AN INTRODU

TO

ELECTROCHEMISTRY

BY

SAMUEL GLASSTONE, D.Sc., Ph.D.

Consultant, United States Atomic Energy Commission

TENTH PRINTING

(AN EAST-WEST EDITION)

AFFILIATED EAST-WEST PRESS PVT. LTD.

NEW DELHI.

First Published May 1942

AFFILIATED EAST-WEST PRESS PVT. LTD.

Price in India Rs. 12 00

Sales Territory : India, Bangla Desh, Burma, Ceylon, Malaysia, Pakistan and Singapore.

This book has been published with the assistance of the joint Indian-American Textbook Programme

Published by K.S. Padmanabhan for AFFILIATED EAST-WEST PRESS PVT. LTD., 9 Nizamuddin East, New Delhi 13, India, and printed by Mohan Makhijani at Rekha Printers, New Delhi.

To V

PREFACE

The object of this book is to provide an introduction to electrochemistry in its present state of development. An attempt has been made to explain the fundamentals of the subject as it stands today, devoting little or no space to the consideration of theories and arguments that have been discarded or greatly modified. In this way it is hoped that the reader will acquire the modern point of view in electrochemistry without being burdened by much that is obsolete. In the opinion of the writer, there have been four developments in the past two decades that have had an important influence on electrochemistry. They are the activity concept, the interionic attraction theory, the proton-transfer theory of acids and bases, and the consideration of electrode reactions as rate processes. These ideas have been incorporated into the structure of the book, with consequent simplification and clarification in the treatment of many aspects of electrochemistry.

This book differs from the author's earlier work, "The Electrochemistry of Solutions," in being less comprehensive and in giving less detail. While the latter is primarily a work of reference, the present book is more suited to the needs of students of physical chemistry, and to those of chemists, physicists and physiologists whose work brings them in contact with a variety of electrochemical problems. As the title implies, the book should also serve as an introductory text for those who intend to specialize in either the theoretical or practical applications of electrochemistry.

In spite of some lack of detail, the main aspects of the subject have been covered, it is hoped impartially and adequately. There has been some tendency in recent electrochemical texts to pay scant attention to the phenomena at active electrodes, such as overvoltage, passivity, corrosion, deposition of metals, and so on. These topics, which are of importance in applied electrochemistry, are treated here at such length as seems reasonable. In addition, in view of the growing interest in electrophoresis, and its general acceptance as a branch of electrochemistry, a chapter on electrokinetic phenomena has been included.

No claim is made to anything approaching completeness in the matter of references to the scientific literature. Such references as are given are generally to the more recent publications, to review articles, and to papers that may, for one reason or another, have some special interest. References are also frequently included to indicate the sources from which data have been obtained for many of the diagrams and tables. Since no effort was made to be exhaustive in this connection, it was felt that an author index would be misleading. This has consequently been omitted, but where certain theories, laws or equations are usually associated with the names of specific individuals, such names have been included in the general index.

In conclusion, attention may be drawn to the problems which are to be found at the end of each chapter. These have been chosen with the object of illustrating particular points; very few are of the kind which involve mere substitution in a formula, and repetition of problems of the same type has been avoided as far as possible. Many of the problems are based on data taken directly from the literature, and their solution should provide both valuable exercise and instruction. The reference to the publication from which the material was taken has been given in the hope that when working the problem the student may become sufficiently interested to read the original paper and thus learn for himself something of the methods and procedures of electrochemical research.

SAMUEL GLASSTONE

NORMAN, OKLAHOMA
March 1942

CONTENTS

CHAPTER I

INTRODUCTION

Properties of Electric Current.—When plates of two dissimilar metals are placed in a conducting liquid, such as an aqueous solution of a salt or an acid, the resulting system becomes a source of electricity; this source is generally referred to as a **voltaic cell** or **galvanic cell,** in honor of Volta and Galvani, respectively, who made the classical discoveries in this field. If the plates of the cell are connected by a wire and a magnetic needle placed near it, the needle will be deflected from its normal position; it will be noted, at the same time, that the wire becomes warm. If the wire is cut and the two ends inserted in a conducting solution, chemical action will be observed where the wires come into contact with the liquid; this action may be in the form of gas evolution, or the liberation of a metal whose salt is present in the solution may be observed. These phenomena, viz., magnetic, heating and chemical effects, are said to be caused by the passage, or flow, of a current of electricity through the wire. Observation of the direction of the deflection of the magnetic needle and the nature of the chemical action, shows that it is possible to associate direction with the flow of electric current. The nature of this direction cannot be defined in absolute terms, and so it is desirable to adopt a convention and the one generally employed is the following: if a man were swimming with the electric current and watching a compass needle, the north-seeking pole of the needle would turn towards his left side. When electricity is passed through a solution, oxygen is generally liberated at the wire at which the positive current enters whereas hydrogen or a metal is set free at the wire whereby the current leaves the solution.

It is unfortunate that this particular convention was chosen, because when the electron was discovered it was observed that a flow of electrons produced a magnetic effect opposite in direction to that accompanying the flow of positive current in the same direction. It was necessary, therefore, to associate a negative charge with the electron, in order to be in harmony with the accepted convention concerning the direction of a current of electricity. Since current is carried through metals by means of electrons only, it means that the flow of electrons is opposite in direction to that of the conventional current flow. It should be emphasized that there is nothing fundamental about this difference, for if the direction of current flow had been defined in the opposite manner, the electron would have been defined as carrying a positive charge and the flow of electrons and of current would have been in the same direction. Although a considerable simplification would result from the change in convention, it is too late in the development of the subject for any such change to be made.

E.M.F., Current and Resistance: Ohm's Law.—If two voltaic cells are connected together so that one metal, e.g., zinc, of one cell is connected to the other metal, e.g., copper, of the second cell, in a manner analogous to that employed by Volta in his electric pile, the magnetic and chemical effects of the current are seen to be increased, provided the same external circuit is employed. The two cells have a greater electrical driving force or pressure than a single one, and this force or pressure * which is regarded as driving the electric current through the wire is called the **electromotive force,** or **E.M.F.** Between any two points in the circuit carrying the current there is said to be a **potential difference,** the total E.M.F. being the algebraic sum of all the potential differences.

By increasing the length of the wire connecting the plates of a given voltaic cell the effect on the magnetic needle and the chemical action are seen to be decreased: the greater length of the wire thus opposes the flow of current. This property of hindering the flow of electricity is called **electrical resistance,** the longer wire having a greater electrical resistance than the shorter one.

It is evident that the current strength in a given circuit, as measured by its magnetic or chemical effect, is dependent on the E.M.F. of the cell producing the current and the resistance of the circuit. The relationship between these quantities is given by **Ohm's law** (1827), which states that the current strength (I) is directly proportional to the applied E.M.F. (E) and inversely proportional to the resistance (R); thus

$$I = \frac{E}{R} \tag{1}$$

is the mathematical expression of Ohm's law. The accuracy of this law has been confirmed by many experiments with conductors of various types: it fails, apparently, for certain solutions when alternating currents of very high frequency are employed, or with very high voltages. The reasons for this failure of Ohm's law are of importance in connection with the theory of solutions (see Chap. III). It is seen from equation (1) that the E.M.F. is equal to the product of the current and the resistance: a consequence of this result is that the potential difference between any two points in a circuit is given by the product of the resistance between those points and the current strength, the latter being the same throughout the circuit. This rule finds a number of applications in electrochemical measurements, as will be evident in due course.

* Electrical force or pressure does not have the dimensions of mechanical force or pressure; the terms are used, however, by analogy with the force or pressure required to produce the flow of a fluid through a pipe.

Electrical Dimensions and Units.—The electrostatic force (F) between two charges ϵ and ϵ' placed at a distance r apart is given by

$$F = \frac{\epsilon\epsilon'}{\kappa r^2},$$

where κ depends on the nature of the medium. Since force has the dimensions mlt^{-2}, where m represents a mass, l length and t time, it can be readily seen that the dimensions of electric charge are $m^{\frac{1}{2}}l^{\frac{3}{2}}t^{-1}\kappa^{\frac{1}{2}}$, the dimensions of κ not being known. The strength of an electric current is defined by the rate at which an electric charge moves along a conductor, and so the dimensions of current are $m^{\frac{1}{2}}l^{\frac{3}{2}}t^{-2}\kappa^{\frac{1}{2}}$. The electromagnetic force between two poles of strength p and p' separated by a distance r is $pp'/\mu r^2$, where μ is a constant for the medium, and so the dimensions of pole strength must be $m^{\frac{1}{2}}l^{\frac{3}{2}}t^{-1}\mu^{\frac{1}{2}}$. It can be deduced theoretically that the work done in carrying a magnetic pole round a closed circuit is proportional to the product of the pole strength and the current, and since the dimensions of work are ml^2t^{-2}, those of current must be $m^{\frac{1}{2}}l^{\frac{1}{2}}t^{-1}\mu^{-\frac{1}{2}}$. Since the dimensions of current should be the same, irrespective of the method used in deriving them, it follows that

$$m^{\frac{1}{2}}l^{\frac{3}{2}}t^{-2}\kappa^{\frac{1}{2}} = m^{\frac{1}{2}}l^{\frac{1}{2}}t^{-1}\mu^{-\frac{1}{2}};$$
$$\therefore \quad \kappa^{\frac{1}{2}}\mu^{\frac{1}{2}} = l^{-1}t.$$

The dimensions $l^{-1}t$ are those of a reciprocal velocity, and it has been shown, both experimentally and theoretically, that the velocity is that of light, i.e., 2.9977×10^{10} cm. per sec., or, with sufficient accuracy for most purposes, 3×10^{10} cm. per sec.

In practice κ and μ are assumed to be unity in vacuum: they are then dimensionless and are called the **dielectric constant** and **magnetic permeability,** respectively, of the medium. Since κ and μ cannot both be unity for the same medium, it is evident that the units based on the assumption that κ is unity must be different from those obtained by taking μ as unity. The former are known as **electrostatic** (e.s.) and the latter as **electromagnetic** (e.m.) **units,** and according to the facts recorded above

$$\frac{1 \text{ e.m. unit of current}}{1 \text{ e.s. unit of current}} = 3 \times 10^{10} \text{ cm. per sec.}$$

It follows, therefore, that if length, mass and time are expressed in centimeters, grams and seconds respectively, i.e., in the c.g.s. system, the e.m. unit of current is 3×10^{10} times as great as the e.s. unit. The e.m. unit of current on this system is defined as that current which flowing through a wire in the form of an arc one cm. long and of one cm. radius exerts a force of one dyne on a unit magnetic pole at the center of the arc.

The product of current strength and time is known as the **quantity of electricity**; it has the same dimensions as electric charge. The e.m. unit of charge or quantity of electricity is thus 3×10^{10} larger than the corresponding e.s. unit. The product of quantity of electricity and potential or E.M.F. is equal to work, and if the same unit of work, or energy, is adopted in each case, the e.m. unit of potential must be smaller than the e.s. unit in the ratio of 1 to 3×10^{10}. When one e.m. unit of potential difference exists between two points, one erg of work must be expended to transfer one e.m. unit of charge, or quantity of electricity, from one point to the other; the e.s. unit of potential is defined in an exactly analogous manner in terms of one e.s. unit of charge.

The e.m. and e.s. units described above are not all of a convenient magnitude for experimental purposes, and so a set of **practical units** have been defined. The practical unit of current, the **ampere,** often abbreviated to "amp.," is one-tenth the e.m. (c.g.s.) unit, and the corresponding unit of charge or quantity of electricity is the **coulomb**; the latter is the quantity of electricity passing when one ampere flows for one second. The practical unit of potential or E.M.F. is the **volt,** defined as 10^8 e.m. units. Corresponding to these practical units of current and E.M.F. there is a unit of electrical resistance; this is called the **ohm,** and it is the resistance of a conductor through which a current of one ampere passes when the potential difference between the ends is one volt. With these units of current, E.M.F. and resistance it is possible to write Ohm's law in the form

$$\text{amperes} = \frac{\text{volts}}{\text{ohms}}. \qquad (2)$$

By utilizing the results given above for the relationships between e.m., e.s. and practical units, it is possible to draw up a table relating the various units to each other. Since the practical units are most frequently employed in electrochemistry, the most useful method of expressing the connection between the various units is to give the number of e.m. or e.s. units corresponding to one practical unit: the values are recorded in Table I.

TABLE I. CONVERSION OF ELECTRICAL UNITS

	Practical Unit	Equivalent in	
		e.m.u.	e.s.u.
Current	Ampere	10^{-1}	3×10^9
Quantity or Charge	Coulomb	10^{-1}	3×10^9
Potential or E.M.F.	Volt	10^8	$(300)^{-1}$

International Units.—The electrical units described in the previous section are defined in terms of quantities which cannot be easily established in the laboratory, and consequently an International Committee (1908) laid down alternative definitions of the practical units of electricity. The **international ampere** is defined as the quantity of electricity which flowing for one second will cause the deposition of 1.11800 milligrams of silver from a solution of a silver salt, while the **international ohm** is the resistance at 0° c. of a column of mercury 106.3 cm. long, of uniform cross-section, weighing 14.4521 g. The **international volt** is then the difference of electrical potential, or E.M.F., required to maintain a current of one international ampere through a system having a resistance of one international ohm. Since the international units were defined it has been found that they do not correspond exactly with those defined above in terms of the c.g.s. system; the latter are thus referred to as **absolute units** to distinguish them from the international units. The international ampere is 0.99986 times the absolute ampere, and the international ohm is 1.00048 times the absolute ohm, so that the international volt is 1.00034 times the absolute practical unit.*

Electrical Energy.—As already seen, the passage of electricity through a conductor is accompanied by the liberation of heat; according to the first law of thermodynamics, or the principle of conservation of energy, the heat liberated must be exactly equivalent to the electrical energy expended in the conductor. Since the heat can be measured, the value of the electrical energy can be determined and it is found, in agreement with anticipation, that the heat liberated by the current in a given conductor is proportional to the quantity of electricity passing and to the difference of potential at the extremities of the conductor. The practical **unit of electrical energy** is, therefore, defined as the energy developed when one coulomb is passed through a circuit by an E.M.F. of one volt; this unit is called the **volt-coulomb,** and it is evident from Table I that the absolute volt-coulomb is equal to 10^7 ergs, or **one joule.** It follows, therefore, that if a current of I amperes is passed for t seconds through a conductor under the influence of a potential of E volts, the energy liberated (Q) will be given by

$$Q = EIt \times 10^7 \text{ ergs}, \tag{3}$$

or, utilizing Ohm's law, if R is the resistance of the conductor,

$$Q = I^2Rt \times 10^7 \text{ ergs}. \tag{4}$$

These results are strictly true only if the ampere, volt and ohm are in absolute units; there is a slight difference if international units are employed, the absolute volt-coulomb or joule being different from the international value. The United States Bureau of Standards has recommended that the unit of heat, the calorie, should be defined as the equivalent of 4.1833 international joules, and hence

$$Q = \frac{EIt}{4.183} \text{ calories}, \tag{5}$$

where E and I are now expressed in international volts and amperes, respectively. Alternatively, it may be stated that one international volt-coulomb is equivalent to 0.2390 standard calorie.

* These figures are obtained from the set of consistent fundamental constants recommended by Birge (1941); slightly different values are given in the International Critical Tables.

Classification of Conductors.—All forms of matter appear to be able to conduct the electric current to some extent, but the conducting powers of different substances vary over a wide range; thus silver, one of the best conductors, is 10^{24} times more effective than paraffin wax, which is one of the poorest conductors. It is not easy to distinguish sharply between good and bad conductors, but a rough division is possible; the systems studied in electrochemistry are generally good conductors. These may be divided into three main categories; they are: (*a*) gaseous, (*b*) metallic and (*c*) electrolytic.

Gases conduct electricity with difficulty and only under the influence of high potentials or if exposed to the action of certain radiations. Metals are the best conductors, in general, and the passage of current is not accompanied by any movement of matter; it appears, therefore, that the electricity is carried exclusively by the electrons, the atomic nuclei remaining stationary. This is in accordance with modern views which regard a metal as consisting of a relatively rigid lattice of ions together with a system of mobile electrons. **Metallic conduction,** or **electronic conduction,** as it is often called, is not restricted to pure metals, for it is a property possessed by most alloys, carbon and certain solid salts and oxides.

Electrolytic conductors, or **electrolytes,** are distinguished by the fact that passage of an electric current through them results in an actual transfer of matter; this transfer is manifested by changes of concentration and frequently, in the case of electrolytic solutions, by the visible separation of material at the points where the current enters and leaves the solution. Electrolytic conductors are of two main types; there are, first, substances which conduct electrolytically in the pure state, such as fused salts and hydrides, the solid halides of silver, barium, lead and some other metals, and the α-form of silver sulfide. Water, alcohols, pure acids, and similar liquids are very poor conductors, but they must be placed in this category. The second class of electrolytic conductors consists of solutions of one or more substances; this is the type of conductor with which the study of electrochemistry is mainly concerned. The most common electrolytic solutions are made by dissolving a salt, acid or base in water; other solvents may be used, but the conducting power of the system depends markedly on their nature. Conducting systems of a somewhat unusual type are lithium carbide and alkaline earth nitrides dissolved in the corresponding hydride, and organic acid amides and nitro-compounds in liquid ammonia or hydrazine.

The distinction between electronic and electrolytic conductors is not sharp, for many substances behave as **mixed conductors**; that is, they conduct partly electronically and partly electrolytically. Solutions of the alkali and alkaline earth metals in liquid ammonia are apparently mixed conductors, and so also is the β-form of silver sulfide. Fused cuprous sulfide conducts electronically, but a mixture with sodium or ferrous sulfide also exhibits electrolytic conduction; a mixture with nickel sulfide is, however, a pure electronic conductor. Although pure metals conduct electronically, conduction in certain liquid alloys involves the transfer of matter and appears to be partly electrolytic in nature. Some materials conduct electronically at one temperature and electrolytically at another; thus cuprous bromide changes its method of conduction between 200° and 300°.

The Phenomena and Mechanism of Electrolysis.—The materials, generally small sheets of metal, which are employed to pass an electric current through an electrolytic solution, are called **electrodes**; the one at which the positive current enters is referred to as the positive electrode or **anode,** whereas the electrode at which current leaves is called the negative electrode, or **cathode.** The passage of current through solutions of salts of such metals as zinc, iron, nickel, cadmium, lead, copper, silver and mercury results in the liberation of these metals at the cathode; from solutions of salts of the very base metals, e.g., the alkali and alkaline earth metals, and from solutions of acids the substance set free is hydrogen gas. If the anode consists of an attackable metal, such as one of those just enumerated, the flow of the current is accompanied by the passage of the metal into solution. When the anode is made of an inert metal, e.g., platinum, an element is generally set free at this electrode; from solutions of nitrates, sulfates, phosphates, etc., oxygen gas is liberated, whereas from halide solutions, other than fluorides, the free halogen is produced. The decomposition of solutions by the electric current, resulting in the liberation of gases or metals, as described above, is known as **electrolysis.**

The first definite proposals concerning the mechanism of electrolytic conduction and electrolysis were made by Grotthuss (1806); he suggested that the dissolved substance consisted of particles with positive and negative ends, these particles being distributed in a random manner throughout the solution. When a potential was applied it was believed that the particles (molecules) became oriented in the form of chains with the positive parts pointing in one direction and the negative parts in the opposite direction (Fig. 1, I). It was supposed that the positive electrode attracts the negative part of one end particle in the chain, resulting in the liberation of the corresponding material, e.g., oxygen in the electrolysis of water. Similarly, the negative electrode attracts the positive portion of the particle, e.g., the hydrogen of water, at the other end of the chain, and sets it free (Fig. 1, II). The residual parts of the end units were then imagined to exchange partners with adjacent molecules, this interchange being carried on until a complete series of new particles

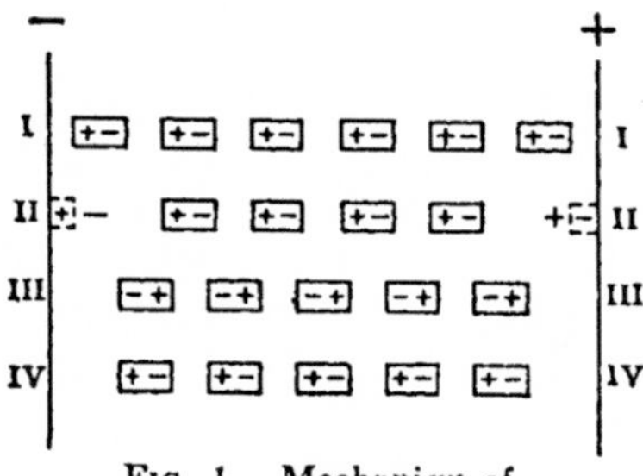

Fig. 1. Mechanism of Grotthuss conduction

is formed (Fig. 1, III). These are now rotated by the current to give the correct orientation (Fig. 1, IV), followed by their splitting up, and so on. The chief objection to the theory of Grotthuss is that it would require a relatively high E.M.F., sufficient to break up the molecules, before any appreciable current was able to flow, whereas many solutions can be electrolyzed by the application of quite small potentials. Although the proposed mechanism has been discarded, as far as most electrolytic conduction is concerned, it will be seen later (p. 66) that a type of Grotthuss conduction occurs in solutions of acids and bases.

In order to account for the phenomena observed during the passage of an electric current through solutions, Faraday (1833) assumed that the flow of electricity was associated with the movement of particles of matter carrying either positive or negative charges. These charged particles were called **ions**; the ions carrying positive charges and moving in the direction of the current, i.e., towards the cathode, were referred to as **cations**, and those carrying a negative charge and moving in the opposite direction, i.e., towards the anode, were called **anions** * (see Fig. 2). The function of the applied E.M.F. is to direct the ions towards the appropriate electrodes where their charges are neutralized and they are set free as atoms or molecules. It may be noted that since hydrogen and metals are discharged at the cathode, the metallic part of a salt or base and the hydrogen of an acid form cations and carry positive charges. The acidic portion of a salt and the hydroxyl ion of a base consequently carry negative charges and constitute the anions.

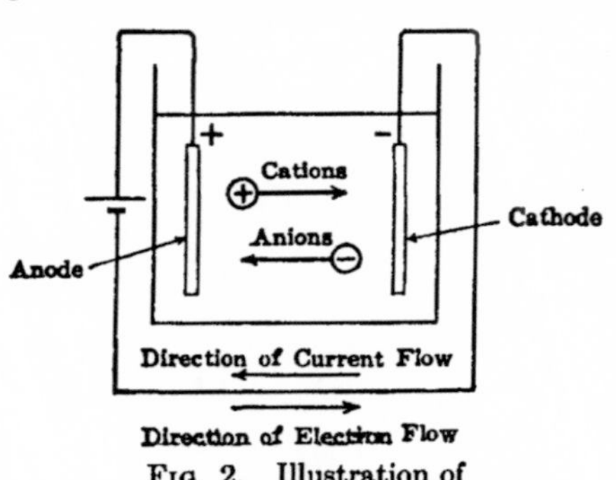

Fig. 2. Illustration of electrochemical terms

Although Faraday postulated the existence of charged material particles, or ions, in solution, he offered no explanation of their origin: it was suggested, however, by Clausius (1857) that the positive and negative parts of the solute molecules were not firmly connected, but were each in a state of vibration that often became vigorous enough to cause the portions to separate. These separated charged parts, or ions, were believed to have relatively short periods of free existence; while free they were supposed to carry the current. According to Clausius, a small fraction only of the total number of dissolved molecules was split into

* The term "ion" is derived from a Greek word meaning "wanderer" or "traveler," the prefixes *ana* and *cata* meaning "up" and "down," respectively; the anion is thus the ion moving *up*, and the cation that moving *down* the potential gradient. These terms, as well as electrode, anode and cathode, were suggested to Faraday by Whewell (1834); see Oesper and Speter, *Scientific Monthly*, 45, 535 (1937).

ions at any instant, but sufficient ions were always available for carrying the current and hence for discharge at the electrodes. Since no electrical energy is required to break up the molecules, this theory is in agreement with the fact that small E.M.F.'s are generally adequate to cause electrolysis to occur; the applied potential serves merely to guide the ions to the electrodes where their charges are neutralized.

The Electrolytic Dissociation Theory.[1]—From his studies of the conductances of aqueous solutions of acids and their chemical activity, Arrhenius (1883) concluded that an electrolytic solution contained two kinds of solute molecules; these were supposed to be "active" molecules, responsible for electrical conduction and chemical action, and inactive molecules, respectively. It was believed that when an acid, base or salt was dissolved in water a *considerable portion*, consisting of the so-called active molecules, was spontaneously split up, or dissociated, into positive and negative ions; it was suggested that these ions are free to move independently and are directed towards the appropriate electrodes under the influence of an electric field. The proportion of active, or dissociated, molecules to the total number of molecules, later called the **"degree of dissociation,"** was considered to vary with the concentration of the electrolyte, and to be equal to unity in dilute solutions.

This **theory of electrolytic dissociation,** or the **ionic theory,** attracted little attention until 1887 when van't Hoff's classical paper on the theory of solutions was published. The latter author had shown that the ideal gas law equation, with osmotic pressure in place of gas pressure, was applicable to dilute solutions of non-electrolytes, but that electrolytic solutions showed considerable deviations. For example, the osmotic effect, as measured by depression of the freezing point or in other ways, of hydrochloric acid, alkali chlorides and hydroxides was nearly twice as great as the value to be expected from the gas law equation; in some cases, e.g., barium hydroxide, and potassium sulfate and oxalate, the discrepancy was even greater. No explanation of these facts was offered by van't Hoff, but he introduced an empirical factor i into the gas law equation for electrolytic solutions, thus

$$\Pi = iRTc,$$

where Π is the observed osmotic pressure of the solution of concentration c; the temperature is T, and R is the gas constant. According to this equation, the van't Hoff factor i is equal to the ratio of the experimental osmotic effect to the theoretical osmotic effect, based on the ideal gas laws, for the given solution. Since the osmotic effect is, at least approximately, proportional to the number of individual molecular particles, a value of two for the van't Hoff factor means that the solution contains about twice the number of particles to be expected. This result

[1] Arrhenius, *J. Chem. Soc.*, 105, 1414 (1914); Walker, *ibid.*, 1380 (1928).

is clearly in agreement with the views of Arrhenius, if the ions are regarded as having the same osmotic effect as uncharged particles.

The concept of "active molecules," which was part of the original theory, was later discarded by Arrhenius as being unnecessary; he suggested that whenever a substance capable of yielding a conducting solution was dissolved in water, it dissociated *spontaneously* into ions, the extent of the dissociation being very considerable with salts and with strong acids and bases, especially in dilute solution. Thus, a molecule of potassium chloride should, according to the theory of electrolytic dissociation, be split up into potassium and chloride ions in the following manner:

$$KCl = K^+ + Cl^-.$$

If dissociation is complete, then each "molecular particle" of solid potassium chloride should give two particles in solution; the osmotic effect will thus approach twice the expected value, as has actually been found. A bi-univalent salt, such as barium chloride, will dissociate spontaneously according to the equation

$$BaCl_2 = Ba^{++} + 2Cl^-,$$

and hence the van't Hoff factor should be approximately 3, in agreement with experiment.

Suppose a solution is made up by dissolving m molecules in a given volume and α is the fraction of these molecules dissociated into ions; if each molecule produces ν ions on dissociation, there will be present in the solution $m(1 - \alpha)$ undissociated molecules and $\nu m\alpha$ ions, making a total of $m - m\alpha + \nu m\alpha$ particles. If the van't Hoff factor is equal to the ratio of the number of molecular particles actually present to the number that would have been in the solution if there had been no dissociation, then

$$i = \frac{m - m\alpha + \nu m\alpha}{m} = 1 - \alpha + \nu\alpha;$$

$$\therefore \quad \alpha = \frac{i - 1}{\nu - 1}. \tag{6}$$

Since the van't Hoff factor is obtainable from freezing-point, or analogous, measurements, the value of α, the so-called degree of dissociation, in the given solution can be calculated from equation (6). An alternative method of evaluating α, using conductance measurements (see p. 51), was proposed by Arrhenius (1887), and he showed that the results obtained by the two methods were in excellent agreement: this agreement was accepted as strong evidence for the theory of electrolytic dissociation, which has played such an important rôle in the development of electrochemistry.

It is now known that the agreement referred to above, which convinced many scientists of the value of the Arrhenius theory, was to a great extent fortuitous; the conductance method for calculating the degree of dissociation is not applicable to salt solutions, and such solutions would, in any case, not be expected to obey the ideal gas law equation. Nevertheless, the theory of electrolytic dissociation, with certain modifications, is now universally accepted; it is believed that when a solute, capable of forming a conducting solution, is dissolved in a suitable solvent, it dissociates spontaneously into ions. If the solute is a salt or a strong acid or base the extent of dissociation is very considerable, it being almost complete in many cases provided the solution is not too concentrated; substances of this kind, which are highly dissociated and which give good conducting solutions in water, are called **strong electrolytes.** Weak acids and weak bases, e.g., amines, phenols, most carboxylic acids and some inorganic acids and bases, such as hydrocyanic acid and ammonia, and a few salts, e.g., mercuric chloride and cyanide, are dissociated only to a small extent at reasonable concentrations; these compounds constitute the **weak electrolytes.*** Salts of weak acids or bases, or of both, are generally strong electrolytes, in spite of the fact that one or both constituents are weak. These results are in harmony with modern developments of the ionic theory, as will be evident in later chapters. As is to be expected, it is impossible to classify all electrolytes as "strong" or "weak," although this forms a convenient rough division which is satisfactory for most purposes. Certain substances, e.g., trichloroacetic acid, exhibit an intermediate behavior, but the number of **intermediate electrolytes** is not large, at least in aqueous solution. It may be noted, too, that the nature of the solvent is often important; a particular compound may be a strong electrolyte, being dissociated to a large extent, in one solvent, but may be only feebly dissociated, and hence is a weak electrolyte, in another medium (cf. p. 13).

Evidence for the Ionic Theory.—There is hardly any branch of electrochemistry, especially in its quantitative aspects, which does not provide arguments in favor of the theory of electrolytic dissociation; without the ionic concept the remarkable systematization of the experimental results which has been achieved during the past fifty years would certainly not have been possible. It is of interest, however, to review briefly some of the lines of evidence which support the ionic theory.

Although exception may be taken to the quantitative treatment given by Arrhenius, the fact of the abnormal osmotic properties of electrolytic solutions still remains; the simplest explanation of the high values can be given by postulating dissociation into ions. This, in conjunction with the ability of solutions to conduct the electric current, is one of the strongest arguments for the ionic theory. Another powerful argument is

* Strictly speaking, the term "electrolyte" should refer to the conducting system as a whole, but it is also frequently applied to the solute; the word "ionogen," i.e., producer of ions, has been suggested for the latter [see, for example, Blum, *Trans. Electrochem. Soc.*, 47, 125 (1925)], but this has not come into general use.

based on the realization in recent years, as a result of X-ray diffraction studies, that the structural unit of solid salts is the ion rather than the molecule. That is to say, salts are actually ionized in the solid state, and it is only the restriction to movement in the crystal lattice that prevents solid salts from being good electrical conductors. When fused or dissolved in a suitable solvent, the ions, which are already present, can move relatively easily under the influence of an applied E.M.F., and conductance is observed. The concept that salts consist of ions held together by forces of electrostatic attraction is also in harmony with modern views concerning the nature of valence.

Many properties of electrolytic solutions are additive functions of the properties of the respective ions; this is at once evident from the fact that the chemical properties of a salt solution are those of its constituent ions. For example, potassium chloride in solution has no chemical reactions which are characteristic of the compound itself, but only those of potassium and chloride ions. These properties are possessed equally by almost all potassium salts and all chlorides, respectively. Similarly, the characteristic chemical properties of acids and alkalis, in aqueous solution, are those of hydrogen and hydroxyl ions, respectively. Certain physical properties of electrolytes are also additive in nature; the most outstanding example is the electrical conductance at infinite dilution. It will be seen in Chap. II that conductance values can be ascribed to all ions, and the appropriate conductance of any electrolyte is equal to the sum of the values for the individual ions. The densities of electrolytic solutions have also been found to be additive functions of the properties of the constituent ions. The catalytic effects of various acids and bases, and of mixtures with their salts, can be accounted for by associating a definite catalytic coefficient with each type of ion; since undissociated molecules often have appreciable catalytic properties due allowance must be made for their contribution.

Certain thermal properties of electrolytes are in harmony with the theory of ionic dissociation; for example, the heat of neutralization of a strong acid by an equivalent amount of a strong base in dilute solution is about 13.7 kcal. at 20° irrespective of the exact nature of the acid or base.[2] If the acid is hydrochloric acid and the base is sodium hydroxide, then according to the ionic theory the neutralization reaction should be written

$$(H^+ + Cl^-) + (Na^+ + OH^-) = (Na^+ + Cl^-) + H_2O,$$

the acid, base and the resulting salt being highly dissociated, whereas the water is almost completely undissociated. Since Na^+ and Cl^- appear on both sides of this equation, the essential reaction is

$$H^+ + OH^- = H_2O,$$

[2] Richards and Rowe, *J. Am. Chem. Soc.*, 44, 684 (1922); see also, Lambert and Gillespie, *ibid.*, 53, 2632 (1931); Rossini, *J. Res. Nat. Bur. Standards*, 6, 847 (1931); Pitzer, *J. Am. Chem. Soc.*, 59, 2365 (1937).

and this is obviously independent of the particular acid or base employed: the heat of neutralization would thus be expected to be constant. It is of interest to mention that the heat of the reaction between hydrogen and hydroxyl ions in aqueous solution has been calculated by an entirely independent method (see p. 344) and found to be almost identical with the value obtained from neutralization experiments. The heat of neutralization of a weak acid or a weak base is generally different from 13.7 kcal., since the acid or base must dissociate completely in order that it may be neutralized and the process of ionization is generally accompanied by the absorption of heat.

Influence of the Solvent on Dissociation.[3]—The nature of the solvent often plays an important part in determining the degree of dissociation of a given substance, and hence in deciding whether the solution shall behave as a strong or as a weak electrolyte. Experiments have been made on solutions of tetraisoamylammonium nitrate in a series of mixtures of water and dioxane (see p. 54). In the water-rich solvents the system behaves like a strong electrolyte, but in the solvents containing relatively large proportions of dioxane the properties are essentially those of a weak electrolyte. In this case, and in analogous cases where the solute consists of units which are held together by bonds that are almost exclusively electrovalent in character, it is probable that the dielectric constant is the particular property of the solvent that influences the dissociation (cf. Chaps. II and III). The higher the dielectric constant of the medium, the smaller is the electrostatic attraction between the ions and hence the greater is the probability of their existence in the free state. Since the dielectric constant of water at 25° is 78.6, compared with a value of about 2.2 for dioxane, the results described above can be readily understood.

It should be noted, however, that there are many instances in which the dielectric constant of the solvent plays a secondary part: for example, hydrogen chloride dissolves in ethyl alcohol to form a solution which behaves as a strong electrolyte, but in nitrobenzene, having a dielectric constant differing little from that of alcohol, the solution is a weak electrolyte. As will be seen in Chap. IX the explanation of this difference lies in the ability of a molecule of ethyl alcohol to combine readily with a bare hydrogen ion, i.e., a proton, to form the ion $C_2H_5OH_2^+$, and this represents the form in which the hydrogen ion exists in the alcohol solution. Nitrobenzene, however, does not form such a combination to any great extent; hence the degree of dissociation of the acid is small and the solution of hydrogen chloride behaves as a weak electrolyte. The ability of oxygen compounds, such as ethers, ketones and even sugars, to accept a proton from a strongly acidic substance, thus forming an ion, e.g., R_2OH^+ or R_2COH^+, accounts for the fact that solutions of such compounds in pure sulfuric acid or in liquid hydrogen fluoride are relatively strong electrolytes.

[3] See, Glasstone, "The Electrochemistry of Solutions," 1937, p. 172.

Another aspect of the formation of compounds and its influence on electrolytic dissociation is seen in connection with substituted ammonium salts of the type R_3NHX; although they are strong electrolytes in hydroxylic solvents, e.g., in water and alcohols, they are dissociated to only a small extent in nitrobenzene, nitromethane, acetone and acetonitrile. It appears that in the salts under consideration the hydrogen atom can act as a link between the nitrogen atom and the acid radical X, so that the molecule $R_3N \cdot H \cdot X$ exists in acid solution. If the solvent S is of such a nature, however, that its molecules tend to form strong hydrogen bonds, it can displace the X^- ions, thus

$$R_3N \cdot H \cdot X + S \rightleftharpoons R_3N \cdot H \cdot S^+ + X^-,$$

so that ionization of the salt is facilitated. Hydroxylic solvents, in virtue of the type of oxygen atom which they contain, form hydrogen bonds more readily than do nitro-compounds, nitriles, etc.; the difference in behavior of the two groups of solvents can thus be understood.

Salts of the type R_4NX function as strong electrolytes in both groups of solvents, since the dielectric constants are relatively high, and the question of compound formation with the solvent is of secondary importance. The fact that salts of different types show relatively little difference of behavior in hydroxylic solvents has led to these substances being called **levelling solvents.** On the other hand, solvents of the other group, e.g., nitro-compounds and nitriles, are referred to as **differentiating solvents** because they bring out the differences between salts of different types. The characteristic properties of the levelling solvents are due partly to their high dielectric constants and partly to their ability to act both as electron donors and acceptors, so that they are capable of forming compounds with either anions or cations.

The formation of a combination of some kind between the ion and a molecule of solvent, known as **solvation,** is an important factor in enhancing the dissociation of a given electrolyte. The solvated ions are relatively large and hence their distance of closest approach is very much greater than the bare unsolvated ions. It will be seen in Chap. V that when the distance between the centers of two oppositely charged ions is less than a certain limiting value the system behaves as if it consisted of undissociated molecules. The effective degree of dissociation thus increases as the distance of closest approach becomes larger; hence solvation may be of direct importance in increasing the extent of dissociation of a salt in a particular solvent. It may be noted that solvation does not necessarily involve a covalent bond, e.g., as is the case in $Cu(NH_3)_4^{++}$ and $Cu(H_2O)_4^{++}$; there is reason for believing that solvation is frequently electrostatic in character and is due to the orientation of solvent molecule dipoles about the ion. A solvent with a large dipole moment will thus tend to facilitate solvation and it will consequently increase the degree of dissociation.

It was mentioned earlier in this chapter that acid amides and nitro-compounds form conducting solutions in liquid ammonia and hydrazine; the ionization in these cases is undoubtedly accompanied by, and is associated with, compound formation between solute and solvent. The same is true of triphenylmethyl chloride which is a fair electrolytic conductor when dissolved in liquid sulfur dioxide; it also conducts to some extent in nitromethane, nitrobenzene and acetone solutions. In chloroform and benzene, however, there is no compound formation and no conductance. The electrolytic conduction of triphenylmethyl chloride in fused aluminum chloride, which is itself a poor conductor, appears to be due to the reaction

$$Ph_3CCl + AlCl_3 = Ph_3C^+ + AlCl_4^-;$$

this process is not essentially different from that involved in the ionization of an acid, where the H^+ ion, instead of a Cl^- ion, is transferred from one molecule to another.

Faraday's Laws of Electrolysis.—During the years 1833 and 1834, Faraday published the results of an extended series of investigations on the relationship between the quantity of electricity passing through a solution and the amount of metal, or other substance, liberated at the electrodes: the conclusions may be expressed in the form of the two following laws.

I. The amount of chemical decomposition produced by a current is proportional to the quantity of electricity passing through the electrolytic solution.

II. The amounts of different substances liberated by the same quantity of electricity are proportional to their chemical equivalent weights.

The first law can be tested by passing a current of constant strength through a given electrolyte for various periods of time and determining the amounts of material deposited, on the cathode, for example; the weights should be proportional to the time in each case. Further, the time may be kept constant and the current varied; in these experiments the quantity of deposit should be proportional to the current strength. The second law of electrolysis may be confirmed by passing the same quantity of electricity through a number of different solutions, e.g., dilute sulfuric acid, silver nitrate and copper sulfate; if a current of one ampere flows for one hour the weights liberated at the respective cathodes should be 0.0379 gram of hydrogen, 4.0248 grams of silver and 1.186 grams of copper. These quantities are in the ratio of 1.008 to 107.88 to 31.78, which is the ratio of the equivalent weights. As the result of many experiments, in both aqueous and non-aqueous media, some of which will be described below, much evidence has been obtained for the accuracy of Faraday's laws of electrolysis within the limits of reasonable experimental error. Apart from small deviations, which can be readily explained by the difficulty of obtaining pure deposits or by similar analytical problems, there are a number of instances of more serious apparent exceptions to the laws of electrolysis. The amount of sodium liberated in the electrolysis of a solution of the metal in liquid ammonia is less than would be expected. It must be remembered, however, that Faraday's laws are applicable only when the whole of the conduction is electrolytic in character; in the sodium solutions in liquid ammonia some of the conduction is electronic in nature. The quantities of metal deposited from solutions of lead or antimony in liquid ammonia containing sodium are in excess of those required by the laws of electrolysis; in these solutions the metals exist in the form of complexes and the ions are quite different from those present in aqueous solution. It is consequently not possible to calculate the weights of the deposits to be expected from Faraday's laws.

The applicability of the laws has been confirmed under extreme conditions: for example, Richards and Stull (1902) found that a given quantity of electricity deposited the same weight of silver, within 0.005 per cent, from an aqueous solution of silver nitrate at 20° and from a solution of this salt in a fused mixture of sodium and potassium nitrates at 260°. The experimental results are quoted in Table II.

TABLE II. COMPARISON OF SILVER DEPOSITS AT 20° AND 260°

Deposit at 20°	Deposit at 260°	Difference
1.14916 g.	1.14919 g.	0.003 per cent
1.12185	1.12195	0.009
1.10198	1.10200	0.002

A solution of silver nitrate in pyridine at − 55° also gives the same weight of silver on the cathode as does an aqueous solution of this salt at ordinary temperatures. Pressures up to 1500 atmospheres have no effect on the quantity of silver deposited from a solution of silver nitrate in water.

Faraday's law holds for solid electrolytic conductors as well as for fused electrolytes and solutions; this is shown by the results of Tubandt and Eggert (1920) on the electrolysis of the cubic form of silver iodide quoted in Table III. The quantities of silver deposited in an ordinary silver coulometer in the various experiments are recorded, together with the amounts of silver gained by the cathode and lost by the anode, respectively, when solid silver iodide was used as the electrolyte.

TABLE III. APPLICATION OF FARADAY'S LAWS TO SOLID SILVER IODIDE

Temp.	Current	Ag deposited in coulometer	Ag deposited on cathode	Ag lost from anode
150°	0.1 amp.	0.8071 g.	0.8072 g.	0.8077 g.
150°	0.1	0.9211	0.9210	0.9217
400°	0.1	0.3997	0.3991	0.4004
400°	0.4	0.4217	0.4218	0.4223

The Faraday and its Determination.—The quantity of electricity required to liberate 1 equiv. of any substance should, according to the second of Faraday's laws, be independent of its nature; this quantity is called the **faraday**; it is given the symbol F and, as will be seen shortly, is equal to 96,500 coulombs, within the limits of experimental error. If e is the equivalent weight of any material set free at an electrode, then 96,500 amperes flowing for one second liberate e grams of this substance; it follows, therefore, from the first of Faraday's laws, that I amperes flowing for t seconds will cause the deposition of w grams, where

$$w = \frac{Ite}{96,500}. \tag{7}$$

If the product It is unity, i.e., the quantity of electricity passed is 1 coulomb, the weight of substance deposited is $e/96,500$; the result is known as the **electrochemical equivalent** of the deposited element. If this quantity is given the symbol $\boldsymbol{e}$, it follows that

$$w = It\boldsymbol{e}. \tag{7a}$$

The electrochemical equivalents of some of the more common elements are recorded in Table IV; * since the value for any given element depends

TABLE IV. ELECTROCHEMICAL EQUIVALENTS IN MILLIGRAMS PER COULOMB

Element	Valence	$\boldsymbol{e}$	Element	Valence	$\boldsymbol{e}$
Hydrogen	1	0.01045	Copper	2	0.3294
Oxygen	2	0.08290	Bromine	1	0.8281
Chlorine	1	0.36743	Cadmium	2	0.5824
Iron	2	0.2893	Silver	1	1.1180
Cobalt	2	0.3054	Iodine	1	1.3152
Nickel	2	0.3041	Mercury	2	1.0394

on the valence of the ions from which it is being deposited, the actual valence for which the results were calculated is given in each case.

The results given above, and equation (7) or (7a), are the quantitative expression of Faraday's laws of electrolysis; they can be employed either to calculate the weight of any substance deposited by a given quantity of electricity, or to find the quantity of electricity passing through a circuit by determining the weight of a given metal set free by electrolysis. The apparatus used for the latter purpose was at one time referred to as a "voltameter," but the name **coulometer,** i.e., coulomb measurer, proposed by Richards and Heimrod (1902), is now widely employed.

The most accurate determinations of the faraday have been made by means of the **silver coulometer** in which the amount of pure silver deposited from an aqueous solution of silver nitrate is measured. The first reliable observations with the silver coulometer were those of Kohlrausch in 1886, but the most accurate measurements in recent years were made by Smith, Mather and Lowry (1908) at the National Physical

* For a complete list of electrochemical equivalents and for other data relating to Faraday's laws, see Roush, *Trans. Electrochem. Soc.*, **73**, 285 (1938).

Laboratory in England, by Richards and Anderegg (1915–16) at Harvard University, and by Rosa and Vinal,[4] and others, at the National Bureau of Standards in Washington, D. C. (1914–16). The conditions for obtaining precise results have been given particularly by Rosa and Vinal (1914): these are based on the necessity of insuring purity of the silver nitrate, of preventing particles of silver from the anode, often known as the "anode slime," from falling on to the cathode, and of avoiding the inclusion of water and silver nitrate in the deposited silver.

The silver nitrate is purified by repeated crystallization from acidified solutions, followed by fusion. The purity of the salt is proved by the absence of the so-called "volume effect," the weight of silver deposited by a given quantity of electricity being independent of the volume of liquid in the coulometer: this means that no extraneous impurities are included in the deposit. The solution of silver nitrate employed for the actual measurements should contain between 10 and 20 g. of the salt in 100 cc.; it should be neutral or slightly acid to methyl red indicator, after removal of the silver by neutral potassium chloride, both at the beginning and end of the electrolysis. The anode should be of pure silver with an area as large as the apparatus permits; the current density at the anode should not exceed 0.2 amp. per sq. cm. To prevent the anode slime

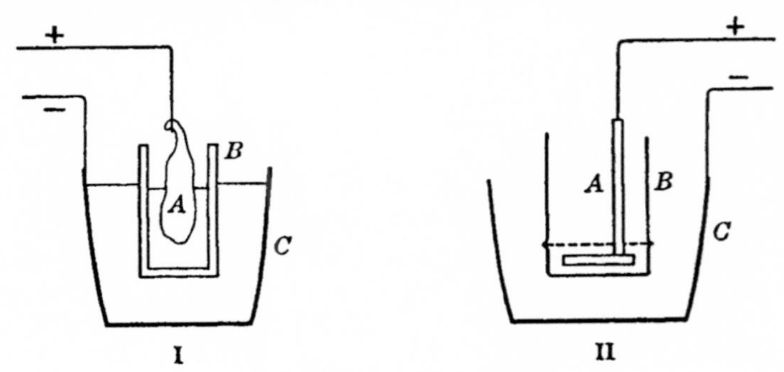

FIG. 3. Silver coulometers

from reaching the cathode, the former electrode (*A* in Fig. 3), is inserted in a cup of porous porcelain, as shown at *B* in Fig. 3, I (Richards, 1900), or is surrounded by a glass vessel, *B* in Fig. 3, II (Smith, 1908). The cathode is a platinum dish or cup (*C*) and its area should be such as to make the cathodic current density less than 0.02 amp. per sq. cm. After electrolysis the solution is removed by a siphon, the deposited silver is washed thoroughly and then the platinum dish and deposit are dried at 150° and weighed. The gain in weight gives the amount of silver deposited by the current; if the conditions described are employed, the impurities should not be more than 0.004 per cent.

[4] Rosa and Vinal, *Bur. Standards Bull.*, 13, 479 (1936); see also, Vinal and Bovard, *J. Am. Chem. Soc.*, 38, 496 (1916); Bovard and Hulett, *ibid.*, 39, 1077 (1917).

If the observations are to be used for the determination of the faraday, it is necessary to know exactly the quantity of electricity passed or the current strength, provided it is kept constant during the experiment. In the work carried out at the National Physical Laboratory the absolute value of the current was determined by means of a magnetic balance, but at the Bureau of Standards the current strength was estimated from the known value of the applied E.M.F., based on the Weston standard cell as 1.01830 international volt at 20° (see p. 193), and the measured resistance of the circuit. According to the experiments of Smith, Mather and Lowry, one absolute coulomb deposits 1.11827 milligrams of silver, while Rosa and Vinal (1916) found that one international coulomb deposits 1.1180 milligrams of silver. The latter figure is identical with the one used for the definition of the international coulomb (p. 4) and since it is based on the agreed value of the E.M.F. of the Weston cell it means that these definitions are consistent with one another within the limits of experimental accuracy. If the atomic weight of silver is taken as 107.88, it follows that

$$\frac{107.88}{0.0011180} = 96{,}494 \text{ international coulombs}$$

are required to liberate one gram equivalent of silver. If allowance is made for the 0.004 per cent of impurity in the deposit, this result becomes 96,498 coulombs. Since the atomic weight of silver is not known with an accuracy of more than about one part in 10,000, the figure is rounded off to 96,500 coulombs. It follows, therefore, that this quantity of electricity is required to liberate 1 gram equivalent of any substance: hence

1 faraday = 96,500 coulombs.

The reliability of this value of the faraday has been confirmed by measurements with the **iodine coulometer** designed by Washburn and Bates, and employed by Bates and Vinal.[5] The apparatus is shown in Fig. 4; it consists of two vertical tubes, containing the anode (*A*) and cathode (*C*) of platinum-iridium foil, joined by a V-shaped portion. A 10 per cent solution of potassium iodide is first placed in the limbs and then

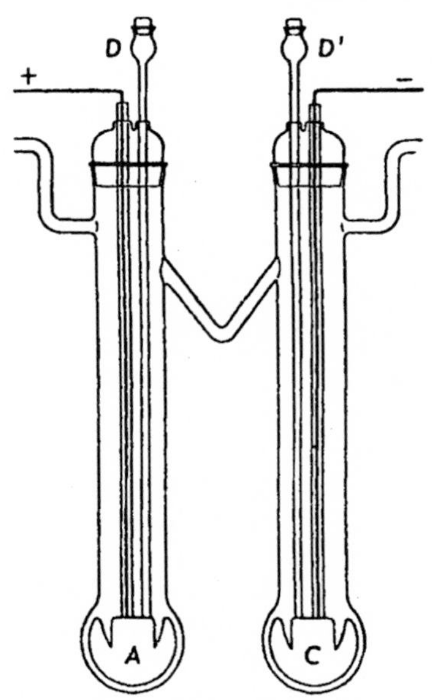

Fig. 4. Iodine coulometer (Washburn and Bates)

[5] Washburn and Bates, *J. Am. Chem. Soc.*, **34**, 1341, 1515 (1912); Bates and Vinal, *ibid.*, **36**, 916 (1914).

by means of the filling tubes D and D' a concentrated solution of potassium iodide is introduced carefully beneath the dilute solution in the anode compartment, and a standardized solution of iodine in potassium iodide is similarly introduced into the cathode compartment. During the passage of current iodine is liberated at the anode while an equivalent amount is reduced to iodide ions at the cathode. After the completion of electrolysis the anode and cathode liquids are withdrawn, through D and D', and titrated with an accurately standardized solution of arsenious acid. In this way the amounts of iodine formed at one electrode and removed at the other can be determined; the agreement between the two results provides confirmation of the accuracy of the measurements. The results obtained by Bates and Vinal in a number of experiments, in which a silver and an iodine coulometer were in series, are given in Table V; the first column records the weight of silver de-

TABLE V. DETERMINATION OF THE FARADAY BY THE IODINE COULOMETER

Silver mg.	Iodine mg.	Coulombs Passed: From Silver deposited	Coulombs Passed: From E.M.F. and Resistance	Milligrams of Iodine per Coulomb	Faraday
4099.03	4822 24	3666.39	3666.65	1.31526	96,498
4397.11	5172.73	3933.01	....	1.31521	96,502
4105.23	4828.51	3671.94	3671.84	1.31498	96,518
4123.10	4849.42	3687.92	...	1.31495	96,521
4104.75	4828.60	3671.51	3671.61	1.31515	96,506
4184.24	4921.30	3742.61	.	1.31494	96,521
4100.27	4822.47	3667.50	3667.65	1.31492	96,523
4105.16	4828.44	3671.88	3671.82	1.31498	96,519
				Mean 1.31502	96,514

posited and the second the mean quantity of iodine liberated or removed; in the third column are the number of coulombs passed, calculated from the data in the first column assuming the faraday to be 96,494 coulombs, and in the fourth are the corresponding values derived from the E.M.F. of the cell employed, that of the Weston standard cell being 1.01830 volt at 25°, and the resistance of the circuit. The agreement between the figures in these two columns shows that the silver coulometer was functioning satisfactorily. The fifth column gives the electrochemical equivalent of iodine in milligrams per coulomb, and the last column is the value of the faraday, i.e., the number of coulombs required to deposit 1 equiv. of iodine, the atomic weight being taken as 126.92.

The faraday, calculated from the work on the iodine coulometer, is thus 96,514 coulombs compared with 96,494 coulombs from the silver coulometer; the agreement is within the limits of accuracy of the known atomic weights of silver and iodine. In view of the small difference between the two values of the faraday given above, the mean figure 96,500 coulombs is probably best for general use.

Measurement of Quantities of Electricity.—Since the magnitude of the faraday is known, it is possible, by means of equation (7), to determine the quantity of electricity passing through any circuit by including in it a coulometer in which an element of known equivalent weight is deposited. Several coulometers, of varying degrees of accuracy and convenience of manipulation, have been described. Since the silver and iodine coulometers have been employed to determine the faraday, these are evidently capable of giving the most accurate results; the iodine coulometer is, however, rarely used in practice because of the difficulty of manipulation. One of the disadvantages of the ordinary form of the silver coulometer is that the deposits are coarse-grained and do not adhere to the cathode; a method of overcoming this is to use an electrolyte made by dissolving silver oxide in a solution of hydrofluoric and boric acids.[6]

In a simplified form of the **silver coulometer,** which is claimed to give results accurate to within 0.1 per cent, the amount of silver dissolved from the anode into a potassium nitrate solution during the passage of current is determined volumetrically.[7]

For general laboratory purposes the **copper coulometer** is the one most frequently employed; [8] it contains a solution of copper sulfate, and the metallic copper deposited on the cathode is weighted. The chief sources of error are attack of the cathode in acid solution, especially in the presence of atmospheric oxygen, and formation of cuprous oxide in neutral solution. In practice slightly acid solutions are employed and the errors are minimized by using cathodes of small area and operating at relatively low temperatures; the danger of oxidation is obviated to a great extent by the presence of ethyl alcohol or of tartaric acid in the electrolyte. The cathode, which is a sheet of copper, is placed midway between two similar sheets which act as anodes; the current density at the cathode should be between 0.002 and 0.02 ampere per sq. cm. At the conclusion of the experiment the cathode is removed, washed with water and dried at 100°. It can be calculated from equation (7) that one coulomb of electricity should deposit 0.3294 milligram of copper.

In a careful study of the copper coulometer, in which electrolysis was carried out at about 0° in an atmosphere of hydrogen, and allowance made for the copper dissolved from the cathode by the acid solution, Richards, Collins and Heimrod (1900) found the results to be within 0.03 per cent of those obtained from a silver coulometer in the same circuit.

The **electrolytic gas coulometer** is useful for the approximate measurement of small quantities of electricity; the total volume of hydrogen and oxygen liberated in the electrolysis of an aqueous solution of sulfuric acid or of sodium, potassium or barium hydroxide can be measured, and from this the quantity of electricity passed can be estimated. If the electrolyte is dilute acid it is necessary to employ platinum electrodes,

[6] von Wartenberg and Schütza, *Z. Elektrochem.*, **36**, 254 (1930).

[7] Kistiakowsky, *Z. Elektrochem.*, **12**, 713 (1906).

[8] Datta and Dhar, *J. Am. Chem. Soc.*, **38**, 1156 (1916); Matthews and Wark, *J. Phys. Chem.*, **35**, 2345 (1931).

but with alkaline electrolytes nickel electrodes are frequently used. One faraday of electricity should liberate one gram equivalent of hydrogen at the cathode and an equivalent of oxygen at the anode, i.e., there should be produced 1 gram of hydrogen and 8 grams of oxygen. Allowing for the water vapor present in the liberated gases and for the decrease in volume of the solution as the water is electrolyzed, the passage of one coulomb of electricity should be accompanied by the formation of 0.174 cc. of mixed hydrogen and oxygen at S.T.P., assuming the gases to behave ideally.

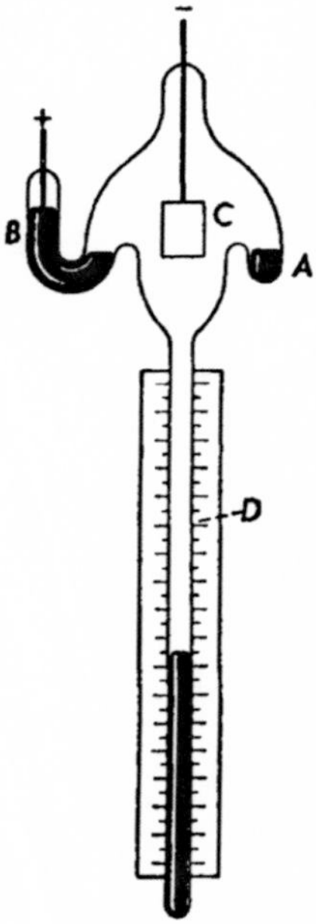

Fig. 5. Mercury coulometer electricity meter

The **mercury coulometer** has been employed chiefly for the measurement of quantities of electricity for commercial purposes, e.g., in electricity meters.[9] The form of apparatus used is shown in Fig. 5; the anode consists of an annular ring of mercury (A) surrounding the carbon cathode (C); the electrolyte is a solution of mercuric iodide in potassium iodide. The mercury liberated at the cathode falls off, under the influence of gravity, and is collected in the graduated tube D. From the height of the mercury in this tube the quantity of electricity passed may be read off directly. When the tube has become filled with mercury the apparatus is inverted and the mercury flows back to the reservoir B. In actual practice a definite fraction only of the current to be measured is shunted through the meter, so that the life of the latter is prolonged. The accuracy of the mercury electricity meter is said to be within 1 to 2 per cent.

A form of mercury coulometer suitable for the measurement of small currents of long duration has also been described.[10]

An interesting form of coulometer, for which an accuracy of 0.01 per cent has been claimed, is the **sodium coulometer**; it involves the passage of sodium ions through glass.[11] The electrolyte is fused sodium nitrate at 340° and the electrodes are tubes of highly conducting glass, electrical contact being made by means of a platinum wire sealed through the glass and dipping into cadmium in the cathode, and cadmium containing some sodium in the anode (Fig. 6). When current is passed, sodium is deposited in the

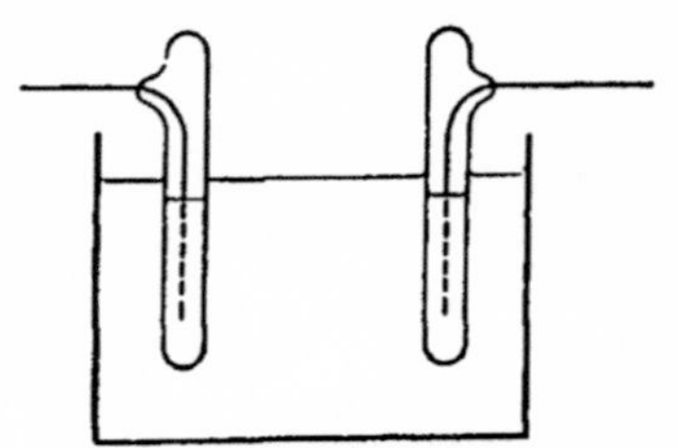

Fig. 6. Sodium coulometer (Stewart)

[9] Hatfield, *Z. Elektrochem.*, **15**, 728 (1909); Schulte, *ibid.*, **27**, 745 (1921).
[10] Lehfeldt, *Phil. Mag.*, **3**, 158 (1902).
[11] Burt, *Phys. Rev.*, **27**, 813 (1926); Stewart, *J. Am. Chem. Soc.*, **53**, 3366 (1931).

glass of the cathode and an equal amount moves out of the anode tube. From the change in weight the quantity of electricity passing may be determined; the anode gives the most reliable results, for with the cathode there is a possibility of the loss of silicate ions from the glass. In spite of the great accuracy that has been reported, it is doubtful if the sodium coulometer as described here will find any considerable application because of experimental difficulties; its chief interest lies in the fact that it shows Faraday's laws hold under extreme conditions.

General Applicability of Faraday's Laws.—The discussion so far has been concerned mainly with the application of Faraday's laws to the material deposited at a cathode, but the laws are applicable to all types of processes occurring at both anode and cathode. The experiments on the iodine coulometer proved that the amount of iodine liberated at the anode was equal to that converted into iodide ions at the cathode, both quantities being in close agreement with the requirements of Faraday's laws. Similarly, provided there are no secondary processes to interfere, the volume of oxygen evolved at an anode in the electrolysis of a solution of dilute acid or alkali is half the volume of hydrogen set free at the cathode.

In the cases referred to above, the anode consists of a metal which is not attacked during the passage of current, but if an attackable metal, e.g., zinc, silver, copper or mercury, is used as the anode, the latter dissolves in amounts exactly equal to that which would be deposited on the cathode by the same quantity of electricity. The results obtained by Bovard and Hulett [12] for the loss in weight of a silver anode and for the amount of silver deposited on the cathode by the same current are given in Table VI; the agreement between the values in the eight experiments shows that Faraday's laws are applicable to the anode as well as to the cathode.

TABLE VI. COMPARISON OF ANODIC AND CATHODIC PROCESSES

Anode loss	Cathode gain	Anode loss	Cathode gain
4.18685 g.	4.18703 g.	4.17651 g.	4.17741 g.
4.13422	4.13422	4.14391	4.14320
4.21204	4.21240	4.08147	4.08097
4.08371	4.08473	4.09386	4.09478

The results obtained at the cathode in the iodine coulometer show that Faraday's laws hold for the reduction of iodine to iodide ions; the laws apply, in fact, to all types of electrolytic reduction occurring at the cathode, e.g., reduction of ferric to ferrous ions, ferricyanide to ferrocyanide, quinone to hydroquinone, etc. The laws are applicable similarly to the reverse process of electrolytic oxidation at the anode. The equivalent weight in these cases is based, of course, on the nature of the oxidation-reduction process.

[12] Bovard and Hulett, *J. Am. Chem. Soc.*, 39, 1077 (1917).

In the discussion hitherto it has been supposed that only one process occurs at each electrode; there are numerous instances, however, of two or more reactions occurring simultaneously. For example, in the electrolysis of nickel salt solutions the deposition of the metal is almost invariably accompanied by the evolution of some hydrogen; when current is passed through a solution of a stannic salt there may be simultaneous reduction of the stannic ions to stannous ions, deposition of tin and liberation of hydrogen at the cathode. Similarly, the electrolysis of a dilute hydrochloric acid solution yields a mixture of oxygen and chlorine at the anode. The conditions which determine the possibility of two or more electrode processes occurring at the same time will be examined in later chapters; in the meantime, it must be pointed out that whenever simultaneous reactions occur, the *total* number of equivalents deposited or reduced at the cathode, or dissolved or oxidized at the anode, are equal to the amount required by Faraday's laws. The passage of one faraday of electricity through a solution of a nickel salt under certain conditions gave a deposit of 25.48 g. of the metal, instead of the theoretical amount 29.34 g.; the number of equivalents of nickel deposited is thus 25.48/29.34, i.e., 0.8684, instead of unity. It follows, therefore, that 0.1316 equiv., i.e., 0.1326 g., of hydrogen is evolved at the same time. The ratio of the actual amount of material deposited, or, in general, the ratio of the actual extent of any electrode reaction, to that expected theoretically is called the **current efficiency** of the particular reaction. In the case under consideration the current efficiency for the deposition of nickel under the given conditions is 0.8684 or 86.84 per cent.

Ions in Two Valence Stages.—A special case of simultaneous electrode processes arises when a given ion can exist in two valence stages, e.g., mercuric (Hg^{++}) and mercurous (Hg_2^{++});* the passage of one faraday then results in the discharge at the cathode or the formation at the anode of a *total* of one gram equivalent of the two ions. An equilibrium exists between a metal and the ions of lower and higher valence; thus, for example,

$$Hg + Hg^{++} \rightleftharpoons Hg_2^{++},$$

and if the law of mass action is applicable to the system, it follows that

$$\frac{\text{Concn. of mercurous ions}}{\text{Concn. of mercuric ions}} = \text{constant},$$

the concentration of the metal being constant. By shaking a simple mercuric salt, e.g., the nitrate, with mercury until equilibrium was established and analyzing the solution, the constant was found to be 120, at room temperatures. When a mercury anode dissolves, the mercurous and mercuric ions are formed in amounts necessary to maintain the

* There is much evidence in favor of the view that the mercurous ion has the formula Hg_2^{++} and not Hg^+ (see p. 264).

equilibrium under consideration; that is, the proportion of mercurous ions is 120 to one part of mercuric ions. It would appear, therefore, that 99.166 per cent of the mercury which dissolves anodically should form mercurous ions: this is true provided no secondary reactions take place in the solution. If the electrolyte is a chloride, the mercurous ions are removed in the form of insoluble mercurous chloride, and in order to maintain the equilibrium between mercury, mercuric and mercurous ions, the anode dissolves almost exclusively in the mercurous form. On the other hand, in a cyanide or iodide solution the mercuric ions are removed by the formation of complex ions, and hence a mercury anode dissolves mainly in the mercuric form. In each case the electrode material passes into solution in such a manner as to establish the theoretical equilibrium, but the existence of subsidiary equilibria in the electrolyte often results in the anode dissolving in the two valence stages in a ratio different from that of the concentrations of the simple ions at equilibrium.

With a copper electrode, the equilibrium is greatly in favor of the cupric ions and so a copper anode normally dissolves virtually completely in the higher valence (cupric) state, i.e., as a bivalent metal. In a cyanide solution, however, cuprous ions are removed as complex cuprocyanide ions; a copper anode then dissolves as a univalent element. Anodes of iron, lead and tin almost invariably dissolve in the lower valence state.

Similar arguments to those given above will apply to the deposition at the cathode; the proportion in which the higher and lower valence ions are discharged is identical with that in which an anode would dissolve in the same electrolyte. Thus, from a solution containing simple mercurous and mercuric ions only, e.g., from a solution of the perchlorates or nitrates, the two ions would be discharged in the ratio of 120 to unity. From a complex cyanide or iodide electrolyte, however, mercuric ions are discharged almost exclusively.

Significance of Faraday's Laws.—Since the discharge at a cathode or the formation at an anode of one gram equivalent of any ion requires the passage of one faraday, it is reasonable to suppose that this represents the charge * carried by a gram equivalent of any ion. If the ion has a valence z, then a mole of these ions, which is equivalent to z equiv., carries a charge of z faradays, i.e., zF coulombs, where F is 96,500. The number of individual ions in a mole is equal to the Avogadro number N, and so the electric charge carried by a single ion is zF/N coulombs. Since z is an integer, viz., one for a univalent ion, two for a bivalent ion, three for a tervalent ion, and so on, it follows that the charge of electricity carried by any single ion is a multiple of a fundamental unit charge equal to F/N. This result implies that electricity, like matter, is atomic in nature and that F/N is the unit or "atom" of electric charge. There are many reasons for identifying this unit charge with the charge of an electron

* It was seen on page 3 that quantity of electricity and electric charge have the same dimensions.

(ϵ), so that

$$\epsilon = \frac{F}{N}. \tag{8}$$

According to these arguments a univalent, i.e., singly charged, cation is formed when an atom loses a single electron, e.g.,

$$Na \rightarrow Na^{+} + \epsilon.$$

A bivalent cation results from the loss of two electrons, e.g.,

$$Cu \rightarrow Cu^{++} + 2\epsilon,$$

and so on. Similarly, a univalent anion is formed when an atom gains an electron, e.g.,

$$Cl + \epsilon \rightarrow Cl^{-}.$$

In general an ion carries the number of charges equal to its valence, and it differs from the corresponding uncharged particle by a number of electrons equal in magnitude to the charge.

Electrons in Electrolysis.—The identification of the unit charge of a single ion with an electron permits a more complete picture to be given of the phenomena of electrolysis. It will be seen from Fig. 2 that the passage of current through a circuit is accompanied by a flow of electrons from anode to cathode, outside the electrolytic cell. If the current is to continue, some process must occur at the surface of the cathode in the electrolyte which removes electrons, while at the anode surface electrons must be supplied: these requirements are satisfied by the discharge and formation of positive ions, respectively, or in other ways. In general. a chemical reaction involving the formation or removal of electrons must always occur when current passes from an electronic to an electrolytic conductor. For example, at a cathode in a solution of silver nitrate, each silver ion takes an electron from the electrode, forming metallic silver; thus

$$Ag^{+} + \epsilon \rightarrow Ag.$$

At a silver anode it is necessary for electrons to be supplied, and this can be achieved by the atoms passing into solutions as ions; thus

$$Ag \rightarrow Ag^{+} + \epsilon.$$

If the anode consisted of an unattackable metal, e.g., platinum, then the electrons must be supplied by the discharge of anions, e.g.,

$$OH^{-} \rightarrow OH + \epsilon,$$

which is followed by

$$2OH = H_2O + \tfrac{1}{2}O_2,$$

resulting in the liberation of oxygen; or

$$Cl^{-} \rightarrow Cl + \epsilon,$$

followed by

$$2Cl = Cl_2,$$

which gives chlorine gas by the discharge of chloride ions. Since the same number of electrons is required by the anode as must be removed from the cathode, it is evident that equivalent amounts of chemical reaction, proportional to the quantity of electricity passing, i.e., to the number of electrons transferred, must take place at both electrodes. The electronic concept, in fact, provides a very simple interpretation of Faraday's laws of electrolysis. It should be clearly understood that although the current is carried through the metallic part of the circuit by the flow of electrons, it is carried through the electrolyte by the ions; the positive ions move in one direction and the negative ions in the opposite direction, the *total* charge of the moving ions being equivalent to the flow of electrons. This aspect of the subject of electrolytic conduction will be considered more fully in Chap. IV.

Equations involving electron transfer, such as those given above, are frequently employed in electrochemistry to represent processes occurring at electrodes, either during electrolysis or in a voltaic cell capable of producing current. It is opportune, therefore, to emphasize their significance at this point: an equation such as

$$Cu \rightarrow Cu^{++} + 2\epsilon$$

means not only that an atom of copper gives up two electrons and becomes a copper ion; it also implies that two faradays are required to cause one gram atom of copper to go into solution forming a mole, or gram-ion, of cupric ions. In general, an electrode process written as involving z electrons requires the passage of z faradays for it to occur completely in terms of moles.

PROBLEMS

1. A constant current, which gave a reading of 25.0 milliamp. on a milliammeter, was passed through a solution of copper sulfate for exactly 1 hour; the deposit on the cathode weighed 0.0300 gram. What is the error of the meter at the 25 milliamp. reading?

2. An average cell, in which aluminum is produced by the electrolysis of a solution of alumina in fused cryolite, takes about 20,000 amps. How much aluminum is produced per day in each cell, assuming a current efficiency of 92 per cent?

3. A current of 0.050 amp. was passed through a silver titration coulometer, and at the conclusion 23 8 cc. of 0.1 N sodium chloride solution were required to titrate the silver dissolved from the anode. How long was the current flowing?

4. What weights of sodium hydroxide and of sulfuric acid are produced at the cathode and anode, respectively, when 1,000 coulombs are passed through a solution of sodium sulfate?

5. Calculate the amount of iodine that would be liberated by a quantity of electricity which sets free 34.0 cc. of gas, at S.T.P., in an electrolytic gas coulometer.

6. In the electrolysis of a solution containing copper (cuprous), nickel and zinc complex cyanides, Faust and Montillon [*Trans. Electrochem. Soc.*, **65**, 361 (1934)] obtained 0.175 g. of a deposit containing 72.8 per cent by weight of copper, 4.3 per cent of nickel and 22.9 per cent of zinc. Assuming no hydrogen was evolved, how many coulombs were passed through the solution?

7. Anthracene can be oxidized anodically to anthraquinone with an efficiency of 100 per cent, according to the reaction $C_{14}H_{10} + 3O = C_{14}H_8O_2 + H_2O$. What weight of anthraquinone is produced by the passage of a current of 1 amp. for 1 hour?

8. A current of 0.10 amp. was passed for two hours through a solution of cuprocyanide and 0.3745 g. of copper was deposited on the cathode. Calculate the current efficiency for copper deposition and the volume of hydrogen, measured at S.T.P., liberated simultaneously.

9. The 140 liters of solution obtained from an alkali-chlorine cell, operating for 10 hours with a current of 1250 amps., contained on the average 116.5 g. of sodium hydroxide per liter. Determine the current efficiency with which the cells were operating.

10. In an experiment on the electrolytic reduction of sodium nitrate solution, Muller and Weber [*Z. Elektrochem.*, **9**, 955 (1903)] obtained 0.0495 g. of sodium nitrite, 0.0173 g. ammonia and 695 cc. of hydrogen at S.T.P., while 2.27 g. of copper were deposited in a coulometer. Evaluate the current efficiency for each of the three products.

11. Oxygen at 25 atm. pressure is reduced cathodically to hydrogen peroxide: from the data of Fischer and Priess [*Ber.*, **46**, 698 (1913)] the following results were calculated for the combined volume of hydrogen and oxygen, measured at S.T.P., liberated in an electrolytic gas coulometer (I) compared with the amount of hydrogen peroxide (II) obtained from the same quantity of electricity.

I.	35.5	200	413	583	1,670 cc. of gas.
II.	34.7	150	265	334	596 mg. of H_2O_2.

Calculate the current efficiency for the formation of hydrogen peroxide in each case, and plot the variation of the current efficiency with the quantity of electricity passed.

12 In the electrolysis of an alkaline sodium chloride solution at 52°, Muller [*Z. anorg. Chem.*, 22, 33 (1900)] obtained the following results:

Active Oxygen as		Copper in
Hypochlorite	Chlorate	Coulometer
0.0015 g.	0.0095 g.	0.115 g.
0.0058	0.0258	0.450
0.105	0.2269	3.110
0.135	0.3185	4.800
0.139	0.4123	7.030

Plot curves showing the variation with the quantity of electricity passed of the current efficiencies for the formation of hypochlorite and of chlorate.

CHAPTER II

ELECTROLYTIC CONDUCTANCE

Specific Resistance and Conductance.—Consider a uniform bar of a conductor of length l cm. and cross-sectional area a sq. cm.; suppose, for simplicity, that the cross section is rectangular and that the whole is divided into cubes of one cm. side, as shown in Fig. 7, I. The resistance

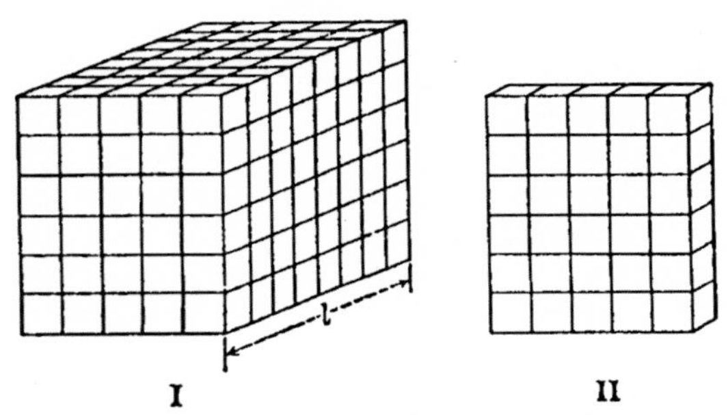

FIG. 7. Calculation of specific resistance

of the bar is seen to be equivalent to that of l layers, such as the one depicted in Fig. 7, II, in *series* with one another; further, each layer is equivalent to a cubes, each of one cm. side, whose resistances are in *parallel*. If ρ is the resistance, in ohms, of a centimeter cube, generally called the **specific resistance** of the substance constituting the conductor, the resistance r of the layer containing the a cubes is given by

$$\frac{1}{r} = \frac{1}{\rho} + \frac{1}{\rho} + \cdots,$$

there being a terms on the right-hand side: it follows, therefore, that

$$r = \frac{\rho}{a}.$$

If R is the resistance of the whole bar, which is equivalent to l layers each of resistance r in series, then

$$R = lr = \rho \frac{l}{a} \text{ ohms.} \tag{1}$$

This equation is applicable to all conductors, electronic or electrolytic, and for uniform conductors of any cross section, not necessarily rectangular.

The **specific conductance** of any conducting material is defined as the reciprocal of the specific resistance; it is given the symbol κ and is stated in reciprocal ohm units, sometimes called "mhos."* Since, by definition, κ is equal to $1/\rho$, it follows from equation (1) that

$$R = \frac{1}{\kappa} \cdot \frac{l}{a} \text{ ohms.} \tag{2}$$

The conductance (C) is the reciprocal of resistance, i.e., $C = 1/R$, and hence

$$C = \kappa \frac{a}{l} \text{ ohms}^{-1}. \tag{3}$$

The physical meaning of the specific conductance may be understood by supposing an E.M.F. of one volt to be applied to a conductor; since $E = 1$, it follows, by Ohm's law, that the current I is equal to $1/R$, and hence to the conductance (C). For a centimeter cube a and l are unity, and so C is equal to κ. It is seen, therefore, that when a potential difference of one volt is applied to a centimeter cube of a conductor, the current in amperes flowing is equal in magnitude to the specific conductance in $\text{ohm}^{-1}\ \text{cm.}^{-1}$ units.

Equivalent Conductance.—For electrolytes it is convenient to define a quantity called the **equivalent conductance** (Λ), representing the conducting power of all the ions produced by 1 equiv. of electrolyte in a given solution. Imagine two large parallel electrodes set 1 cm. apart, and suppose the whole of the solution containing 1 equiv. is placed between these electrodes; the area of the electrodes covered will then be v sq. cm., where v cc. is the volume of solution containing the 1 equiv. of solute. The conductance of this system, which is the equivalent conductance Λ, may be derived from equation (3), where a is equal to v sq. cm. and l is 1 cm.; thus

$$\Lambda = \kappa v, \tag{4}$$

where v is the "dilution" of the solution in cc. per equiv. If c is the concentration of the solution, in *equivalents per liter*, then v is equal to 1000/c, so that equation (4) becomes

$$\Lambda = 1000 \frac{\kappa}{c} \cdot \tag{5}$$

The equivalent conductance of any solution can thus be readily derived from its specific conductance and concentration. Since the units of κ are $\text{ohm}^{-1}\ \text{cm.}^{-1}$, those of Λ are seen from equation (4) or (5) to be $\text{ohm}^{-1}\ \text{cm.}^{2}$

*** It will be apparent from equation (1) or (2) that if R is in ohms, and l and a are in cm. and sq. cm. respectively, the units of κ are $\text{ohm}^{-1}\ \text{cm}^{-1}$. This exact notation will be used throughout the present book.**

In some cases the **molecular conductance** (μ) is employed; it is the conductance of 1 mole of solute, instead of 1 equiv. If v_m is the volume in cc. containing a mole of solute, and c is the corresponding concentration in *moles per liter,** then

$$\mu = \kappa v_m = 1000 \frac{\kappa}{c}. \qquad (6)$$

For an electrolyte consisting of two univalent ions, e.g., alkali halides, the values of Λ and μ are, of course, identical.

Determination of Resistance.—The measurement of resistance is most frequently carried out with some form of Wheatstone bridge circuit, the principle of which may be explained with the aid of Fig. 8. The four arms of the bridge, viz., *ab, ac, bd* and *cd*, have resistances R_1, R_2, R_3 and R_4, respectively; a source of current S is connected across the bridge between b and c, and a current detector D is connected between a and d. Let E_1, E_2, E_3 and E_4 be the fall of potential across the four arms, corresponding to the resistances R_1, R_2, R_3 and R_4, respectively, and suppose the currents in these arms are I_1, I_2, I_3 and I_4, then by Ohm's law:

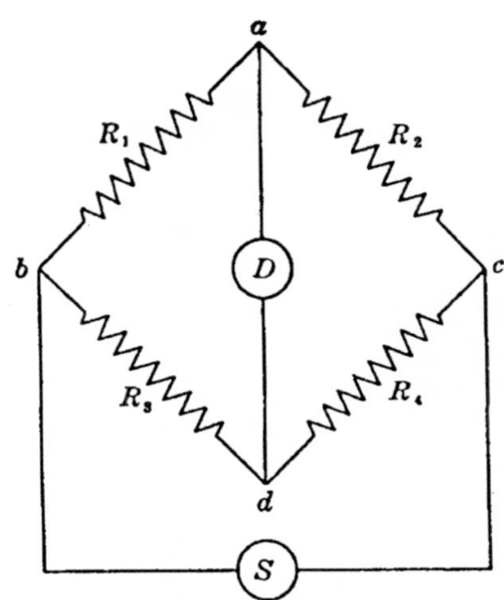

Fig. 8. Wheatstone bridge circuit

$$E_1 = I_1R_1, \qquad E_2 = I_2R_2,$$

$$E_3 = I_3R_3 \quad \text{and} \quad E_4 = I_4R_4.$$

If the resistances are adjusted so that there is no flow of current through the detector D, that is to say, when the bridge is "balanced," the potential at a must be the same as that at d. Since the arms *ab* and *bd* are joined at b and the potentials are the same at a and d, it follows that the fall of potential across *ab*, i.e., E_1, must equal that across *bd*, i.e., E_3. Similarly, the fall of potential across *ac* must be the same as that across *cd*, i.e., E_2 and E_4 are equal. Introducing the values of the various E's given above, it is seen that

$$I_1R_1 = I_3R_3 \quad \text{and} \quad I_2R_2 = I_4R_4,$$

$$\therefore \frac{I_1R_1}{I_2R_2} = \frac{I_3R_3}{I_4R_4}.$$

Since no current passes through *ad* when the bridge is balanced, it is

* In accordance with the practice adopted by a number of writers, the symbol **c** is used to represent concentrations in equivalents and *c* in moles, per liter.

evident that the current flowing in the arm *ab* must be the same as that in *ac*, i.e., $I_1 = I_2$, while that passing through *bd* must be identical with that in *cd*, i.e., $I_3 = I_4$. It follows, therefore, that at the balance point

$$\frac{R_1}{R_2} = \frac{R_3}{R_4}, \tag{7}$$

and so if the resistances of three of the arms of the bridge are known, that of the fourth can be readily evaluated. In practice, R_1 is generally the unknown resistance, and R_2 is a resistance box which permits various known resistances to be used; the so-called "ratio arms" R_3 and R_4 may be a uniform wire (*bdc*) on which the position of *d* is adjusted until the bridge is balanced, as shown by the absence of current in *D*. The ratio of the lengths of the two parts of the uniform wire, corresponding to *bd* and *dc*, gives the ratio R_3/R_4.

Resistance of Electrolytes: Introduction.—In the earliest attempts to determine the resistance of electrolytic solutions the results were so erratic that it was considered possible that Ohm's law was not applicable to electrolytic conductors. The erratic behavior was shown to be due to the use of direct current in the measurement, and when the resulting errors were eliminated it became evident that Ohm's law held good for electrolytic as well as for metallic systems. The passage of direct current through an electrolyte is, as seen in Chap. I, accompanied by changes in composition of the solution and frequently by the liberation of gases at the electrodes. The former alter the conductance and the latter set up an E.M.F. of "polarization" (see Chap. XIII) which tends to oppose the flow of current. The difficulties may be overcome by the use of non-polarizable electrodes and the employment of such small currents that concentration changes are negligible; satisfactory conductance measurements have been made in this way with certain electrolytes by the use of direct current, as will be seen later (p. 47).

The great majority of the work with solutions has, however, been carried out with a rapidly alternating current of low intensity, following the suggestion made by Kohlrausch in 1868. The underlying principle of the use of an alternating current is that as a result of the reversal of the direction of the current about a thousand times per second, the polarization produced by each pulse of the current is completely neutralized by the next, provided the alternations are symmetrical. There is also exact compensation of any concentration changes which may occur. Kohlrausch used an induction coil as a source of alternating current (abbreviated to A.C.) and in his early work a bifilar galvanometer acted as detector; later (1880) he introduced the telephone earpiece, and this, with some improvements, is still the form of A.C. detector most frequently employed in electrolytic conductance measurements. The electrolyte was placed in a cell and its resistance measured by a Wheatstone bridge arrangement shown schematically in Fig. 9. The cell *C* is in the arm *ab* and a resistance box *R* constitutes the arm *ac*; the source of A.C. is represented by *S*, and *D* is the telephone earpiece detector. In the simplest form of bridge, frequently employed for ordinary laboratory purposes, the arms *bd* and *dc* are in the form of a uniform wire, preferably of platinum-iridium, stretched along a meter scale, i.e., the so-called "meter bridge," or suitably wound round a slate cylinder. The point *d* is a sliding contact which is moved back and forth until no sound can be heard in the detector; the bridge is then balanced. If the wire *bc* is uniform, the ratio of the resistances of the two arms is equal to the ratio of the lengths, *bd* and *dc*, as seen above. If the resistance taken from the box *R* is adjusted so as to be approximately equal to that of the electrolyte in the cell *C*, the balance point *d* will be roughly midway between *b* and *c*; a small error in the setting of *d* will then cause the least discrepancy in the final value for the resistance of *C*. If somewhat greater accuracy is desired, two variable resistance boxes may be used for *bd* and *dc*, i.e., R_3 and R_4 (cf. Fig. 8), the resistance taken from each being adjusted until the bridge is balanced. Alternatively, two resistance boxes or coils may be joined by a wire, whose resistance is known in terms of that of the boxes or coils, for the purpose of making the final adjustment.

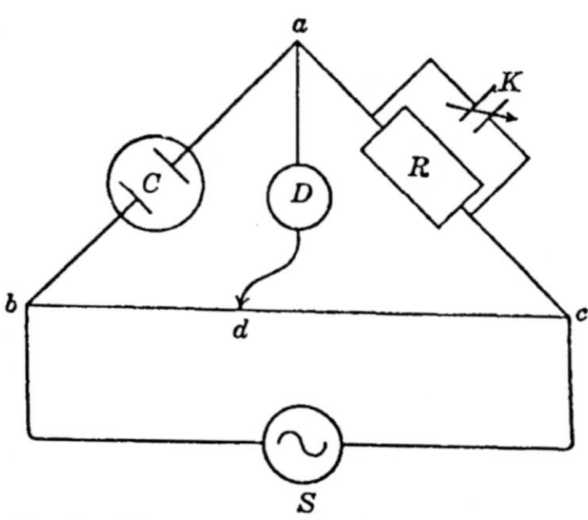

Fig. 9. Measurement of resistance of electrolyte

It will be seen shortly that for precision measurements of electrolytic conductance it is necessary to take special precautions to obviate errors due to inductance and capacity in the bridge circuit. One immediate effect of these factors is to make the minimum sound in the telephone earpiece difficult to detect; for most general purposes this source of error can be overcome by using a good resistance box, in which the coils are wound in such a manner as to eliminate self-induction, and to use a straight-wire bridge, if a special non-inductive bridge is not available. Further, a variable condenser *K* is connected across the resistance box and adjusted until the telephone earpiece gives a sharply defined sound minimum; in this way the unavoidable capacity of the conductance cell may be balanced to some extent.

A.C. Sources and Detectors.—Although the induction coil suffers from being noisy in operation and does not give a symmetrical alternating current, it is still often employed in conductance measurements where great accuracy is not required. A mechanical high-frequency

A.C. generator was employed by Washburn (1913), and Taylor and Acree (1916) recommended the use of the Vreeland oscillator, which consists of a double mercury-arc arrangement capable of giving a symmetrical sine-wave alternating current of constant frequency variable at will from 160 to 4,200 cycles per second. These costly instruments have been displaced in recent years by some kind of vacuum-tube oscillator, first employed in conductance work by Hall and Adams.[1] Several types of suitable oscillators have been described and others are available commercially; the essential circuit of one form of oscillator is shown in Fig. 10. The grid circuit of the thermionic vacuum tube T contains a grid coil L_1 of suitable inductance which is connected to the oscillator coil L_2 in parallel with the variable condenser C. The output coil L_3, which is coupled inductively with L_2, serves to convey the oscillations to the conductance bridge.

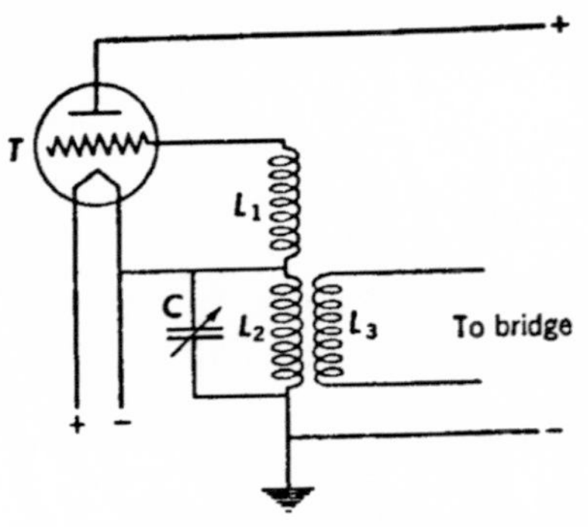

Fig. 10. Vacuum-tube oscillator

The chief advantages of the vacuum-tube oscillator are that it is relatively inexpensive, it is silent in operation and gives a symmetrical sinusoidal alternating current of constant frequency; by suitable adjustment of inductance and capacity the frequency of the oscillations may be varied over the whole audible range, but for conductance work frequencies of 1,000 to 3,000 cycles per sec. are generally employed. The disadvantage of this type of oscillator is that it is liable to introduce stray capacities into the bridge circuit which can be a serious source of error in precision work. The difficulty may be overcome, however, by the use of special grounding devices (see p. 42).

If properly tuned to the frequency of the A.C., the telephone earpiece can be used to detect currents as small as 10^{-9} amp.; it is still regarded as the most satisfactory instrument for conductance measurements. The sensitivity of the telephone can be greatly increased by the addition of a vacuum-tube (low frequency) amplifier; this is particularly valuable when working with very dilute solutions having a high resistance, for it is then possible to determine the balance point of the bridge with greater precision than without the amplifier. The basic circuit of a simple type of audio-frequency amplifier is shown in Fig. 11, in which the conductance bridge is connected to the primary coil of an iron-cored transformer (P); T is a suitable vacuum tube and C is a condenser. The use of a vacuum-tube amplifier introduces the possibility of errors

[1] Hall and Adams, *J. Am. Chem. Soc.*, **41**, 1515 (1919); see also, Jones and Josephs, *ibid.*, **50**, 1049 (1928); Luder, *ibid.*, **62**, 89 (1940); Jones, Mysels and Juda, *ibid.*, **62**, 2919 (1940).

due to capacity and interaction effects, but these can be largely eliminated by suitable grounding and shielding (see p. 42).

If results of a low order of accuracy are sufficient, as, for example, in conductance measurements for analytical or industrial purposes, the A.C. supply mains, of frequency about 60 cycles per sec., can be employed as a source of current; in this case an A.C. galvanometer is a satisfactory detector. A combination of a vacuum-tube, or other form of A.C. rectifier, and a direct current galvanometer has been employed,

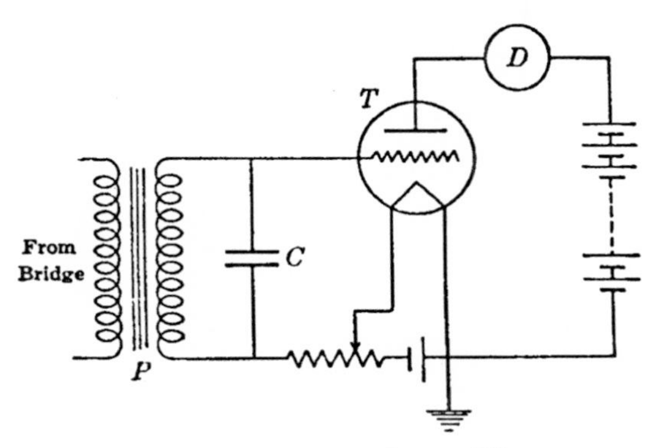

Fig. 11. Vacuum-tube amplifier

and in some cases the thermal effect of the alternating current has been used, in conjunction with a thermocouple and a sensitive galvanometer, for detection purposes.

Electrodes for Conductance Measurements.—For the determination of electrolytic conductance it is the general practice to use two parallel sheets of stout platinum foil, that do not bend readily; their relative positions are fixed by sealing the connecting tubes into the sides of the measuring cell (cf. Fig. 12). In order to aid the elimination of polarization effects by the alternating current, Kohlrausch (1875) coated the electrodes with a layer of finely divided platinum black; these are called **platinized platinum** electrodes. The platinization is carried out by electrolysis of a solution containing about 3 per cent of chloroplatinic acid and 0.02 to 0.03 per cent of lead acetate; the lead salt apparently favors the formation of the platinum deposit in a finely-divided, adherent form. The large surface area of the finely divided platinum appears to catalyze the union of the hydrogen and oxygen which tend to be liberated by the successive pulses of the current; the polarization E.M.F. is thus eliminated.

In some cases the very properties which make the platinized platinum electrodes satisfactory for the reduction of polarization are a disadvantage. The finely-divided platinum may catalyze the oxidation of organic compounds, or it may adsorb appreciable quantities of the solute present in the electrolyte and so alter its concentration. Some workers have overcome this disadvantage of platinized electrodes by heating them to redness and so obtaining a gray surface; the resulting electrode is probably not so effective in reducing polarization, but it adsorbs much less solute than does the black deposit. Others have employed electrodes covered with very thin layers of platinum black, and sometimes smooth electrodes have been used. By making measurements with smooth platinum electrodes at various frequencies and extrapolating the results to infinite frequency, conductance values have been obtained which are in agreement with those given by platinized electrodes; this method is thus available when platinum black must not be used. For the great majority of solutions of simple salts and of inorganic acids and bases it is the practice to employ electrodes coated with a thin layer of platinum black obtained by electrolysis as already described.

Conductance Cells: The Cell Constant.—The cells for electrolytic conductance measurements are made of highly insoluble glass, such as Pyrex, or of quartz; they should be very carefully washed and steamed before use. For general laboratory requirements the simple cell designed by Ostwald (Fig. 12, I) is often employed, but for industrial purposes the "dipping cell" (Fig. 12, II) or the pipette-type of cell (Fig. 12, III) have been found convenient. By means of the two latter cells, samples obtained at various stages in a chemical process can be readily tested.

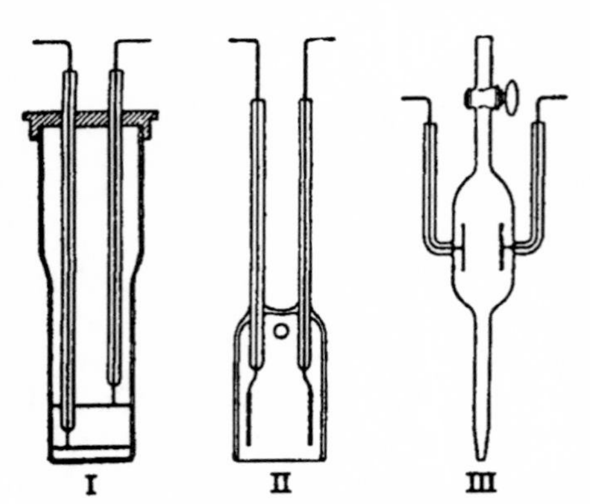

FIG. 12. Types of conductance cells

The resistance (R) of the solution in the cell can be measured, as already explained, and hence the specific conductance (κ) is given by equation (2) as

$$\kappa = \frac{l}{aR},$$

where l is the distance between the electrodes and a is the area of cross section of the electrolyte through which the current passes. For a given cell with fixed electrodes l/a is a constant, called the **cell constant**; if this is given the symbol K cm.$^{-1}$, it follows that

$$\kappa = \frac{K}{R}. \tag{8}$$

It is neither convenient nor desirable, with the cells in general use, to measure l and a with any degree of accuracy, and so an indirect method is employed for the evaluation of the cell constant. If a solution whose specific conductance is known accurately, from other measurements, is placed in the experimental cell and its resistance R is measured, it is possible to obtain K for the given cell directly, by means of equation (8). The electrolyte almost invariably used for this purpose is potassium chloride, its specific conductance having been determined with high precision in cells calibrated by measurement with a concentrated solution of sulfuric acid, the resistance of which has been compared in another cell with that of mercury; the specific conductance of the latter is known accurately from the definition of the international ohm as 10629.63 $\text{ohms}^{-1}\ \text{cm.}^{-1}$ at 0°.

The potassium chloride solutions employed in the most recent work contain 1.0, 0.1 or 0.01 mole in a cubic decimeter of solution at 0°, i.e., 0.999973 liter; these solutions, designated as 1.0 D, 0.1 D and 0.01 D, where D stands for "demal," contain 76.627, 7.4789 and 0.74625 grams of potassium chloride to 1000 grams of water, respectively. The specific conductances of these solutions at 0°, 18° and 25° are quoted in Table VII;[2] the particular solution chosen for calibrating a given cell depends

TABLE VII. SPECIFIC CONDUCTANCES OF POTASSIUM CHLORIDE SOLUTIONS IN $\text{OHM}^{-1}\ \text{CM.}^{-1}$

Temp.	1.0 D	0.1 D	0.01 D
0°	0.065176	0.0071379	0.00077364
18°	0.097838	0.0111667	0.00122052
25°	0.111342	0.0128560	0.00140877

on the range of conductances for which it is to be employed. The values recorded in this table do not include the conductance of the water; when carrying out a determination of the constant of a given conductance cell allowance must be made for this quantity.

Design of Cells.—In the design of conductance cells for precision measurements a number of factors must be taken into consideration. Kohlrausch showed theoretically that the error resulting from polarization was determined by the quantity $P^2/\omega R^2$, where P is the polarization E.M.F., R is the resistance of the electrolyte in the cell and ω is the frequency of the alternating current. It is evident that the error can be made small by adjusting the experimental conditions so that ωR^2 is much greater than P^2; this can be done by making either ω or R, or both, as large as is reasonably possible. There is a limit to the increase in the frequency of the A.C. because the optimum range of audibility of the telephone earpiece is from 1,000 to 4,000 cycles per sec., and so it is desirable to make the resistance high. If this is too high, however, the current strength may fall below the limit of satisfactory audibility, and it is not possible to determine the balance point of the bridge. The

[2] Jones and Bradshaw, *J. Am. Chem. Soc.*, 55, 1780 (1933); see also, Jones and Prendergast, *ibid.*, 59, 731 (1937); Bremner and Thompson, *ibid.*, 59, 2371 (1937); Davies, *J. Chem. Soc.*, 432, 1326 (1937).

highest electrolytic resistances which can be measured with accuracy, taking advantage of the properties of the vacuum-tube audio-amplifier, are about 50,000 ohms. In order to measure low resistances the polarization P should be reduced by adequate platinization of the electrodes, but there is a limit to which this can be carried and experiments show that resistances below 1,000 ohms cannot be measured accurately. The resistances which can be determined in a given cell, therefore, cover a ratio of about 50 to unity. The observed specific conductances of electrolytes in aqueous solution range from approximately 10^{-1} to 10^{-7} ohms^{-1} cm.$^{-1}$, and so it is evident that at least three cells of different dimensions, that is with different cell constants, must be available.

Another matter which must be borne in mind in the design of a conductance cell is the necessity of preventing a rise of temperature in the electrolyte due to the heat liberated by the current. This can be achieved either by using a relatively large volume of solution or by making the cell in the form of a long narrow tube which gives good thermal contact with the liquid in the thermostat.

Two main types of cell have been devised for the accurate measurement of electrolytic conductance; there is the "pipette" type, used by Washburn (1916), and the flask type, introduced by Hartley and Barrett (1913). In the course of a careful study of cells of the pipette form, Parker (1923) found that with solutions of high resistance, for which the polarization error is negligible, there was an apparent decrease of the cell constant with increasing resistance. This phenomenon, which became known as the "Parker effect," was confirmed by other workers; it was at first attributed to adsorption of the electrolyte by the platinized electrode, but its true nature was elucidated by Jones and Bollinger.[3] The pipette type of cell (Fig. 13, I) is electrically equivalent to the circuit depicted in Fig. 13, II; the resistance R_0 is that of the solution contained between the electrodes in the cell, and this is in parallel with the resistance (R_p) of the electrolyte in the filling tube at the right and a capacity (C_p). The latter is equivalent to the distributed capacity between the electrolyte in the body of the cell and the mercury in the contact tube, on the one hand, and the solution in the filling tube, on the other hand; the glass walls of the tubes and the thermostat liquid act as the dielectric medium. An analysis of the effect of shunting the resistance R_0 by a capacity C_p and a resistance R_p shows that, provided

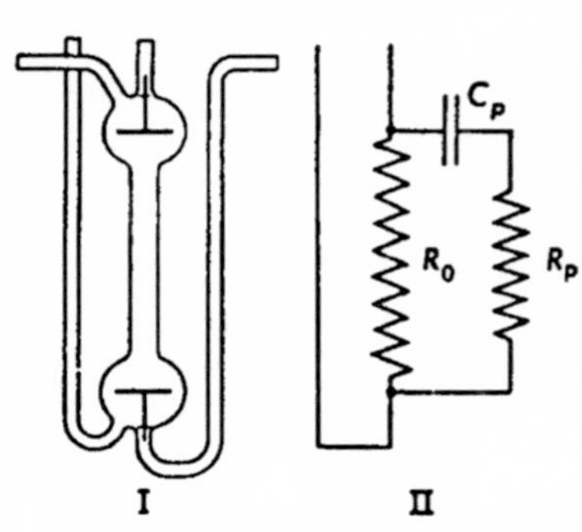

FIG. 13. Illustration of the "Parker effect"

[3] Jones and Bollinger, *J. Am. Chem. Soc.*, **53**, 411 (1931); cf., Washburn, *ibid.*, **38**, 2431 (1916).

the cell is otherwise reasonably well designed, the error ΔR in the measured resistance is given by

$$- \Delta R \propto R_0^3 \omega^2 C_p, \tag{9}$$

where ω, as before, is the frequency of the alternating current. According to equation (9) the apparent cell constant will decrease with increasing resistance R_0, as found in the Parker effect. In order to reduce this source of error, it is necessary that R_0, ω and C_p should be small; as already seen, however, R_0 and ω must be large to minimize the effect of polarization, and so the shunt capacity C_p should be negligible if the Parker effect is to be eliminated. Since most of the shunt capacity lies between the filling tube and the portions of the cell of opposite polarity (cf. Fig. 13, I) it is desirable that these should be as far as possible from each other. This principle is embodied in the cells shown in Fig. 14, designed by Jones and Bollinger; the wider the tube and the closer the electrodes, the smaller the cell constant. These cells exhibit no appreciable Parker effect: the cell constants are virtually independent of the frequency of the A.C. and of the resistance of the electrolyte within reasonable limits.

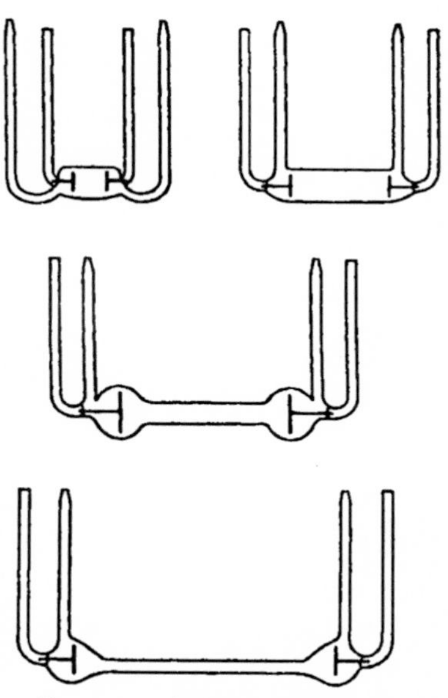

Fig. 14. Cells for accurate conductance measurements (Jones and Bollinger)

The Parker effect is absent from cells with dipping electrodes, such as in cells of the flask type; there are other sources, of error, however, as was pointed out by Shedlovsky.[4] In the cell represented diagrammatically in Fig. 15, I, the true resistance of the solution between the electrodes is R_0, and there is a capacity C_1 between the contact tubes above the electrolyte, and a capacity C_2 in series with a resistance r between those parts immersed in the liquid; the equivalent electrical circuit is shown by Fig. 15, II. When the cell is placed in the arm of a Wheatstone bridge it is found necessary to insert a resistance R and a capacity C in parallel in the opposite arm in order to obtain a balance (cf. p. 33); it can be shown from the theory of alternating currents that

$$\frac{1}{R} = \frac{1}{R_0}\left(1 + \frac{1}{a} - \frac{1}{a^3\omega^2 C_2^2 R^2} + \cdots\right), \tag{10}$$

[4] Shedlovsky, *J. Am. Chem. Soc.*, **54**, 1411 (1932).

where r is taken as proportional to R_0, the constant a being equal to r/R_0. It follows, therefore, that if the cell is balanced by a resistance and a capacity in parallel, no error results if part of the current through the cell is shunted by a pure capacity such as C_1, since the quantity C_1 does

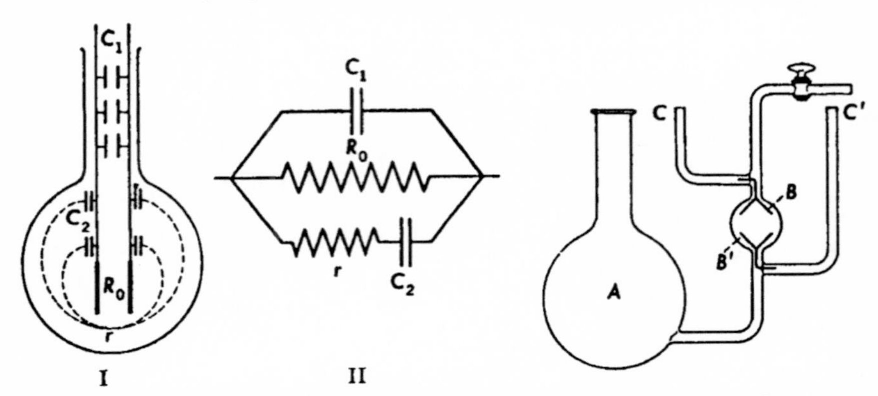

FIG. 15. Equivalent resistance and capacity of flask cell

FIG. 16. Shedlovsky flask cell

not appear in equation (10). On the other hand, parasitic currents resulting from a series resistance-capacity path, i.e., involving r and C_2, will introduce errors, since the apparent resistance R will be different from the true resistance R_0. In order to eliminate parasitic currents, yet retaining the advantages of the flask type of cell for work with a series of solutions of different concentrations, Shedlovsky designed the cell depicted in Fig. 16; the experimental solution contained in the flask A is forced by gas pressure through the side tube into the bulb containing the electrodes B and B'. These consist of perforated platinum cones fused to the walls of the bulb; the contact tubes C and C' are kept apart in order to diminish the capacity between them. The Shedlovsky cell has been used particularly for accurate determination of the conductances of a series of dilute solutions of strong electrolytes.

Temperature Control.—The temperature coefficient of conductance of electrolytes is relatively high, viz., about 2 per cent per degree; in order to obtain an accuracy of two parts in 10,000, which is desirable for accurate work, the temperature should be kept constant within 0.01°. The use of water in the thermostat is not recommended; this liquid has an appreciable conductance and there is consequently a danger of current leakage leading to errors in the measurement, as explained below. The thermostatic liquid should, therefore, be a hydrocarbon oil which is a non-conductor.

Design of the A.C. Bridge.—Strictly speaking the condition of balance of the Wheatstone bridge given by equation (7) is applicable for alternating current only if R_1, R_2, R_3 and R_4 are pure resistances. It is unlikely that the resistance coils will be entirely free from inductance and capacity and, in addition, the conductance cell and its connecting tubes are equivalent to a resistance shunted by a condenser. One consequence of this fact is that the alternating currents in the two arms (R_1 and R_2) of the bridge are not in phase and it is found impossible to obtain any adjustment of the bridge which gives complete silence in the telephone earpiece. For most purposes, this difficulty may be overcome by the use of the condenser K in parallel with the resistance box R_2, as suggested on page 33.

For precision work it is necessary, however, to consider the problem in further detail. For alternating current, Ohm's law takes the form $E = IZ$, where Z is the impedance of the circuit, i.e., Z^2 is equal to $R^2 + X^2$, the quantities R and X being the resistance and reactance, respectively. The condition for balance of a Wheatstone bridge circuit with alternating current is, consequently,

$$\frac{I_1 Z_1}{I_2 Z_2} = \frac{I_3 Z_3}{I_4 Z_4}.$$

If there is no leakage of current from the bridge network to ground, or from one part of the bridge to any other part, and there is no mutual inductance between the arms, I_1 is equal to I_2, and I_3 to I_4, so that

$$\frac{Z_1}{Z_2} = \frac{Z_3}{Z_4} \tag{11}$$

at balance. It follows, therefore, that in a Wheatstone A.C. bridge, under the conditions specified, the impedances, rather than the resistances, are balanced. It can be shown that if the resistances are also to be balanced, i.e., for R_1/R_2 to be equal to R_3/R_4, at the same time as Z_1/Z_2 is equal to Z_3/Z_4, it is necessary that

$$\frac{X_1}{R_1} = \frac{X_2}{R_2} \quad \text{and} \quad \frac{X_3}{R_3} = \frac{X_4}{R_4}. \tag{12}$$

The fraction X/R for any portion of an A.C. circuit is equal to $\tan \theta$, where θ is the phase angle between the voltage and current in the given conductor. It is seen, therefore, that the conditions for the simple Wheatstone bridge relationship between *resistances*, i.e., for equation (7), to be applicable when alternating current is used, are (*a*) that there should be no leakage currents, and (*b*) that the phase angles should be the same in the two pairs of adjacent arms of the bridge.

These requirements have been satisfied in the A.C. bridge designed for electrolytic conductance measurements by Jones and Josephs;[5] the second condition is met by making the two ratio arms (R_3 and R_4, Fig. 8) as nearly as possible identical in resistance and construction, so that any

[5] Jones and Josephs, *J. Am. Chem. Soc.*, 50, 1049 (1928); see also, Luder, *ibid.*, 62, 89 (1940).

reactance, which is deliberately kept small, is the same in each case. In this way X_3/R_3 is made equal to X_4/R_4. It may be noted that this condition is automatically obtained when a straight bridge wire is employed. The reactance of the measuring cell, i.e., X_1, should be made small, but as it cannot be eliminated it should be balanced by a variable condenser in parallel with the resistance box R_2; in this way X_1/R_1 can be made equal to X_2/R_2.

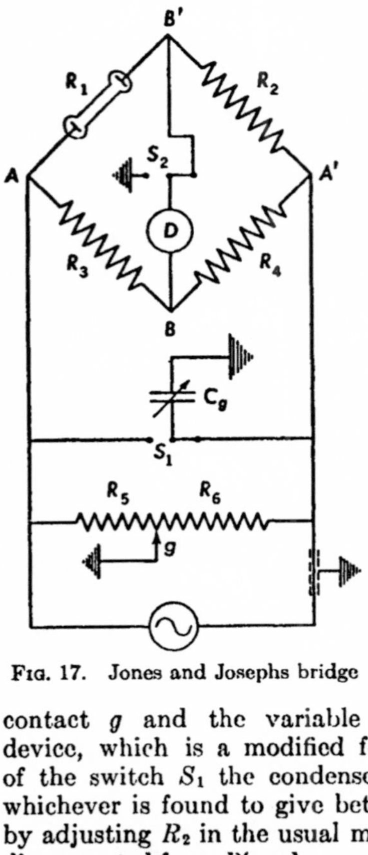

It has often been the practice in conductance work to ground certain parts of the bridge network for the purpose of improving the sharpness of the sound minimum in the detector at the balance point; unless this is done with care it is liable to introduce errors because of the existence of leakage currents to earth. The telephone earpiece must, however, be at ground potential, otherwise the capacity between the telephone coils and the observer will result in a leakage of current. Other sources of leakage are introduced by the use of vacuum-tube oscillator and amplifier, and by various unbalanced capacities to earth, etc.

FIG. 17. Jones and Josephs bridge

The special method of grounding proposed by Jones and Josephs is illustrated in Fig. 17. The bridge circuit consists essentially of the resistances R_1, R_2, R_3 and R_4, as in Fig. 8; the resistances R_5 and R_6, with the movable contact g and the variable condenser C_g, constitute the earthing device, which is a modified form of the Wagner ground. By means of the switch S_1 the condenser C_g is connected either to A or to A', whichever is found to give better results. The bridge is first balanced by adjusting R_2 in the usual manner;* the telephone detector D is then disconnected from B' and connected to ground by means of the switch S_2. The position of the contact g and the condenser C_g are adjusted until there is silence in the telephone, thus bringing B to ground potential. The switch S_2 is now returned to its original position, and R_2 is again adjusted so as to balance the bridge. If the changes from the original positions are appreciable, the process of adjusting g and C_g should be repeated and the bridge again balanced.

Shielding the A.C. Bridge.—In order to eliminate the electrostatic influence between parts of the bridge on one another, and also that due

* This adjustment includes that of a condenser (not shown) in parallel, as explained above; see also page 33 and Fig. 9.

to outside disturbances, grounded metallic shields have sometimes been placed between the various parts of the bridge, or the latter has been completely surrounded by such shields. It has been stated that this form of shielding may introduce more error than it eliminates, on account of the capacity between the shield and the bridge; it has been recommended, therefore, that the external origin of the disturbance, rather than the bridge, should be shielded. According to Shedlovsky[6] the objection to the use of electrostatic screening is based on unsymmetrical shielding which introduces unbalanced capacity effects to earth; further, it is pointed out that it is not always possible to shield the disturbing source. A bridge has, therefore, been designed in which the separate arms of each pair are screened symmetrically; the shields surrounding the cell and the variable resistance (R_2) are grounded, while those around the ratio arms (R_3 and R_4) are not. The leads connecting the oscillator and detector to the bridge are also screened and grounded. In this way mutual and external electrostatic influences on the bridge are eliminated. By means of a special type of twin variable condenser, connected across R_1 and R_2, the reactances in these arms can be compensated so as to give a sharp minimum in the telephone detector and also the correct conditions for R_1/R_2 to be equal to R_3/R_4. It is probable that the screened bridge has advantages over the unscreened bridge when external disturbing influences are considerable.

Preparation of Solvent: Conductance Water.—Distilled water is a poor conductor of electricity, but owing to the presence of impurities such as ammonia, carbon dioxide and traces of dissolved substances derived from containing vessels, air and dust, it has a conductance sufficiently large to have an appreciable effect on the results in accurate work. This source of error is of greatest importance with dilute solutions or weak electrolytes, because the conductance of the water is then of the same order as that of the electrolyte itself. If the conductance of the solvent were merely superimposed on that of the electrolyte the correction would be a comparatively simple matter. The conductance of the electrolyte would then be obtained by subtracting that of the solvent from the total; this is possible, however, for a limited number of solutes. In most cases the impurities in the water can influence the ionization of the electrolyte, or vice versa, or chemical reaction may occur, and the observed conductance of the solution is not the sum of the values of the constituents. It is desirable, therefore, to use water which is as free as possible from impurities; such water is called **conductance water,** or **conductivity water.**

The purest water hitherto obtained was prepared by Kohlrausch and Heydweiller (1894) who distilled it forty-two times under reduced pressure; this water had a specific conductance of 0.043×10^{-6} ohm^{-1} cm.$^{-1}$

[6] Shedlovsky, *J. Am. Chem. Soc.*, 52, 1793 (1930).

at 18°.* Water of such a degree of purity is extremely tedious to prepare, but the so-called "ultra-pure" water, with a specific conductance of 0.05 to 0.06 × 10^{-6} ohm^{-1} cm.$^{-1}$ at 18°, can be obtained without serious difficulty.[7] The chief problem is the removal of carbon dioxide and two principles have been adopted to achieve this end; either a rapid stream of pure air is passed through the condenser in which the steam is being condensed in the course of distillation, or a small proportion only of the vapor obtained by heating ordinary distilled water is condensed, the gaseous impurities being carried off by the uncondensed steam. Ultra-pure water will maintain its low conductance only if air is rigidly excluded, but as such water is not necessary except in special cases, it is the practice to allow the water to come to equilibrium with the carbon dioxide of the atmosphere. The resulting "equilibrium water" has a specific conductance of 0.8 × 10^{-6} ohm^{-1} cm.$^{-1}$ and is quite satisfactory for most conductance measurements.

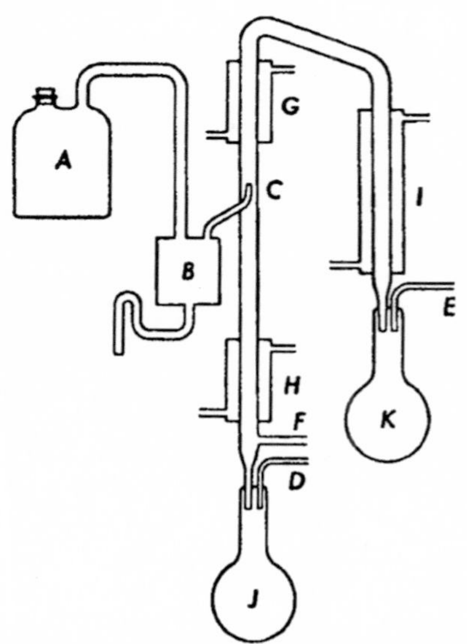

Fig. 18. Apparatus for preparation of conductance water (Vogel and Jeffery)

The following brief outline will indicate the method[8] used for the ready preparation of water having a specific conductance of 0.8 × 10^{-6} ohm^{-1} cm.$^{-1}$; it utilizes both the air-stream and partial condensation methods of purification. The 20-liter boiler *A* (Fig. 18) is of copper, while the remainder of the apparatus should be made of pure tin or of heavily tinned copper. Distilled water containing sodium hydroxide and potassium permanganate is placed in the boiler and the steam passes first through the trap *B*, which collects spray, and then into the tube *C*. A current of purified air, drawn through the apparatus by connecting *D* and *E* to a water pump, enters at *F*; a suction of about 8 inches of water is employed. The temperature of the condenser *G* is so arranged (about 80°) that approximately half as much water is condensed in *H* as in *I*; the best conductance

* Calculations based on the known ionization product of water and the conductances of the hydrogen and hydroxyl ions at infinite dilution (see p. 340) show that the specific conductance of perfectly pure water should be 0.038 × 10^{-6} ohm^{-1} cm.$^{-1}$ at 18°.

[7] Kraus and Dexter, *J. Am. Chem. Soc.*, **44**, 2468 (1922); Bencowitz and Hotchkiss, *J. Phys. Chem.*, 29, 705 (1925); Stuart and Wormwell, *J. Chem. Soc.*, 85 (1930).

[8] Vogel and Jeffery, *J. Chem. Soc.*, 1201 (1931).

water collects in the Pyrex flask *J*, while a somewhat inferior quality is obtained in larger amount at *K*.

For general laboratory measurements water of specific conductance of about 1×10^{-6} ohm^{-1} cm.$^{-1}$ at 18° is satisfactory; this can be obtained by distilling good distilled water, to which a small quantity of permanganate or Nessler's solution is added. A distilling flask of resistance glass is used and the vapor is condensed either in a block-tin condenser or in one of resistance glass. If corks are used they should be covered with tin foil to prevent direct contact with water or steam.

Non-aqueous solvents should be purified by careful distillation, special care being taken to eliminate all traces of moisture. Not only are conductances in water appreciably different from those in non-aqueous media, but in certain cases, particularly if the electrolytic solution contains hydrogen, hydroxyl or alkoxyl ions, small quantities of water have a very considerable effect on the conductance. Precautions should thus be taken to prevent access of water, as well as of carbon dioxide and ammonia from the atmosphere.

Solvent Corrections.—The extent of the correction which must be applied for the conductance of the solvent depends on the nature of the electrolyte; [9] although not all workers are in complete agreement on the subject, the following conclusions are generally accepted. If the solute is a neutral salt, i.e., the salt of a strong acid and a strong base, the ionization and conductance of the carbonic acid, which is the main impurity in water, are not affected to any great extent; the whole of the conductance of the solvent should then be subtracted from that of the solution. With such electrolytes the particular kind of conductance water employed is not critical. Strictly speaking the change in ionic concentration due to the presence of the salt does affect the conductance of the carbonic acid to some extent; when the solvent correction is a small proportion of the total, e.g., in solutions of neutral salts more concentrated than about 10^{-3} N, the alteration is negligible. For more dilute solutions, however, it is advisable to employ ultra-pure water, precautions being taken to prevent the access of carbon dioxide.

Salts of weak bases or weak acids are hydrolyzed in aqueous solution (see Chap. XI) and they behave as if they contained excess of strong acid and strong base, respectively. According to the law of mass action the presence of one acid represses the ionization of a weaker one, so that the effective conductance of the water, which is due mainly to carbonic acid, is diminished. The solvent correction in the case of a salt of a weak base and a strong acid should thus be somewhat less than the total conductance of the water. For solutions of salts of a weak acid and a strong base, which react alkaline, the correction is uncertain, but methods of calculating it have been described; they are based on the assumption

[9] Kolthoff, *Rec. trav. chim.*, **48**, 664 (1929); Davies, *Trans. Faraday Soc.*, **25**, 129 (1929); "The Conductivity of Solutions," 1933, Chap. IV.

that the impurity in the water is carbonic acid.[10] If ultra-pure water is used, the solvent correction can generally be ignored, provided the solution is not too dilute.

If the solution being studied is one of a strong acid of concentration greater than 10^{-4} N, the ionization of the weak carbonic acid is depressed to such an extent that its contribution towards the total conductance is negligible. In these circumstances no water correction is necessary; at most, the value for pure water, i.e., about 0.04×10^{-6} ohm^{-1} cm.$^{-1}$ at ordinary temperatures, may be subtracted from the total. If the concentration of the strong acid is less than 10^{-4} N, a small correction is necessary and its magnitude may be calculated from the dissociation constant of carbonic acid.

The specific conductance of a 10^{-4} N solution of a strong acid, which represents the lowest concentration for which the solvent correction may be ignored, is about 3.5×10^{-5} ohm^{-1} cm.$^{-1}$ Similarly, with weak acids the correction is unnecessary provided the specific conductance exceeds this value. For more dilute solutions the appropriate correction may be calculated, as mentioned above.

The solvent correction to be applied to the results obtained with solutions of bases is very uncertain; the partial neutralization of the alkali by the carbonic acid of the conductance water results in a *decrease* of conductance, and so the solvent correction should be *added*, rather than subtracted. A method of calculating the value of the correction has been suggested, but it would appear to be best to employ ultra-pure water in conductance work with bases.

With non-aqueous solvents of a hydroxylic type, such as alcohols, the corrections are probably similar to those for water; other solvents must be considered on their own merits. In general, the solvent should be as pure as possible, so that the correction is, in any case, small; as indicated above, access of atmospheric moisture, carbon dioxide and ammonia should be rigorously prevented. Since non-hydroxylic solvents such as acetone, acetonitrile, nitromethane, etc., have very small conductances when pure, the correction is generally negligible.

Preparation of Solutions.—When the conductances of a series of solutions of a given electrolyte are being measured, it is the custom to determine the conductance of the water first. Some investigators recommend that measurements should then commence with the most concentrated solution of the series, in order to diminish the possibility of error resulting from the adsorption of solute from the more dilute solutions by the finely divided platinum on the electrodes. When working with cells of the flask type it is the general practice, however, to fill the cell with a known amount of pure solvent, and then to add successive small quantities of a concentrated solution of the electrolyte, of known con-

[10] Davies, *Trans. Faraday Soc.*, 28, 607 (1932); MacInnes and Shedlovsky, *J. Am. Chem. Soc.*, 54, 1429 (1932); Jeffery, Vogel and Lowry, *J. Chem. Soc.*, 1637 (1933); 166 (1934); 21 (1935).

centration, from a weight burette. When cells of other types are used it is necessary to prepare a separate solution for each measurement; this procedure must be adopted in any case if the solute is relatively insoluble.

Direct Current Methods.—A few measurements of electrolytic conductance have been made with direct current and non-polarizable electrodes; the electrodes employed have been mercury-mercurous chloride in chloride solutions, mercury-mercurous sulfate in sulfate solutions, and hydrogen electrodes in acid electrolytes.* Two main principles have been applied: in the first, the direct current is passed between two electrodes whose nature is immaterial; the two non-polarizable electrodes are then inserted at definite points in the electrolyte and the fall of potential between them is measured. The current strength is calculated by determining the potential difference between two ends of a wire of accurately known resistance placed in the circuit. Knowing the potential difference between the two non-polarizable electrodes and the current passing, the resistance of the column of solution separating these electrodes is obtained immediately by means of Ohm's law. The second principle which has been employed is to use the non-polarizable electrodes for leading direct current into and out of the solution in the normal manner and to determine the resistance of the electrolyte by means of a Wheatstone bridge network. A sensitive mirror galvanometer is used as the null instrument and no special precautions need be taken to avoid inductance, capacity and leakage effects, since these do not arise with direct current.[11]

The cells used in the direct current measurements are quite different from those employed with alternating current; there is nothing critical about their design, and they generally consist of horizontal tubes with the electrodes inserted either at the ends or at definite intermediate positions. The constants of the cells are determined either by direct measurement of the tubes, by means of mercury, or by using an electrolyte whose specific conductance is known accurately from other sources. It is of interest to record that where data are available for both direct and alternating current methods, the agreement is very satisfactory, showing that the use of alternating current does not introduce any inherent error. The direct current method has the disadvantage of being applicable only to those electrolytes for which non-polarizable electrodes can be found.

The following simple method for measuring the resistance of solutions of very low specific conductance has been used.[12] A battery of storage

* The nature of these electrodes will be understood better after Chap. VI has been studied.

[11] Eastman, *J. Am. Chem. Soc.*, **42**, 1648 (1920); Brønsted and Nielsen, *Trans. Faraday Soc.*, **31**, 1478 (1935); Andrews and Martin, *J. Am. Chem. Soc.*, **60**, 871 (1938).

[12] LaMer and Downes, *J. Am. Chem. Soc.*, **53**, 888 (1931); Fuoss and Kraus, *ibid.*, **55**, 21 (1933); Bent and Dorfman, *ibid.*, **57**, 1924 (1935).

cells, having an E.M.F. of about 150 volts, is applied to the solution whose resistance exceeds 100,000 ohms; the strength of the current which passes is then measured on a calibrated mirror galvanometer. From a knowledge of the applied voltage and the current strength, the resistance is calculated with the aid of Ohm's law. In view of the high E.M.F. employed, relative to the polarization E.M.F., the error due to polarization is very small; further, since only minute currents flow, the influence of electrolysis and heating is negligible.

Conductance Determinations at High Voltage and High Frequency.—The electrolytic conductances of solutions with alternating current of very high frequency or of high voltage have acquired special interest in connection with modern theories of electrolytic solutions. Under these

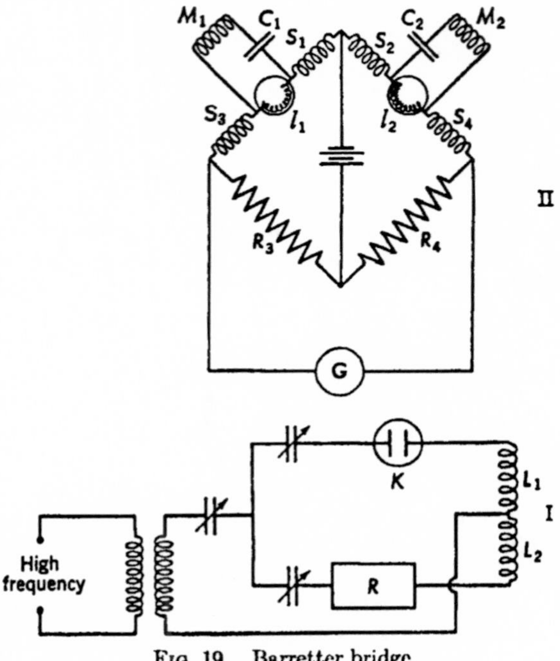

Fig. 19. Barretter bridge

extreme conditions the simple Wheatstone bridge method cannot be used, and other experimental procedures have been described. The chief difficulty lies in the determination of the balance point, and in this connection the "barretter bridge" has been found to be particularly valuable. A form of this bridge is shown in Fig. 19, II; it is virtually a Wheatstone bridge, one arm containing the choke inductances S_1 and S_3, and a small fine-wire filament "barretter" lamp (l_1), across which is shunted a coupling inductance M_1 and a condenser C_1; the corresponding arm of the bridge contains the chokes S_2 and S_4, and the barretter tube l_2, which is carefully matched with l_1, shunted by the coupling inductance M_2 and the condenser C_2. The ratio arms of the bridge consist of the variable resistances R_3 and R_4. The actuating direct current voltage for the bridge is supplied by a direct current battery, and the detecting instrument is the galvanometer G; an inductance in series with the latter prevents induced currents from passing through it. At the beginning of the experiment the resistances R_3 and R_4 are adjusted until the bridge is balanced.

The actual resistance circuit is depicted in Fig. 19, I; K is the conductance cell and R is a variable resistance which are coupled to the barretter circuit by means of the inductances L_1 and L_2. The high frequency or high voltage is applied to the terminals of this circuit, and the currents induced in the bridge are restricted to the barretters l_1 and l_2 by the pairs of inductances S_1-S_3 and S_2-S_4, respectively. The heating effect of these currents causes a change of resistance of the barretters, and if the currents in L_1 and L_2 are different, the bridge will be thrown out of balance. The resistance R is then adjusted until the bridge remains balanced when the current is applied to the cell circuit. The cell K is now replaced by a standard variable resistance and, keeping R constant, this is adjusted until the bridge is again balanced; the value of this resistance is then equal to that of the cell K.[13]

Results of Conductance Measurements.—The results recorded here refer to measurements made at A.C. frequencies and voltages that are not too high, i.e., one to four thousand cycles per sec. and a few volts per cm., respectively. Under these conditions the electrolytic conductances are independent of the voltage, i.e., Ohm's law is obeyed, and of frequency, provided polarization is eliminated. Although the property of a solution that is actually measured is the specific conductance at a given concentration, this quantity is not so useful for comparison purposes as is the equivalent conductance; the latter gives a measure of the conducting power of the ions produced by one equivalent of the electrolyte at the given concentration and is invariably employed in electrochemical work. The equivalent conductance is calculated from the measured specific conductance by means of equation (5).

A large number of conductance measurements of varying degrees of accuracy have been reported; the most reliable results for some electrolytes in aqueous solution at 25° are recorded in Table VIII, the concentrations being expressed in equivalents per liter.[14]

These data show that the equivalent conductance, and hence the conducting power of the ions in one gram equivalent of any electrolyte, increases with decreasing concentration. The figures appear to approach

[13] Malsch and Wien, *Ann. Physik*, **83**, 305 (1927); Neese, *ibid.*, **8**, 929 (1931); Wien, *ibid.*, **11**, 429 (1931); Schiele, *Physik. Z.*, **35**, 632 (1934).

[14] For a critical compilation of recent accurate data, see MacInnes, "The Principles of Electrochemistry," 1939, p. 339; for other data International Critical Tables, Vol. VI, and the Landolt-Bornstein Tabellen should be consulted.

TABLE VIII. EQUIVALENT CONDUCTANCES AT 25° IN OHMS^{-1} CM.[2]

Concn.	HCl	KCl	NaI	NaOH	$AgNO_3$	$\frac{1}{2}BaCl_2$	$\frac{1}{2}NiSO_4$	$\frac{1}{3}LaCl_3$	$\frac{1}{4}K_4Fe(CN)_6$
0.0005 N	422.74	147.81	125.36	246	131.36	135.96	118.7	139.6	—
0.001	421.36	146.95	124.25	245	130.51	134.34	113.1	137.0	167.24
0.005	415.80	143.55	121.25	240	127.20	128.02	93.2	127.5	146.09
0.01	412.00	141.27	119.24	237	124.76	123.94	82.7	121.8	134.83
0.02	407.24	138.34	116.70	233	121.41	119.09	72.3	115.3	122.82
0.05	399.09	133.37	112.79	227	115.24	111.48	59.2	106.2	107.70
0.10	391.32	128.96	108.78	221	109.14	105.19	50.8	99.1	97.87

a limiting value in very dilute solutions; this quantity is known as the **equivalent conductance at infinite dilution** and is represented by the symbol Λ_0.

An examination of the results of conductance measurements of many electrolytes of different kinds shows that the variation of the equivalent conductance with concentration depends to a great extent on the *type* of electrolyte, rather than on its actual nature. For strong uni-univalent electrolytes, i.e., with univalent cation and anion, such as hydrochloric acid, the alkali hydroxides and the alkali halides, the decrease of equivalent conductance with increasing concentration is not very large. As the valence of the ions increases, however, the falling off is more marked; this is shown by the curves in Fig. 20 in which the equivalent conduct-

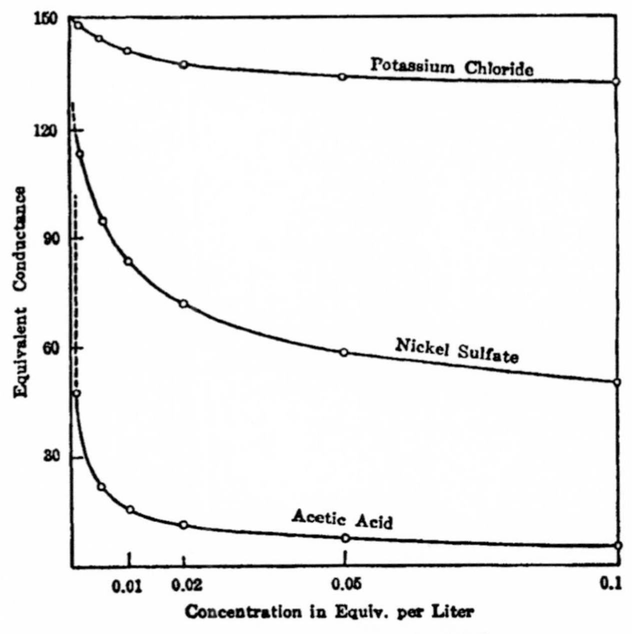

FIG. 20. Conductances of electrolytes of different types

ances of potassium chloride, a typical uni-univalent strong electrolyte, and of nickel sulfate, a bi-bivalent electrolyte, are plotted as functions of the concentration. Electrolytes of an intermediate valence type, e.g., potassium sulfate, a uni-bivalent electrolyte, and barium chloride, which is a bi-univalent salt, behave in an intermediate manner.

The substances referred to in Table VIII are all strong, or relatively strong, electrolytes, but weak electrolytes, such as weak acids and bases, exhibit an apparently different behavior. The results for acetic acid, a typical weak electrolyte, at 25° are given in Table IX.

TABLE IX. EQUIVALENT CONDUCTANCE OF ACETIC ACID AT 25°

Concn.	0.0001	0.001	0.005	0.01	0.02	0.05	0.10 N
Λ	131.6	48.63	22.80	16.20	11.57	7.36	5.20 $ohms^{-1}$ $cm.^2$

It is seen that at the higher concentrations the equivalent conductance is very low, which is the characteristic of a weak electrolyte, but in the more dilute solutions the values rise with great rapidity; the limiting equivalent conductance of acetic acid is known from other sources to be 390.7 $ohms^{-1}$ $cm.^2$ at 25°, and so there must be an increase from 131.6 to this value as the solution is made more dilute than 10^{-4} equiv. per liter. The plot of the results for acetic acid, shown in Fig. 20, may be regarded as characteristic of a weak electrolyte. As mentioned in Chap. I, it is not possible to make a sharp distinction between electrolytes of different classes, and the variation of the equivalent conductance of an intermediate electrolyte, such as trichloroacetic, cyanoacetic and mandelic acids, lies between that for a weak electrolyte, e.g., acetic acid, and a moderately strong electrolyte, e.g., nickel sulfate (cf. Fig. 20).

The Conductance Ratio.—The ratio of the equivalent conductance (Λ) at any concentration to that at infinite dilution (Λ_0)* has played an important part in the development of electrochemistry; it is called the **conductance ratio,** and is given the symbol α, thus

$$\alpha = \frac{\Lambda}{\Lambda_0}. \tag{13}$$

In the calculations referred to on page 10, Arrhenius assumed the conductance ratio to be equal to the degree of dissociation of the electrolyte; this appears to be approximately true for weak electrolytes, but not for salts and strong acids and bases. Quite apart from any theoretical significance which the conductance ratio may have, it is a useful empirical quantity because it indicates the extent to which the equivalent conductance at any specified concentration differs from the limiting value. The change of conductance ratio with concentration gives a measure of the corresponding falling off of the equivalent conductance. In accordance with the remarks made previously concerning the connection be-

* For the methods of extrapolation of conductance data to give the limiting value, see p. 54.

tween the variation of equivalent conductance with concentration and the valence type of the electrolyte, a similar relationship should hold for the conductance ratio. In dilute solutions of strong electrolytes, other than acids, the conductance ratio is in fact almost independent of the nature of the salt and is determined almost entirely by its valence type. Some mean values, derived from the study of a number of electrolytes at room temperatures, are given in Table X; the conductance ratio at any

TABLE X. CONDUCTANCE RATIO AND VALENCE TYPE OF SALT

Valence Type	0.001	0.01	0.1 N
Uni-uni	0.98	0.93	0.83
Uni-bi / Bi-uni	0.95	0.87	0.75
Bi-bi	0.85	0.65	0.40

given concentration is seen to be smaller the higher the valence type. For weak electrolytes the conductance ratios are obviously very much less, as is immediately evident from the data in Table IX.

As a general rule increase of temperature increases the equivalent conductance both at infinite dilution and at a definite concentration; the conductance ratio, however, usually decreases with increasing temperature, the effect being greater the higher the concentration. These conclusions are supported by the results for potassium chloride solutions in Table XI taken from the extensive measurements of Noyes and his

TABLE XI. VARIATION OF CONDUCTANCE RATIO OF POTASSIUM CHLORIDE SOLUTIONS WITH TEMPERATURE

	18°	100°	150°	218°	306°
0.01 N	0.94	0.91	0.90	0.90	0.81
0.08 N	0.87	0.83	0.80	0.77	0.64

collaborators.[15] The falling off is more marked for electrolytes of higher valence type, and especially for weak electrolytes. A few cases are known in which the conductance ratio passes through a maximum as the temperature is increased; this effect is probably due to changes in the extent of dissociation of relatively weak electrolytes.

Equivalent Conductance Minima.—Provided the dielectric constant of the medium is greater than about 30, the conductance behavior in that medium is usually similar to that of electrolytes in water; the differences are not fundamental and are generally differences of degree only. With solvents of low dielectric constant, however, the equivalent conductances often exhibit distinct abnormalities. It is frequently found, for example, that with decreasing concentration, the equivalent conductance *decreases* instead of increasing; at a certain concentration, however, the value passes through a minimum and the subsequent variation is normal. In other cases, e.g., potassium iodide in liquid sulfur dioxide and tetra-

[15] Noyes *et al.*, *J. Am. Chem. Soc.*, **32**, 159 (1910); see also, Kraus, "Electrically Conducting Systems," 1922, Chap. VI.

propylammonium iodide in methylene chloride, the equivalent conductances pass through a maximum and a minimum with decreasing concentration. The problem of the minimum equivalent conductance was investigated by Walden [16] who concluded that there was a definite relationship between the concentration at which such a minimum could be observed and the dielectric constant of the solvent. If $c_{min.}$ is the concentration for the minimum equivalent conductance, and D is the dielectric constant of the medium, then Walden's conclusion may be represented as

$$c_{min.} = kD^3, \tag{14}$$

where k is a constant for the given electrolyte. It is evident from this equation that in solvents of high dielectric constant the minimum should be observed only at extremely high concentrations; even if such solutions could be prepared, it is probable that other factors would interfere under these conditions. It will be seen later that equation (14) has a theoretical basis.

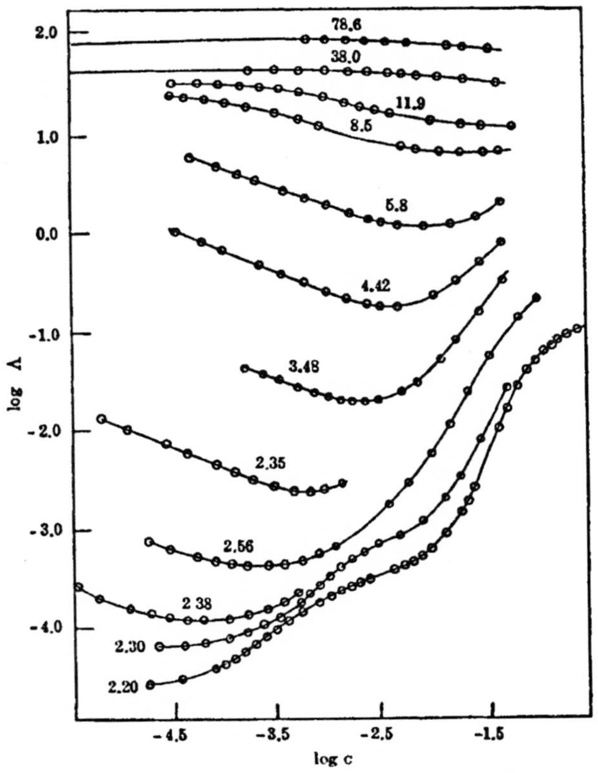

FIG. 21. Influence of dielectric constant on conductance (Fuoss and Kraus)

[16] Walden, *Z. physik. Chem.*, 94, 263 (1920); 100, 512 (1922).

The influence of dielectric constant on the variation of equivalent conductance with concentration has been demonstrated in a striking manner by the measurements made by Fuoss and Kraus[17] on tetra-isoamylammonium nitrate at 25° in a series of mixtures of water and dioxane, with dielectric constant varying from 78.6 to 2.2. The results obtained are depicted graphically in Fig. 21, the dielectric constant of the medium being indicated in each case; in view of the large range of conductances and concentrations the figure has been made more compact by plotting log Λ against log c. It is seen that as the dielectric constant becomes smaller, the falling off of equivalent conductance with increasing concentration is more marked. At sufficiently low dielectric constants the conductance minimum becomes evident; the concentration at which this occurs decreases with decreasing dielectric constant, in accordance with the Walden equation. The theoretical implication of these results will be considered more fully in Chap. V.

Equivalent Conductance at Infinite Dilution.—A number of methods have been proposed at various times for the extrapolation of experimental equivalent conductances to give the values at infinite dilution. Most of the procedures described for strong electrolytes are based on the use of a formula of the type

$$\Lambda = \Lambda_0 - ac^n, \tag{15}$$

where Λ is the equivalent conductance measured at concentration c; the quantities a and n are constants, the latter being approximately 0.5, as required by the modern theoretical treatment of electrolytes. If data for sufficiently dilute solutions are available, a reasonably satisfactory value for Λ_0 may be obtained by plotting the experimental equivalent conductances against the square-root of the concentration and performing a linear extrapolation to zero concentration. It appears doubtful, from recent accurate work, if an equation of the form of (15) can represent completely the variation of equivalent conductance over an appreciable range of concentrations; it follows, therefore, that no simple extrapolation procedure can be regarded as entirely satisfactory. An improved method[18] is based on the theoretical Onsager equation (p. 90), i.e.,

$$\Lambda = \Lambda_0' - (A + B\Lambda_0')\sqrt{c},$$

$$\therefore \quad \Lambda_0' = \frac{\Lambda + A\sqrt{c}}{1 - B\sqrt{c}},$$

where A and B are constants which may be evaluated from known properties of the solvent. The results for Λ_0' derived from this equation for solutions of appreciable concentration are not constant, and hence the prime has been added to the symbol for the equivalent conductance.

[17] Fuoss and Kraus, *J. Am. Chem. Soc.*, **55**, 21 (1933).
[18] Shedlovsky, *J. Am. Chem. Soc.*, **54**, 1405 (1932).

For many strong electrolytes Λ_0' is a linear function of the concentration, thus

$$\Lambda_0' = \Lambda_0 + ac,$$

so that if the values of Λ_0' are plotted against the concentration c, the equivalent conductance at infinite dilution may be obtained by linear extrapolation. The data for sodium chloride and hydrochloric acid at 25° are shown in Fig. 22; the limiting equivalent conductances at zero concentration are 126.45 and 426.16 ohm^{-1}cm.^{2}, respectively.

For weak electrolytes, no form of extrapolation is satisfactory, as will be evident from an examination of Fig. 20. The equivalent conductance at infinite dilution can then be obtained only from the values of the individual ions, as will be described shortly. For electrolytes exhibiting intermediate behavior, e.g., solutions of salts in media of relatively low dielectric constant, an extrapolation method based on theoretical considerations can be employed (see p. 167).

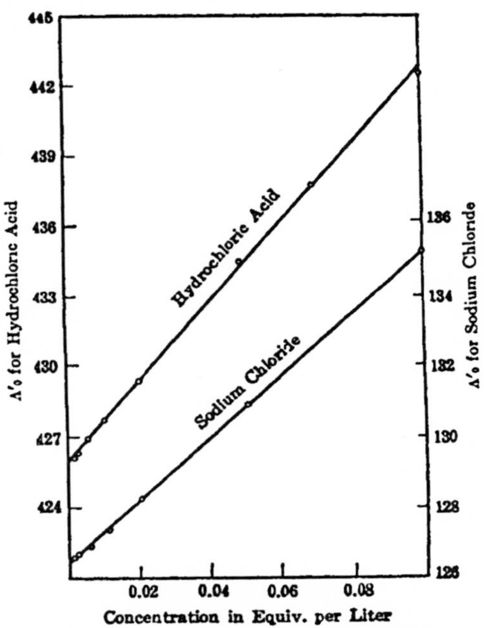

FIG. 22. Extrapolation to infinite dilution

The Independent Migration of Ions.—A survey of equivalent conductances at infinite dilution of a number of electrolytes having an ion in common will bring to light certain regularities; the data in Table XII,

TABLE XII. COMPARISON OF EQUIVALENT CONDUCTANCES AT INFINITE DILUTION

Electrolyte	Λ_0	Electrolyte	Λ_0	Difference
KCl	130.0	NaCl	108.9	21.1
KNO_3	126.3	$NaNO_3$	105.2	21.1
K_2SO_4	133.0	Na_2SO_4	111.9	21.1

for example, are for corresponding sodium and potassium salts at 18°. The difference between the conductances of a potassium and a sodium salt of the same anion is seen to be independent of the nature of the latter. Similar results have been obtained for other pairs of salts with an anion or a cation in common, both in aqueous and non-aqueous solvents. Observations of this kind were first made by Kohlrausch (1879, 1885) by comparing equivalent conductances at high dilutions; he ascribed them to the fact that under these conditions every ion makes a definite contribution towards the equivalent conductance of the electrolyte, irrespective of the nature of the other ion with which it is associated in the solution. The value of the equivalent conductance at infinite dilution may thus be regarded as made up of the sum of two independent factors, one characteristic of each ion; this result is known as Kohlrausch's **law of independent migration of ions.** The law may be expressed in the form

$$\Lambda_0 = \lambda_+^0 + \lambda_-^0, \tag{16}$$

where λ_+^0 and λ_-^0 are known as the **ion conductances,** of cation and anion, respectively, at infinite dilution. The ion conductance is a definite constant for each ion, in a given solvent, its value depending only on the temperature.

It will be seen later that the ion conductances at infinite dilution are related to the speeds with which the ions move under the influence of an applied potential gradient. Although it is possible to derive their values from the equivalent conductances of a number of electrolytes by a method of trial and error, a much more satisfactory procedure is based on the use of accurate transference number data; these transference numbers are determined by the relative speeds of the ions present in the electrolyte and hence are related to the relative ion conductances. The determination of transference numbers will be described in Chap. IV and the method of evaluating ion conductances will be given there; the results will, however, be anticipated and some of the best values for ion conductances in water at 25° are quoted in Table XIII.[19] It should be noted that since these are

TABLE XIII. ION CONDUCTANCES AT INFINITE DILUTION AT 25° IN OHMS^{-1} CM.^{2}

Cation	λ_+^0	$\alpha \times 10^2$	Anion	λ_-^0	$\alpha \times 10^2$
H^+	349.82	1.42	OH^-	198	1.60
Tl^+	74.7	1.87	Br^-	78.4	1.87
K^+	73.52	1.89	I^-	76.8	1.86
NH_4^+	73.4	1.92	Cl^-	76.34	1.88
Ag^+	61.92	1.97	NO_3^-	71.44	1.80
Na^+	50.11	2.09	ClO_4^-	68.0	—
Li^+	38.69	2.26	HCO_3^-	44.5	—
$\frac{1}{2}Ba^{++}$	63.64	2.06	$\frac{1}{2}SO_4^{--}$	79.8	1.96
$\frac{1}{2}Ca^{++}$	59.50	2.11	$\frac{1}{3}Fe(CN)_6^{---}$	101.0	—
$\frac{1}{2}Sr^{++}$	59.46	2.11	$\frac{1}{4}Fe(CN)_6^{----}$	110.5	—
$\frac{1}{2}Mg^{++}$	53.06	2.18			

actually *equivalent* conductances, symbols such as $\frac{1}{2}Ba^{++}$ and $\frac{1}{3}Fe(CN)_6^{---}$ are employed. (The quantities recorded in the columns headed α are approximate temperature coefficients; their significance will be explained on page 61.)

In the results recorded in Table XIII, there appears to be no connection between ionic size and conductance; for a number of ions be-

[19] See MacInnes, *J. Franklin Inst.*, **225**, 661 (1938); "The Principles of Electrochemistry," 1939, p. 342.

longing to a homologous series, as for example the ions of normal fatty acids, a gradual decrease of conductance is observed and a limiting value appears to be approached with increasing chain length. The data for certain fatty acid anions are known accurately, but others are approximate only; the values in Table XIV, nevertheless, show the definite trend

TABLE XIV. ION CONDUCTANCES OF FATTY ACID IONS AT 25°

Anion	Formula	λ_-^0
Formate	HCO_2^-	~52 ohms^{-1} cm.^{2}
Acetate	$CH_3CO_2^-$	40.9
Propionate	$CH_3CH_2CO_2^-$	35.8
Butyrate	$CH_3(CH_2)_2CO_2^-$	32.6
Valerianate	$CH_3(CH_2)_3CO_2^-$	~29
Caproate	$CH_3(CH_2)_4CO_2^-$	~28

towards a constant ion conductance. A similar tendency has been observed in connection with the conductances of alkylammonium ions.

A large number of ion conductances, of more or less accuracy, have been determined in non-aqueous solvents; reference to these will be made shortly in the section dealing with the relationship between the conductance of a given ion in various solvents and the viscosities of the latter.

Application of Ion Conductances.—An important use of ion conductances is to determine the equivalent conductance at infinite dilution of certain electrolytes which cannot be, or have not been, evaluated from experimental data. For example, with a weak electrolyte the extrapolation to infinite dilution is very uncertain, and with sparingly soluble salts the number of measurements which can be made at appreciably different concentrations is very limited. The value of Λ_0 can, however, be obtained by adding the ion conductances. For example, the equivalent conductance of acetic acid at infinite dilution is the sum of the conductances of the hydrogen and acetate ions; the former is derived from a study of strong acids and the latter from measurements on acetates. It follows, therefore, that at 25°

$$\begin{aligned}\Lambda_{0(CH_3CO_2H)} &= \lambda_{H^+}^0 + \lambda_{CH_3CO_2^-}^0\\ &= 349.8 + 40.9 = 390.7 \text{ ohms}^{-1}\text{ cm.}^2\end{aligned}$$

The same result can be derived in another manner which is often convenient since it avoids the necessity of separating the conductance of an electrolyte into the contributions of its constituent ions. The equivalent conductance of any electrolyte MA at infinite dilution $\Lambda_{0(MA)}$ is equal to $\lambda_{M^+}^0 + \lambda_{A^-}^0$, where $\lambda_{M^+}^0$ and $\lambda_{A^-}^0$ are the ion conductances of the ions M^+ and A^- at infinite dilution; it follows, therefore, that

$$\Lambda_{0(MA)} = \Lambda_{0(MCl)} + \Lambda_{0(NaA)} - \Lambda_{0(NaCl)},$$

where $\Lambda_{0(MCl)}$, $\Lambda_{0(NaA)}$ and $\Lambda_{0(NaCl)}$ are the equivalent conductances at infinite dilution of the chloride of the metal M, i.e., MCl, of the sodium salt of the anion A, i.e., NaA, and of sodium chloride, respectively. Any convenient anion may be used instead of the chloride ion, and similarly the sodium ion may be replaced by another metallic cation or by the hydrogen ion. For example, if M^+ is the hydrogen ion and A^- is the acetate ion, it follows that

$$\begin{aligned}\Lambda_{0(CH_3CO_2H)} &= \Lambda_{0(HCl)} + \Lambda_{0(CH_3CO_2Na)} - \Lambda_{0(NaCl)} \\ &= 426.16 + 91.0 - 126.45 \\ &= 390.71 \text{ ohms}^{-1} \text{ cm.}^2 \text{ at } 25°.\end{aligned}$$

In order to determine the equivalent conductance of a sparingly soluble salt it is the practice to add the conductances of the constituent ions; thus for silver chloride and barium sulfate the results are as follows:

$$\begin{aligned}\Lambda_{0(AgCl)} &= \lambda^0_{Ag^+} + \lambda^0_{Cl^-} \\ &= 61.92 + 76.34 = 138.3 \text{ ohms}^{-1} \text{ cm.}^2 \text{ at } 25°, \\ \Lambda_{0(\frac{1}{2}BaSO_4)} &= \lambda^0_{\frac{1}{2}Ba^{++}} + \lambda^0_{\frac{1}{2}SO_4^{--}} \\ &= 63.64 + 79.8 = 143.4 \text{ ohms}^{-1} \text{ cm.}^2 \text{ at } 25°.\end{aligned}$$

Absolute Ionic Velocities: Ionic Mobilities.—The approach of the equivalent conductances of all electrolytes to a limiting value at very high dilutions may be ascribed to the fact that under these conditions all the ions that can be derived from one gram equivalent are taking part in conducting the current. At high dilutions, therefore, solutions containing one equivalent of various electrolytes will contain equivalent numbers of ions; the total charge carried by all the ions will thus be the same in every case. The ability of an electrolyte to transport current, and hence its conductance, is determined by the product of the number of ions and the charge carried by each, i.e., the total charge, and by the actual speeds of the ions. Since the total charge is constant for equivalent solutions at high dilution, the limiting equivalent conductance of an electrolyte must depend only on the ionic velocities: it is the difference in the speeds of the ions which is consequently responsible for the different values of ion conductances. The speed with which a charged particle moves is proportional to the potential gradient, i.e., the fall of potential per cm., directing the motion, and so the speeds of ions are specified under a potential gradient of unity, i.e., one volt per cm. These speeds are known as the **mobilities** of the ions.

If u^0_+ and u^0_- are the actual velocities of positive and negative ions of a given electrolyte at infinite dilution under unit potential gradient, i.e., the respective mobilities, then the equivalent conductance at infinite dilution must be proportional to the sum of these quantities; thus

$$\Lambda_0 = k(u^0_+ + u^0_-) = ku^0_+ + ku^0_-, \tag{17}$$

where k is the proportionality constant which must be the same for all electrolytes. The equivalent conductance, as seen above, is the sum of the ion conductances, i.e.,

$$\Lambda_0 = \lambda^0_+ + \lambda^0_-,$$

and since λ_+^0 and u_+^0 are determined only by the nature of the positive ion, while λ_-^0 and u_-^0 are determined only by the negative ion, it follows that

$$\lambda_+^0 = ku_+^0 \quad \text{and} \quad \lambda_-^0 = ku_-^0. \tag{18}$$

Imagine a very dilute solution of an electrolyte, at a concentration c equiv. per liter, to be placed in a cube of 1 cm. side with square electrodes of 1 sq. cm. area at opposite faces, and suppose an E.M.F. of 1 volt to be applied. By definition, the specific conductance (κ) is the conductance of a centimeter cube, and the equivalent conductance of the given dilute solution, which is virtually that at infinite dilution, is $1000\ \kappa/c$ [see equation (5)], so that

$$1000\frac{\kappa}{c} = \Lambda_0 = \lambda_+^0 + \lambda_-^0,$$

$$\therefore \quad \kappa = \frac{c(\lambda_+^0 + \lambda_-^0)}{1000}.$$

It was shown on page 30 that when a potential difference of 1 volt is applied to a 1 cm. cube, the current in amperes is numerically equal to the specific conductance, i.e.,

$$I = \kappa = \frac{c(\lambda_+^0 + \lambda_-^0)}{1000},$$

and this represents the number of coulombs flowing through the cube per second.

Since the mobilities u_+^0 and u_-^0 are the ionic velocities in cm. per sec. under a fall of potential of 1 volt per cm., all the cations within a length of u_+^0 cm. will pass across a given plane in the direction of the current in 1 sec., while all the anions within a length of u_-^0 cm. will pass in the opposite direction. If the plane has an area of 1 sq. cm., all the cations in a volume u_+^0 cc. and all the anions in u_-^0 cc. will move in opposite directions per sec.; since 1 cc. of the solution contains c/1000 equiv., it follows that a total of $(u_+^0 + u_-^0)c/1000$ equiv. of cations and anions will be transported by the current in 1 sec. Each equivalent of any ion carries one faraday (F) of electricity; hence the total quantity carried per sec. will be $F(u_+^0 + u_-^0)c/1000$ coulombs. It has been seen that the quantity of electricity flowing per sec. through the 1 cm. cube is equal to I as given above; consequently,

$$\frac{F(u_+^0 + u_-^0)c}{1000} = \frac{c(\lambda_+^0 + \lambda_-^0)}{1000},$$

$$\therefore \quad F(u_+^0 + u_-^0) = \lambda_+^0 + \lambda_-^0. \tag{19}$$

It follows, therefore, that the constant k in equation (17) is equal to F, and hence by equation (18),

$$\lambda_+^0 = Fu_+^0 \quad \text{and} \quad \lambda_-^0 = Fu_-^0. \tag{20}$$

The absolute velocity of any ion in cm. per sec. under a potential gradient of 1 volt per cm. can thus be obtained by dividing the ion conductance in $ohms^{-1}$ $cm.^2$ by the value of the faraday in coulombs, i.e., 96,500. Since the velocity is proportional to the potential gradient, as a consequence of the applicability of Ohm's law to electrolytes, the speed of an ion can be evaluated for any desired fall of potential. It should be pointed out that equation (20) gives the ionic velocity at infinite dilution; the values decrease with increasing concentration, especially for strong electrolytes.

The ion conductances in Table XIII have been used to calculate the mobilities of a number of ions at infinite dilution at 25°; the results are recorded in Table XV. It will be observed that, apart from hydrogen

TABLE XV. CALCULATED IONIC MOBILITIES AT 25°

Cation	Mobility cm./sec.	Anion	Mobility cm./sec.
Hydrogen	36.2×10^{-4}	Hydroxyl	20.5×10^{-4}
Potassium	7.61	Sulfate	8.27
Barium	6.60	Chloride	7.91
Sodium	5.19	Nitrate	7.40
Lithium	4.01	Bicarbonate	4.61

and hydroxyl ions, most ions have velocities of about 5×10^{-4} cm. per sec. at 25° under a potential gradient of unity. The influence of temperature on ion conductance, and hence on ionic speeds, is discussed below.

Experimental Determination of Ionic Velocities.—An attempt to measure the speeds of ions directly was made by Lodge (1886) who made use of some characteristic property of the ion, e.g., production of color with an indicator or formation of a precipitate, to follow its movement under an applied field. In Lodge's apparatus the vessels containing the anode and cathode, respectively, were joined by a tube 40 cm. long filled with a conducting gelatin gel in which the indicating material was dissolved. For example, in determining the velocity of barium and chloride ions the gel contained acetic acid as conductor and a trace of silver sulfate as indicator; barium chloride was used in both anode and cathode vessels and the electrodes were of platinum. On passing current the barium and chloride ions moved into the gel, in opposite directions, producing visible precipitates of barium sulfate and silver chloride, respectively: the rates of forward movement of the precipitates gave the speeds of the respective ions under the particular potential gradient employed.

Although the results obtained by Lodge in this manner were of the correct order of magnitude, they were generally two or three times less than those calculated from ion conductances by the method described above. The discrepancies were shown by Whetham (1893) to be due to a non-uniform potential gradient and to lack of precautions to secure sharp boundaries. Taking these factors into consideration, Whetham devised an apparatus for observing the movement of the boundary between a colorless and a colored ion, or between two colored ions, without the use of a gel. The values for the velocities of ions obtained in this manner were in satisfactory agreement with those calculated, especially when allowance was made for the fact that the latter refer to infinite dilution. The principle employed by Whetham is almost identical with that used in the modern "moving boundary" method for determining transference numbers and this is described in Chap. IV.

Influence of Temperature on Ion Conductances.—Increase of temperature invariably results in an increase of ion conductance at infinite dilution; the variation with temperature may be expressed with fair accuracy by means of the equation

$$\lambda_t^0 = \lambda_{25}^0[1 + \alpha(t - 25) + \beta(t - 25)^2], \tag{21}$$

where λ_t^0 is the ion conductance at infinite dilution at the temperature t, and λ_{25}^0 is the value at 25°. The factors α and β are constants for a given ion in the particular solvent; for a narrow temperature range, e.g., about 10° on either side of 25°, the constant β may be neglected, and approximate experimental values of α are recorded in Table XIII above. It will be apparent that, except for hydrogen and hydroxyl ions, the temperature coefficients α are all very close to 0.02 at 25°.

Since the conductance of an ion depends on its rate of movement, it seems reasonable to treat conductance in a manner analogous to that employed for other processes taking place at a definite rate which increases with temperature. If this is the case, it is possible to write

$$\lambda^0 = Ae^{-E/RT}, \tag{22}$$

where A is a constant, which may be taken as being independent of temperature over a relatively small range; E is the activation energy of the process which determines the rate of movement of the ions, R is the gas constant and T is the absolute temperature. Differentiation of equation (22) with respect to temperature, assuming A to be constant, gives

$$\frac{d \ln \lambda^0}{dT} = \frac{1}{\lambda^0} \cdot \frac{d\lambda^0}{dT} = \frac{E}{RT^2}. \tag{23}$$

Further, differentiation of equation (21) with respect to temperature, the factor β being neglected, shows that for a narrow temperature range

$$\frac{1}{\lambda^0} \cdot \frac{d\lambda^0}{dT} = \alpha$$

and hence the activation energy is given by

$$E = \alpha RT^2.$$

Since α is approximately 0.02 for all ions, except hydrogen and hydroxyl ions, at 25°, it is seen that for conductance in water the activation energy is about 3.60 kcal. in every case.

Ion Conductance and Viscosity: Temperature and Pressure Effects.— It is an interesting fact that the activation energy for electrolytic conductance is almost identical with that for the viscous flow of water, viz., 3.8 kcal. at 25°; hence, it is probable that ionic conductance is related to the viscosity of the medium. Quite apart from any question of mechanism, however, equality of the so-called activation energies means that the positive temperature coefficient of ion conductance is roughly equal to the negative temperature coefficient of viscosity. In other words, the product of the conductance of a given ion and the viscosity of water at a series of temperatures should be approximately constant. The results in Table XVI give the product of the conductance of the acetate ion at

TABLE XVI. CONDUCTANCE-VISCOSITY PRODUCT OF THE ACETATE ION

Temperature	0°	18°	25°	59°	75°	100°	128°	156°
$\lambda_0\eta_0$	0.366	0.368	0.366	0.368	0.369	0.368	0.369	0.369

infinite dilution (λ_0) and the viscosity of water (η_0), i.e., $\lambda_0\eta_0$, at temperatures between 0° and 156°; the results are seen to be remarkably constant. It is true that such constancy is not always obtained, but the conductance-viscosity product for infinite dilution is, at least, approximately independent of temperature for a number of ions in water. The data for non-aqueous media are less complete, but it appears that in general the product of the ionic conductance and the viscosity in such media is also approximately constant over a range of temperatures.*

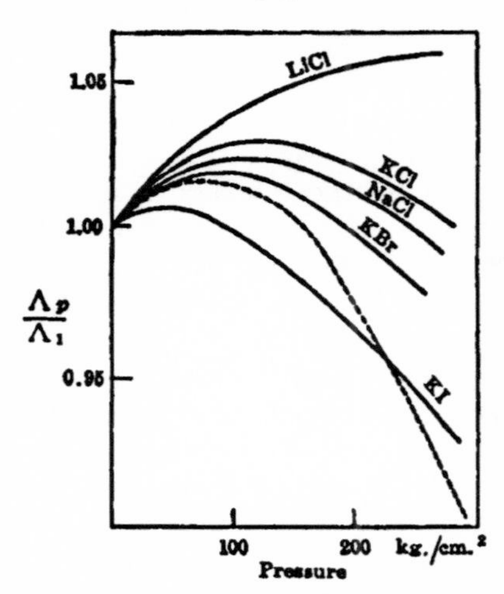

Fig. 23. Variation of conductance with pressure

Another fact which points to a relationship between ionic mobility and viscosity is the effect of pressure on electrolytic conductance. Data are not available for infinite dilution, but the results of measurements on a number of electrolytes at a concentration of 0.01 N in water at 20° are shown in Fig. 23; the ordinates give the ratio of the equivalent conductance at a pressure p to that at unit pressure, i.e., Λ_p/Λ_1, while the abscissae represent the pressures in kg. per sq. cm.[20] The dotted line

* It should be emphasized that the conductance-viscosity product constancy is, on the whole, not applicable to solutions of appreciable concentration.

[20] Data mainly from Körber, *Z. physik. Chem.*, 67, 212 (1909); see also, Adams and Hall, *J. Phys. Chem.*, 35, 2145 (1931); Zisman, *Phys. Rev.*, 39, 151 (1932).

indicates the variation with pressure of the fluidity, i.e., the reciprocal of the viscosity, of water relative to that at unit pressure. The existence of a maximum in both the conductance and fluidity curves suggests that there is some parallelism between these quantities: exact agreement would be expected only at infinite dilution, for other factors which are influenced by pressure may be important in solutions of appreciable concentration.

The relationship between viscosity and ion conductance has been interpreted in at least two ways; some writers have suggested that the constancy of the product $\lambda_0\eta_0$ proves the applicability of Stokes's law to ions in solution. According to this law

$$f = 6\pi\eta r u, \tag{24}$$

where u is the steady velocity with which a particle of radius r moves through a medium of viscosity η when a force f is applied. For a particular ion, r may be regarded as constant, and since the conductance is proportional to the speed of the ion under the influence of a definite applied potential (see p. 58), it follows that according to Stokes's law $\lambda_0\eta_0$ should be constant, as found experimentally. Another suggestion that has been made to explain this fact is that the ion in solution is so completely surrounded by solvent molecules which move with it, that is to say, it is so extensively "solvated," that its motion through the medium is virtually the same as the movement of solvent molecules past one another in viscous flow of the solvent.

It is not certain, however, that either of these conclusions can be legitimately drawn from the results. Since the activation energies for ionic mobility and viscous flow are approximately equal, it is reasonable to suppose that the rate-determining stage in the movement of an ion under the influence of an applied electric field and that involved in the viscous flow of the medium are the same. It has been suggested that in the latter process the slow stage is the jump of a solvent molecule from one equilibrium position to another, and this must also be rate-determining for ionic conductance. It appears, therefore, that when an electric field is applied to a solution containing ions, the latter can move forward only if a solvent molecule standing in its path moves in the opposite direction. The actual rate of movement of an ion will depend to a great extent on its effective size in the given solvent, but the temperature coefficient should be determined almost entirely by the activation energy for viscous flow.

Influence of Solvent on Ion Conductance.—In the course of his investigation of the conductance of tetraethylammonium iodide in various solvents, Walden (1906) noted that the product of the equivalent conductance at infinite dilution and the viscosity of the solvent was approximately constant and independent of the nature of the latter;[21] this conclusion, known as **Walden's rule,** may be expressed as

$$\Lambda_0\eta_0 \approx \text{constant}, \tag{25}$$

for a given electrolyte in any solvent. The values of $\Lambda_0\eta_0$ for the aforementioned salt, obtained by Walden and others, in a variety of media are given in Table XVII; the viscosities are in poises, i.e., dynes per sq. cm.

TABLE XVII. VALUES OF $\Lambda_0\eta_0$ FOR TETRAETHYLAMMONIUM IODIDE IN VARIOUS SOLVENTS

Solvent	CH_3OH	CH_3COCH_3	CH_3CN	$C_2H_4Cl_2$	CH_3NO_2	$C_6H_5NO_2$	C_6H_5OH
$\Lambda_0\eta_0$	0.63	0.66	0.64	0.60	0.69	0.67	0.63

The results were generally obtained at 25°, but since $\lambda_0\eta_0$ is approximately independent of temperature, as seen above, it is evident that $\Lambda_0\eta_0$ will also not vary appreciably.

If Walden's rule holds for other electrolytes, it follows, since Λ_0 is the sum of the conductances of the constituent ions, that $\lambda_0\eta_0$ should be approximately constant for a given ion in all solvents. The extent to which this is true may be seen from the conductance-viscosity products for a number of ions collected in Table XVIII; the data for hydrogen

TABLE XVIII. ION CONDUCTANCE-VISCOSITY PRODUCTS IN VARIOUS SOLVENTS AT 25°

Solvent	Na^+	K^+	Ag^+	$N(C_2H_5)_4^+$	I^-	ClO_4^-	Picrate
H_2O	0.460	0.670	0.563	0.295	0.685	0.606	0.276
CH_3OH	0.250	0.293	0.274	0.338	0.334	0.387	0.255
C_2H_5OH	0.204	0.235	0.195	0.310	0.290	0.340	0.292
CH_3COCH_3	0.253	0.259	—	0.284	0.366	0.366	0.275
CH_3CN	0.241	0.296	—	0.296	0.347	0.359	0.268
CH_3NO_2	0.364	0.383	0.326	0.310	0.403	—	0.276
$C_6H_5NO_2$	—	—	0.330	0.322	—	0.366	0.277
NH_3 (−33°)	0.333	0.430	0.297	—	0.437	—	—

and hydroxyl ions are deliberately excluded from Table XVIII, for reasons which will appear later. The results show that, for solvents other than water, the conductance-viscosity product of a given ion is approximately constant, thus confirming the approximate validity of Walden's rule. If Stokes's law were obeyed, the value of $\lambda_0\eta_0$ would be constant only if the effective radius of the ion were the same in the different media; since there are reasons for believing that most ions are solvated in solution, the dimensions of the moving unit will undoubtedly

[21] Walden *et al.*, *Z. physik. Chem.*, **107**, 219 (1923); **114**, 297 (1925); **123**, 429 (1926); "Salts, Acids and Bases," 1929; Ulich, *Fortschritte der Chemie, Physik and phys. Chem.*, **18**, No. 10 (1926); *Trans. Faraday Soc.*, **23**, 388 (1927); Barak and Hartley, *Z. phys. Chem.*, **165**, 273 (1933); Coates and Taylor, *J. Chem. Soc.*, 1245, 1495 (1936); see also Longsworth and MacInnes, *J. Phys. Chem.*, **43**, 239 (1939).

vary to some extent and exact constancy of the conductance-viscosity product is not to be expected. It should be pointed out, also, that the deduction of Stokes's law is based on the assumption of a spherical particle moving in a continuous medium, and this condition can be approximated only if the moving particle is large in comparison with the molecules of the medium. It is of interest to note in this connection that for large ions, such as the tetraethylammonium and picrate ions, the $\lambda_0\eta_0$ values are much more nearly constant than is the case with other ions; further, the behavior of such ions in water is not exceptional. Stokes's law is presumably applicable to these large ions, and since they are probably solvated to a small extent only, they will have the same size in all solvents; the constancy of the conductance-viscosity product is thus to be expected. For small ions the value of $\lambda_0\eta_0$ will depend to some extent on the fundamental properties of the solvent, as well as on the effective size of the ion: for such ions, too, Stokes's law probably does not hold, and so exact constancy of the conductance-viscosity product is not to be expected.

An interesting test of the validity of the Walden rule is provided by the conductance measurements, made by LaMer and his collaborators, of various salts in a series of mixtures of light water (H_2O) and heavy water (D_2O). The results indicate that, although the rule holds approximately, it is by no means exact.[22]

Although no actual tabulation has been made here of the ion conductances of various ions in different solvents, it may be pointed out that these values are implicit in Table XVIII; knowing the viscosity of the solvent, the ion conductance at infinite dilution can be calculated.

Abnormal Ion Conductances.—An inspection of the conductance-viscosity products for the hydrogen ion recorded in Table XIX immediately reveals the fact that the values in the hydroxylic solvents, and particularly in water, are abnormally high. It might appear, at first sight, that the high conductance-viscosity product of the hydrogen ion in water could be explained by its small size. In view of the high free energy of hydration of the proton (cf. p. 308), however, in aqueous solution the reaction

TABLE XIX. CONDUCTANCE-VISCOSITY PRODUCT OF THE HYDROGEN ION

Solvent	H_2O	CH_3OH	C_2H_5OH	CH_3COCH_3	CH_3NO_2	$C_6H_5NO_2$	NH_3
$\lambda_0\eta_0$	3.14	0.774	0.641	0.277	0.395	0.401	0.359

$$H^+ + H_2O = H_3O^+,$$

where H^+ represents a **proton** or "bare" hydrogen ion, must go to virtual completion. The hydrogen ion in water cannot, therefore, consist of a

[22] LaMer *et al.*, *J. Chem. Phys.*, 3, 406 (1935); 9, 265 (1941); *J. Am. Chem. Soc.*, 58, 1642 (1936); 59, 2425 (1937); see also, Longsworth and MacInnes, *ibid.*, 59, 1666 (1937).

bare ion, but must be combined with at least one molecule of water. The hydrogen ion in water is thus probably to be represented by H_3O^+, and its effective size and conducting power should then be approximately the same as that of the sodium ion; it is, however, actually many times greater, as the figures in Table XIX show. It is of interest to note that in acetone, nitromethane, nitrobenzene, liquid ammonia, and probably in other non-hydroxylic solvents, the conductance-viscosity product, and hence the conductance, of the hydrogen ion, which is undoubtedly solvated, is almost the same as that of the sodium ion. It is doubtful, therefore, if the high conductance of the hydrogen ion in hydroxylic solvents can be explained merely by its size.

The suggestion has been frequently made that the high conductance is due to a type of Grotthuss conduction (p. 7), and this view has been developed by a number of workers in recent years.[23] It is supposed, as already indicated, that the hydrogen ion in water is H_3O^+ with three hydrogen atoms attached to the central oxygen atom. When a potential gradient is applied to an aqueous solution containing hydrogen ions, the latter travel to some extent by the same mechanism as do other ions, but there is in addition another mechanism which permits of a more rapid ionic movement. This second process is believed to involve the transfer of a proton (H^+) from a H_3O^+ ion to an adjacent water molecule; thus

$$\underset{\oplus}{\mathrm{H{-}\overset{\overset{\displaystyle H}{|}}{O}{-}H}} + \mathrm{\overset{\overset{\displaystyle H}{|}}{O}{-}H} \rightarrow \mathrm{H{-}\overset{\overset{\displaystyle H}{|}}{O}} + \underset{\oplus}{\mathrm{H{-}\overset{\overset{\displaystyle H}{|}}{O}{-}H}}.$$

The resulting H_3O^+ ion can now transfer a proton to another water molecule, and in this way the positive charge will be transferred a considerable distance in a short time. It has been calculated from the known structure of water that the proton has to jump a distance of 0.86×10^{-8} cm. from a H_3O^+ ion to a water molecule, but as a result the positive charge is effectively transferred through 3.1×10^{-8} cm. The electrical conductance will thus be much greater than that due solely to the normal mechanism. It will be observed that after the proton has passed from the H_3O^+ ion to the water molecule, the resulting water molecule, i.e., the one shown on the right-hand side, is oriented in a different manner from that to which the proton was transferred, i.e., the one on the left-hand side. If the process of proton jumping is to continue, each water molecule must rotate after the proton has passed on, so that it may be ready to receive another proton coming from the same direction. The combination of proton transfer and rotation of the water molecule, which has some features in common with the Grotthuss mechanism for conduc-

[23] Hückel, *Z. Elektrochem.*, **34**, 546 (1928); Bernal and Fowler, *J. Chem. Phys.*, 1, 515 (1933); Wannier, *Ann. Physik*, **24**, 545, 569 (1935); Stearn and Eyring, *J. Chem. Phys.*, **5**, 113 (1937); see also, Glasstone, Laidler and Eyring, "The Theory of Rate Processes," 1941, Chap. X.

tion, is sufficient to account for the high conductance of the hydrogen ion in aqueous solution.

The abnormal conductance of the hydrogen ion in methyl and ethyl alcohols, which is somewhat less than in water, can also be accounted for by a proton transfer analogous to that suggested for water; thus, if the hydrogen ion in an alcohol ROH is represented by ROH_2^+, the process is

$$\underset{\oplus}{\mathrm{H{-}\overset{\overset{\displaystyle R}{|}}{O}{-}H}} + \mathrm{\overset{\overset{\displaystyle R}{|}}{O}{-}H} \rightarrow \mathrm{H{-}\overset{\overset{\displaystyle R}{|}}{O}} + \underset{\oplus}{\mathrm{H{-}\overset{\overset{\displaystyle R}{|}}{O}{-}H}},$$

followed by rotation of the alcohol molecule. To account for the dependence of abnormal conductance on the nature of R, it must be supposed that the transfer of a proton from one alcohol molecule to another involves the passage over an energy barrier whose height increases as R is changed from hydrogen to methyl to ethyl. The Grotthuss type of conduction, therefore, diminishes in this order. It is probable that the effect decreases steadily with increasing chain length of the alcohol.

Abnormal Conductances of Hydroxyl and Other Ions.—The conductance of the hydroxyl ion in water is less than that of the hydrogen ion; it is nevertheless three times as great as that of most other anions (cf. Table XIII). It is probable that the abnormal conductance is here also due to the transfer of a proton, in this case from a water molecule to a hydroxyl ion, thus

$$\underset{\ominus}{\mathrm{\overset{\overset{\displaystyle H}{|}}{O}}} + \mathrm{H{-}\overset{\overset{\displaystyle H}{|}}{O}} \rightarrow \mathrm{\overset{\overset{\displaystyle H}{|}}{O}{-}H} + \underset{\ominus}{\mathrm{\overset{\overset{\displaystyle H}{|}}{O}}},$$

followed by rotation of the resulting water molecule. If this is the case, it might be expected that the anion RO^- should possess abnormal conductance in the corresponding alcohol ROH; such abnormalities, if they exist at all, are very small, for the conductances of the CH_3O^- and $C_2H_5O^-$ ions in methyl and ethyl alcohol, respectively, are almost the same as that of the chloride ion which exhibits normal conductance only. The energy barriers involved in the abnormal mobility process must therefore be considerably higher than for water.

These results emphasize the fact that ions produced by self-ionization of the solvent, e.g., H_3O^+ and OH^- in water, ROH_2^+ and RO^- in alcohols, and NH_4^+ and NH_2^- in liquid ammonia, do not of necessity possess abnormal conductance, although they frequently do so. It is seen from Table XIX that the conductance of the hydrogen ion in liquid ammonia, i.e., NH_4^+, is normal; the same is true for the NH_2^- ion. The anilinium and pyridinium ions also have normal conductances in the corresponding solvents. The conductance of the HSO_4^- ion in sulfuric acid as solvent is, however, abnormally high; it is probable that a Grotthuss type of conduction, involving proton transfer, viz.,

$$HSO_4^- + H_2SO_4 = H_2SO_4 + HSO_4^-,$$

is responsible for the abnormal conductance.[24]

Influence of Traces of Water.—The change in the equivalent conductance of a strong electrolyte, other than an acid, in a non-aqueous solvent resulting from the addition of small amounts of water, generally corresponds to the alteration in the viscosity. With strong acids, however, there is an initial decrease of conductance in an alcoholic solvent which is much greater than is to be expected from the change in viscosity; this is subsequently followed by an increase towards the value in water. When acetone is the solvent, however, the conductance in the presence of water runs parallel with the viscosity of the medium. It should be noted that the abnormal behavior is observed in solvents in which the hydrogen ion manifests the Grotthuss type of conduction. The hydrogen ion in alcoholic solution is ROH_2^+ and the addition of water results in the occurrence of the reversible reaction

$$ROH_2^+ + H_2O \rightleftharpoons ROH + H_3O^+.$$

The equilibrium of this system lies well to the right, and so a large proportion of the ROH_2^+ ions will be converted into H_3O^+ ions. Although the former possess abnormal conductance in the alcohol solution, the latter do not, since the proton must pass from H_3O^+ to ROH, and the position of the equilibrium referred to above shows that this process must be slow. The result of the addition of small quantities of water to an alcoholic solution of an acid is to replace an ion capable of abnormal conduction by one which is able to conduct in a normal manner only; the equivalent conductance of the system must consequently decrease markedly. As the amount of water present is increased it will become increasingly possible for the proton to pass from H_3O^+ to a molecule of water, and so there is some abnormal contribution to the conductance; the conductance thus eventually increases towards the usual value for the acid in pure water, which is higher than that in the alcohol.

From the initial decrease in conductance accompanying the addition of small amounts of water to a solution of hydrochloric acid in ethyl alcohol, it is possible to evaluate the conductance of the H_3O^+ ion in the alcohol. The value has been found to be 16.8 $ohms^{-1}$ $cm.^2$ at 25°, which may be compared with 18.7 $ohms^{-1}$ $cm.^2$ for the sodium ion in the same solvent. It is evident, therefore, that the H_3O^+ ion possesses only normal conductance in ethyl alcohol.[25]

Determination of Solubilities of Sparingly Soluble Electrolytes.—If a slightly soluble electrolyte dissociates in a simple manner, it is possible to calculate the saturation solubility from conductance measure-

[24] Hammett and Lowenheim, *J. Am. Chem. Soc.*, **56**, 2620 (1934).

[25] Goldschmidt, *Z. physik. Chem.*, **89**, 129 (1914).

ments. If s is the solubility, in equivalents per liter, of a given substance and κ is the specific conductance of the saturated solution, then the equivalent conductance of the solution is given by

$$\Lambda = 1000 \frac{\kappa}{s}. \tag{26}$$

In general, the solution will be sufficiently dilute for the equivalent conductance to be little different from the value at infinite dilution: the latter can be obtained, as already seen, from the ion conductances of the constituent ions. It follows, therefore, since Λ is known and κ for the saturated solution can be determined experimentally, that it is possible to evaluate the solubility s by means of equation (26).

From Kohlrausch's measurements on the conductance of saturated solutions of pure silver chloride the specific conductance at 25° may be estimated as 3.41×10^{-6} ohm^{-1} cm.$^{-1}$; the specific conductance of the water used was 1.60×10^{-6} ohm^{-1} cm.$^{-1}$, and so that due to the salt may be obtained by subtraction as 1.81×10^{-6} ohm^{-1} cm.$^{-1}$ This is the value of κ to be employed in equation (26). From Table XIII the equivalent conductance of silver chloride at infinite dilution is 138.3 ohms^{-1} cm.^{2} at 25°, and so if this is assumed to be the equivalent conductance in the saturated solution of the salt, it follows from equation (26) that

$$s = 1000 \frac{\kappa}{\Lambda} = \frac{1000 \times 1.81 \times 10^{-6}}{138.3}$$

$$= 1.31 \times 10^{-5} \text{ equiv. per liter at } 25°.$$

By means of this first approximation for the concentration of the saturated solution of silver chloride, it is possible to make a more exact estimate of the actual equivalent conductance by means of the Onsager equation (p. 89); a more precise value of the solubility may then be determined. In the particular case of silver chloride, however, the difference is probably within the limits of the experimental error.

It should be realized that the method described actually gives the *ionic* concentration in the saturated solution, and it is only when dissociation is virtually complete that the result is identical with the solubility. This fact is brought out by the data for thallous chloride: the solubility at 18° calculated from Kohlrausch's conductance measurements is 1.28×10^{-2} equiv. per liter, but the value obtained by direct solubility measurement is 1.32×10^{-2} equiv. per liter. The discrepancy, which is not very large in this instance, is probably to be ascribed to incomplete dissociation of the salt in the saturated solution; the degree of dissociation appears to be 128/132, i.e., 0.97.

If the sparingly soluble salt does not undergo simple dissociation, the solubility obtained by the conductance method may be seriously in error. For example, the value found for lanthanum oxalate in water at 25° is

6.65×10^{-6} equiv. per liter, but direct determination gives 2.22×10^{-5} equiv. per liter. The difference is partly due to incomplete dissociation and partly to the formation of complex ions. In other words, the lanthanum oxalate does not ionize to yield simple La^{+++} and $C_2O_4^{--}$ ions, as is assumed in the conductance method for determining the solubility; in addition complex ions, containing both lanthanum and oxalate, are present to an appreciable extent in the saturated solution. It is necessary, therefore, to exercise caution in the interpretation of the results obtained from conductance measurements with saturated solutions of sparingly soluble electrolytes.

Determination of Basicity of Acids.—From an examination of the conductances of the sodium salts of a number of acids, Ostwald (1887) discovered the empirical relation

$$\Lambda_{1024} - \Lambda_{32} \approx 11b, \tag{27}$$

where Λ_{1024} and Λ_{32} are the equivalent conductances of the salt at 25° at dilutions of 1024 and 32 liters per equivalent, respectively, and b is the basicity of the acid. The data in Table XX are taken from the work of

TABLE XX. BASICITY OF ACID AND EQUIVALENT CONDUCTANCE OF SALT

Sodium salt of:	Λ_{1024}	Λ_{32}	Difference	Basicity
Nicotinic acid	85.0	73.8	11.2	1
Quinolinic acid	104.9	83.4	21.5	2
1 : 2 : 4-Pyridine tricarboxylic acid	121.0	88.8	32.2	3
1 : 2 : 3 : 4-Pyridine tetracarboxylic acid	131.1	87.3	43.8	4
Pyridine pentacarboxylic acid	138 1	83.9	54.2	5

Ostwald, recalculated so as to give the equivalent conductances in ohm^{-1} cm.^{2} units, instead of reciprocal Siemens units; they show that the equation given above is approximately true, and hence it may be employed to determine the basicity of an acid. The method fails when applied to very weak acids whose salts are considerably hydrolyzed in solution. The results quoted in Table XX are perhaps exceptionally favorable, for the agreement with equation (27) is not always as good as these figures would imply. The Ostwald rule is, nevertheless, an expression of the facts already discussed, viz., that substances of the same valence type have approximately the same conductance ratios at equivalent concentrations and that the values diminish with increasing valence of one or both ions. The rule has been extended by Bredig (1894) to include electrolytes of various types.

Mode of Ionization of Salts.—Most ions, with the exception of hydrogen, hydroxyl and long-chain ions, have ion conductances of about 60 ohms^{-1} cm.^{2} at 25°, and this fact may be utilized to throw light on the mode of ionization of electrolytes. It has been found of particular value, in connection with the Werner co-ordination compounds, to determine whether a halogen atom, or other negative group, is attached in a covalent or an electrovalent manner.

Since the mode of ionization of the salt is not known, it is not possible to determine the equivalent weight and hence the equivalent conductance cannot be calculated; it is necessary, therefore, to make use of the *molar* conductance, as defined on p. 31. In the simple case of a series of salts all of which have one univalent ion, either the cation or anion, whereas the other ion has a valence of z, the gram molecule contains z gram equivalents; the molar conductance is thus z-times the equivalent conductance. If the mean equivalent conductance of all ions is taken as 60, the equivalent conductance of any salt is 120 $ohms^{-1}$ $cm.^2$, and the molar conductance is 120 z $ohms^{-1}$ $cm.^2$ The approximate results for a number of salts of different valence types with one univalent ion at 25° are given in Table XXI. The observed molar conductances of the platinosammine

TABLE XXI. APPROXIMATE MOLAR CONDUCTANCES OF SALTS OF DIFFERENT VALENCE TYPES

Type	Molar Conductance
Uni-uni	120 $ohms^{-1}$ $cm.^2$
Uni-bi or bi-uni	240
Uni-ter or ter-uni	360
Uni-tetra or tetra-uni	480

series, at a concentration of 0.001 M, are in general agreement with expectation, as the following data show:

$[Pt(NH_3)_4]^{++}2Cl^-$	$[Pt(NH_3)_3Cl]^+Cl^-$
260	116
$K^+[Pt(NH_3)Cl_3]^-$	$2K^+[PtCl_4]^{--}$
107	267 $ohms^{-1}$ $cm.^2$

The other member of this group, $Pt(NH_3)_2Cl_2$, is a non-electrolyte and so produces no ions in solution; the two chlorine atoms are thus held to the central platinum atom by covalent forces.

Conductometric Titration: (*a*) *Strong Acids.*—When a strong alkali, e.g., sodium hydroxide, is added to a solution of a strong acid, e.g., hydrochloric acid, the reaction

$$(H^+ + Cl^-) + (Na^+ + OH^-) = Na^+ + Cl^- + H_2O$$

occurs, so that the highly conducting hydrogen ions initially present in the solution are replaced by sodium ions having a much lower conductance. In other words, the salt formed has a smaller conductance than the strong acid from which it was made. The addition of the alkali to the acid solution will thus be accompanied by a decrease of conductance. When neutralization is complete the further addition of alkali results in an increase of conductance, since the hydroxyl ions are no longer used up in the chemical reaction. At the neutral point, therefore, the conductance of the system will have a minimum value, from which the equivalence-point of the reaction can be estimated. When the specific conductance of the acid solution is plotted against the volume of alkali added, the result will be of the form of Fig. 24. If the initial solution is relatively dilute and there is no appreciable change in volume in the course of the titration, the specific conductance will be approximately proportional to the concentration of unneutralized acid or free alkali present at any instant. The specific conductance during the course of the titration of an acid by an alkali under these conditions will consequently be linear with the amount of alkali added. It is seen, therefore, that Fig. 24 will consist of two straight lines which intersect at the equivalence-point.

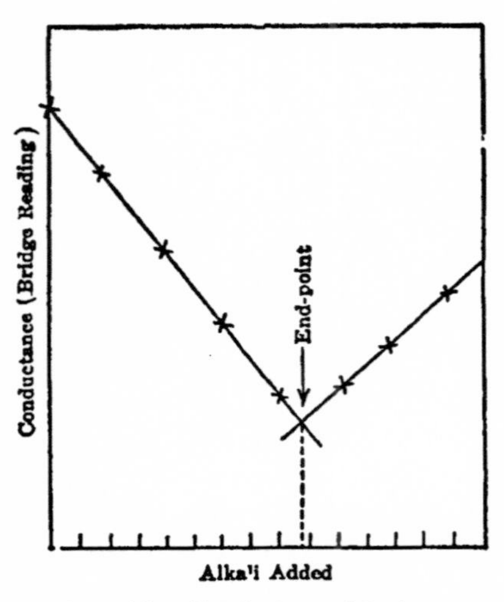

Fig. 24. Conductance titration of strong acid and alkali

If the strong acid is titrated with a weak base, e.g., an aqueous solution of ammonia, the first part of the conductance-titration curve, representing the neutralization of the acid and its replacement by a salt, will be very similar to the first part of Fig. 24, since both salts are strong electrolytes. When the equivalence-point is passed, however, the conductance will remain almost constant since the free base is a weak electrolyte and consequently has a very small conductance compared with that of the acid or salt.

The determination of the end-point of a titration by means of conductance measurements is known as **conductometric titration.**[26] For practical purposes it is not necessary to know the actual specific conductance of the solution; any quantity proportional to it, as explained below, is satisfactory. The conductance readings corresponding to various added amounts of titrant are plotted against the latter, as in Fig. 24. The titrant should be at least ten times as concentrated as the solution being titrated, in order to keep the volume change small; if necessary the titrated solution may be diluted in order to satisfy this condition, for the method can be applied to solutions of strong acids as dilute as 0.0001 N. Since the variation of conductance is linear, it is sufficient to obtain six or eight readings covering the range before and after the end-point, and to draw two straight lines through them, as seen in Fig. 24; the intersection of the lines gives the required end-point. The method of con-

[26] Kolthoff, *Ind. Eng. Chem.* (*Anal. Ed.*), 2, 225 (1930); Davies, "The Conductivity of Solutions," 1933, Chap. XIX; Glasstone, *Ann. Rep. Chem. Soc.*, 30, 294 (1933); Britton, "Conductometric Analysis," 1934; Jander and Pfundt, Böttger's "Physikalische Methoden der analytischen Chemie," 1935, Part II.

ductometric titration is capable of considerable accuracy provided there is good temperature control and a correction is applied for the volume change during titration. It can be used with very dilute solutions, as mentioned above, but in that case it is essential that extraneous electrolytes should be absent; in the presence of such electrolytes the change of conductance would be a very small part of the total conductance and would be difficult to measure with accuracy.

(*b*) *Weak Acids.*—If a moderately weak acid, such as acetic acid, is titrated with a strong base, e.g., sodium hydroxide, the form of the conductance-titration curve is as shown in Fig. 25, I. The initial solution of the weak acid has a low conductance and the addition of alkali may at first result in a further decrease, in spite of the formation of a salt, e.g., sodium acetate, with a high conducting power. The reason for this is that the common anion, i.e., the acetate ion, represses the dissociation of the acetic acid. With further addition of alkali, however, the conductance of the highly ionized salt soon exceeds that of the weak acid which it replaces, and so the specific conductance of the solution increases. After the equivalence-point there is a further increase of conductance because of the excess free alkali; the curve is then parallel to the corresponding part of Fig. 24.

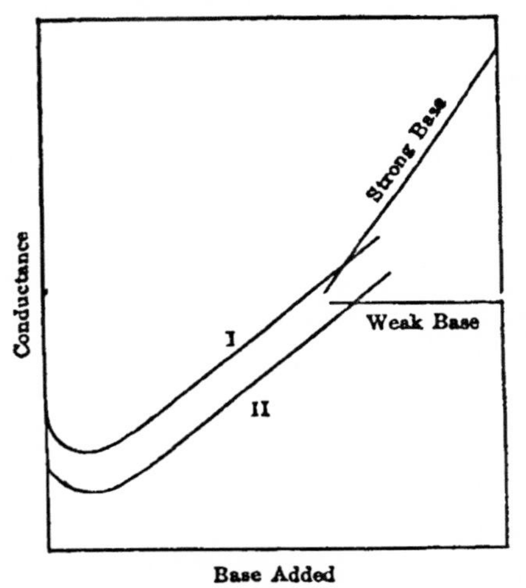

Fig. 25. Conductance titration of weak acid

When a weak acid is titrated with a weak base the initial portion of the conductance-titration curve is similar to that for a strong base, since the salt is a strong electrolyte in spite of the weakness of the acid and base. Beyond the equivalence-point, however, there is no change in conductance because of the small contribution of the free weak base. The complete conductance-titration curve is shown in Fig. 25, II. It will be observed that the intersection is sharper than in Fig. 25, I, for titration with a strong base; it is thus possible to determine the end-point of the titration of a moderately weak acid by the conductometric method if a moderately weak, rather than a strong, base is employed. As long as there is present an excess of acid or base the extent of hydrolysis of the salt is repressed, but in the vicinity of the equivalence-point the salt of the weak acid and weak base is extensively split up by the water; the conductance measurements do not then fall on the lines shown, but these readings can be ignored in the graphical estimation of the end-point.

If the acid is very weak, e.g., phenol or boric acid, or a very dilute solution of a moderately weak acid is employed, the initial conductance is extremely small and the addition of alkali is not accompanied by any decrease of conductance, such as is shown in Fig. 25. The conductance of the solution increases from the commencement of the neutralization as the very weak acid is replaced by its salt which is a strong electrolyte. After the equivalence-point the conductance shows a further increase if a strong base is used, and so the end-point can be found in the usual manner. Owing to the extensive hydrolysis of the salt of a weak base and a very weak acid, even when excess of acid is still present, the titration by a weak base cannot be employed to give a conductometric end-point.

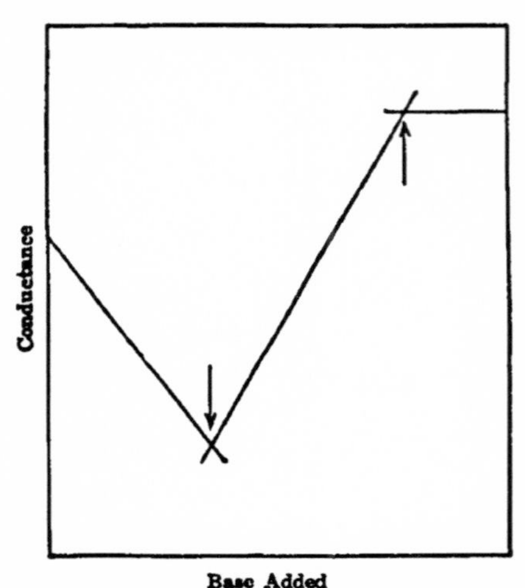

FIG. 26. Conductance titration of mixture of strong and weak acid

One of the valuable features of the conductance method of analysis is that it permits the analysis of a mixture of a strong and a weak acid in one titration. The type of conductance-titration curve using a weak base is shown in Fig. 26; the initial decrease is due to the neutralization of the strong acid, and this is followed by an increase as the weak acid is replaced by its salt. When the neutralization is complete there is little further change of conductance due to the excess weak base. The first point of intersection gives the amount of strong acid in the mixture and the difference between the first and second is equivalent to the amount of weak acid.

(c) *Strong and Weak Bases.*—The results obtained in the titration of a base by an acid are very similar to those just described for the reverse process. When a strong base is neutralized the highly conducting hydroxyl ion is replaced by an anion with a smaller conductance; the conductance of the solution then decreases as the acid is added. When the end-point is passed, however, there is an increase of conductance, just as in Fig. 24, if a strong acid is used for titration purposes, but the value remains almost constant if a weak or very weak acid is employed. With an acid of intermediate strength there will be a small increase of conductance beyond the equivalence-point. In any case the intersections are relatively sharp and, provided carbon dioxide from the air can be excluded, the best method of titrating acids of any degree of weakness conductometrically is to add the acid solution to that of a standard strong alkali.

The conductometric titration of weak bases and those of intermediate strength is analogous to the titration of the corresponding acids. Similarly, a mixture of a strong and a weak base can be titrated quantitatively by means of a weak acid; the results are similar to those depicted in Fig. 26.

(*d*) *Displacement Reactions.*—The titration of the salt of a weak acid, e.g., sodium acetate, by a strong acid, e.g., hydrochloric acid, in which the weak acid is displaced by the strong acid, e.g.,

$$(CH_3CO_2^- + Na^+) + (H^+ + Cl^-) = CH_3CO_2H + Na^+ + Cl^-,$$

can be followed conductometrically. In this reaction the highly ionized sodium acetate is replaced by highly ionized sodium chloride and almost un-ionized acetic acid. Since the chloride ion has a somewhat higher conductance than does the acetate ion, the conductance of the solution increases slowly at first, in this particular case, although in other instances the conductance may decrease somewhat or remain almost constant; in general, therefore, the change in conductance is small. After the end-point is passed, however, the free strong acid produces a marked increase, and its position can be determined by the intersection of the two straight lines. The salt of a weak base and a strong acid, e.g., ammonium chloride, may be titrated by a strong base, e.g., sodium hydroxide, in an analogous manner. It is also possible to carry out conductometrically the titration of a mixture of a salt of a weak acid, e.g., sodium acetate, and weak base, e.g., ammonia, by a strong acid; the first break corresponds to the neutralization of the base and the second to the completion of the displacement reaction. Similarly, it is possible to titrate a mixture of a weak acid and the salt of a weak base by means of a strong base.

(*e*) *Precipitation Reactions.*—In reactions of the type

$$(K^+ + Cl^-) + (Ag^+ + NO_3^-) = AgCl + K^+ + NO_3^-$$

and

$$(Mg^{++} + SO_4^{--}) + 2(Na^+ + OH^-) = Mg(OH)_2 + 2Na^+ + SO_4^{--},$$

where a precipitate is formed, one salt is replaced by an equivalent amount of another, e.g., potassium chloride by potassium nitrate, and so the conductance remains almost constant in the early stages of the titration. After the equivalence-point is passed, however, the excess of the added salt causes a sharp rise in the conductance (Fig. 27, I); the end-point of the reaction can thus be determined.

If both products of the reaction are sparingly soluble, as for example in the titration of sulfates by barium hydroxide, viz.,

$$(Mg^{++} + SO_4^{--}) + (Ba^{++} + 2OH^-) = Mg(OH)_2 + BaSO_4,$$

the conductance of the solution decreases right from the commencement, but increases after the end-point because of the free barium hydroxide (Fig. 27, II).

Precipitation reactions cannot be carried out conductometrically with such accuracy as can the other reactions considered above; this is due to slow separation of the precipitate, with consequent supersaturation of the solution, to removal of titrated solute by adsorption on the precipitate, and to other causes.[27] The best results have been obtained by working with dilute solutions in the presence of a relatively large amount of alcohol; the latter causes a diminution of the solubility of the precipitate and there is also less adsorption.

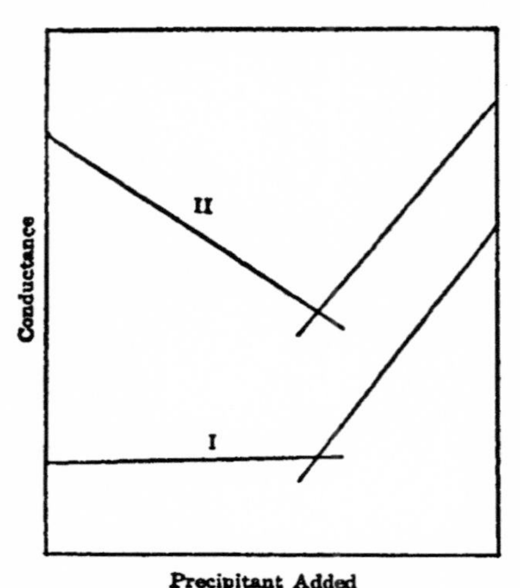

Fig. 27. Conductance titration of precipitation reactions

Conductometric Titration: Experimental Methods.—The titration cell may take any convenient form, the electrodes being arranged vertically so as to permit mixing of the liquids being titrated (see Fig. 28). The conventional Wheatstone bridge, or other simple method of measuring conductance, may be employed. If the form of Fig. 9 is used and the resistance R is kept constant, the specific conductance of the solution in the measuring cell can be readily shown to be proportional to dc/bd. An alternative procedure is to make the ratio arms equal, i.e., $R_3 = R_4$ in Fig. 8 or $bd = dc$ in Fig. 9; the resistance of the cell is then equal to that taken from the box R_2 in Fig. 8 or R in Fig. 9 when the bridge is balanced. If two boxes, or other standard resistances, one for coarse and the other for fine adjustment, are used in series, it is possible to read off directly the resistance of the cell; the reciprocal of this reading is proportional to the specific conductance and is plotted in the titration-conductance curve.

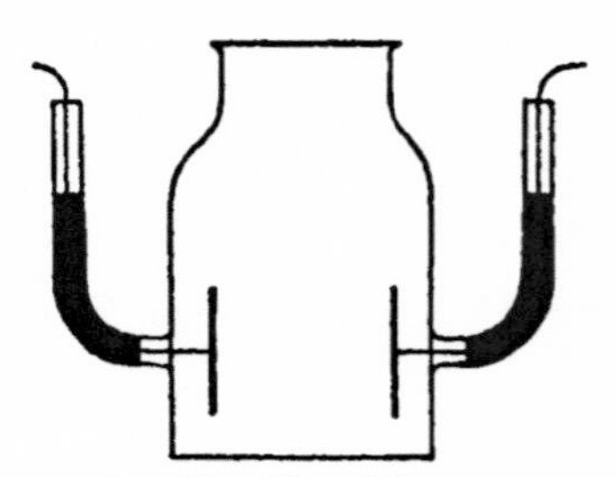
Fig. 28. Vessel for conductometric titration

Since for most titration purposes it is unnecessary to have results of high precision, a certain amount of accuracy has been sacrificed to convenience in various forms of conductometric apparatus.[28] In some cases the Wheatstone bridge arrangement is retained, but a form of visual

[27] van Suchtelen and Itano, *J. Am. Chem. Soc.*, **36**, 1793 (1914); Harned, *ibid.*, **39**, 252 (1917); Freak, *J. Chem. Soc.*, **115**, 55 (1919); Lucasse and Abrahams, *J. Chem. Ed.*, **7**, 341 (1930); Kolthoff and Kameda, *Ind. Eng. Chem.* (*Anal. Ed.*), **3**, 129 (1931).

[28] Treadwell and Paoloni, *Helv. Chim. Acta*, **8**, 89 (1925); Callan and Horrobin, *J. Soc. Chem. Ind.*, **47**, 329T (1928).

detector replaces the telephone earpiece (see p. 35). In other simplified conductance-titration procedures the alternating current is passed directly through the cell and its magnitude measured by a suitable instrument in series; if the applied voltage is constant, then, by Ohm's law, the current is proportional to the conductance of the circuit. For analytical purposes all that is required is the *change* of conductance during the course of the titration, and this is equivalent to knowing the change of current at constant voltage. The type of apparatus employed is shown in Fig. 29; the source of current is the alternating-current supply mains

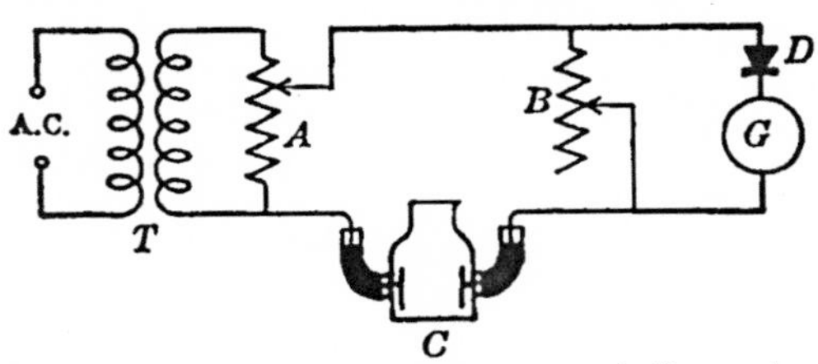

FIG. 29. Conductometric titration using A.C. supply mains

(A.C.), which is reduced to about 3 to 5 volts by means of the transformer T. The secondary of this transformer forms part of the circuit containing the titration cell and also a direct current galvanometer G and a rectifier D; the 400-ohm resistances A and B are used as shunts for the purpose of adjusting the current to a value suitable for the measuring instrument. The rectifier D may be a rectifying crystal, a copper-copper oxide rectifier or a suitable vacuum-tube circuit giving rectification and amplification; alternatively, D and G may be combined in the form of a commercial A.C. microammeter. The solution to be titrated is placed in the vessel C, the resistances A and B are adjusted and then the current on G is noted: the titration is now carried out and the galvanometer readings are plotted against the volume of titrant added. The end-point is determined, as already explained, from the point of intersection of the two parts of the titration curve.

PROBLEMS

1. A conductance cell has two parallel electrodes of 1.25 sq. cm. area placed 10.50 cm. apart; when filled with a solution of an electrolyte the resistance was found to be 1995.6 ohms. Calculate the cell constant of the cell and the specific conductance of the solution.

2. Jones and Bradshaw [*J. Am. Chem. Soc.*, 55, 1780 (1933)] found the resistance of a conductance cell (Z_4) when filled with mercury at 0° to be 0.999076 ohm when compared with a standard ohm. The cell Z_4 and another cell Y_1 were filled with sulfuric acid, and the ratio of the resistances Y_1/Z_4 was 0.107812. The resistance of a third cell N_2 to that of Y_1, i.e., N_2/Y_1, was found to be 0.136564. Evaluate the cell constant of N_2, calculating the specific resistance of mercury at 0° from the data on page 4. (It may be mentioned that the result is 0.014 per cent too high, because of a difference in the current lines in the cell Z_4 when filled with mercury and sulfuric acid, respectively.)

3. A conductance cell having a constant of 2.485 cm.$^{-1}$ is filled with 0.01 N potassium chloride solution at 25°; the value of Λ for this solution is 141.3 ohms^{-1} cm.^{2} If the specific conductance of the water employed as solvent is 1.0×10^{-6} ohm^{-1} cm.$^{-1}$, what is the measured resistance of the cell containing the solution?

4. The measured resistance of a cell containing a 0.1 demal solution of potassium chloride at 25°, in water having a specific conductance of 0.8×10^{-6} ohm^{-1} cm.$^{-1}$, was found to be 3468.86 ohms. A 0.1 N solution of another salt, dissolved in the same conductance water, had a resistance of 4573.42 ohms in the same cell. Calculate the specific conductance of the given solution at 25°.

5. A conductance cell containing 0.01 N potassium chloride was found to have a resistance of 2573 ohms at 25°. The same cell when filled with a solution of 0.2 N acetic acid had a resistance of 5085 ohms. Calculate (*a*) the cell constant, (*b*) the specific resistances of the potassium chloride and acetic acid solutions, (*c*) the conductance ratio of 0.2 N acetic acid, utilizing data given in Chap. II. (The conductance of the water may be neglected.)

6. Use the data in Tables X and XIII to estimate the equivalent conductance of 0.1 N sodium chloride, 0.01 N barium nitrate and 0.001 N magnesium sulfate at 25°. (Compare the results with the values in Table VIII.)

7. The following values for the resistance were obtained when 100 cc. of a solution of hydrochloric acid were titrated with 1.045 N sodium hydroxide:

0	1.0	2.0	3.0	4.0	5.0 cc. NaOH
2564	3521	5650	8065	4831	3401 ohms

Determine the concentration of the acid solution.

8. A 0.01 N solution of hydrochloric acid ($\Lambda = 412.0$) was placed in a cell having a constant of 10.35 cm.$^{-1}$, and titrated with a more concentrated solution of sodium hydroxide. Assuming the equivalent conductance of each electrolyte to depend only on the total ionic concentration of the solution, plot the variation of the cell conductance resulting from the addition of 25, 50, 75, 100, 125 and 150 per cent of the amount of sodium hydroxide required for complete neutralization. The equivalent conductance of the sodium chloride may be taken as 118.5 ohms^{-1} cm.^{2}; the change in volume of the solution during titration may be neglected.

9. The following values were obtained by Shedlovsky [*J. Am. Chem. Soc.*, 54, 1405 (1932)] for the equivalent conductance of potassium chloride at various concentrations at 25°:

0.1	0.05	0.02	0.01	0.005	0.001 N
128.96	133.37	138.34	141.27	143.55	146.95 ohms^{-1} cm.^{2}

Evaluate the equivalent conductance of the salt at infinite dilution by the method described on page 54; the values of B and A may be taken as 0.229 and 60.2, respectively.

10. A potential of 5.6 volts is applied to two electrodes placed 9.8 cm. apart: how far would an ammonium ion be expected to move in 1 hour in a dilute solution of an ammonium salt at 25°?

11. A saturated solution of silver chloride when placed in a conductance cell whose constant is 0.1802 had a resistance of 67,953 ohms at 25°. The resistance of the water used as solvent was found to be 212,180 ohms in the same cell. Calculate the solubility of the salt at 25°, assuming it to be completely dissociated in its saturated solution in water.

CHAPTER III

THE THEORY OF ELECTROLYTIC CONDUCTANCE

Variation of Ionic Speeds.—It has been seen (p. 58) that the equivalent conductance of an electrolyte depends on the number of ions, on the charge carried by each ionic species and on their speeds. For a given solute the charge is, of course, constant, and so the variation of equivalent conductance with concentration means that there is either a change in the number of ions present or in their velocities, or in both. In the early development of the theory of electrolytic dissociation, Arrhenius made the tacit assumption that the ionic speeds were independent of the concentration of the solution; the change of equivalent conductance would then be due to the change in the number of ions produced from the one equivalent of electrolyte as a result of the change of concentration. In other words, the change in the equivalent conductance should then be attributed to the change in the degree of dissociation. All electrolytes are probably completely dissociated into ions at infinite dilution, and so, if the speeds of the ions do not vary with the concentration of the solution, it is seen that the ratio of the equivalent conductance Λ at any concentration to that (Λ_0) at infinite dilution, i.e., Λ/Λ_0, should be equal to the degree of dissociation of the electrolyte. For many years, therefore, following the original work of Arrhenius, this quantity, which is now given the non-committal name of "conductance ratio" (p. 51), was identified with the degree of dissociation.

There are good reasons for believing that the speeds of the ions do vary as the concentration of the solution of electrolyte is changed, and so the departure of the conductance ratio (Λ/Λ_0) from unity with increasing concentration cannot be due merely to a decrease in the degree of dissociation. For strong electrolytes, in which the ionic concentration is high, the mutual interaction of the oppositely charged ions results in a considerable decrease in the velocities of the ions as the concentration of the solution is increased; the fraction Λ/Λ_0 under these conditions bears no relation to the degree of dissociation. In solutions of weak electrolytes the number of ions in unit volume is relatively small, and hence so also is the interionic action which reduces the ionic speeds. The latter, consequently, do not change greatly with concentration, and the conductance ratio gives a reasonably good value of the degree of dissociation; some correction should, however, be made for the influence of interionic forces, as will be seen later.

The Degree of Dissociation.—An expression for the degree of dissociation which will be found useful at a later stage is based on a consideration of the relationship between the equivalent conductance of a solution and the speeds of the ions. It was deduced on page 59 that the speed of an ion at infinite dilution under a potential gradient of 1 volt per cm. is equal to λ^0/F, the derivation being based on the assumption that the electrolyte is completely dissociated. A consideration of the arguments presented shows that they are of general applicability to solutions of any concentration; the only change is that if the electrolyte is not completely dissociated, an allowance must be made in calculating the actual *ionic* concentration. If α is the true degree of dissociation and c is the total (stoichiometric) concentration of the electrolyte, the ionic concentration αc equiv. per liter must be employed in evaluating the quantity of electricity carried by the ions; the total concentration c is still used, however, for calculating the equivalent conductance. The result of making this change is that equation (19) on page 59 becomes

$$\alpha F(u_+ + u_-) = \lambda_+ + \lambda_- = \Lambda, \tag{1}$$

where λ_+ and λ_- are the actual ion conductances and Λ the equivalent conductance of the solution; u_+ and u_- are the mobilities of the ions in the same solution and α is the degree of dissociation at the given concentration. It follows, therefore, that

$$\frac{\alpha F(u_+ + u_-)}{F(u_+^0 + u_-^0)} = \frac{\Lambda}{\Lambda_0},$$

$$\therefore \quad \alpha = \frac{\Lambda}{\Lambda_0} \cdot \frac{u_+^0 + u_-^0}{u_+ + u_-}. \tag{2}$$

For a weak electrolyte the sum $u_+^0 + u_-^0$, for infinite dilution, does not differ greatly from $u_+ + u_-$ in the actual solution, and so the degree of dissociation is approximately equal to the conductance ratio, as stated above.

If equation (1) is divided into its constituent parts, for positive and negative ions, it is seen that

$$\alpha F u_i = \lambda_i \tag{3}$$

for each ion; hence

$$\frac{\alpha F u_i}{F u_i^0} = \frac{\lambda_i}{\lambda_i^0},$$

$$\therefore \quad \alpha = \frac{\lambda_i}{\lambda_i^0} \cdot \frac{u_i^0}{u_i}, \tag{4}$$

where λ_i and u_i are the equivalent conductance and mobility of the *i*th ion in the actual solution.

Interionic Attraction: The Ionic Atmosphere.—The possibility that the attractive forces between ions might have some influence on electrolytic conductance, especially with strong electrolytes, was considered by Noyes (1904), Sutherland (1906), Bjerrum (1909), and Milner (1912) among others, but the modern quantitative treatment of this concept is due mainly to the work of Debye and Hückel and its extension chiefly by Onsager and by Falkenhagen.[1] The essential postulate of the **Debye-Hückel theory** is that every ion may be considered as being surrounded by an **ionic atmosphere** of opposite sign: this atmosphere can be regarded as arising in the following manner. Imagine a positive ion situated at the point A in Fig. 30, and consider a small volume element dv at the end of a radius vector r; the distance r is supposed to be of the order of less than about one hundred times the diameter of an ion. As a result of thermal movements of the ions, there will sometimes be an excess of positive and sometimes an excess of negative ions in the volume element dv; if a time-average is taken, however, it will be found to have, as a consequence of electrostatic attraction by the positive charge at A, a negative charge density. In other words, the probability of finding ions of opposite sign in the space surrounding a given ion is greater than the probability of finding ions of the same sign; every ion may thus be regarded as being associated with an ionic atmosphere of opposite sign. The net charge of the atmosphere is, of course, equal in magnitude but opposite in sign to that of the central ion: the charge density will obviously be greater in the immediate vicinity of the latter and will fall off with increasing distance. It is possible, nevertheless, to define an effective thickness of the ionic atmosphere, as will be explained shortly.

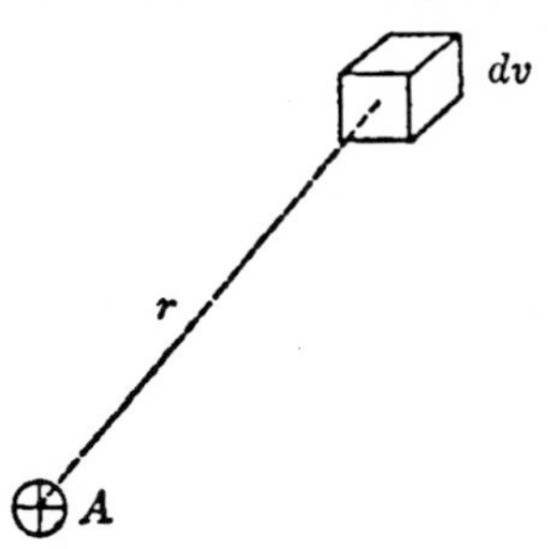

FIG. 30. The ionic atmosphere

Suppose the time-average of the electrical potential in the center of the volume element dv in Fig. 30 is ψ; the work required to bring a positive ion from infinity up to this point is then $z_+\epsilon\psi$ and to bring up a negative ion it is $-z_-\epsilon\psi$, where z_+ and z_- are the *numerical* values of the valences of the positive and negative ions, respectively, and ϵ is the unit charge, i.e., the electronic charge. If the Boltzmann law of the distribution of particles in a field of varying potential energy is applicable to ions, the time-average numbers of positive ions (dn_+) and of negative ions (dn_-) present in the volume element dv are given by

$$dn_+ = n_+e^{-(z_+\epsilon\psi/kT)}dv$$

and

$$dn_- = n_-e^{-(-z_-\epsilon\psi/kT)}dv,$$

where n_+ and n_- are the total numbers of positive and negative ions,

[1] Debye and Hückel, *Physik. Z.*, **24**, 185, 305 (1923); **25**, 145 (1924); for reviews, see Falkenhagen and Williams, *Chem. Revs.*, **6**, 317 (1929); Williams, *ibid.*, **8**, 303 (1931); Hartley *et al.*, *Ann. Rep. Chem. Soc.*, **27**, 326 (1930); Falkenhagen, *Rev. Modern Phys.*, **3**, 412 (1931); "Electrolytes" (Translated by Bell), 1934; MacInnes *et al.*, *Chem. Revs.*, **13**, 29 (1933); *Trans. Electrochem. Soc.*, **66**, 237 (1934); *J. Franklin Inst.*, **225**, 661 (1938).

respectively, in unit volume of the solution; k is the Boltzmann constant, i.e., the gas constant per single molecule, and T is the absolute temperature. The electrical density ρ, i.e., the net charge per unit volume, in the given volume element is therefore given by

$$\rho = \frac{\epsilon(z_+dn_+ - z_-dn_-)}{dv}$$
$$= \epsilon(n_+z_+e^{-z_+\epsilon\psi/kT} - n_-z_-e^{z_-\epsilon\psi/kT}). \tag{5}$$

For a uni-univalent electrolyte z_+ and z_- are unity, and n_+ and n_- must be equal, because of electrical neutrality; hence equation (5) becomes

$$\rho = n\epsilon(e^{-\epsilon\psi/kT} - e^{\epsilon\psi/kT}), \tag{6}$$

where n is the number of either kind of ion in unit volume. Expanding the two exponential series, and writing x in place of $\epsilon\psi/kT$, equation (6) becomes

$$\rho = -\frac{\epsilon^2\psi}{kT}2n\left(1 + \frac{x^2}{3!} + \frac{x^4}{5!} + \cdots\right),$$

and if it is assumed that x, i.e., $\epsilon\psi/kT$, is small in comparison with unity, all terms beyond the first in the parentheses may be neglected, so that

$$\rho = -\frac{\epsilon^2\psi}{kT}2n. \tag{7}$$

In the general case, when z_+ and z_- are not necessarily unity, if the assumption is made that $z\epsilon\psi/kT$ is much less than unity in each case, the corresponding expression for the electrical density is

$$\rho = -\frac{\epsilon^2\psi}{kT}\sum_i n_iz_i^2, \tag{8}$$

where n_i and z_i represent the number (per unit volume) and valence of the ions of the ith kind. The summation is taken over all the types of ions present in the solution, and equation (8) is applicable irrespective of the number of different kinds of ions.

In order to solve for ψ it is necessary to have another relationship between ρ and ψ, and this may be obtained by introducing Poisson's equation, which is equivalent to assuming that Coulomb's law of force between electrostatic charges also holds good for ions. This equation in rectangular coordinates is

$$\frac{\partial^2\psi}{\partial x^2} + \frac{\partial^2\psi}{\partial y^2} + \frac{\partial^2\psi}{\partial z^2} = -\frac{4\pi\rho}{D}; \tag{9}$$

x, y and z are the coordinates of the point in the given volume element, and D is the dielectric constant of the medium. Converting to polar coordinates, and making use of the fact that the terms containing $\partial\psi/\partial\theta$ and $\partial\psi/\partial\phi$ will be zero, since the distribution of potential about any point in the electrolyte must be spherically symmetrical, and consequently independent of the angles θ and ϕ, equation (9) becomes

$$\frac{1}{r^2}\cdot\frac{\partial}{\partial r}\left(r^2\frac{\partial\psi}{\partial r}\right) = -\frac{4\pi\rho}{D}. \tag{10}$$

If the value of ρ given by equation (8) is inserted, this becomes

$$\begin{aligned}\frac{1}{r^2}\cdot\frac{\partial}{\partial r}\left(r^2\frac{\partial\psi}{\partial r}\right) &= \frac{4\pi\epsilon^2}{DkT}\psi\sum_i n_i z_i^2\\ &= \kappa^2\psi,\end{aligned} \tag{11}$$

where the quantity κ (not to be confused with specific conductance) is defined by

$$\kappa = \left(\frac{4\pi\epsilon^2}{DkT}\sum_i n_i z_i^2\right)^{\frac{1}{2}}. \tag{12}$$

The differential equation (11) can be solved, and the solution has the general form

$$\psi = \frac{Ae^{-\kappa r}}{r} + \frac{A'e^{\kappa r}}{r}, \tag{13}$$

where A and A' are constants which can be evaluated in the following manner. Since ψ must approach zero as r increases, because the potential at an infinite distance from a given point in the solution must be zero, it follows that the constant A' must be zero; equation (13) consequently becomes

$$\psi = \frac{Ae^{-\kappa r}}{r}. \tag{14}$$

For a very dilute solution $\Sigma n_i z_i^2$ is almost zero, and hence so also is κ, as may be seen from equation (12); the value of the potential at the point under consideration will then be A/r, according to equation (14). In such a dilute solution the potential in the neighborhood of any ion will be due to that ion alone, since other ions are too far away to have any influence: further, if the ion is regarded as being a point charge, the potential at small distances will be $z_i\epsilon/Dr$. It follows, therefore, that

$$\frac{A}{r} = \frac{z_i\epsilon}{Dr},$$

$$\therefore\quad A = \frac{z_i\epsilon}{D},$$

and insertion of this result in equation (14) gives

$$\psi = \frac{z_i\epsilon}{D}\cdot\frac{e^{-\kappa r}}{r}. \tag{15}$$

This equation may be written in the form

$$\psi = \frac{z_i \epsilon}{Dr} - \frac{z_i \epsilon}{Dr}(1 - e^{-\kappa r}),$$

and if the solution is dilute, so that κ is small and $1 - e^{-\kappa r}$ is practically equal to κr, this becomes

$$\psi = \frac{z_i \epsilon}{Dr} - \frac{z_i \epsilon \kappa}{D}. \tag{16}$$

The first term on the right of equation (16) is the potential at a distance r due to a given point ion when there are no surrounding ions; the second term must, therefore, represent the potential arising from the ionic atmosphère. It is seen, therefore, that ψ_i, the potential due to the ionic atmosphere, is given by

$$\psi_i = -\frac{z_i \epsilon \kappa}{D} \tag{17}$$

for a dilute solution. Since this expression is independent of r, it may be assumed to hold when r is zero, so that the potential *on the ion itself*, due to its surrounding atmosphere, is given by equation (17). If the whole of the charge of the ionic atmosphere which is $-z_i\epsilon$, since it is equal in magnitude and opposite in sign to that of the central ion itself, were placed at a distance $1/\kappa$ from the ion the potential produced at it would be $-z_i\epsilon\kappa/D$, which is identical with the value given by equation (17). It is seen, therefore, that the effect of the ion atmosphere is equivalent to that of a single charge, of the same magnitude, placed at a distance $1/\kappa$ from the ion; the quantity $1/\kappa$ can thus be regarded as a measure of the **thickness of the ion atmosphere** in a given solution.

According to the definition of κ, i.e., equation (12), the thickness of the ionic atmosphere will depend on the number of ions of each kind present in unit volume and on their valence. If c_i is the concentration of the ions of the ith kind expressed in *moles* (gram-ions) per liter, then

$$n_i = c_i \frac{N}{1000},$$

where N is the Avogadro number; hence, from equation (12), after making a slight rearrangement,

$$\frac{1}{\kappa} = \left(\frac{DT}{\Sigma c_i z_i^2} \cdot \frac{1000k}{4\pi\epsilon^2 N}\right)^{\frac{1}{2}}. \tag{18}$$

The values of the universal constants are as follows: k is 1.38×10^{-16} erg per degree, ϵ is 4.802×10^{-10} e.s. unit, and N is 6.025×10^{23}; hence

$$\frac{1}{\kappa} = 2.81 \times 10^{-10}\left(\frac{DT}{\Sigma c_i z_i^2}\right)^{\frac{1}{2}} \text{ cm.}$$

For water as solvent at 25°, D is 78.6 and T is 298°, so that

$$\frac{1}{\kappa} = \frac{4.31 \times 10^{-8}}{(\Sigma c_i z_i^2)^{\frac{1}{2}}} \text{ cm.} \tag{19}$$

The thickness of the ionic atmosphere is thus seen to be of the order of 10^{-8} cm.; it decreases with increasing concentration and increasing valence of the ions present in the electrolyte, and increases with increasing dielectric constant of the solvent and with increasing temperature. The value of $1/\kappa$ in Ångström units for solutions of various types of electrolytes at concentrations of 0.1, 0.01 and 0.001 *moles* per liter in water at 25° are given in Table XXII.

TABLE XXII. THICKNESS OF THE IONIC ATMOSPHERE IN WATER AT 25°

	Concentration of Solution		
Valence Type	0.10 M	0.01 M	0.001 M
Uni-uni	9.64Å	30.5Å	96.4Å
Uni-bi and bi-uni	5.58	19.3	55.8
Bi-bi	4.82	15.3	48.2
Uni-ter and ter-uni	3.94	13.6	39.4

Time of Relaxation of Ionic Atmosphere.—As long as the ionic atmosphere is "stationary," that is to say, it is not exposed to an applied electrical field or to a shearing force tending to cause movement of the ion with respect to the solvent, it has spherical symmetry. When the ion is made to move under the influence of an external force, however, e.g., by the application of an electrical field, the symmetry of the ionic atmosphere is disturbed. If a particular kind of ion moves to the right, for example, each ion will constantly have to build up its ionic atmosphere to the right, while the charge density to the left gradually decays. The rate at which the atmosphere to the right forms and that to the left dies away is expressed in terms of a quantity called the **time of relaxation** of the ionic atmosphere. The decay of the ionic atmosphere occurs exponentially, and so the return to random distribution is asymptotic in nature; it follows, therefore, that the time required for the ionic atmosphere to fall actually to zero is, theoretically, infinite. It has been shown, however, that, after the removal of the central ion, the surrounding atmosphere falls virtually to zero in the time $4q\theta$, where θ is the time of relaxation of the ionic atmosphere and q is defined by

$$q = \frac{z_+ z_-}{z_+ + z_-} \cdot \frac{\lambda_+ + \lambda_-}{z_+\lambda_- + z_-\lambda_+}; \tag{20}$$

z is the valence, excluding the sign, and λ is the ion conductance, of the respective ions. For a binary electrolyte, i.e., one yielding only two ions, z_+ and z_- are equal and q is 0.5; the time for the ionic atmosphere to decay virtually to zero is then 2θ.

When an ion of valence z is moving with a steady velocity through a solution, under the influence of an electrical force ϵzV, where V is the applied potential gradient, this force must balance the force due to resistance represented by Ku; K is the resultant coefficient of frictional resistance and u is the steady velocity of the ion. It follows, therefore, that

$$\epsilon z V = Ku,$$

$$\therefore \quad K = \frac{\epsilon z V}{u}.$$

If the potential gradient is 1 volt per cm., then V is 1/300 e.s. unit; further the velocity u is then given, according to equation (20), Chap. II, by λ/F, where F is 96,500, and hence

$$K = \frac{\epsilon z F}{300\lambda} = 15.4 \times 10^{-8} \frac{z}{\lambda}, \tag{21}$$

since ϵ is 4.802×10^{-10} e.s. unit. It has been shown by Debye and Falkenhagen [2] that the relaxation time is related to the frictional coefficients K_+ and K_- of the two ions constituting a *binary* electrolyte by the expression

$$\theta = \frac{2K_+K_-}{K_+ + K_-} \cdot \frac{1}{kT\kappa^2} \text{ sec.}, \tag{22}$$

where κ has the same significance as before. Utilizing equation (21) and remembering that z_+ is equal to z_- for a binary electrolyte and that $\lambda_+ + \lambda_-$ is equal to Λ, the equivalent conductance of the electrolyte, equation (22) becomes

$$\theta = 30.8 \times 10^{-8} \frac{z}{\Lambda} \cdot \frac{1}{kT\kappa^2} \text{ sec.} \tag{23}$$

Introducing the value of $1/\kappa$ for aqueous solutions at 25°, given by equation (19), into equation (23), the result is

$$\theta = \frac{71.3 \times 10^{-10}}{cz\Lambda} \text{ sec.}, \tag{24}$$

where c is the concentration of the solution in moles per liter. For most solutions other than acids and bases, Λ is about 120 ohms^{-1} cm.^{2} at 25°, so that

$$\theta \approx \frac{0.6 \times 10^{-10}}{cz} \text{ sec.}$$

The time of relaxation of the ionic atmosphere for a binary electrolyte is thus seen to be inversely proportional to the concentration of the solution and to the valence of the ions. The approximate relaxation times for 0.1, 0.01 and 0.001 N solutions of a uni-univalent electrolyte are 0.6×10^{-9}, 0.6×10^{-8} and 0.6×10^{-7} sec., respectively.

[2] Debye and Falkenhagen, *Physik. Z.*, 29, 121, 401 (1928); Falkenhagen and Williams, *Z. physik. Chem.*, 137, 399 (1928); *J. Phys. Chem.*, 33, 1121 (1929).

Mechanism of Electrolytic Conductance.—The existence of a finite time of relaxation means that the ionic atmosphere surrounding a moving ion is not symmetrical, the charge density being greater behind than in front; since the net charge of the atmosphere is opposite to that of the central ion, there will be an excess charge of the opposite sign behind the moving ion. The asymmetry of the ionic atmosphere, due to the time of relaxation, will thus result in a retardation of the ion moving under the influence of an applied field. This influence on the speed of an ion is called the **relaxation effect** or **asymmetry effect.**

Another factor which tends to retard the motion of an ion in solution is the tendency of the applied potential to move the ionic atmosphere, with its associated solvent molecules, in a direction opposite to that in which the central ion, with its solvent molecules (cf. p. 114), is moving. An additional retarding influence, equivalent to an increase in the viscous resistance of the solvent, is thus exerted on the moving ion; this is known as the **electrophoretic effect,** since it is analogous to the resistance acting against the movement of a colloidal particle in an electrical field (cf. p. 530).

An attempt to calculate the magnitude of the forces opposing the motion of an ion through a solution was made by Debye and Hückel: they assumed the applicability of Stokes's law and derived the following expression for the electrophoretic force on an ion of the ith kind:

$$\text{Electrophoretic Force} = \frac{\epsilon z_i \kappa}{6\pi\eta} K_i V, \tag{25}$$

where ϵ, z_i and κ have their usual significance, the latter being taken as equal to the reciprocal of the thickness of the ionic atmosphere; η is the viscosity of the medium, K_i is the coefficient of frictional resistance of the solvent opposing the motion of the ion of the ith kind, and V is the applied potential gradient.* The same result was derived in an alternative manner by Onsager,[3] who showed that it is not necessary for Stokes's law to be strictly applicable in the immediate vicinity of an ion.

In the first derivation of the relaxation force Debye and Hückel did not take into account the natural Brownian movement of the ions; allowance for this was made by Onsager who deduced the equation:

$$\text{Relaxation Force} = \frac{\epsilon^3 z_i \kappa}{6DkT} wV, \tag{26}$$

* The coefficient K_i given here differs somewhat from that (K) employed on page 86; the latter is defined as the resultant frictional coefficient, based on the tacit assumption that all the forces opposing the motion of the ion in a solution of appreciable concentration are frictional in nature. An attempt is made here to divide these forces into the true frictional force due to the solvent, for which the coefficient K_i is employed, and the electrophoretic and relaxation forces due to the presence of other ions. At infinite dilution, K and K_i are, of course, identical.

[3] Onsager, *Physik. Z.*, 27, 388 (1926); 28, 277 (1927); *Trans. Faraday Soc.*, 23, 341 (1927).

where D is the dielectric constant of the medium and w is defined by

$$w \equiv z_+ z_- \frac{2q}{1 + q^{\frac{1}{2}}}, \tag{27}$$

the value of q being given by equation (20).

It is now possible to equate the forces acting on an ion of the ith kind when it is moving through a solution with a steady velocity u_i; the driving force due to the applied electrical field is $\epsilon z_i V$, and this is opposed by the frictional force of the solvent, equal to $K_i u_i$, together with the electrophoretic and relaxation forces; hence

$$\epsilon z_i V = K_i u_i + \frac{\epsilon z_i \kappa}{6\pi\eta} K_i V + \frac{\epsilon^3 z_i \kappa}{6DkT} wV. \tag{28}$$

On dividing through by $K_i V$ and rearranging, this becomes

$$\frac{u_i}{V} = \frac{\epsilon z_i}{K_i} - \frac{\epsilon z_i \kappa}{6\pi\eta} - \frac{\epsilon^3 z_i \kappa}{6DkT} \cdot \frac{w}{K_i}.$$

If the field strength, or potential gradient, is taken as 1 volt per cm., i.e., V is 1/300, then

$$u_i = \frac{\epsilon z_i}{300 K_i} - \frac{\epsilon \kappa}{300} \left(\frac{z_i}{6\pi\eta} + \frac{\epsilon^2 z_i}{6DkT} \cdot \frac{w}{K_i} \right). \tag{29}$$

At infinite dilution κ is zero, and so under these conditions this equation becomes

$$u_i^0 = \frac{\epsilon z_i}{300 K_i}$$

and since $F u_i^0$ is equal to λ_i^0, it follows that

$$\frac{\epsilon z_i}{300 K_i} = \frac{\lambda_i^0}{F}. \tag{30}$$

Further, according to equation (3), u_i is equal to $\lambda_i / \alpha F$, where α is the degree of dissociation; and if this result and that of equation (30) are introduced into (29) the latter becomes

$$\frac{\lambda_i}{\alpha F} = \frac{\lambda_i^0}{F} - \frac{\epsilon \kappa}{300} \left(\frac{z_i}{6\pi\eta} + \frac{\epsilon}{6DkT} \cdot \frac{\epsilon z_i}{K_i} w \right). \tag{31}$$

For simplicity, *the assumption is now made that the electrolyte is completely dissociated*, that is to say, α is assumed to be unity; this, as will be evident shortly, is true for solutions of strong electrolytes at quite appreciable concentrations. Equation (31) can then be put in the form

$$\lambda_i = \lambda_i^0 - \frac{\epsilon \kappa}{300} \left(\frac{z_i}{6\pi\eta} F + \frac{300 \epsilon}{6DkT} \lambda_i^0 w \right), \tag{32}$$

making use of equation (30) to replace $\epsilon z_i/K_i$ by $300\lambda_i^0/F$. Introducing the expression for κ given by equation (12), and utilizing the standard values of ϵ, k and N (p. 84), equation (32) becomes

$$\lambda_i = \lambda_i^0 - \left[\frac{29.15 z_i}{(DT)^{\frac{1}{2}}\eta} + \frac{9.90 \times 10^5}{(DT)^{\frac{3}{2}}}\lambda_i^0 w\right]\sqrt{c_+ z_+^2 + c_- z_-^2}. \qquad (33)$$

The quantities c_+ and c_- represent the concentrations of the ions in moles per liter; these may be replaced by the corresponding concentrations c in equivalents per liter, where c, which is the same for both ions, is equal to $c_i z_i$; hence

$$\lambda_i = \lambda_i^0 - \left[\frac{29.15 z_i}{(DT)^{\frac{1}{2}}\eta} + \frac{9.90 \times 10^5}{(DT)^{\frac{3}{2}}}\lambda_i^0 w\right]\sqrt{\mathrm{c}(z_+ + z_-)}. \qquad (34)$$

The equivalent conductance of an electrolyte is equal to the sum of the conductances of the constituent ions, and so it follows from equation (34) that

$$\Lambda = \Lambda_0 - \left[\frac{29.15(z_+ + z_-)}{(DT)^{\frac{1}{2}}\eta} + \frac{9.90 \times 10^5}{(DT)^{\frac{3}{2}}}\Lambda_0 w\right]\sqrt{\mathrm{c}(z_+ + z_-)}. \qquad (35)$$

In the simple case of a uni-univalent electrolyte, z_+ and z_- are unity, and w is $2 - \sqrt{2}$; equation (35) then reduces to

$$\Lambda = \Lambda_0 - \left[\frac{82.4}{(DT)^{\frac{1}{2}}\eta} + \frac{8.20 \times 10^5}{(DT)^{\frac{3}{2}}}\Lambda_0\right]\sqrt{c}, \qquad (36)$$

the concentration c, in equivalents, being replaced by c, in moles, since both are now identical. This equation and equations (33), (34) and (35) represent forms of the **Debye-Hückel-Onsager conductance equation;** these relationships, based on the assumption that dissociation of the electrolyte is complete, attempt to account for the falling off of the equivalent conductance at appreciable concentrations in terms of a decrease in ionic velocity resulting from interionic forces. The decrease of conductance due to these forces is represented by the quantities in the square brackets; the first term in the brackets gives the effect due to the electrophoretic force and the second term represents the influence of the relaxation, or asymmetry, force. It will be apparent from equation (35) that, for a given solvent at a definite temperature, the magnitude of the interionic forces increases, as is to be anticipated, with increasing valence of the ions and with increasing concentration of the electrolyte.

Before proceeding with a description of the experiments that have been made to test the validity of the Onsager equation, attention may be called to the concentration term c (or c) which appears in the equations (33) to (36). This quantity arises from the expression for κ [equation (12)], and in the latter it represents strictly the *actual* ionic concentration. As long as dissociation is complete, as has been assumed above, this is equal to the stoichiometric concentration, but when cases of incomplete dissociation are considered it must be remembered that the actual ionic concentration is αc, and this should be employed in the Onsager equation.

Validity of the Debye-Hückel-Onsager Equation.—For a uni-univalent electrolyte, the Onsager equation (36), assuming complete dissociation, may be written in the form

$$\Lambda = \Lambda_0 - (A + B\Lambda_0)\sqrt{c}, \tag{37}$$

where A and B are constants dependent only on the nature of the solvent and the temperature; thus

$$A = \frac{82.4}{(DT)^{\frac{1}{2}}\eta}$$

and

$$B = \frac{8.20 \times 10^5}{(DT)^{\frac{3}{2}}}.$$

The values of A and B for a number of common solvents at 25° are given in Table XXIII.

TABLE XXIII. VALUES OF THE ONSAGER CONSTANTS FOR UNI-UNIVALENT ELECTROLYTES AT 25°

Solvent	D	$\eta \times 10^3$	A	B
Water	78.5	8.95	60.20	0.229
Methyl alcohol	31.5	5.45	156.1	0.923
Ethyl alcohol	24.3	10.8	89.7	1.33
Acetone	21.2	3.16	32.8	1.63
Acetonitrile	36.7	3.44	22.9	0.716
Nitromethane	37.0	6.27	125.1	0.708
Nitrobenzene	34.8	18.3	44.2	0.776

(*a*) *Aqueous Solutions.*—In testing the validity of equation (37), it is not sufficient to show that the equivalent conductance is a linear function of the square-root of the concentration, as is generally found to be the case (cf. p. 54); the important point is that the slope of the line must be numerically equal to $A + B\Lambda_0$, where A and B have the values given in Table XXIII. It must be realized, further, that the Onsager equation is to be regarded as a limiting expression applicable to very dilute solutions only; the reason for this is that the identification of the ionic atmosphere with $1/\kappa$, where κ is defined by equation (12), involves simplifications resulting from the assumption of point charges and dilute solutions. It is necessary, therefore, to have reliable data of conductances for solutions of low concentration in order that the accuracy of the Onsager equation may be tested. Such data have become available in recent years, particularly for aqueous solutions of a few uni-univalent electrolytes, e.g., hydrochloric acid, sodium and potassium chlorides and silver nitrate. The experimental results for these solutions at 25° are indicated by the points in Fig. 31, in which the observed equivalent conductances are plotted against the square-roots of the corresponding concentrations.[4] The theoretical slopes of the straight lines to be expected from the Onsager equation, calculated from the values of A and B in Table XXIII in conjunction with an estimated equivalent conductance

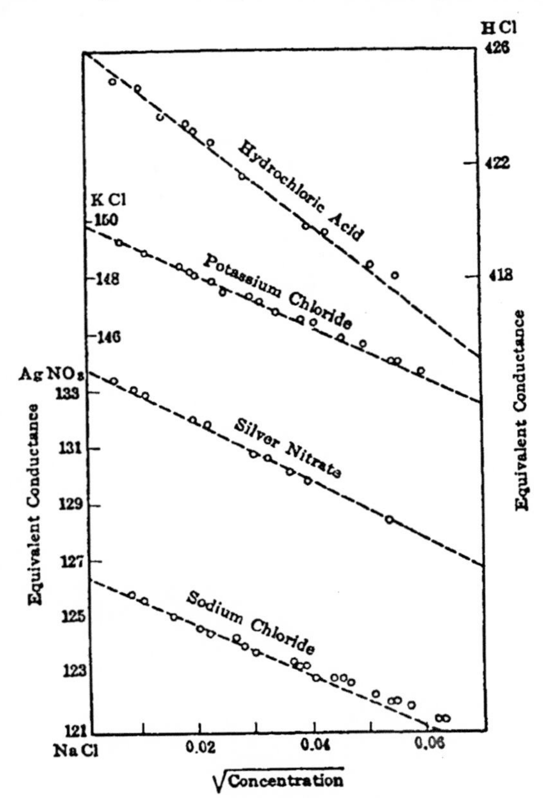

Fig. 31. Test of the Onsager equation

at infinite dilution, are shown by the dotted lines. It is evident from Fig. 31 that for aqueous solutions of the uni-univalent electrolytes for which data are available, the Onsager equation is very closely obeyed at concentrations up to about 2×10^{-3} equiv. per liter.

For electrolytes of unsymmetrical valence types, i.e., z_+ and z_- are different, the verification of the Debye-Hückel-Onsager equation is more difficult since the evaluation of the factor w in equation (35) requires a knowledge of the mobilities of the individual ions at infinite dilution; for this purpose it is necessary to know the transference numbers of the

[4] Shedlovsky, *J. Am. Chem. Soc.*, **54**, 1411 (1932); Shedlovsky, Brown and MacInnes, *Trans. Electrochem. Soc.*, **66**, 165 (1934); Krieger and Kilpatrick, *J. Am. Chem. Soc.*, **59**, 1878 (1937).

ions constituting the electrolyte (see Chap. IV). The requisite data for dilute aqueous solutions at 25° are available for calcium and lanthanum chlorides, i.e., $CaCl_2$ and $LaCl_3$, and in both instances the results are in close agreement with the requirements of the theoretical equation at concentrations up to 4×10^{-5} equiv. per liter.[5] It is apparent that the higher the valence type of the electrolyte the lower is the limit of concentration at which the Onsager equation is applicable.

Less accurate measurements of the conductances of aqueous solutions of various electrolytes have been made, and in general the results bear out the validity of the Onsager equation.[6] A number of values of the experimental slopes are compared in Table XXIV with those calculated

TABLE XXIV. COMPARISON OF OBSERVED AND CALCULATED ONSAGER SLOPES IN AQUEOUS SOLUTIONS AT 25°

Electrolyte	Observed Slope	Calculated Slope
$LiCl$	81.1	72.7
$NaNO_3$	82.4	74.3
KBr	87.9	80.2
$KCNS$	76.5	77.8
$CsCl$	76.0	80.5
$MgCl_2$	144.1	145.6
$Ba(NO_3)_2$	160.7	150.5
K_2SO_4	140.3	159.5

theoretically; the agreement is seen to be fairly good, but it may be even better than would at first appear, owing to the lack of data in sufficiently dilute solutions. It is of interest to record in this connection that the experimental slope of the Λ versus $\sqrt{c}$ curve for silver nitrate was given at one time as 88.2, compared with the calculated value 76.5 at 18°; more recent work on very dilute solutions has shown much better agreement than these results would imply (see Fig. 31).

Further support for the Onsager theory is provided by conductance measurements of a number of electrolytes made at 0° and 100°. At both temperatures the observed slope of the plot of Λ against $\sqrt{c}$ agrees with the calculated result within the limits of experimental error. The slope of the curve for potassium chloride changes from 47.3 to 313.4 within the temperature range studied.

The data recorded above indicate that the Onsager equation represents in a satisfactory manner the dependence on the concentration of the equivalent conductances of uni-univalent and uni-bi- (or bi-uni-) valent electrolytes. With bi-bivalent solutes, however, very marked discrepancies are observed; in the first place the plot of the equivalent con-

[5] Jones and Bickford, *J. Am. Chem. Soc.*, **56**, 602 (1934); Shedlovsky and Brown, *ibid.*, **56**, 1066 (1934).

[6] See, Davies, "The Conductivity of Solutions," 1933, Chap. V; Hartley *et al.*, *Ann. Rep. Chem. Soc.*, **27**, 341 (1930); *J. Chem. Soc.*, 1207 (1933); *Z. physik. Chem.*, **165A**, 272 (1933).

ductance against the square-root of the concentration is not a straight line, but is concave to the axis of the latter parameter (Fig. 32). Further, the slopes at appreciable concentrations are much greater than those calculated theoretically. It is probable that these results are to be explained by incomplete dissociation at the experimental concentrations: the shapes of the curves do in fact indicate that in sufficiently dilute solutions the slopes would probably be very close to the theoretical Onsager values.

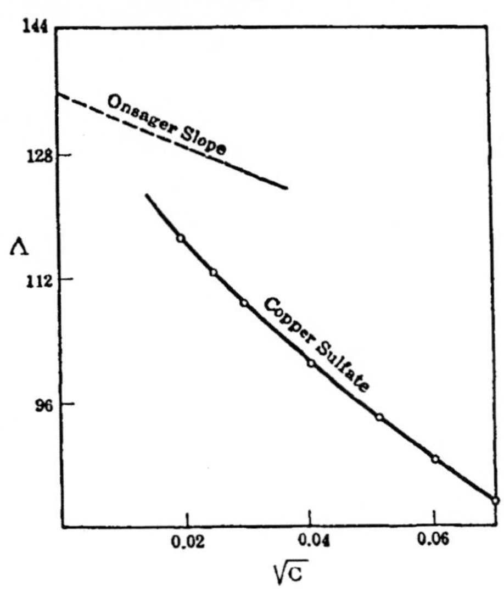

Fig. 32. Deviation from Onsager equation

(*b*) *Non-Aqueous Solutions.*—A number of cases of satisfactory agreement with theoretical requirements have been found in methyl alcohol solutions; this is particularly the case for the chlorides and thiocyanates of the alkali metals.[7] Other electrolytes, such as nitrates, tetralkylammonium salts and salts of higher valence types, however, exhibit appreciable deviations. These discrepancies become more marked the lower the dielectric constant of the medium, especially if the latter is non-hydroxylic in character. The conductance of potassium iodide has been determined in a number of solvents at 25° and the experimental and calculated slopes of the plots of Λ against $\sqrt{c}$ are quoted in Table XXV,

TABLE XXV. OBSERVED AND CALCULATED ONSAGER SLOPES FOR POTASSIUM IODIDE AT 25°

Solvent	*D*	Onsager Slope Observed	Onsager Slope Calculated
Water	78.6	73	80
Methyl alcohol	31.5	260	268
Ethyl cyanoacetate	27.7	115	63
Ethyl alcohol	25.2	209	153
Benzonitrile	25.2	263	142
Acetone	20.9	1000	638

together with the dielectric constant of the medium in each case. At still lower dielectric constants, and for other electrolytes, even greater discrepancies have been recorded: in many cases substances which are strong electrolytes, and hence almost completely dissociated in water, behave as weak, incompletely dissociated electrolytes in solvents of low

[7] Hartley *et al.*, *Proc. Roy. Soc.*, **127A**, 228 (1930); **132A**, 427 (1931); *J. Chem. Soc.*, 2488 (1930).

dielectric constant. It is not surprising, therefore, to find departures from the theoretical Onsager behavior.

Deviations from the Onsager Equation.—Two main types of deviation from the Onsager equation have been observed: the first type is exhibited by a number of salts in aqueous solution which give conductances that are too large at relatively high concentrations, although the values are in excellent agreement with theory in the more dilute solutions. This effect can be seen from the results plotted in Fig. 31; it is probably to be ascribed to the approximations made in the derivation of the Onsager equation which, as already explained, can only be expected to hold for point ions in dilute solution. An empirical correction, involving c and $\log c$, has been applied to allow for these approximations in the following manner. Solving equation (37), for a uni-univalent electrolyte, for Λ_0 it is found that

$$\Lambda_0 = \frac{\Lambda + A\sqrt{c}}{1 - B\sqrt{c}}, \tag{38}$$

according to the simple Onsager theory, and after applying the corrections proposed by Shedlovsky,[8] this becomes

$$\Lambda_0 = \frac{\Lambda + A\sqrt{c}}{1 - B\sqrt{c}} - Cc - Dc \log c + Ec^2, \tag{39}$$

where C, D and E are empirical constants. In some cases D and E are very small and equation (39) reduces to the form

$$\Lambda_0 = \frac{\Lambda + A\sqrt{c}}{1 - B\sqrt{c}} - Cc, \tag{40}$$

which was employed on page 55 to calculate equivalent conductances at infinite dilution. Its validity is confirmed by the results depicted in Fig. 22. In general, the Shedlovsky equation (39) adequately represents the behavior of a number of electrolytes in relatively concentrated solutions; it reduces to the simple Onsager equation at high dilutions when c is small. It is of interest to call attention to the fact that if the term in equation (39) involving $\log c$ is small, as it often is, and can be neglected, this equation can be written in the form of the power series

$$\Lambda = \Lambda_0 - A'c^{\frac{1}{2}} + B'c - C'c^{\frac{3}{2}} + D'c^2 - E'c^{\frac{5}{2}}, \tag{41}$$

where A', B', etc., are constants for the given solute and solvent.

For many electrolytes the plot of the equivalent conductance against the square-root of the concentration is linear, or slightly concave to the concentration axis, but the experimental slopes are numerically greater

[8] Shedlovsky, *J. Am. Chem. Soc.*, **54**, 1405 (1932); Shedlovsky and Brown, *ibid.*, **56**, 1066 (1934); cf., Onsager and Fuoss, *J. Phys. Chem.*, **36**, 2689 (1932). See, however, Jones and Bickford, *J. Am. Chem. Soc.*, **56**, 602 (1934).

than those expected theoretically; this constitutes the second type of deviation from the Onsager equation, instances of which are given in Table XXV. In these cases the conductance is less than required by the theory and the explanation offered for the discrepant behavior, as indicated above, is that dissociation of the electrolyte is incomplete: the number of ions available for carrying the current is thus less than would be expected from the stoichiometric concentration. It will be seen from the treatment on page 89 that, strictly speaking, the left-hand side of equation (32), and hence of all other forms of the Onsager equation, should include a factor $1/\alpha$, where α is the degree of dissociation of the electrolyte; further, it was noted on page 90 that the concentration term should really be αc. It follows, therefore, that for a uni-univalent electrolyte the correct form of equation (37), which makes allowance for incomplete dissociation, is

$$\Lambda = \alpha[\Lambda_0 - (A + B\Lambda_0)\sqrt{\alpha c}]. \tag{42}$$

This equation is sometimes written as

$$\Lambda = \alpha\Lambda', \tag{43}$$

where Λ', defined by

$$\Lambda' \equiv \Lambda_0 - (A + B\Lambda_0)\sqrt{\alpha c}, \tag{44}$$

is the equivalent conductance of 1 equiv. of *free* ions at the concentration αc equiv. per liter, i.e., at the *actual ionic concentration* in the solution.

It is not evident from equation (42) that the plot of Λ against $\sqrt{c}$ will be a straight line, since α varies with the concentration; but as α is less than unity, it is clear that the observed values of the equivalent conductance will be appreciably less than is to be expected from the simple Onsager equation. The second type of deviation, which occurs particularly with salts of high valence types and in media of low dielectric constant, can thus be accounted for by incomplete dissociation of the solute. It is seen from equation (43) that the degree of dissociation α is numerically equal to Λ/Λ', instead of to Λ/Λ_0 as proposed by Arrhenius. It is apparent from equation (44) that for all electrolytes, and especially those which are relatively strong, Λ' is considerably smaller than Λ_0; the true degree of dissociation (Λ/Λ') is thus appreciably closer to unity than is the value assumed to be equal to the conductance ratio (Λ/Λ_0). For a weak electrolyte, the degree of dissociation is in any case small, and αc will also be small; the difference between Λ' and Λ_0 is thus not large and the degree of dissociation will be approximately equal to the conductance ratio. The values for the degree of dissociation obtained in this way are, however, in all circumstances too small, the difference being greater the more highly ionized the electrolyte.

The fact that the type of deviation from Onsager's equation under discussion is not observed, at least up to relatively high concentrations, with many simple electrolytes, e.g., the alkali halides in both aqueous and methyl alcohol solutions, shows that these substances are completely or almost completely dissociated under these conditions. At appreciable concentrations the degree of dissociation probably falls off from unity, but the value of α is undoubtedly much greater than the conductance ratio at the same concentration.

Significance of the Degree of Dissociation.—The quantity α, referred to as the degree of dissociation, represents the fraction of the solute which is free to carry current at a given concentration. The departure of the value of α from unity may be due to two causes which are, however, indistinguishable as far as conductance is concerned. Although many salts probably exist in the ionic form even in the solid state, so that they are probably to be regarded as completely or almost completely ionized at all reasonable concentrations, the ions are not necessarily free to move independently. As a result of electrostatic attraction, ions of opposite sign may form a certain proportion of **ion-pairs**; although any particular ion-pair has a temporary existence only, for there is a continual interchange between the various ions in the solution, nevertheless, at any instant a number of ions are made unavailable in this way for the transport of current. In cases of this kind the electrolyte may be completely *ionized*, but not necessarily completely *dissociated*. At high dilutions, when the simple Onsager equation is obeyed, the solute is both ionized and dissociated completely.

In addition to the reason for incomplete dissociation just considered, there are some cases, e.g., weak acids and many salts of the transition and other metals, in which the electrolyte is not wholly ionized. These substances exist to some extent in the form of un-ionized molecules; a weak acid, such as acetic acid, provides an excellent illustration of this type of behavior. The solution contains un-ionized, covalent molecules, quite apart from the possibility of ion-pairs. With sodium chloride, and similar electrolytes, on the other hand, there are probably no actual covalent molecules of sodium chloride in solution, although there may be ion-pairs in which the ions are held together by forces of electrostatic attraction.

The quantity which has been called the "degree of dissociation" represents the fraction of the electrolyte present as *free* ions capable of carrying the current, the remainder including both un-ionized and undissociated portions. Neither of the latter is able to transport current under normal conditions, and so the ordinary conductance treatment is unable to differentiate between them.

The experimental data show that the deviations from the Onsager equation which may be attributed to incomplete dissociation occur more readily the smaller the ions, the higher their valence and the lower the dielectric constant of the medium. This generalization, as far as ionic size is concerned, appears at first sight not to hold for the salts of the alkali metals, for the deviations from the Onsager equation become more marked as the atomic weight of the metal increases; owing to the effect of hydration, however, the *effective* size of the ion in solution decreases with increasing atomic weight. It is consequently the radius of the ion as it exists in solution, i.e. together with its associated solvent molecules, and not the size of the bare ion, that determines the extent of dissociation of the salt.

According to the concept of **ion association,** developed by Bjerrum (see p. 155), small size and high valence of the ions and a medium of low dielectric constant are just the factors that would facilitate the formation of ion-pairs. The observed results are thus in general agreement with the theory of incomplete dissociation due to the association of ions in pairs held together by electrostatic forces. The theory of Bjerrum leads to the expectation that the extent of association of an electrolyte consisting of small or high-valence ions in a solvent of low dielectric constant would only become inappreciable, and hence the degree of dissociation becomes equal to unity, at very high dilutions. It follows, therefore, that the simple Onsager equation could only be expected to hold at very low concentrations; under these conditions, however, the experimental results would not be sufficiently accurate to provide an adequate test of the equation.

Determination of the Degree of Dissociation.—The determination of the degree of dissociation involves the evaluation of the quantity Λ' at the given concentration, as defined by equation (44); as seen previously, Λ' is the equivalent conductance the electrolyte would have if the solute were completely dissociated at the same ionic concentration as in the experimental solution. Since the definition of Λ' involves α, whereas Λ' is required in order to calculate α, it is evident that the former quantity can be obtained only as the result of a series of approximations. Two of the methods that have been used will be described here.

If Kohlrausch's law of independent ionic migration is applicable to solutions of appreciable concentration, as well as to infinite dilution, as actually appears to be the case, the equivalent conductance of an electrolyte MA may be represented by an equation similar to the one on page 57, viz.,

$$\Lambda_{MA} = \Lambda_{MCl} + \Lambda_{NaA} - \Lambda_{NaCl}, \tag{45}$$

where the various equivalent conductances refer to solutions at the *same ionic concentration.* If MCl, NaA and NaCl are strong electrolytes, they may be regarded as completely dissociated, provided the solutions are not too concentrated; the equivalent conductances in equation (45) consequently refer to the same stoichiometric concentration in each case. If MA is a weak or intermediate uni-univalent electrolyte, however, the value of Λ_{MA} derived from equation (45) will be equivalent to Λ'_{MA}, the corresponding ionic concentration being αc, where α is the degree of dissociation of MA at the total concentration c moles per liter.[9]

[9] MacInnes and Shedlovsky, *J. Am. Chem. Soc.*, **54**, 1429 (1932).

The equivalent conductances of the three strong electrolytes may be written in the form of the power series [cf. equation (41)],

$$\Lambda = \Lambda_0 - A'c^{\frac{1}{2}} + B'c - C'c^{\frac{3}{2}} + \cdots, \tag{41a}$$

where c is the *actual* ionic concentration, which in these instances is identical with the stoichiometric concentration. Combining the values of Λ_{MCl}, Λ_{NaA} and Λ_{NaCl} expressed in this form, it is possible by adding Λ_{MCl} and Λ_{NaA} and subtracting Λ_{NaCl} to derive an equation for Λ'_{MA}; thus

$$\Lambda'_{MA} = \Lambda_{0(MA)} + A''(\alpha c)^{\frac{1}{2}} + B''(\alpha c) - C''(\alpha c)^{\frac{3}{2}} + \cdots, \tag{46}$$

the c terms being replaced by αc to give the actual ionic concentration of the electrolyte MA. Since $\Lambda_{0(MA)}$ is known, and A'', B'', C'', etc., are derived from the A', B', C', etc. values for MCl, NaA and NaCl, it follows that Λ'_{MA} could be calculated if α were available. An approximate estimate is first made by taking α as equal to Λ/Λ_0 for MA, and in this way a preliminary value for Λ'_{MA} is derived from equation (46); α can now be obtained more accurately as $\Lambda_{MA}/\Lambda'_{MA}$, and the calculations are repeated until there is no change in Λ'_{MA}. The method may be illustrated with special reference to the determination of the dissociation of acetic acid. The conductances of hydrochloric acid (MCl), sodium acetate (NaA) and sodium chloride (NaCl) can be expressed in the form of equation (41a): thus, at 25°,

$$\begin{aligned}
\Lambda_{(HCl)} &= 426.16 - 156.62\sqrt{c} + 169.0c\ (1 - 0.2273\sqrt{c}),\\
\Lambda_{(CH_3CO_2Na)} &= 91.00 - 80.46\sqrt{c} + 90.0c\ (1 - 0.2273\sqrt{c}),\\
\Lambda_{(NaCl)} &= 126.45 - 88.52\sqrt{c} + 95.8c\ (1 - 0.2273\sqrt{c}),\\
\therefore\ \Lambda'_{(CH_3CO_2H)} &= 390.7 - 148.56\sqrt{\alpha c} + 163.2\alpha c\ (1 - 0.2273\sqrt{\alpha c}).
\end{aligned}$$

At a concentration of 1.0283×10^{-3} equiv. per liter, for example, the observed equivalent conductance of acetic acid is 48.15 ohms^{-1} cm.^{2} and since Λ_0 is 390.7 ohms^{-1} cm.^{2}, the value of α, as a first approximation, is 48.15/390.7, i.e., 0.1232; inserting this result in the expression for $\Lambda'_{(CH_3CO_2H)}$, the latter is found to be 389.05. As a second approximation, α is now taken as 48.15/389.05, i.e., 0.1238; repetition of the calculation produces no appreciable change in the value of Λ', and so 0.1238 may be taken as being the correct degree of dissociation of acetic acid at the given concentration. The difference between this result and the conductance ratio, 0.1232, is seen to be relatively small in this instance; for stronger electrolytes, however, the discrepancy is much greater.

If there are insufficient data for the equivalent conductances to be expressed analytically in the form of equation (41a), the calculations described above can be carried out in the following manner.[10] As a first approximation the value of α is taken as equal to the conduct-

[10] Sherrill and Noyes, *J. Am. Chem. Soc.*, **48**, 1861 (1926); MacInnes, *ibid.*, **48**, 2068 (1926).

ance ratio and from this the ionic concentration αc is estimated. By graphical interpolation from the conductance data the equivalent conductances of MCl, NaA and NaCl are found at this stoichiometric concentration, which in these cases is the same as the ionic concentration, and from them a preliminary result for $\Lambda'_{(MA)}$ is obtained. With this a more accurate value of α is derived and the calculation of $\Lambda'_{(MA)}$ is repeated; this procedure is continued until the latter quantity remains unchanged. The final result is utilized to derive the correct degree of dissociation. This method of calculation is, of course, identical in principle with that described previously; the only difference lies in the fact that in the one case the interpolation to give the value of Λ' at the ionic concentration is carried out graphically while in the other it is achieved analytically.

In the above procedure for determining the degree of dissociation, the correction for the change in ionic speeds due to interionic forces is made empirically by utilizing the experimental conductance data: the necessary correction can, however, also be applied with the aid of the Onsager equation.[11] Since Λ/Λ' is equal to α, equation (44) can be written as

$$\Lambda' = \Lambda_0 - k\sqrt{\Lambda c/\Lambda'}, \tag{47}$$

where k represents $A + B\Lambda_0$ and is a constant for the given solute in a particular solvent at a definite temperature. The value of Λ_0 for the electrolyte under consideration can, in general, be obtained from the ion conductances at infinite dilution or from other conductance data (see p. 54); it may, therefore, be regarded as known. As a first approximation, Λ' in the term $\sqrt{\Lambda c/\Lambda'}$ is taken as equal to Λ_0, which is equivalent to identifying the degree of dissociation with the conductance ratio, and a preliminary value for Λ' can be derived from equation (47) by utilizing the experimental equivalent conductance Λ at the concentration c. This result for Λ' is inserted under the square-root sign, thus introducing a better value for α, and Λ' is again computed by means of equation (47). The procedure is continued until there is no further change in Λ' and this may be taken as the correct result from which the final value of α is calculated.

Conductance Ratio and the Onsager Equation.—Equation (42) can be written in the form

$$\frac{\Lambda}{\Lambda_0} = \alpha\left[1 - \left(\frac{A}{\Lambda_0} + B\right)\sqrt{\alpha c}\right], \tag{48}$$

which is an expression for the conductance ratio, Λ/Λ_0; the values, clearly, decrease steadily with increasing concentration. For weak electrolytes, the degree of dissociation decreases with increasing temperature, since

[11] Davies, *Trans. Faraday Soc.*, 23, 351 (1927); "The Conductivity of Solutions," 1933, p. 101; see also, Banks, *J. Chem. Soc.*, 3341 (1931).

these substances generally possess a positive heat of ionization. It is apparent, therefore, from equation (48), that the conductance ratio will also decrease as the temperature is raised. For strong electrolytes, α being virtually unity, equation (48) becomes

$$\frac{\Lambda}{\Lambda_0} = 1 - \left(\frac{A}{\Lambda_0} + B\right)\sqrt{c}; \tag{49}$$

the influence of temperature on the conductance ratio is consequently determined by the quantity in the parentheses, viz., $(A/\Lambda_0) + B$. In general, this quantity increases with increasing temperature. That this is the case, at least with water as the solvent, is shown by the data in Table XXVI, for potassium chloride and tetraethylammonium picrate in

TABLE XXVI. INFLUENCE OF TEMPERATURE ON CONDUCTANCE RATIO

Temp.	Potassium Chloride		Tetraethylammonium Picrate	
	Λ_0	$\frac{A}{\Lambda_0} + B$	Λ_0	$\frac{A}{\Lambda_0} + B$
0°	81.8	0.54	31.2	1.16
18°	129.8	0.61	53.2	1.17
100°	406.0	0.77	196.5	1.30

aqueous solution. It follows, therefore, that the conductance ratio for strong electrolytes should decrease with increasing temperature, as found experimentally (p. 52). It will be evident from equation (49) that the decrease should be greater the more concentrated the solution, and this also is in agreement with observation. It may be noted that the quantity $(A/\Lambda_0) + B$ is equal to $(A + B\Lambda_0)/\Lambda_0$, in which the numerator is a measure of the decrease in equivalent conductance due to the diminution of ionic speeds by interionic forces (p. 89): it follows, therefore, that as a general rule the interionic forces increase with increasing temperature.

Introducing the expressions for A and B given on page 90, it is seen that

$$\frac{A}{\Lambda_0} + B = \frac{82.4}{(DT)^{1/2}\eta\Lambda_0} + \frac{8.20 \times 10^5}{(DT)^{3/2}},$$

and since $\eta\Lambda_0$ is approximately constant for a given electrolyte in different solvents (cf. p. 64), this result may be written in the form

$$\frac{A}{\Lambda_0} + B = \frac{a}{(DT)^{1/2}} + \frac{b}{(DT)^{3/2}}, \tag{50}$$

where a and b are numerical constants. It is at once evident, therefore, that the smaller the dielectric constant of the solvent, at constant temperature, the greater will be the value of $(A/\Lambda_0) + B$, and hence the smaller the conductance ratio. The increase of ion association which accompanies the decrease of dielectric constant will also result in a decrease of the conductance ratio.

The discussion so far has referred particularly to uni-univalent electrolytes; it is evident from equation (35) that the valences of the ions are important in determining the decrease of conductance due to interionic forces and hence they must also affect the conductance ratio. The general arguments concerning the effect of concentration, temperature and dielectric constant apply to electrolytes of all valence types; in order to investigate the effect of valence, equation (35) for a strong electrolyte may be written in the general form

$$\frac{\Lambda}{\Lambda_0} = 1 - [A'(z_+ + z_-) + B'\Lambda_0 w]\sqrt{c(z_+ + z_-)}, \tag{51}$$

where A' and B' are constants for the solvent at a definite temperature. It is clear that for a given concentration the conductance ratio decreases with increasing valence of the ions, since the factors $z_+ + z_-$ and w both increase. It was seen in Chap. II that the equivalent conductances of most electrolytes, other than acids or bases, at infinite dilution are approximately the same; in this event it is apparent from equation (51) that for electrolytes of a given valence type the conductance ratio will depend only on the concentration of the solution (cf. p. 52).

In the foregoing discussion the Onsager equation has been used for the purpose of drawing a number of qualitative conclusions which are in agreement with experiment. The equation could also be used for quantitative purposes, but the results would be expected to be correct only in very dilute solutions. At appreciable concentrations additional terms must be included, as in the Shedlovsky equation, to represent more exactly the variation of conductance with concentration; the general arguments presented above would, however, remain unchanged.

Dispersion of Conductance at High Frequencies.—An important consequence of the existence of the ionic atmosphere, with a finite time of relaxation, is the variation of conductance with frequency at high frequencies, generally referred to as the **dispersion of conductance** or the **Debye-Falkenhagen effect.** If an alternating potential of high frequency is applied to an electrolyte, so that the time of oscillation is small in comparison with the relaxation time of the ionic atmosphere, the unsymmetrical charge distribution generally formed around an ion in motion will not have time to form completely. In fact, if the oscillation frequency is high enough, the ion will be virtually stationary and its ionic atmosphere will be symmetrical. It follows, therefore, that the retarding force due to the relaxation or assymmetry effect will thus disappear partially or entirely as the frequency of the oscillations of the current is increased. At sufficiently high frequencies, therefore, the conductance of a solution should be greater than that observed with low-frequency alternating or with direct current. The frequency at which the increase of conductance might be expected will be approximately $1/\theta$, where θ is the relaxation time; according to equation (24) the relaxation time for a binary electrolyte is $71.3 \times 10^{-10}/cz\Lambda$ sec., and so the limiting frequency ν above which abnormal conductance is to be expected is given by

$$\nu \approx \frac{cz\Lambda}{71.3} \times 10^{10} \text{ oscillations per second.}$$

The corresponding wave length in centimeters is obtained by dividing the velocity of light, i.e., 3×10^{10} cm. per sec. by this frequency; the result may be divided by 100 to give the value in meters, thus

$$\lambda \approx \frac{2.14}{cz\Lambda} \text{ meters.}$$

For most electrolytes, other than acids and bases, in aqueous solutions Λ is about 120 at 25°, and hence

$$\lambda \approx \frac{2 \times 10^{-2}}{cz} \text{ meters.}$$

If the electrolyte is of the uni-univalent type and has a concentration of 0.001 molar, the Debye-Falkenhagen effect should become evident with high-frequency oscillations of wave length of about 20 meters or less. The higher the valence of the ions and the more concentrated the solution the smaller the wave length, and hence the higher the frequency, of the oscillations required for the effect to become apparent.

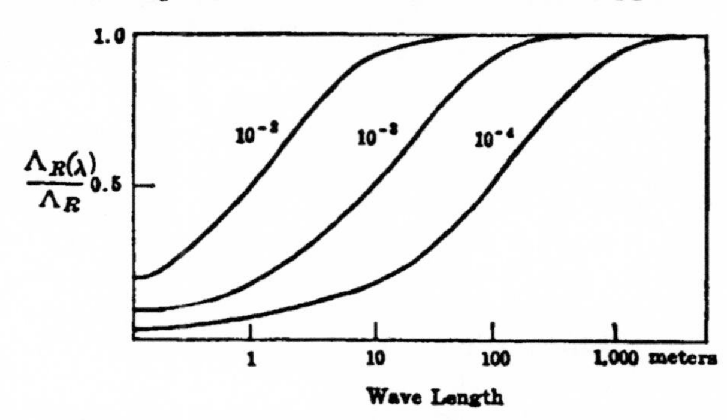

FIG. 33. High frequency conductance dispersion of potassium chloride

The dispersion of conductance at high frequencies was predicted by Debye and Falkenhagen,[12] who developed the theory of the subject; the phenomena were subsequently observed by Sack and others.[13] The

[12] Debye and Falkenhagen, *Physik. Z.*, **29**, 121, 401 (1928); Falkenhagen and Williams, *Z. physik. Chem.*, **137**, 399 (1928); *J. Phys. Chem.*, **33**, 1121 (1929); Falkenhagen, *Physik. Z.*, **39**, 807 (1938).

[13] Sack *et al.*, *Physik. Z.*, **29**, 627 (1928); **30**, 576 (1929); **31**, 345, 811 (1930); Brendel, *ibid.*, **32**, 327 (1931); Debye and Sack, *Z. Elektrochem.*, **39**, 512 (1933); Arnold and Williams, *J. Am. Chem. Soc.*, **58**, 2613, 2616 (1936).

nature of the results to be expected will be evident from an examination of Figs. 33 and 34, in which the calculated ratio of the decrease of conductance due to the relaxation effect * at a short wave length λ, i.e., $\Lambda_{R(\lambda)}$, to that at long wave length Λ_R, i.e., at low frequency, is plotted as

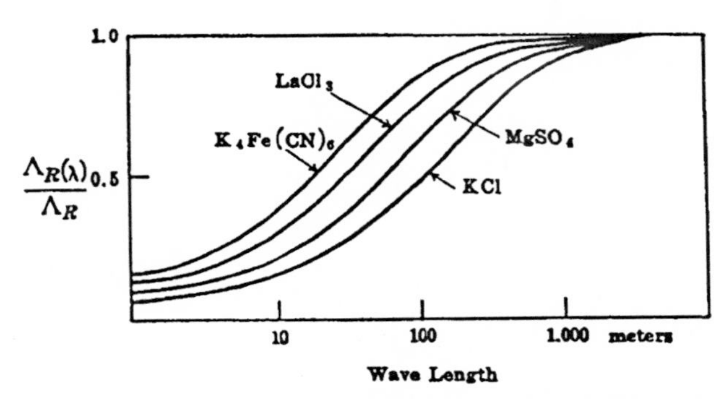

FIG. 34. High frequency conductance dispersion of salts at 10^{-4} mole per liter

ordinate against the wave length as abscissa. The values for potassium chloride at concentrations of 10^{-2}, 10^{-3}, and 10^{-4} mole per liter are plotted in Fig. 33, and those for potassium chloride, magnesium sulfate, lanthanum chloride and potassium ferrocyanide at 10^{-4} mole per liter in water at 18° are shown in Fig. 34. It is seen that, in general, the decrease of conductance caused by the relaxation or asymmetry effect decreases with decreasing wave length or increasing frequency; the actual conductance of the solution thus increases correspondingly. The effect is not noticeable, however, until a certain low wave length is reached, which, as explained above, is smaller the higher the concentration. The influence of the valence of the ions is represented by the curves in Fig. 34; the higher the valence the smaller the relative conductance change at a given high frequency.

The measurements of the Debye-Falkenhagen effect are generally made with reference to potassium chloride; the results for a number of electrolytes of different valence types have been found to be in satisfactory agreement with the theoretical requirements. Increase of temperature and decrease of the dielectric constant of the solvent necessitates the use of shorter wave lengths for the dispersion of conductance to be observed; these results are also in accordance with expectation from theory.

Conductance with High Potential Gradients.—When the applied potential is of the order of 20,000 volts per cm., an ion will move at a speed of about 1 meter per sec., and so it will travel several times the thickness of the effective ionic atmosphere in the time of relaxation.

* At low frequencies this quantity is equal to the second term in the brackets in equation (35), multiplied by $\sqrt{c}(z_+ + z_-)$.

As a result, the moving ion is virtually free from an oppositely charged ion atmosphere, since there is never time for it to be built up to any extent. In these circumstances both asymmetry and electrophoretic effects will be greatly diminished and at sufficiently high voltages should

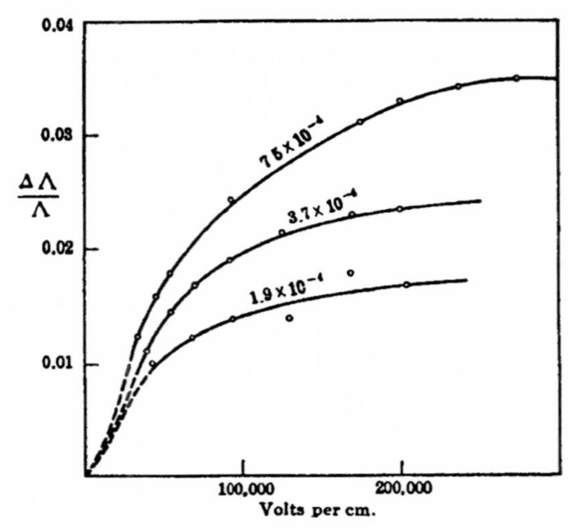

FIG. 35. Wien effect for potassium ferricyanide

disappear. Under the latter conditions the equivalent conductance at any appreciable concentration should be greater than the value at low voltages. The increase in conductance of an electrolyte at high potential gradients was observed by Wien [14] before any theoretical interpretation had been given, and it is consequently known as the **Wien effect.**

It is to be expected that the Wien effect will be most marked under such conditions that the influence of the interionic forces resulting from the existence of an ionic atmosphere is abnormally large; this would be the case for concentrated solutions of high-valence ions. The experimental results shown in Figs. 35 and 36 confirm these expectations; those in Fig. 35 are for solutions of containing potassium ferricyanide at concentrations of 7.5, 3.7 and 1.9×10^{-4} mole per liter, respectively, and the curves in Fig. 36 are for electrolytes of various valence types in solutions having equal low voltage conductances. The quantity $\Delta\Lambda$ is the increase of equivalent conductance resulting from the application of a potential gradient represented by the abscissa.

[14] Wien, *Ann. Physik*, **83**, 327 (1927); **85**, 795 (1928); **1**, 400 (1929); *Physik. Z.*, **32**, 545 (1931); Falkenhagen, *ibid.*, **32**, 353 (1931); Schiele, *Ann. Physik*, **13**, 811 (1932); Debye, *Z. Elektrochem.*, **39**, 478 (1933); Mead and Fuoss, *J. Am. Chem. Soc.*, **61**, 2047, 3257, 3589 (1939); **62**, 1720 (1940); for review, see Eckstrom and Schmelzer, *Chem. Revs.*, **24**, 367 (1939).

It will be observed that the values of $\Delta\Lambda$ tend towards a limit at very high potentials; the relaxation and electrophoretic effects are then virtually entirely eliminated. For an incompletely dissociated electrolyte the measured equivalent conductance under these conditions should be $\alpha\Lambda_0$, where α is the true degree of dissociation; since Λ_0 is known, determinations of conductance at high voltages would seem to provide a method of obtaining the degree of dissociation at any concentration. It has been found, however, that the Wien effect for weak acids and bases, which are known to be dissociated to a relatively small extent, is several times greater than is to be expected; the discrepancy increases as the voltage is raised. It is very probable that in these cases the powerful electrical fields produce a temporary dissociation into ions of the molecules of weak acid or base; this phenomenon, referred to as the **dissociation field effect,** invalidates the proposed method for calculating the degree of dissociation. With strong electrolytes, which are believed to be completely dissociated, the conductances observed at very high potential gradients are close to the values for infinite dilution, in agreement with anticipation.

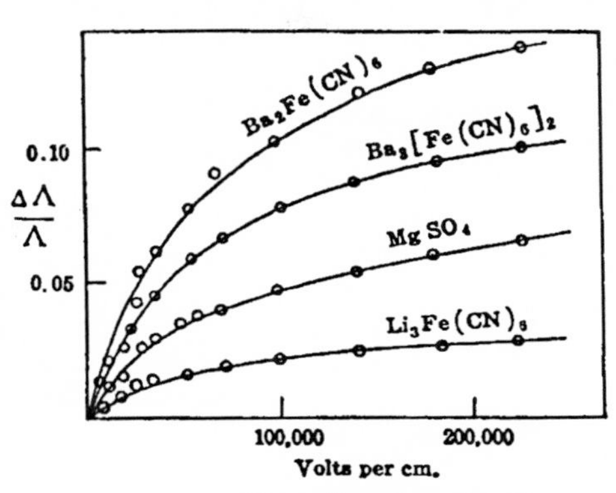

FIG. 36. Wien effect for salts of different valence types

It may be pointed out in conclusion that the conductance phenomena with very high frequency currents and at high potential gradients provide striking evidence for the theory of electrolytic conductance, based on the existence of an ionic atmosphere surrounding every ion, proposed by Debye and Hückel and described in this chapter. Not only does the theory account qualitatively for conductance results of all types, but it is also able to predict them quantitatively provided the solutions are not too concentrated.

PROBLEMS

1. Calculate the thickness of the ionic atmosphere in 0.1 N solutions of a uni-univalent electrolyte in the following solvents: nitrobenzene ($D = 34.8$); ethyl alcohol ($D = 24.3$); and ethylene dichloride ($D = 10.4$).

2. Utilize the results obtained in the preceding problem to calculate the relaxation times of the ionic atmospheres and the approximate minimum frequencies at which the Debye-Falkenhagen effect is to be expected. It may be assumed that $\Lambda_0\eta_0$ has a constant value of 0.6. The viscosities of the solvents are as follows: nitrobenzene (0.0183 poise); ethyl alcohol (0.0109); and ethylene dichloride (0.00785).

3. The viscosity of water at 0° is 0.01793 poise and at 100° it is 0.00284; the corresponding dielectric constants are 87.8 and 56. Calculate the values of the Onsager constants A and B for a uni-univalent electrolyte at these temperatures. Make an approximate comparison of the slopes of the plots of Λ against $\sqrt{c}$ at the two temperatures for an electrolyte for which Λ_0 is 100 ohms^{-1} cm.^{2} at 0°, assuming Walden's rule to be applicable.

4. Make an approximate comparison, by means of the Onsager equation, of the conductance ratios at 25° of 0.01 N solutions of a strong uni-univalent electrolyte in water and in ethyl alcohol; it may be assumed that $\Lambda_0\eta_0$ has the constant value of 0.6 in each case.

5. The following values were obtained by Martin and Tartar [*J. Am. Chem. Soc.*, 59, 2672 (1937)] for the equivalent conductance of sodium lactate at various concentrations at 25°:

$c \times 10^3$	0.1539	0.3472	0.6302	1.622	2.829	4.762
Λ	87.89	87.44	86.91	85.80	84.87	83.78

Plot the values of Λ against $\sqrt{c}$ and determine the slope of the line; estimate Λ_0 and compare the experimental slope with that required by the Onsager equation.

6. Calculate the limiting theoretical slope for the plot of Λ against $\sqrt{c}$ for lanthanum chloride ($LaCl_3$) in water at 25°; Λ_0 for this salt is 145.9 ohms^{-1} cm.^{2} and λ^0_- for the chloride ion is 76.3 ohms^{-1} cm.^{2}

7. Saxton and Waters [*J. Am. Chem. Soc.*, 59, 1048 (1937)] gave the ensuing expressions for the equivalent conductances in water at 25° of hydrochloric acid, sodium chloride and sodium α-crotonate (Naα-C.):

$$\Lambda_{HCl} = 426.28 - 156.84\sqrt{c} + 169.7c\ (1 - 0.2276\sqrt{c})$$
$$\Lambda_{NaCl} = 126.47 - 88.65\sqrt{c} + 94.8c\ (1 - 0.2276\sqrt{c})$$
$$\Lambda_{Na\alpha\text{-}C.} = 83.30 - 78.84\sqrt{c} + 97.27c\ (1 - 0.2276\sqrt{c}).$$

The equivalent conductances of α-crotonic acid at various concentrations were as follows:

$c \times 10^3$	Λ	$c \times 10^3$	Λ
0.95825	51.632	7.1422	19.861
1.7050	39.473	14.511	14.053
3.2327	29.083	22.512	11.318
4.9736	23.677	33.246	9.317

Calculate the degree of dissociation of the crotonic acid at each concentration, making due allowance for interionic attraction. Compare the values obtained with the corresponding conductance ratios.

(The results of this problem are required for Problem 8 of Chap. V.)

8. Employ the data of the preceding problem to calculate the degree of dissociation of α-crotonic acid at the various concentrations using the method of Davies described on page 99.

CHAPTER IV

THE MIGRATION OF IONS

Transference Numbers.—The quantity of electricity q_i carried through a certain volume of an electrolytic solution by ions of the ith kind is proportional to the number in unit volume, i.e., to the concentration c_i in gram-ions or moles per liter, to the charge z_i carried by each ion, and to the mobility u_i, i.e., the velocity under unit potential gradient (cf. p. 58); thus

$$q_i = kc_iz_iu_i, \tag{1}$$

where k is the proportionality constant, which includes the time. The total quantity of electricity Q carried by all the ions present in the electrolyte is thus the sum of the q_i terms for each species; that is

$$Q = kc_1z_1u_1 + kc_2z_2u_2 + kc_3z_3u_3 + \cdots \tag{2}$$

$$= k\Sigma c_iz_iu_i, \tag{2a}$$

the proportionality constant being the same for all the ions. It follows, therefore, that the fraction of the total current carried by an ion of the ith kind is given by

$$t_i = \frac{q_i}{Q} = \frac{c_iz_iu_i}{\sum_i c_iz_iu_i}. \tag{3}$$

This fraction is called the **transference number,** or **transport number,** of the given ion in the particular solution and is designated by the symbol t_i; the sum of the transference numbers of all the ions present in the solution is clearly equal to unity. In the simplest case of a single electrolyte yielding two ions, designated by the suffixes $_+$ and $_-$, the corresponding transference numbers are given, according to equation (3), by

$$t_+ = \frac{c_+z_+u_+}{c_+z_+u_+ + c_-z_-u_-} \quad \text{and} \quad t_- = \frac{c_-z_-u_-}{c_+z_+u_+ + c_-z_-u_-}.$$

The quantities c_+z_+ and c_-z_-, which represent the equivalent concentrations of the ions, are equal, and hence for this type of electrolyte, which has been most frequently studied,

$$t_+ = \frac{u_+}{u_+ + u_-} \quad \text{and} \quad t_- = \frac{u_-}{u_+ + u_-}, \tag{4}$$

and

$$t_+ + t_- = 1.$$

The speed of an ion in a solution at any concentration is proportional to the conductance of the ion at that concentration (p. 80), and so the transference number may be alternatively expressed in the form

$$t_+ = \frac{\lambda_+}{\Lambda} \quad \text{and} \quad t_- = \frac{\lambda_-}{\Lambda}, \tag{5}$$

where the values of the ion conductances λ_+ and λ_-, and the equivalent conductance Λ of the solution, are those at the particular concentration to which the transference numbers are applicable. These values are, of course, different from those at infinite dilution, and so it is not surprising to find, as will be seen shortly, that transference numbers vary with the concentration of the solution; they approach a limiting value, however, at infinite dilution.

Three methods have been generally employed for the experimental determination of transference numbers: the first, based on the procedure originally proposed by Hittorf (1853), involves measurement of changes of concentration in the vicinity of the electrodes; in the second, known as the "moving boundary" method, the rate of motion of the boundary between two solutions under the influence of current is studied (cf. p. 116); the third method, which will be considered in Chap. VI, is based on electromotive force measurements of suitable cells.

Faraday's Laws and Ionic Velocities.—It may appear surprising, at first sight, that equivalent quantities of different ions are liberated at the two electrodes in a given solution, as required by Faraday's laws,

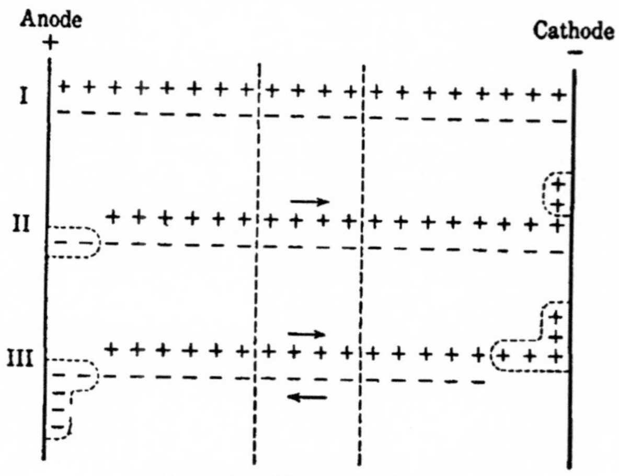

FIG. 37. Migration of ions

in spite of the possible difference in the speeds of the ions moving towards the respective electrodes. The situation can, however, be understood by reference to the diagram in Fig. 37; this represents an electrolytic cell in which there are an equivalent number of positive and negative ions, indicated by plus and minus signs. The condition of the system at the commencement of electrolysis is shown in Fig. 37, I. Suppose that the cations *only* are able to move under the influence of an applied potential, and that two of these ions move from left to right; the condition attained will then be as at Fig. 37, II. At each electrode there are two ions unpaired and these must be considered to be discharged; the two electrons given up by the negative ions at the anode may be imagined to travel through the external circuit and discharge the two positive ions at the cathode. It is seen, therefore, that although only the positive ions are able to move, equivalent amounts of the two ions are discharged at the respective electrodes. A condition of this kind actually arises in certain solid and fused electrolytes, where all the current is carried by the cations.

If while the two cations are moving in one direction, three anions are carrying electricity in the opposite direction, so that the ionic velocities are in the ratio of 2 to 3, the result will be as in Fig. 37, III. Five ions are seen to be discharged at each electrode, in spite of the difference in speeds of the two ions. There is thus no difficulty in correlating Faraday's laws with the fact that the oppositely charged ions in a solution may have different velocities. Incidentally it will be noted that the conclusions to be drawn from Fig. 37 are in harmony with the results derived above, e.g., equation (4); the fraction of the total current carried by each ion, i.e., its transference number, is proportional to its speed. In the condition of Fig. 37, III, the total quantity of electricity passing may be taken as five faradays, since five ions are discharged; of these five faradays, two are carried by the cations in one direction and three by the anions in the opposite direction.

Attention may be called here to a matter which will receive further discussion in Chap. XIII; the ions that carry the current through the solution are not necessarily those to be discharged at the electrodes. This is assumed to be the case here, however, for the sake of simplicity.

The Hittorf Method.—Suppose an electric current is passed through a solution of an electrolyte which yields the ions M^+ and A^-; these ions are not necessarily univalent, although a single + or − sign is used for the sake of simplicity of representation. The fraction of the total current carried by the cations is t_+ and that carried by the anions is t_-; hence when one faraday of electricity is passed through the solution, t_+ faradays are carried in one direction by t_+ equivalents of M^+ ions and t_- faradays are carried in the other direction by t_- equivalents of A^- ions. At the same time one equivalent of each ion is discharged at the appropriate electrode. The migration of the ions and their discharge under the influence of the current bring about changes of concentration in the vicinity of the electrodes, and from these changes it is possible to calculate the transference numbers.

Imagine the cell containing the electrolyte to be divided into three compartments by means of two hypothetical partitions; one compartment surrounds the cathode, another the anode, and the third is a middle compartment in which there is no resultant change of concentration. The effect of passing one faraday of electricity through the solution of the electrolyte MA can then be represented in the following manner.

Cathode Compartment (I)	Middle Compartment	Anode Compartment (II)
1 equiv. of M^+ is discharged t_+ equiv. of M^+ migrate in t_- equiv. of A^- migrate out	t_+ equiv. of M^+ migrate to I t_- equiv. of A^- migrate from I t_+ equiv. of M^+ migrate from II t_- equiv. of A^- migrate to II	1 equiv. of A^- is discharged t_- equiv. of A^- migrate in t_+ equiv. of M^+ migrate out
Net Result:		
Loss of $1 - t_+ = t_-$ equiv. of M^+ Loss of t_- equiv. of A^- ∴ Net loss is t_- equiv. of MA	No change of concentration	Loss of $1 - t_- = t_+$ equiv. of A^- Loss of t_+ equiv. of M^+ ∴ Net loss is t_+ equiv. of MA

It follows, therefore, if the discharged ions may be regarded as being completely removed from the system and the electrodes are not attacked, as is tacitly assumed in the above tabulation, that

$$\frac{\text{Equiv. of electrolyte lost from anode compartment}}{\text{Equiv. of electrolyte lost from cathode compartment}} = \frac{t_+}{t_-}.$$

The total decrease in amount of the electrolyte MA in both compartments of the experimental cell is equal to the number of equivalents deposited on each electrode; if a coulometer (p. 17) is included in the circuit, then by Faraday's laws the same number of equivalents of material, no matter what its nature, will be deposited. It follows, therefore, that

$$\frac{\text{Equiv. of electrolyte lost from anode compartment}}{\text{Equiv. deposited on each electrode of cell or in coulometer}} = t_+, \quad (6)$$

and

$$\frac{\text{Equiv. of electrolyte lost from cathode compartment}}{\text{Equiv. deposited on each electrode of cell or in coulometer}} = t_-. \quad (7)$$

By measuring the fall in concentration of electrolyte in the vicinity of anode and cathode of an electrolytic cell, and at the same time determining the amount of material deposited on the cathode of the cell or of a coulometer in the circuit, it is possible to evaluate the transference numbers of the ions present in solution. Since the sum of t_+ and t_- must be unity, it is not necessary to measure the concentration changes in both anode and cathode compartments, except for confirmatory purposes; similarly, if the changes in both compartments are determined it is not strictly necessary to employ a coulometer in the circuit. It is, however, more accurate to evaluate the total amount of material deposited by the current by means of a coulometer than from the concentration changes.

Chemical Changes at the Electrodes.—Although the discharge of a cation generally leads to the deposition of metal on the cathode and its consequent removal from the system, this is not true for anions. If the anode consists of an attackable metal which does not form an insoluble compound with the anions present in the solution, these ions are not removed on discharge but an equivalent amount of the anode material passes into solution. In these circumstances the concentration of the anode solution actually increases instead of decreasing, but allowance can be readily made for the amount of dissolved material. In the simplest case the anode metal is the same as that of the cations in the electrolyte, e.g., a silver anode in silver nitrate solution; the changes in the anode compartment resulting from the passage of one faraday of electricity are as follows:

1 equiv. of M^+ dissolves from the electrode
t_- equiv. of A^- migrate in
t_+ equiv. of M^+ migrate out
Net *gain* is t_- equiv. of MA.

It is thus possible to determine the transference number of the cation from the increase in concentration of the anode compartment. An alternative way of treating the results is to subtract from the observed gain in amount of electrolyte the number of equivalents of M^+ dissolved from the anode; the net result is a *loss* of $1 - t_-$, i.e., t_+, equiv. of MA per faraday, as would have been the case if the anions had been completely removed on discharge and the anode had not dissolved. It should be noted that the general results derived are applicable even if the anode material consists of a metal M′ which differs from M; the increase or decrease of concentration now refers to the total number of equivalents of MA and M′A, but the presence of the extraneous ions will affect the transference numbers of the M^+ and A^- ions.

When working with a solution of an alkali or alkaline-earth halide, the anode is generally made of silver coated with the same metal in a finely-divided state, and the cathode is of silver covered with silver halide. In this case the discharged halogen at the anode combines with the silver to form the insoluble silver halide, and so is effectively removed from the anode compartment. At the cathode, however, the silver halide is reduced to metallic silver and halide ions pass into solution; there is consequently a gain in the concentration of the cathode compartment for which allowance must be made.

Hittorf Method: Experimental Procedure.—In Hittorf's original determination of transference numbers short, wide electrolysis tubes were used in order to reduce the electrical resistance, and porous partitions were inserted to prevent mixing by diffusion and convection. These partitions are liable to affect the results and so their use has been avoided in recent work, and other precautions have been taken to minimize errors due to mixing. Many types of apparatus have been devised for the determination of transference numbers by the Hittorf method. One form, which was favored by earlier investigators and is still widely used for ordinary laboratory purposes, consists of an H-shaped tube, as shown in Fig. 38, or a tube of this form in which the limbs are separated by a U-tube. The vertical tubes, about 1.5 to 2 cm. in width and 20 to 25 cm. approximately in length, contain the anode and cathode, respectively. If the electrolyte being studied is the salt of a metal, such as silver or copper, which is capable of being deposited on the cathode with 100 per cent efficiency, the metal itself may be used as anode and cathode. Transference numbers can be calculated from the concentration changes in one electrode compartment only; if this procedure is adopted the nature of the electrolyte and of the electrode in the other compartment is immaterial. With certain solutions, e.g., acids, alkali hydroxides and alkali halides, there is a possibility that gases may be liberated at one or both electrodes; the mixing thus caused and the acid or alkali set free will vitiate the experiment. Cadmium electrodes have been employed to avoid the liberation of chlorine at the anode, and cathodes of mercury covered with concentrated solutions of zinc chloride or copper nitrate have been used to prevent the evolution of hydrogen. In the latter cases the change in the concentration of the experimental electrolyte in the anode compartment only can be utilized for the calculation of the transference numbers, as indicated above. For alkali halides the best electrodes are finely divided silver as anode and silver coated with silver halide by electrolysis (p. 234) as cathode; the behavior of these electrodes has been explained previously.

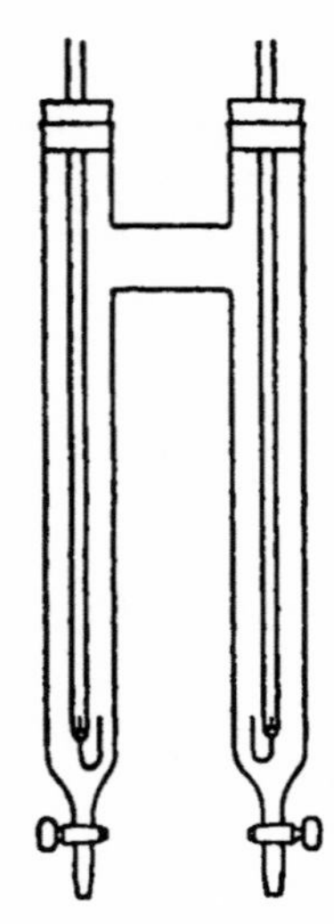

FIG. 38. Simple apparatus for transference numbers.

The apparatus is filled with the experimental solution whose *weight concentration* is known, and the electrodes are connected in series with a copper or silver voltameter; a current of 0.01 to 0.02 ampere is then passed for two to three hours. Too long a time must not be used, otherwise the results will be vitiated by diffusion, etc., and too large a current will produce mixing by convection due to heating. If both the time and current are too small, however, the concentration changes will not be appreciable. At the conclusion of the experiment a sufficient quantity of solution, believed to contain all that has changed in concentration during the electrolysis, is run off slowly from each limb, so as to avoid mixing, and analyzed. A further portion of liquid is removed from each limb; these represent the "middle compartment" and should have the same concentration as the original solution. The amount of metal deposited in the coulometer during the electrolysis is determined and sufficient data are now available for the calculation of the transference numbers.

Since the gain or loss of electrolyte near the electrode is accompanied by changes of density and hence in the volume of the solution, the concentration changes resulting from the passage of current must be determined with reference to a definite weight of solvent present at the conclusion of the electrolysis. Thus, if analysis of x grams of the anode solution showed it to contain y grams of the electrolyte *at the end* of the experiment, then the latter was associated with $x - y$ grams of water. The amount of electrolyte, say z grams, associated with this same amount of water at the beginning, is calculated from the known weight composition of the original solution. The decrease of electrolyte in the anode compartment, assuming due allowance has been made for the amount, if any, of anode material that has dissolved, is thus $z - y$ grams or $(z - y)/e$ equivalents, where e is the equivalent weight of the experimental substance. If c is the number of equivalents of material deposited in the coulometer during the electrolysis, it follows from equation (6) that the transference number of the cation (t_+) is given by

$$t_+ = \frac{z - y}{ec}. \tag{8}$$

The transference number of the anion (t_-) is of course equal to $1 - t_+$.

Improved Apparatus for the Hittorf Method.—Recent work on transference number determinations of alkali and alkaline-earth chlorides by the Hittorf method has been made with a form of apparatus of which the principle is illustrated by Fig. 39.[1] It consists of two parts, each of which contains a stopcock of the same bore as the main tubes; the anode is inserted at A and the cathode at C, the parts of the apparatus being connected by the ground joint at B. The possibility of mixing between the anode and cathode solutions is obviated by introducing right-angle bends below the anode, above the cathode and in the vertical tube between the two portions of the apparatus. For the study of alkali and alkaline-earth chlorides the anode is a coiled silver wire and the cathode is covered with silver chloride. In these cases the anode solution becomes more dilute and tends to rise, while the cathode solution increases in concentration during the course of the electrolysis and has a tendency to sink; the consequent danger of mixing is avoided by placing the anode at a higher level than the cathode, as shown in Fig. 39.

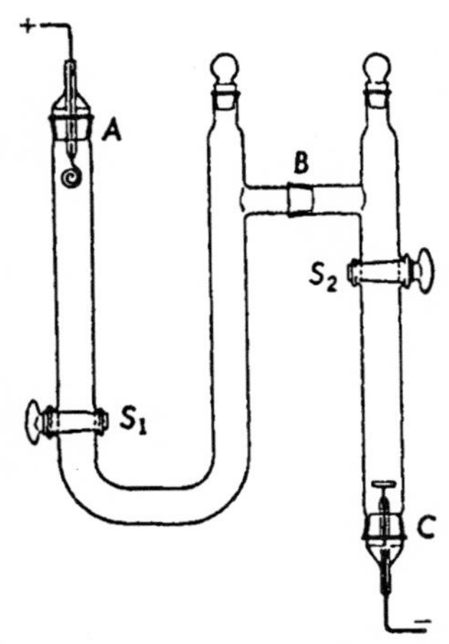

Fig. 39. Apparatus for application of Hittorf method

[1] Jones and Dole, *J. Am. Chem. Soc.*, 51, 1073 (1929); MacInnes and Dole, *ibid.*, 53, 1357 (1931); Jones and Bradshaw, *ibid.*, 54, 138 (1932).

When carrying out a measurement the two parts of the apparatus, with the electrodes in position and the stopcocks open, are fitted together, placed in a thermostat, and filled with the experimental solution. A silver coulometer is connected in series with each electrode to insure the absence of leakage currents. A quantity of electricity, depending in amount on the concentration of the solution, is passed through the circuit, and the stopcocks are then closed. The liquid isolated above S_1 is the anode solution and that below S_2 is the cathode solution; these are removed and analyzed. Quantities of liquid are withdrawn from the intermediate portion between S_1 and S_2 by inserting pipettes through the openings shown; these should have the same concentration as the original electrolyte.

Although the Hittorf method is simple in principle, accurate results are difficult to obtain; it is almost impossible to avoid a certain amount of mixing as the result of diffusion, convection and vibration. Further, the concentration changes are relatively small and any attempt to increase them, by prolonged electrolysis or large currents, results in an enhancement of the sources of error just mentioned. In recent years, therefore, the Hittorf method for the determination of transference numbers has been largely displaced by the moving boundary method, to be described later.

True and Apparent Transference Numbers.—The fundamental assumption of the Hittorf method for evaluating transference numbers from concentration changes is that the water remains stationary. There is ample evidence, however, that ions are solvated in solution and hence they carry water molecules with them in their migration through the electrolyte; this will result in concentration changes which affect the measured or "apparent" transference number. Suppose that each cation and anion has associated with it w_+ and w_- molecules of water, respectively; let T_+ and T_- be the "true" transference numbers, i.e., the actual fraction of current carried by cations and anions, respectively. For the passage of one faraday of electricity the cations will carry w_+T_+ moles of water in one direction and the anions will transport w_-T_- moles in the opposite direction; there will consequently be a resultant transfer of

$$w_+T_+ - w_-T_- = x \tag{9}$$

moles of water from the anode to the cathode compartment. The transference number t_+ is equal to the apparent number of equivalents of electrolyte leaving the anode compartment, for the passage of one faraday, whereas T_+ is the true number of equivalents; the difference between these two quantities is equal to the change of concentration resulting from the transfer to the cathode compartment of x moles of water. If the original solution contained N_s equiv. of salt associated with N_w moles of water, then the removal of x moles of water from the anode compartment, for the passage of one faraday, will increase the amount of salt by

$(N_s/N_w)x$ equiv. The apparent transference number of the cation will thus be smaller than the true value by this amount; that is,

$$T_+ = t_+ + \frac{N_s}{N_w} x. \tag{10}$$

In exactly the same way it may be shown that the water transported by the ions will cause a decrease of concentration in the cathode compartment; hence the transference number will be larger * than the true value, viz.,

$$T_- = t_- - \frac{N_s}{N_w} x. \tag{11}$$

If the net amount of water (x) transported were known, it would thus be possible to evaluate the true and apparent transference numbers from the results obtained by the Hittorf method.

The suggestion was made by Nernst (1900) that the value of x could be determined by adding to the electrolyte solution an indifferent "reference substance," e.g., a sugar, which did not move with the current; if there were no resultant transfer of water by the ions, the concentration of the reference substance would remain unchanged, but if there were such a transfer, there would be a change in the concentration. From this change the amount of water transported could be calculated. The earliest attempts to apply this principle did not yield definite results, but later investigators, particularly Washburn,[2] were more successful. At one time the sugar raffinose was considered to be the best reference substance, since its concentration could be readily determined from the optical rotation of the solution; more recently urea has been employed as the reference material, its amount being determined by chemical methods.[3]

The mean values of x obtained for approximately 1.3 N solutions of a number of halides at 25° are quoted in Table XXVII, together with the

TABLE XXVII. TRUE AND APPARENT TRANSFERENCE NUMBERS IN 1.3 N SOLUTIONS AT 25°

Electrolyte	x	t_+	T_+
HCl	0.24	0.820	0.844
LiCl	1.5	0.278	0.304
NaCl	0.76	0.366	0.383
KCl	0.60	0.482	0.495
CsCl	0.33	0.485	0.491

apparent transference numbers (t_+) of the cations and the corrected values (T_+) derived from equation (10). The difference between the

* The terms "smaller" and "larger" are used here in the algebraic sense; they also refer to the numerical values if x is positive.

[2] Washburn, *J. Am. Chem. Soc.*, **31**, 322 (1909); Washburn and Millard, *ibid.*, **37**, 694 (1915).

[3] Taylor *et al.*, *J. Chem. Soc.*, 2095 (1929); 2497 (1932); 902 (1937).

transference numbers t_+ and T_+ in the relatively concentrated solutions employed is quite appreciable; it will be apparent from equation (10) that, provided x does not change greatly with concentration, the difference between true and apparent transference numbers will be much less in the more dilute solutions, that is when N_s is small.

Another procedure for determining the net amount of water transported during electrolysis is to separate the anode and cathode compartments by means of a parchment membrane and to measure the change in volume accompanying the passage of current. This is achieved by using closed vessels as anode and cathode compartments and observing the movement of the liquid in a capillary tube connected with each vessel (Fig. 40). After making corrections for the volume changes at the electrodes due to chemical reactions, the net change is attributed to the transport of water by the ions.[4] The results may be affected to some extent by electro-osmosis (see p. 521) through the membrane separating the compartments, especially in the more concentrated solutions, but on the whole they are in fair agreement with those given in Table XXVII.

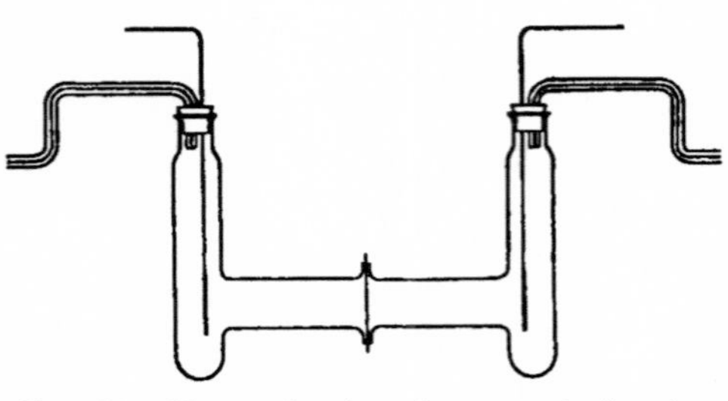

Fig. 40. Determination of transport of water

The Moving Boundary Method.—The moving boundary method for measuring transference numbers involves a modification and improvement of the idea employed by Lodge and by Whetham (cf. p. 60) for the study of the speeds of ions. On account of its relative simplicity and the accuracy of which it is capable, the method has been used in recent years for precision measurements.[5]

If it is required to determine the transference numbers of the ions constituting the electrolyte MA, e.g., potassium chloride, by the moving boundary method, it may be supposed that two other electrolytes, designated by M′A and MA′, e.g., lithium chloride and potassium acetate, each having an ion in common with the experimental solute MA, are available to act as "indicators." Imagine the solution of MA to be placed between the indicator solutions so as to form sharp boundaries at a and b, as shown in Fig. 41; the anode is inserted in the solution of M′A and the cathode in that of MA′. In order that the boundaries

[4] Remy, *Z. physik. Chem.*, **89**, 529 (1915); **118**, 161 (1925); **124**, 394 (1926); *Trans. Faraday Soc.*, **33**, 381 (1927); Baborovský *et al.*, *Rec. trav. chim.*, **42**, 229, 553 (1923); *Z. physik. Chem.*, **129**, 129 (1927); **131**, 129 (1927); **163A**, 122 (1933); *Trans. Electrochem. Soc.*, **75**, 283 (1939); Hepburn, *Phil. Mag.*, **25**, 1074 (1938).

[5] MacInnes and Longsworth, *Chem. Revs.*, **11**, 171 (1932); Longsworth, *J. Am. Chem. Soc.*, **54**, 2741 (1932); **57**, 1185 (1935).

between the solutions may remain distinct during the passage of the current, the first requirement is that the speed of the indicator ion M′ shall be less than that of M, and that the speed of A′ shall be less than that of the A ions. If these conditions hold, as well as another to be considered shortly, the M′ ions do not overtake the M ions at *a*, and neither do the A′ ions overtake the A ions at *b*; the boundaries consequently do not become blurred. In view of the slower speeds of the indicator ions, they are sometimes referred to as "following ions." Under the influence of an electric field the boundary *a* moves to *a*′, while at the same time *b* moves to *b*′; the distances *aa*′ and *bb*′ depend on the speeds of the ions M and A, and since there is a uniform potential gradient through the central solution MA, these will be proportional to the ionic velocities u_+ and u_-. It follows, therefore, from equation (4) that

$$\frac{aa'}{aa' + bb'} = \frac{u_+}{u_+ + u_-} = t_+$$

and

$$\frac{bb'}{aa' + bb'} = \frac{u_-}{u_+ + u_-} = t_-,$$

so that the transference numbers can be determined from observations on the movements of the boundaries *a* and *b*.

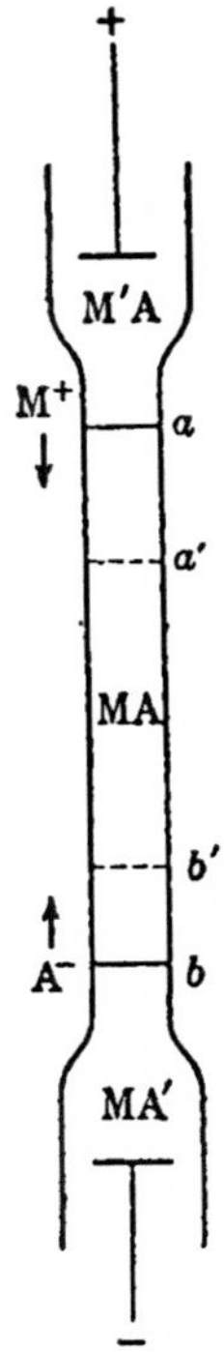

FIG. 41. Principle of the moving boundary method

In the practical application of the moving boundary method one boundary only is observed, and so the necessity of finding two indicator solutions is obviated; the method of calculation is as follows. If one faraday of electricity passes through the system, t_+ equiv. of the cation must pass any given point in one direction; if c equiv. per unit volume is the concentration of the solution in the vicinity of the boundary formed by the M ions, this boundary must sweep through a volume t_+/c while one faraday is passing. The volume ϕ swept out by the cations for the passage of Q coulombs is thus

$$\phi = \frac{Qt_+}{Fc}, \tag{12}$$

where F is one faraday, i.e., 96,500 coulombs. If the cross section of the tube in which the boundary moves is a sq. cm., and the distance through which it moves during the passage of Q coulombs is l cm., then ϕ is equal to la, and hence from equation (12)

$$t_+ = \frac{laFc}{Q}. \tag{13}$$

Since the number of coulombs passing can be determined, the transference number of the ion may be calculated from the rate of movement of one boundary.

In accurate work a correction must be applied for the change in volume occurring as a result of chemical reactions at the electrodes and because of ionic migration. If Δv is the consequent increase of volume of the cathode compartment for the passage of one faraday, equation (12) becomes

$$\phi + \frac{Q}{F}\Delta v = \frac{Qt_{corr.}}{F\text{c}},$$

$$\therefore \quad t_{corr.} = t_{obs.} + \text{c}\Delta v, \tag{14}$$

where $t_{corr.}$ is the corrected transference number and $t_{obs.}$ is the value given by equation (13); the difference is clearly only of importance in concentrated solutions.

The Kohlrausch Regulating Function.—An essential requirement for a sharp boundary is that the cations M and M′, present on the two sides of the boundary, should move with exactly the same speed under the conditions of the experiment. It can be deduced that the essential requirement for this equality of speed is given by the **Kohlrausch regulating function,** viz.,

$$\frac{t_+}{\text{c}} = \frac{t'_+}{\text{c}'}, \tag{15}$$

where t_+ and c are the transference number and equivalent concentration, respectively, of the ion M in the solution of MA, and t'_+ and c′ are the corresponding quantities for the ion M′ in the solution of M′A; the solutions are those constituting the two sides of the boundary. The equivalent concentration of each electrolyte at the boundary, i.e., of MA and M′A should be proportional to the transference number of its cation. Similarly, at the boundary between the salts MA and MA′, the concentrations should be proportional to the transference numbers of the respective anions. The reason for this condition may be seen in an approximate way from equation (3): the transference number divided by the equivalent concentration of the ion, which is equal to cz, is proportional to the speed of the ion; hence, when t/c is the same for both ions the speeds will be equal.

The indicator concentration at the boundary should, theoretically, adjust itself automatically during the passage of current so as to satisfy the requirement of the Kohlrausch regulating function. Suppose the indicator were more concentrated than is necessary according to equation (15); the potential gradient in this solution would then be lower than is required to make the ion M′ travel at the same speed as M. The M′ ions would thus lag behind and their concentration at the boundary would fall; the potential gradient in this region would thus increase until the velocity of the M′ ions was equal to that of the leading ion. Similar automatic adjustment would be expected if the bulk of the indicator solution were more dilute than necessary to satisfy equation (15).

It would appear, therefore, that the actual concentration of the indicator solution employed in transference measurements is immaterial: experiments show, however, that automatic attainment of the Kohlrausch regulating condition is not quite complete, for the transference numbers have been found to be dependent to some extent on the concentration of the bulk of the indicator solution. This is shown by the results in Fig. 42 for the observed transference number of the potassium ion in 0.1 N potassium chloride, with lithium chloride of various concentrations as indicator solution. The concentration of the latter required to satisfy equation (15) is 0.064 N, and hence it appears, from the constancy of the transference number over the range of 0.055 to 0.075 N lithium chloride, that automatic adjustment occurs only when the actual concentration of the indicator solution is not greatly different from the Kohlrausch value. The failure of the adjustment to take place is probably due to the disturbing effects of convection resulting from temperature and density gradients in the electrolyte.[6]

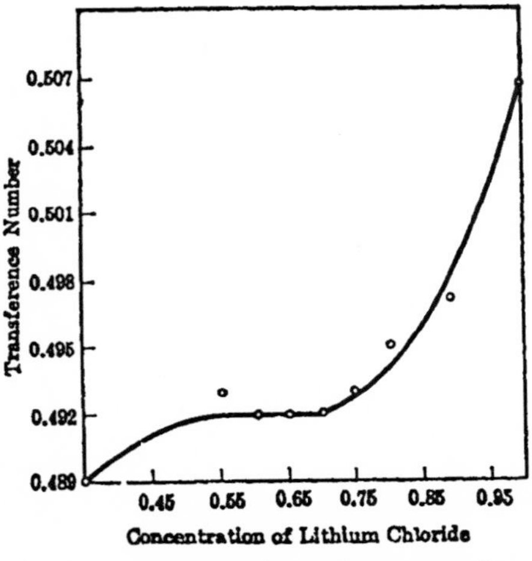

Fig. 42. Variation of transference number with concentration of indicator solution

When carrying out a transference number measurement by the moving boundary method the bulk concentration of the indicator solution is chosen so as to comply with equation (15), as far as possible, using approximate transference numbers for the purpose of evaluating c'. The experiment is then repeated with a somewhat different concentration of indicator solution until a constant value for the transference number is obtained; this value is found to be independent of the applied potential and hence of the current strength.

Experimental Methods.—One of the difficulties experienced in performing transference number measurements by the moving boundary method was the establishment of sharp boundaries; recent work, chiefly by MacInnes and his collaborators, has resulted in such improvements of technique as to make this the most accurate method for the determination of transference numbers. Since the earlier types of apparatus

[6] MacInnes and Smith, *J. Am. Chem. Soc.*, **45**, 2246 (1923); MacInnes and Longsworth, *Chem. Revs.*, **11**, 171 (1932); Hartley and Moilliet, *Proc. Roy. Soc.*, **140A**, 141 (1933).

have been largely superseded, these will not be described here; reference will be made to the more modern forms only.

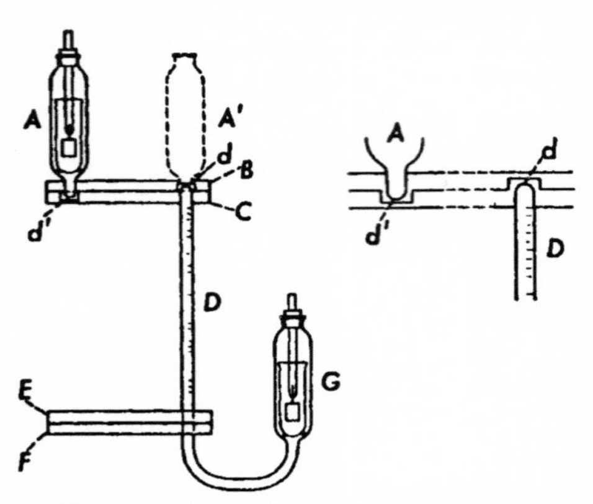

FIG. 43. Sheared boundary apparatus (MacInnes and Brighton)

The apparatus used in the **sheared boundary method** is shown diagrammatically in Fig. 43.[7] The electrode vessel *A* is fitted into the upper of a pair of accurately ground discs, *B* and *C*, which can be rotated with respect to each other. Into the lower disc is fixed the graduated tube *D* in which the boundary is to move, and this is attached by a similar pair of discs, *E* and *F*, to the other electrode vessel *G*. The vessel *A* is filled with the indicator solution and a drop is allowed to protrude below the disc *B*, while the experimental solution is placed in the vessel *G* and the tube *D* so that a drop protrudes above the top of *C*; the discs are so arranged that the protruding drops *d* and *d'* are accommodated in the small holes, as shown in the enlarged diagram at the right of Fig. 43. The disc *B* is now rotated, with the result that the electrode vessel *A* fits exactly over *D*, as shown by the dotted lines at *A'*; in the process the protruding drops of liquid are sheared off and a sharp boundary is formed. The above procedure is employed for a falling boundary, moving down the tube *D* under the influence of current, i.e., when the indicator solution has a lower density than the experimental solution. If the reverse is the case, a rising boundary must be used, and this is formed in a similar manner between the two lower discs *E* and *F*; the indicator solution is now placed in *G* and the experimental solution in *A* and *D*.

If the ions of a metal, such as cadmium or silver, which forms an attackable anode, are suitable as indicator cations, it is possible to use the device of the **autogenic boundary.**[8] No special indicator solution is required, but a block of the metal serves as the anode and the experimental solution is placed in a vertical tube above it. For example, with nitrate solutions a silver anode can be used, and with chloride solutions one of cadmium can be employed; the silver nitrate or cadmium chloride, respectively, that is formed as the anode dissolves acts as indicator solution. It is claimed that there is automatic adjustment of the concentration in accordance with the Kohlrausch regulating function, and a sharp boundary is formed and maintained throughout the experiment.

[7] MacInnes and Brighton, *J. Am. Chem. Soc.*, **47**, 994 (1925).

[8] Cady and Longsworth, *J. Am. Chem. Soc.*, **51**, 1656 (1929); Longsworth, *ibid.*, **57**, 1698 (1935); *J. Chem. Ed.*, **11**, 420 (1934).

The method is capable of giving results of considerable accuracy, although its application is limited to those cases for which a suitable anode material can be found.

An alternative, somewhat simple~ but less accurate, procedure for measuring transference numbers by the moving boundary principle, utilizes the **air-lock method** of establishing the boundary.[9] The apparatus for a rising boundary is shown in Fig. 44; the graduated measuring tube *A* has a bore of about 7 mm., whereas *E* and *F* are fine capillaries; the top of the latter is closed by rubber tubing with two pinchcocks. The electrodes are placed in the vessels *B* and *C*. With electrode *B* in position, and the upper pinchcock at the top of *F* closed, the apparatus is filled with the experimental solution. By closing the lower pinchcock a small column of air *G* is forced into the tube where *F* joins *A*, thus separating the solutions in *A* and *D*. The solution in *CDE* is then emptied by suction through *C* and *E*, care being taken not to disrupt the air column *G*. The tube *CDE* is now filled with the indicator solution, the electrode is inserted in *C*, and the lower pinchcock at the top of *F* is adjusted so that the air column *G* is withdrawn sufficiently to permit a boundary to form between the indicator solution in *CDE* and the experimental solution in *A*. Even if the boundary is not initially sharp, it is soon sharpened by the current.

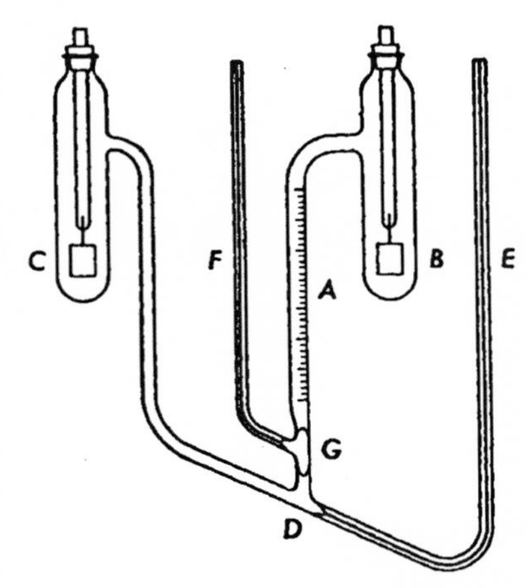

FIG. 44. Air-lock method for establishing boundary (Hartley and Donaldson)

In following the movement of the boundary, no matter how it is formed, use is made of the difference in the refractive indices of the indicator and experimental solutions; if the boundary is to be clearly visible, this difference should be appreciable. If the distance (l) moved in a given time and the area of cross section (a) of the tube are measured, and the equivalent concentration (c) of the experimental solution is known, it is only necessary to determine the number of coulombs (Q) passed for the transference number to be calculated by equation (13). The quantity of electricity passing during the course of a moving boundary experiment is generally too small to be measured accurately in a coulometer. It is the practice, therefore, to employ a current of known strength for a measured period of time; the constancy of the current can be ensured by means of automatic devices which make use of the properties of vacuum tubes.

[9] Hartley and Donaldson, *Trans. Faraday Soc.*, 33, 457 (1937).

Results of Transference Number Measurements.—Provided the measurements are made with great precision, the results obtained by the Hittorf and moving boundary methods agree within the limits of experimental error; this is shown by the most accurate values for various solutions of potassium chloride at 25° as recorded in Table XXVIII.

TABLE XXVIII. TRANSFERENCE NUMBERS OF POTASSIUM CHLORIDE SOLUTIONS AT 25°

Concentration	0.02	0.05	0.10	0.50	1.0 N
Hittorf method	0.489	0.489	0.490	0.490	0.487
Moving boundary method	0.490	0.490	0.490	0.490	0.488

It is probable that, on the whole, transference numbers derived from moving boundary measurements are the more reliable.

It may be noted that the values obtained by the moving boundary method, like those given by the Hittorf method, are the so-called "apparent" transference numbers (p. 114), because the transport of water by the ions will affect the volume through which the boundary moves. It is the practice, however, to record observed transference numbers without applying any correction, since much uncertainty is attached to the determination of the transport of water during the passage of current. Further, in connection with the study of certain types of voltaic cell, it is the "apparent" rather than the "true" transference number that is involved (cf. p. 202).

Some of the most recent data of the transference numbers of the cations of various salts at a number of concentrations at 25°, mainly obtained by the moving boundary method, are given in Table XXIX;[10]

TABLE XXIX. TRANSFERENCE NUMBERS OF CATIONS IN AQUEOUS SOLUTIONS AT 25°

Concn.	HCl	LiCl	NaCl	KCl	KNO_3	$AgNO_3$	$BaCl_2$	K_2SO_4	$LaCl_3$
0.01 N	0.8251	0.3289	0.3918	0.4902	0.5084	0.4648	0.440	0.4829	0.4625
0.02	0.8266	0.3261	0.3902	0.4901	0.5087	0.4652	0.4375	0.4848	0.4576
0.05	0.8292	0.3211	0.3876	0.4899	0.5093	0.4664	0.4317	0.4870	0.4482
0.1	0.8314	0.3168	0.3854	0.4898	0.5103	0 4682	0.4253	0.4890	0.4375
0.2	0.8337	0.3112	0.3821	0.4894	0.5120	—	0.4162	0.4910	0.4233
0.5	—	0.300	—	0.4888	—	—	0.3986	0.4909	0.3958
1.0	—	0.287	—	0.4882	—	—	0.3792	—	—

the corresponding anion transference numbers may be obtained in each case by subtracting the cation transference number from unity.

Influence of Temperature on Transference Numbers.—The extent of the variation of transference numbers with temperature will be evident from the data for the cations of a number of chlorides at a concentration of 0.01 N recorded in Table XXX; these figures were obtained by the Hittorf method and, although they may be less accurate than those in Table XXIX, they are consistent among themselves. The transference

[10] Longsworth, *J. Am. Chem. Soc.*, **57**, 1185 (1935); **60**, 3070 (1938).

TABLE XXX. INFLUENCE OF TEMPERATURE ON CATION TRANSFERENCE NUMBERS IN 0.01 N SOLUTIONS

Temperature	HCl	NaCl	KCl	$BaCl_2$
0°	0.846	0.387	0.493	0.437
18°	0.833	0.397	0.496	—
30°	0.822	0.404	0.498	0.444
50°	0.801	—	—	0.475

numbers of the ions of potassium chloride vary little with temperature, but in sodium chloride solution, and particularly in hydrochloric acid, the change is appreciable. It has been observed, at least for uni-univalent electrolytes, that if the transference number of an ion is greater than 0.5, e.g., the hydrogen ion, there is a decrease as the temperature is raised. It appears, therefore, in general that transference numbers measured at appreciable concentrations tend to approach 0.5 as the temperature is raised; in other words, the ions tend towards equal speeds at high temperatures.

Transference Number and Concentration: The Onsager Equation.— It will be observed from the results in Table XXIX that transference numbers generally vary with the concentration of the electrolyte, and the following relationship was proposed to represent this variation, viz.,

$$t = t_0 - A\sqrt{c}, \tag{16}$$

where t and t_0 are the transference numbers of a given ion in a solution of concentration c and that extrapolated to infinite dilution, respectively, and A is a constant. Although this equation is applicable to dilute solutions, it does not represent the behavior of barium chloride and other electrolytes at appreciable concentrations. A better expression, which holds up to relatively high concentrations, is

$$t = \frac{t_0 + 1}{1 + B\sqrt{c}} - 1, \tag{17}$$

where B is a constant for the given electrolyte.[11] This equation may be written in the form

$$\frac{t_0 + 1}{t + 1} = 1 + B\sqrt{c}, \tag{18}$$

so that the plot of $1/(t + 1)$ against $\sqrt{c}$ should be a straight line, as has been found to be true in a number of instances. Equation (17) can also be expressed as a power series, thus

$$t = t_0 - (t_0 + 1)Bc^{\frac{1}{2}} + (t_0 + 1)B^2c + (t_0 + 1)B^3c^{\frac{3}{2}} + \cdots,$$

and when c is small, i.e., for dilute solutions, so that all terms beyond that involving $c^{\frac{1}{2}}$ can be neglected, this reduces to equation (16) since $(t_0 + 1)B$ is a constant.

[11] Jones and Dole, *J. Am. Chem. Soc.*, 51, 1073 (1929); Jones *et al.*, *ibid.*, 54, 138 (1932); 58, 1476 (1936); Dole, *J. Phys. Chem.*, 35, 3647 (1931).

The Onsager equation for the equivalent conductance λ_i of an ion may be written in the form [cf. equation (34), p. 89]

$$\lambda_i = \lambda_i^0 - A_i\sqrt{c}, \tag{19}$$

where λ_i^0 is the ion conductance at infinite dilution and A_i is a constant. Introducing the expression for the transference number given by equation (5), that is $t_i = \lambda_i/\Lambda$, where Λ is the equivalent conductance of the electrolyte at the experimental concentration, it follows that

$$t_i = \frac{\lambda_i^0 - A_i\sqrt{c}}{\Lambda}. \tag{20}$$

The value of Λ can be expressed in terms of Λ_0 by an equation similar to (19), and then equation (20) can be written in the form

$$t_i = \frac{A}{1 - B\sqrt{c}} + D, \tag{21}$$

where A, B and D are constants. This equation derived from the Debye-Hückel-Onsager theory of conductance is of the same form as the empirical equation (17), and hence is in general agreement with the facts; the constants A, B and D, however, which are required to satisfy the experimental results differ from those required by theory. This discrepancy is largely due to the fact that the transference measurements were made in solutions which are too concentrated for the simple Onsager equation to be applicable.

Since the Onsager equation is, strictly speaking, a limiting equation, it is more justifiable to see if the variation of transference number with concentration approaches the theoretical behavior with increasing dilution. The equivalent conductance of a univalent ion can be expressed in the form of equation (37), page 90, viz.,

$$\lambda_i = \lambda_i^0 - (\tfrac{1}{2}A + B\lambda_i^0)\sqrt{c}, \tag{22}$$

where A and B as used here are the familiar Onsager values (Table XXIII, p. 90); the transference number (t_+) of the cation in a uni-univalent electrolyte can then be represented by

$$t_+ = \frac{\lambda_+}{\lambda_+ + \lambda_-} = \frac{\lambda_+^0 - (\tfrac{1}{2}A + B\lambda_+^0)\sqrt{c}}{\Lambda_0 - (A + B\Lambda_0)\sqrt{c}}, \tag{23}$$

where A and B are the same for both ions. Differentiating equation (23) with respect to $\sqrt{c}$, and introducing the condition that c approaches zero, it is found that

$$\left(\frac{dt_+}{d\sqrt{c}}\right)_{c\to 0} = \frac{2t_+^0 - 1}{2\Lambda_0}A. \tag{24}$$

It follows, therefore, that the slope of the plot of the transference number of an ion against the square-root of the concentration should attain a limiting value, equal to $(2t_+^0 - 1)A/2\Lambda_0$ as infinite dilution is approached; the results in Fig. 45, in which the full lines are drawn through the experimental cation transference numbers in aqueous solution at 25° and the

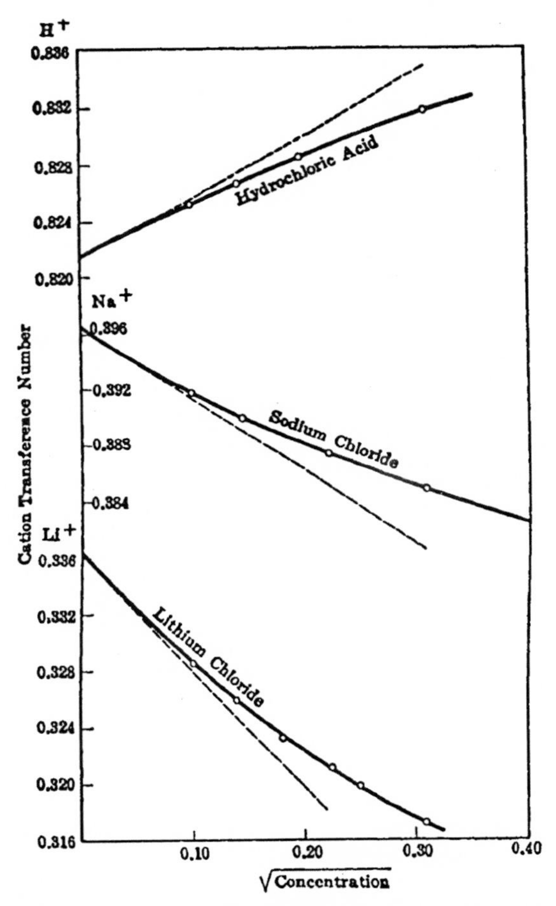

Fig. 45. Transference numbers and the Onsager equation (Longsworth)

dotted lines represent the limiting slopes, are seen to be in good agreement with the requirements of the inter-ionic attraction theory.[12]

Equivalent Conductances of Ions.—Since transference numbers and equivalent conductances at various concentrations are known, it should be possible, by utilizing the expression $\lambda_i = t_i\Lambda$, to extrapolate the re-

[12] Longsworth, *J. Am. Chem. Soc.*, **57**, 1185 (1935); see also, Hartley and Donaldson, *Trans. Faraday Soc.*, **33**, 457 (1937); Samis, *ibid.*, **33**, 469 (1937).

sults to give ion conductances at infinite dilution. Two methods of extrapolating the data are possible. In the first place, the equivalent conductances and the transference numbers may be extrapolated separately to give the respective values at infinite dilution; the product of these quantities would then be equal to the ionic conductance at infinite dilution. The data from which the conductance of the chloride ion can be evaluated are given in Table XXXI; the mean value of the conduct-

TABLE XXXI. CALCULATION OF CHLORIDE ION CONDUCTANCE AT 25°

Electrolyte	$t^0_{Cl^-}$	Λ_0	$\lambda^0_{Cl^-}$
HCl	0.1790	426.16	76.28
LiCl	0.6633	115.03	76.30
NaCl	0.6035	126.45	76.31
KCl	0.5097	149.86	76.40

ance of the chloride ion at infinite dilution at 25°, derived from measurements on solutions of the four chlorides, is thus found to be 76.32 ohms^{-1} cm.^{2} The results in the last column are seen to be virtually independent of the nature of the chloride, in agreement with Kohlrausch's law of the independent migration of ions.

The second method of extrapolation is to obtain the values of λ_i at various concentrations and to extrapolate the results to infinite dilution. The equivalent conductances of the chloride ion at several concentrations obtained from transference and conductance measurements, on the four chlorides to which the data in Table XXXI refer, are given in Table XXXII. These results can be plotted against the square-root of the

TABLE XXXII. EQUIVALENT CONDUCTANCES OF CHLORIDE ION AT 25°

Electrolyte	0.01	0.02	0.05	0.10 N
HCl	72.06	70.62	68.16	65.98
LiCl	72.02	70.52	67.96	65.49
NaCl	72.05	70.54	67.92	65.58
KCl	72.07	70.56	68.03	65.79

concentration and extrapolated to infinite dilution, thus giving 76.3 ohms^{-1} cm.^{2} for the ion conductance, but a more precise method is similar to that described on page 54, based on the use of the Onsager equation. The conductance of a single univalent ion, assuming complete dissociation of the electrolyte, is given by equation (22), the values of A and B being known; if the experimental data for λ_i at various concentrations, as given in Table XXXII, are inserted in this equation, the corresponding results for λ^0_i can be obtained. If the solutions were sufficiently dilute for the Onsager equation to be strictly applicable, the values of λ^0_i would all be the same; on account of the incomplete nature of this equation in its simple form, however, they actually increase with increasing concentration (cf. p. 55). By plotting the results against the concentration and extrapolating to infinite dilution, the equivalent conductance of the chloride ion in aqueous solution has been found to be 76.34 ohms^{-1} cm.^{2}

at 25°; this is the best available datum for the conductance of the chloride ion.[13]

Since the ion conductance of the chloride ion is now known accurately, that of the hydrogen, lithium, sodium, potassium and other cations can be derived by subtraction from the equivalent conductances at infinite dilution of the corresponding chloride solutions; from these results the values for other anions, and hence for further cations, can be obtained. The data recorded in Table XIII, page 56, were calculated in this manner.

It is of interest to note from Table XXXII that the equivalent conductance of the chloride ion is almost the same in all four chloride solutions at equal concentrations, especially in the more dilute solutions. This fact supports the view expressed previously that Kohlrausch's law of the independent migration of ions is applicable to dilute solutions of strong electrolytes at equivalent concentrations, as well as at infinite dilution.

Transference Numbers in Mixtures.—Relatively little work has been done on the transference numbers of ions in mixtures, although both Hittorf and moving boundary methods have been employed. In the former case, it follows from equation (3) that the transference number of any ion in a mixture is equal to the number of equivalents of that ion migrating from the appropriate compartment divided by the total number of equivalents deposited in a coulometer. It is possible, therefore, to derive the required transference numbers by analysis of the anode and cathode compartments before and after electrolysis.

The moving boundary method has been used to study mixtures of alkali chlorides and hydrochloric acid, a cadmium anode being employed to form an "autogenic" boundary. After electrolysis has proceeded for some time two boundaries are observed; the leading boundary is due to the high mobility of the hydrogen ion and is formed between the mixture of hydrochloric acid and the alkali chloride on the one side, and a solution of the alkali chloride from which the hydrogen ion has completely migrated out on the other side. The rate of movement of this boundary gives the transference number of the hydrogen ion in the mixture of electrolytes. The slower boundary is formed between the pure alkali chloride solution and the cadmium chloride indicator solution, and gives no information concerning transference numbers in the mixture. The transference number of the alkali metal ion cannot be determined directly from the movement of the boundaries, and so the transference number of the chloride ion in the mixed solution is obtained from a separate experiment with an anion boundary using a mixture of potassium iodate and iodic acid as indicator. Since the transference numbers of the three

[13] Longsworth, *J. Am. Chem. Soc.*, **54**, 2741 (1932); MacInnes, *J. Franklin Inst.*, **225**, 661 (1938); see also, Owen, *J. Am. Chem. Soc.*, **57**, 2441 (1935).

ions must add up to unity, the value for the alkali metal can now be derived.[14]

Abnormal Transference Numbers.—In certain cases, particularly with solutions of cadmium iodide, the transference number varies markedly with concentration, and the values may become zero or even apparently negative; the results for aqueous solutions of cadmium iodide at 18° are quoted in Table XXXIII. At concentrations greater than 0.5 N,

TABLE XXXIII. CATION TRANSFERENCE NUMBERS IN CADMIUM IODIDE AT 18°

Concn.	0.0005	0.01	0.02	0.05	0.1	0.2	0.5 N
t_+	0.445	0.444	0.442	0.396	0.296	0.127	0.003

the transference number of cadmium apparently becomes negative: this means that in relatively concentrated solutions of cadmium iodide, the cadmium is being carried by the current in a direction *opposite* to that in which positive electricity moves through the solution. In other words, cadmium must form part of the negative ion present in the electrolyte. A reasonable explanation of the results is that in dilute solution cadmium iodide ionizes to yield simple ions; thus

$$CdI_2 \rightleftharpoons Cd^{++} + 2I^-,$$

and so the transference number of the cadmium ion, in solutions containing less than 0.02 equiv. per liter, is normal. As the concentration is increased, however, the iodide ions combine with unionized molecules of cadmium iodide to form complex CdI_4^{--} ions, thus,

$$CdI_2 + 2I^- \rightleftharpoons CdI_4^{--},$$

with the result that appreciable amounts of cadmium are present in the anions and hence are transferred in the direction opposite to that of the flow of positive current. The apparent transference number of the cadmium ion is thus observed to decrease; if equal quantities of electricity are carried in opposite directions by Cd^{++} and CdI_4^{--} ions the transference number will appear to be zero. The proportion of CdI_4^{--} ions increases with increasing concentration and eventually almost the whole of the iodine will be present as CdI_4^{--} ions; the current is then carried almost exclusively by Cd^{++} and CdI_4^{--} ions. If the speed of the latter is greater than that of the former, as appears actually to be the case, the apparent cation transference number will be negative. A similar variation of the cation transference number with concentration has been observed in solutions of cadmium bromide and this may be attributed to the existence of the analogous $CdBr_4^{--}$ ion. Less marked changes of transference number have been observed with other electrolytes; these are also probably to be ascribed to the presence of complex ions in concentrated solutions.

[14] Longsworth, *J. Am. Chem. Soc.*, 52, 1897 (1930).

PROBLEMS

1. MacInnes and Dole [*J. Am. Chem. Soc.*, 53, 1357 (1931)] electrolyzed a 0.5 N solution of potassium chloride, containing 3.6540 g. of salt per 100 g. solution, at 25° using an anode of silver and a cathode of silver coated with silver chloride. After the passage of a current of about 0.018 amp. for approximately 26 hours, 1.9768 g. of silver were deposited in a coulometer in the circuit and on analysis the 119.48 g. of anode solution were found to contain 3.1151 g. potassium chloride per 100 g. solution, while the 122.93 g. of cathode solution contained 4.1786 g. of salt per 100 g. Calculate the values of the transference number of the potassium ion obtained from the anode and cathode solutions, respectively.

2. Jones and Bradshaw [*J. Am. Chem. Soc.*, 54, 138 (1932)] passed a current of approximately 0.025 amp. for 8 hours through a solution of lithium chloride, using a silver anode and a silver chloride cathode; 0.73936 g. of silver was deposited in a coulometer. The original electrolyte contained 0.43124 g. of lithium chloride per 100 g. of water, and after electrolysis the anode portion, weighing 128.615 g., contained 0.35941 g. of salt per 100 g. water, while the cathode portion, weighing 123.074 g., contained 0.50797 g. of salt per 100 g. of water. Calculate the transference number of the chloride ion from the separate data for anode and cathode solutions.

3. In a moving boundary experiment with 0.1 N potassium chloride, using 0.065 N lithium chloride as indicator solution, MacInnes and Smith [*J. Am. Chem. Soc.*, 45, 2246 (1923)] passed a constant current of 0.005893 amp. through a tube of 0.1142 sq. cm. uniform cross section and observed the boundary to pass the various scale readings at the following times:

Scale reading	0.5	5.50	5.80	6.10	6.40	6.70	7.00 cm.
Time	0	1900	2016	2130	2243	2357	2472 sec.

Calculate the mean transference number of the potassium ion. The potential gradient was 4 volts per cm.; evaluate the mobility of the potassium ion for unit potential gradient.

4. The following results were recorded by Jahn and his collaborators [*Z. physik. Chem.*, 37, 673 (1901)] in experiments on the transference number of cadmium in cadmium iodide solutions using a cadmium anode:

Original solution Cd per cent *	Anode solution after electrolysis Weight	Cd per cent	Silver deposited in coulometer
2.5974	138.073	2.8576	0.7521 g.
1.3565	395.023	1.4863	0.9538
0.8820	300.798	1.0096	0.9963
0.4500	289.687	0.5654	0.9978
0.2311	305.750	0.3264	0.9604
0.1390	301.700	0.1868	0.5061

* The expression "Cd per cent" refers to the number of grams of Cd per 100 g. of solution.

Evaluate the apparent transference number of the cadmium ion at the different concentrations, and plot the results as a function of concentration.

5. A0.2 N solution of sodium chloride was found to have a specific conductance of 1.75×10^{-2} ohm^{-1} cm.$^{-1}$ at 18°; the transference number of the

cation in this solution is 0.385. Calculate the equivalent conductance of the sodium and chloride ions.

6. A solution contains 0.04 N sodium chloride, 0.02 N hydrochloric acid and 0.04 N potassium sulfate; calculate, approximately, the fraction of the current carried by each of the ionic species, Na^+, K^+, H^+, Cl^- and SO_4^{--}, in this solution. Utilize the data in Tables X and XIII, and assume that the conductance of each ion is the same as in a solution of concentration equal to the total equivalent concentration of the given solution.

7. The equivalent conductances and cation transference numbers of ammonium chloride at several concentrations at 25° are as follows [Longsworth, *J. Am. Chem. Soc.*, 57, 1185 (1935)]:

c	0.01	0.02	0.05	0.10 N	
Λ	141.28	138.33	133.29	128.75	ohms^{-1} cm.^{2}
t_+	0.4907	0.4906	0.4905	0.4907	

Utilize the results to evaluate the equivalent conductance of the ammonium and chloride ions at infinite dilution by the method described on page 126.

8. Use the results of the preceding problem to calculate the limiting slope, according to the Onsager equation, of the plot of the transference number of the ammonium ion in ammonium chloride against the square-root of the concentration.

9. Hammett and Lowenheim [*J. Am. Chem. Soc.*, 56, 2620 (1934)] electrolyzed, with inert electrodes, a solution of $Ba(HSO_4)_2$ in sulfuric acid as solvent; 1 g. of this solution contained 0.02503 g. $BaSO_4$ before electrolysis. After the passage of 4956 coulombs, 41 cc. of the anode solution and 39 cc. of the cathode solution, each having a density of 1.9, were run off; they were found on analysis to contain 0.02411 and 0.02621 g. of $BaSO_4$ per gram of solution, respectively. Calculate the transference number of the cation.

10. A solution, 100 g. of which contained 2.9359 g. of sodium chloride and 0.58599 g. urea, was electrolyzed with a silver anode and a silver chloride cathode; after the passage of current which resulted in the deposition of 4.5025 g. of silver in a coulometer, Taylor and Sawyer [*J. Chem. Soc.*, 2095 (1929)] found 141.984 g. of anode solution to contain 3.2871 g. sodium chloride and 0.84277 g. urea, whereas 57.712 g. of cathode solution contained 2.5775 g. sodium chloride and 0.32872 g. urea. Calculate the "true" and "apparent" transference numbers of the ions of sodium chloride in the experimental solution.

CHAPTER V

FREE ENERGY AND ACTIVITY

Partial Molar Quantities.[1]—The thermodynamic functions, such as heat content, free energy, etc., encountered in electrochemistry have the property of depending on the temperature, pressure and volume, i.e., the state of the system, and on the amounts of the various constituents present. For a given mass, the temperature, pressure and volume are not independent variables, and so it is, in general, sufficient to express the function in terms of two of these factors, e.g., temperature and pressure. If X represents any such extensive property, i.e., one whose magnitude is determined by the state of the system and the amounts, e.g., number of moles, of the constituents, then the **partial molar** value of that property, for any constituent i of the system, is defined by

$$\bar{X}_i = \left(\frac{\partial X}{\partial n_i}\right)_{p,\,T,\,n_1,\,n_2,\,\cdots} \tag{1}$$

and is indicated by writing a bar over the symbol for the property. The partial molar quantity is consequently the increase in the particular property X resulting from the addition, at constant temperature and pressure, of one mole of the constituent i to such a large quantity of the system that there is no appreciable change in its composition.

If a small change is made in the system at constant temperature and pressure, such that the number of moles of the constituent 1 is increased by dn_1, of 2 by dn_2, or, in general, of the constituent i by dn_i, the total change dX in the value of the property X is given by

$$(dX)_{T,\,p} = \bar{X}_1 dn_1 + \bar{X}_2 dn_2 + \cdots \bar{X}_i dn_i + \cdots \tag{2}$$

In estimating dX from equation (2) it is, of course, necessary to insert a minus sign before the $\bar{X}dn$ term for any constituent whose amount is decreased as a result of the change in the system.

Partial Molar Free Energy: Chemical Potential.—The partial molal free energy is an important thermodynamic property in connection with the study of electrolytes; it can be represented either as $\bar{G}$, where G is employed for the Gibbs, or Lewis, free energy,* or by the symbol μ, when it is referred to as the **chemical potential**; thus the appropriate form

[1] Lewis and Randall, "Thermodynamics and the Free Energy of Substances," 1923, Chap. IV; Glasstone, "Text-book of Physical Chemistry," 1940, Chap. III.

* Electrochemical processes are almost invariably carried out at constant temperature and pressure; under these conditions G is the appropriate thermodynamic function. The symbol F has been generally used to represent the free energy, but in order to avoid confusion with the symbol for the faraday, many writers now adopt G instead.

of equation (2), for the increase of free energy accompanying a change in a given system at constant temperature and pressure, is then

$$(dG)_{T,p} = \mu_1 dn_1 + \mu_2 dn_2 + \cdots \mu_i dn_i + \cdots \tag{3}$$

One of the thermodynamic conditions of equilibrium is that $(dG)_{T,p}$ is zero; it follows, therefore, that *for a system in equilibrium at constant temperature and pressure*

$$\mu_1 dn_1 + \mu_2 dn_2 + \cdots \mu_i dn_i + \cdots = \sum_i \mu_i dn_i = 0. \tag{4}$$

The partial molal volume of the constituent i in a mixture of *ideal gases*, which do not react, is equal to its molar volume v_i in the system, since there is no volume change on mixing; if p_i is the partial pressure of the constituent, then v_i is equal to RT/p_i, where R is the gas constant per mole and T is the absolute temperature. It can be shown by means of thermodynamics that the partial molal volume ($\bar{v}_i$) is related to the chemical potential by the equation

$$\bar{v}_i = \left(\frac{\partial \mu_i}{\partial p}\right)_{T,n_1,n_2,\cdots} \tag{5}$$

and so it follows that, for an ideal gas mixture,

$$\left(\frac{\partial \mu_i}{\partial p}\right)_{T,n_1,n_2,\cdots} = \frac{RT}{p_i}. \tag{6}$$

Integration of equation (6) then gives the chemical potential of the gas i in the mixture, thus

$$\mu_i = \mu_i^0 + RT \ln p_i, \tag{7}$$

where μ_i^0 is a constant depending only on the nature of the gas and on the temperature of the system. It is evident that μ_i^0 is equal to the chemical potential of the ideal gas at unit partial pressure.

Activity and Activity Coefficient.[2]—When a pure liquid or a mixture is in equilibrium with its vapor, the chemical potential of any constituent in the liquid must be equal to that in the vapor; this is a consequence of the thermodynamic requirement that for a system at equilibrium a small change at constant temperature and pressure shall not be accompanied by any change of free energy, i.e., $(\partial G)_{T,p}$ is zero. It follows, therefore, that if the vapor can be regarded as behaving ideally, the chemical potential of the constituent i of a *solution* can be written in the same form as equation (7), where p_i is now the partial pressure of the component in the vapor in equilibrium with the solution. If the vapor is not ideal, the partial pressure should be replaced by an ideal pressure, or "fugacity," but this correction need not be considered further. According to Raoult's

[2] Lewis and Randall, *J. Am. Chem. Soc.*, 43, 1112 (1921); "Thermodynamics and the Free Energy of Substances," 1923, Chaps. XXII to XXVIII; Glasstone, "Textbook of Physical Chemistry," 1940, Chap. IX.

law the partial vapor pressure of any constituent of an ideal solution is proportional to its mole fraction (x_i) in the solution, and hence it follows that the chemical potential in the liquid is given by

$$\begin{aligned}\mu_i &= \mu_p^0 + RT \ln p_i \\ &= \mu_x^0 + RT \ln x_i.\end{aligned} \tag{8}$$

The constant μ_x^0 for the particular constituent of the solution is independent of the composition, but depends on the temperature *and pressure,* for the relationship between the mole fraction and the vapor pressure is dependent on the total pressure of the system.

If the solution under consideration is not ideal, as is generally the case, especially for solutions of electrolytes, equation (8) is not applicable, and it is modified arbitrarily by writing

$$\mu_i = \mu_x^0 + RT \ln x_i f_i, \tag{9}$$

where f_i is a correction factor known as the **activity coefficient** of the constituent i in the given solution. The product xf is called the **activity** of the particular component and is represented by the symbol a, so that

$$\mu_i = \mu_x^0 + RT \ln a_i. \tag{10}$$

As may be seen from equations (8) and (10), the activity in this particular case may thus be regarded as an idealized mole fraction of the given constituent. A comparison of equations (8) and (9) shows that for an ideal solution the activity coefficient f is unity; in general, the difference between unity and the actual value of the activity coefficient in a given solution is a measure of the departure from ideal behavior in that solution.

For a system consisting of a solvent, designated by the suffix 1, and a solute, indicated by the suffix 2, the respective chemical potentials are

$$\mu_1 = \mu_{x(1)}^0 + RT \ln x_1 f_1 \tag{11}$$

and

$$\mu_2 = \mu_{x(2)}^0 + RT \ln x_2 f_2. \tag{12}$$

It is known that a solution tends towards ideal behavior more closely the greater the dilution; hence, it follows that f_2 approaches unity as x_2 approaches zero, and f_1 approaches unity as x_1 attains unity. It is convenient, therefore, to adopt the definitions

$$f_1 \rightarrow 1 \quad \text{as} \quad x_1 \rightarrow 1 \quad \text{and} \quad f_2 \rightarrow 1 \quad \text{as} \quad x_2 \rightarrow 0.$$

Since f_1 and x_1 become unity at infinite dilution, i.e., for the pure solvent, it follows from equation (11) that the chemical potential of a pure liquid becomes equal to $\mu_{x(1)}^0$, and hence is a constant at a given temperature and pressure. By considering the equilibrium between a solid and its vapor, it can be readily shown that the same rule is applicable to a pure solid.

Forms of the Activity Coefficient.—The equations given above are satisfactory for representing the behavior of liquid solutes, but for solid solutes, especially electrolytes, a modified form is more convenient. In a very dilute solution the mole fraction of solute is proportional both to its concentration (c), i.e., moles per liter of solution, and to its molality (m), i.e., moles per 1000 g. of solvent; hence for such solutions, which are known to approach ideal behavior, it is possible to write either

$$\mu = \mu_x^0 + RT \ln x, \tag{13a}$$

or

$$\mu = \mu_c^0 + RT \ln c, \tag{13b}$$

or

$$\mu = \mu_m^0 + RT \ln m, \tag{13c}$$

where μ_x^0, μ_c^0 and μ_m^0 are constants whose relationship to each other depends on the factors connecting x, c and m in dilute solutions. Since solutions of appreciable concentration do not behave ideally, it is necessary to include the appropriate activity coefficients; thus

$$\mu = \mu_x^0 + RT \ln xf_x = \mu_x^0 + RT \ln a_x, \tag{14a}$$

$$\mu = \mu_c^0 + RT \ln cf_c = \mu_c^0 + RT \ln a_c, \tag{14b}$$

and

$$\mu = \mu_m^0 + RT \ln mf_m = \mu_m^0 + RT \ln a_m, \tag{14c}$$

where the a terms are the respective activities.

It is evident from the equations (14) that the activity of a constituent of a solution can be expressed only in terms of a ratio * of two chemical potentials, viz., μ and μ^0, and so it is the practice to choose a **reference state,** or **standard state,** for each constituent in which *the activity is arbitrarily taken as unity.* It can be readily seen from the equations given above that in the standard state the chemical potential μ is equal to the corresponding value of μ^0. The activity of a component in any solution is thus invariably expressed as the ratio of its value to that in the arbitrary standard state. The actual standard state chosen differs, of course, according to which form of equation (14) is employed to define the activity.

At infinite dilution, when a solution behaves ideally, the three activity coefficients of the solute, viz., f_x, f_c and f_m, are all unity, but at appreciable concentrations the values diverge from this figure and they are no longer equal. It is possible, however, to derive a relationship between them in the following manner. The mole fraction x, concentration c, and molality m of a solute can be readily shown to be related thus

$$x = \frac{0.001\,cM_1}{\rho - 0.001\,cM_2 + 0.001\,cM_1} = \frac{0.001\,mM_1}{1 + 0.001\,mM_1}, \tag{15}$$

* It is a ratio, rather than a difference, because in equations (14) the activity appears in a logarithmic term.

where ρ is the density of the solution, and M_1 and M_2 are the molecular weights of solvent and solute, respectively. In very dilute solutions the three related quantities are x_0, c_0 and m_0, and the density is ρ_0, which is virtually that of the pure solvent; since the quantities 0.001 cM_1, 0.001 cM_2 and 0.001 mM_1 are then negligibly small, it follows from equation (15) that

$$x_0 = \frac{0.001\, c_0 M_1}{\rho_0} = 0.001\, m_0 M_1. \tag{16}$$

Incidentally this relationship proves the statement made above that in very dilute solutions the mole fraction, concentration and molality are proportional to each other.

If μ_0 is the chemical potential of a given solute in a very dilute solution, to which the terms x_0, c_0 and m_0 apply, the three activity coefficients are all unity; further, if μ is the chemical potential in some other solution, whose concentration is represented by x, c or m, it follows from the three forms of equation (14) that $\mu - \mu_0$ may be written in three ways, thus

$$RT \ln \frac{xf_x}{x_0} = RT \ln \frac{cf_c}{c_0} = RT \ln \frac{mf_m}{m_0},$$

$$\therefore \frac{xf_x}{x_0} = \frac{cf_c}{c_0} = \frac{mf_m}{m_0}. \tag{17}$$

Combination of equations (15), (16) and (17) then gives the relationship between the three activity coefficients for the solute in the given solution:

$$f_x = f_c \frac{\rho - 0.001\, cM_2 + 0.001\, cM_1}{\rho_0} = f_m(1 + 0.001\, mM_1). \tag{18}$$

It is evident from this expression that f_c and f_m must be almost identical in dilute solutions, and that f_x cannot differ appreciably from the other coefficients for solutions more dilute than about 0.1 N.

The arguments given above are applicable to a single molecular species as solute, but for electrolytes it is the common practice to employ a mean activity coefficient (see p. 138); in this event it is necessary to introduce into the terms 0.001 cM_1 and 0.001 mM_1 the factor ν which is equal to the number of ions produced by one molecule of electrolyte when it ionizes. The result is then

$$f_x = f_c \frac{\rho - 0.001\, cM_2 + 0.001\, \nu cM_1}{\rho_0} = f_m(1 + 0.001\, \nu mM_1). \tag{19}$$

The activity coefficient f_x is sometimes called the **rational activity coefficient,** since it gives the most direct indication of the deviation from the ideal behavior required by Raoult's law. It is, however, not often used in connection with measurements on solutions of electrolytes, and so the coefficients f_c and f_m, which are commonly employed, are described as the **practical activity coefficients.** The coefficient f_c, from which the suffix is dropped, is generally used in the study of electrolytic equilibria to represent the activity of a particular ionic species; thus, the activity of ions of the ith kind is equal to $c_i f_i$, where c_i is the *actual* ionic concentration, due allowance being made for incomplete dissociation if necessary. On the other hand f_m, which is given the symbol γ, is almost invariably used in connection with the thermodynamics of voltaic cells; the activity of an ion is expressed as $m\gamma_i$, where m is the *total* molality of the ionic constituent of the electrolyte with no correction for incomplete dissociation. For this reason γ is sometimes called the **stoichiometric activity coefficient.**

Equilibrium Constant and Free Energy Changes.—If a system involving the reversible chemical process

$$aA + bB + \cdots \rightleftharpoons lL + mM + \cdots$$

is in a state of equilibrium, it can be readily shown, by means of equations (4) and (14), that

$$\frac{a_L^l a_M^m \cdots}{a_A^a a_B^b \cdots} = K, \tag{20}$$

where K is the **equilibrium constant** for the system under consideration. Equation (20) is the exact form of the **law of mass action** applicable to any system, ideal or not. Writing fc or γm in place of the activity a, the following equations for the equilibrium constant, which are frequently employed in electrochemistry, are obtained, viz.,

$$K_c = \frac{c_L^l c_M^m \cdots}{c_A^a c_B^b \cdots} \cdot \frac{f_L^l f_M^m \cdots}{f_A^a f_B^b \cdots}. \tag{21a}$$

and

$$K_m = \frac{m_L^l m_M^m \cdots}{m_A^a m_B^b \cdots} \cdot \frac{\gamma_L^l \gamma_M^m \cdots}{\gamma_A^a \gamma_B^b \cdots}, \tag{21b}$$

If the components of the system under consideration are at their equilibrium concentrations, or activities, the free energy change resulting from the transfer from reactants to resultants is zero. If, however, the various substances are present in arbitrary concentrations, or activities, the transfer process is accompanied by a definite change of free energy; thus, if a moles of A, b moles of B, etc., at arbitrary activities are transferred to l moles of L, m moles of M, etc., under such conditions that the concentrations are not appreciably altered, the increase of free energy (ΔG) at constant temperature is given by the expression

$$-\Delta G = RT \ln K - RT \ln \frac{a_L^l a_M^m \cdots}{a_A^a a_B^b \cdots}; \tag{22}$$

this equation is a form of the familiar **reaction isotherm.** If the arbitrary activities of reactants and resultants are chosen as the respective standard states, i.e., the a's in equation (22) are all unity, it follows that

$$-\Delta G^0 = RT \ln K, \tag{23}$$

where ΔG^0 is the **standard free energy change** of the process.

Activities of Electrolytes.—When the solute is an electrolyte, the standard states for the ions are chosen, in the manner previously indicated, as a hypothetical ideal solution of unit activity; in this solution the thermodynamic properties of the solute, e.g., the partial molal heat content, heat capacity, volume, etc., will be those of a real solution at infinite dilution, i.e., when it behaves ideally. With this definition of the standard state the activity of an ion becomes equal to its concentration at infinite dilution.

For the undissociated part of the electrolyte it is convenient to define the standard state in such a way as to make its chemical potential equal to the sum of the values for the ions in their standard states. Consider, for example, the electrolyte $M_{\nu_+}A_{\nu_-}$ which ionizes thus

$$M_{\nu_+}A_{\nu_-} \rightleftharpoons \nu_+M^+ + \nu_-A^-$$

to yield the number ν_+ of M^+ ions and ν_- of A^- ions. The chemical potentials of these ions are given by the general equations

$$\mu_{M^+} = \mu_+^0 + RT \ln a_+ \tag{24a}$$

and

$$\mu_{A^-} = \mu_-^0 + RT \ln a_-, \tag{24b}$$

where a_+ and a_- are the activities of the ions M^+ and A^- respectively. If μ_2 is the chemical potential of the undissociated portion of the electrolyte in a given solution and μ_2^0 is the value in the standard state, then by the definition given above,

$$\mu_2^0 = \nu_+\mu_+^0 + \nu_-\mu_-^0. \tag{25}$$

When the system of undissociated molecules and free ions in solution is in equilibrium, a small change at constant temperature and pressure produces no change in the free energy of the system; since one molecule of electrolyte produces ν_+ positive and ν_- negative ions, it is seen that [cf. equation (4)]

$$\nu_+(\mu_+^0 + RT \ln a_+) + \nu_-(\mu_-^0 + RT \ln a_-) = \mu_2^0 + RT \ln a_2. \tag{26}$$

Introducing equation (25) it follows, on the basis of the particular standard states chosen, that

$$\nu_+RT \ln a_+ + \nu_-RT \ln a_- = RT \ln a_2,$$

$$\therefore a_+^{\nu_+}a_-^{\nu_-} = a_2. \tag{27}$$

If the total number of ions produced by a molecule of electrolyte, i.e., $\nu_+ + \nu_-$, is represented by ν, then the **mean activity** $a_\pm$ of the electrolyte is defined by

$$a_{\pm} \equiv (a_+^{\nu_+} a_-^{\nu_-})^{1/\nu}, \tag{28}$$

and hence, according to equation (27),

$$a_{\pm} = (a_2)^{1/\nu} \quad \text{or} \quad a_2 = a_{\pm}^{\nu}. \tag{29}$$

The activity of each ion may be written as the product of its activity coefficient and concentration, so that

$$a_+ = \gamma_+ m_+ \quad \text{and} \quad a_- = \gamma_- m_-,$$

$$\therefore \gamma_+ = \frac{a_+}{m_+} \quad \text{and} \quad \gamma_- = \frac{a_-}{m_-}.$$

The **mean activity coefficient** $\gamma_{\pm}$ of the electrolyte, defined by

$$\gamma_{\pm} \equiv (\gamma_+^{\nu_+} \gamma_-^{\nu_-})^{1/\nu}, \tag{30}$$

can consequently be represented by

$$\gamma_{\pm} = \left(\frac{a_+^{\nu_+} a_-^{\nu_-}}{m_+^{\nu_+} m_-^{\nu_-}} \right)^{1/\nu}. \tag{31}$$

If m is the molality of the electrolyte, m_+ is equal to $m\nu_+$ and m_- is equal to $m\nu_-$, so that equation (31) may be written as

$$\gamma_{\pm} = \frac{a_{\pm}}{m(\nu_+^{\nu_+} \nu_-^{\nu_-})^{1/\nu}}. \tag{32}$$

The **mean molality** $m_{\pm}$ of the electrolyte is defined, in an analogous manner, by

$$m_{\pm} \equiv (m_+^{\nu_+} m_-^{\nu_-})^{1/\nu} = m(\nu_+^{\nu_+} \nu_-^{\nu_-})^{1/\nu},$$

so that it is possible to write equation (32) as

$$\gamma_{\pm} = \frac{a_{\pm}}{m_{\pm}}. \tag{33}$$

Relationships similar to those given above may, of course, be derived for the other activity coefficients.

Values of Activity Coefficients.—Without entering into details, it is evident from the foregoing discussion that activities and activity coefficients are related to chemical potentials or free energies; several methods, both direct and indirect, are available for determining the requisite differences of free energy so that activities, relative to the specified standard states, can be evaluated. In the study of the activity coefficients of electrolytes the procedures generally employed are based on measurements of either vapor pressure, freezing point, solubility or electromotive force.[3] The results obtained by the various methods are

[3] See references on page 132, also pages 200 and 203. For a valuable summary of data and other information on activity coefficients, see Robinson and Harned, *Chem. Revs.*, **28**, 419 (1941).

TABLE XXXIV. MEAN ACTIVITY COEFFICIENTS OF ELECTROLYTES IN AQUEOUS SOLUTION AT 25°

Molality	HCl	NaCl	KCl	HBr	NaOH	$CaCl_2$	$ZnCl_2$	H_2SO_4	$ZnSO_4$	$LaCl_3$	$In_2(SO_4)_3$
0.001	0.966	0.966	0.966	—	—	0.888	0.881	—	0.734	0.853	—
0.005	0.930	0.928	0.927	0.930	—	0.789	0.767	0.643	0.477	0.716	0.16
0.01	0.906	0.903	0.902	0.906	0.899	0.732	0.708	0.545	0.387	0.637	0.11
0.02	0.878	0.872	0.869	0.879	0.860	0.669	0.642	0.455	0.298	0.552	0.08
0.05	0.833	0.821	0.816	0.838	0.805	0.584	0.556	0.341	0.202	0.417	0.035
0.10	0.798	0.778	0.770	0.805	0.759	0.524	0.502	0.266	0.148	0.356	0.025
0.20	0.768	0.732	0.719	0.782	0.719	0.491	0.448	0.210	0.104	0.298	0.021
0.50	0.769	0.679	0.652	0.790	0.681	0.510	0.376	0.155	0.063	0.303	0.014
1.00	0.811	0.656	0.607	0.871	0.667	0.725	0.325	0.131	0.044	0.387	—
1.50	0.898	0.655	0.586	—	0.671	—	0.290	—	0.037	0.583	—
2.00	1.011	0.670	0.577	—	0.685	1.554	—	0.125	0.035	0.954	—
3.00	1.31	0.719	0.572	—	—	3.384	—	0.142	0.041	—	—

in good agreement with each other and hence they may be regarded as reliable. Although the description of the principles on which the determinations of activity coefficients are based will be considered later, it will be convenient to summarize in Table XXXIV some actual values of the mean activity coefficients at 25° obtained for a number of electrolytes of several valence types in aqueous solution at various molalities. Some of the results are also depicted by the curves in Fig. 46; it will be observed that the activity coefficients may deviate appreciably from unity. The values always decrease at first as the concentration is increased, but they generally pass through a minimum and then increase again. At high concentrations the activity coefficients often exceed unity, so that the mean activity of the electrolyte is actually greater than the concentration; the deviations from ideal behavior are now in the opposite direction to those which occur at low concentrations. An examination of Table XXXIV brings to light other important facts: it is seen, in the first place, that electrolytes of the same valence type, e.g., sodium and potassium chlorides, etc., or calcium and zinc chlorides, etc.,

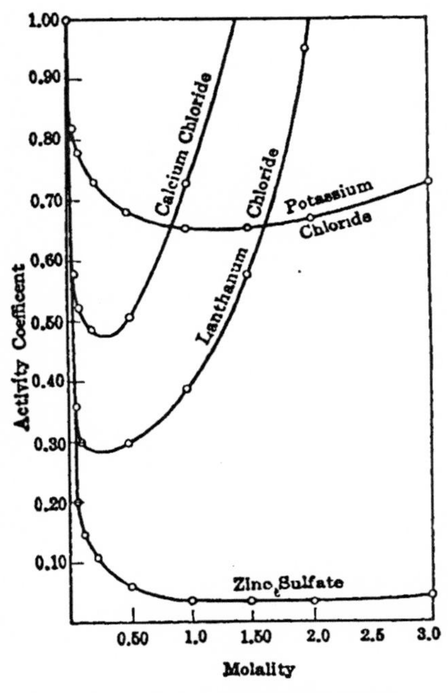

FIG. 46. Activity coefficients of electrolytes of different valence types

have almost identical activity coefficients in dilute solutions. Secondly, the deviation from ideal behavior at a given concentration is greater the higher the product of the valences of the ions constituting the electrolyte.

The Ionic Strength.—In order to represent the variation of activity coefficient with concentration, especially in the presence of added electrolytes, Lewis and Randall introduced the quantity called the **ionic strength,** which is a measure of the intensity of the electrical field due to the ions in a solution.[4] It is given the symbol μ and is defined as half the sum of the terms obtained by multiplying the molality, or concentration, of each ion present in the solution by the square of its valence; that is

$$\mu = \tfrac{1}{2}\sum_i m_i z_i^2. \tag{34}$$

In calculating the ionic strength it is necessary to use the *actual* ionic concentration or molality; for a weak electrolyte this would be obtained by multiplying its concentration by the degree of dissociation.

Although the importance of the ionic strength was first realized from empirical considerations, it is now known to play an important part in the theory of electrolytes. It will be observed that equation (12) on page 83, which gives the reciprocal of the thickness of the ionic atmosphere according to the theory of Debye and Hückel, contains the quantity $\Sigma n_i z_i^2$, where n_i is the number of ions of the ith kind in unit volume and hence is proportional to the concentration. This quantity is clearly related to the ionic strength of the solution as defined above; it will be seen shortly that it plays a part in the theoretical treatment of activity coefficients.

It was pointed out by Lewis and Randall that, in dilute solutions, the activity coefficient of a given strong electrolyte is approximately the same in all solutions of a given ionic strength. The particular ionic strength may be due to the presence of other salts, but their nature does not affect the activity coefficient of the electrolyte under consideration. This generalization, to which further reference will be made later, holds only for solutions of relatively low ionic strength; as the concentration is increased the specific influence of the added electrolyte becomes manifest.

The Debye-Hückel Theory.—The first successful attempt to account for the departure of electrolytes from ideal behavior was made by Milner (1912), but his treatment was very complicated; the ideas were essentially the same as those which were developed in a more elegant manner by Debye and Hückel. The fundamental ideas have already been given on page 81 in connection with the theory of electrolytic conductance, and the application of the Debye-Hückel theory to the problem of activity coefficients will be considered here.[5]

[4] Lewis and Randall, *J. Am. Chem. Soc.*, **43**, 1112 (1921).

[5] Debye and Hückel, *Physik. Z.*, **24**, 185, 334 (1923); **25**, 97 (1924); for reviews, see LaMer, *Trans. Electrochem. Soc.*, **51**, 507 (1927); Falkenhagen, "Electrolytes" (translated by Bell), 1934; Williams, *Chem. Revs.*, **8**, 303 (1931); Schingnitz, *Z. Elektrochem.*, **36**, 861 (1930).

According to equation (16), page 84, the potential ψ due to ions of the ith kind may be represented by

$$\psi = \frac{z_i\epsilon}{Dr} - \frac{z_i\epsilon\kappa}{D}, \tag{35}$$

where the first term is the potential at a distance r from the central ion when there are no surrounding ions, and the second term is the contribution of the ionic atmosphere; κ is defined by equation (12), page 83. Suppose that all the ions are discharged and that successive small charges are brought up to the ions from infinity in such a way that at any instant all the ions have the same fraction λ of their final charge $z_i\epsilon$. It follows, therefore, from equation (35), that at any stage during the charging process the potential ψ_λ due to ions of the ith kind is given by

$$\psi_\lambda = \frac{\lambda z_i\epsilon}{Dr} - \frac{\lambda z_i\epsilon\kappa_\lambda}{D}, \tag{36}$$

where κ_λ is the value of the quantity κ at this stage. It can be seen, from the definition of κ, that since the charge on the ion is then $\lambda z_i\epsilon$, the value of κ_λ will be a fraction λ of the final value; the term κ_λ in equation (36) may thus be replaced by $\lambda\kappa$. Making this substitution, equation (36) becomes

$$\psi_\lambda = \frac{z_i\epsilon}{Dr}\lambda - \frac{z_i\epsilon\kappa}{D}\lambda^2. \tag{37}$$

If $z_i\epsilon d\lambda$ is the magnitude of the small charge brought up to each ion of the ith kind, the corresponding work done is $z_i\epsilon d\lambda \times \psi_\lambda$, and hence the total electrical work (W_i) done in charging completely, i.e., from $\lambda = 0$ to $\lambda = 1$, an ion of the ith kind is

$$\begin{aligned} W_i &= \int_{\lambda=0}^{\lambda=1} z_i\epsilon\psi_\lambda d\lambda \\ &= \frac{z_i^2\epsilon^2}{Dr}\int_0^1 \lambda d\lambda - \frac{z_i^2\epsilon^2\kappa}{D}\int_0^1 \lambda^2 d\lambda \\ &= \frac{z_i^2\epsilon^2}{2Dr} - \frac{z_i^2\epsilon^2\kappa}{3D}. \end{aligned} \tag{38}$$

If N_i is the *total number* of ions of the ith kind,* the total electrical work (W) done in charging completely *all the ions* of the solution is obtained by multiplying equation (38) by N_i and summing over all the ions, thus

$$W = \sum_i \frac{N_i z_i^2\epsilon^2}{2Dr} - \sum_i \frac{N_i z_i^2\epsilon^2\kappa}{3D}. \tag{39}$$

* This should not be confused with n_i, the number of these ions in *unit volume*.

At infinite dilution there is no ionic atmosphere, and so κ is zero and the second term on the right-hand side of equation (39) disappears; since the dielectric constant is that of the pure solvent, i.e., D_0, the electrical work (W_0) done in charging the ions at infinite dilution is

$$W_0 = \sum_i \frac{N_i z_i^2 \epsilon^2}{2D_0 r}. \tag{40}$$

Provided the solution is not too concentrated, D and D_0 are approximately equal, and hence the difference in the electrical work of charging the same ions at a definite concentration and at infinite dilution is given by

$$W - W_0 = -\sum_i \frac{N_i z_i^2 \epsilon^2 \kappa}{3D}. \tag{41}$$

The volume change accompanying the charging process at constant pressure is negligible, and so $W - W_0$ may be identified with the difference between the electrical free energy of an ionic solution at a definite concentration and at infinite dilution.

The free energy (G) of a solution containing ions may be regarded as being made up of two parts: first, that corresponding to the value for an ideal solution at the same concentration as the ionic solution (G_0), and second, an amount due to the electrical interaction of the ions ($G_{el.}$); thus

$$G = G_0 + G_{el.}, \tag{42}$$

where $G_{el.}$ may be taken as being equal to $W - W_0$, as given by equation (41). Differentiating with respect to N_i, the *number of ions* of the ith kind, at constant temperature and pressure, the result is

$$\frac{\partial G}{\partial N_i} = \frac{\partial G_0}{\partial N_i} + \frac{\partial G_{el.}}{\partial N_i},$$

or

$$\mu_i = \mu_{i(0)} + \mu_{i(el.)}. \tag{43}$$

According to the definition of the chemical potential μ_i, which now applies to a *single ion*, instead of to a g.-ion,

$$\begin{aligned}\mu_i &= \mu_i^0 + kT \ln a_i \\ &= \mu_i^0 + kT \ln x_i + kT \ln f_i,\end{aligned} \tag{44}$$

where k is the Boltzmann constant, i.e., the gas constant per single molecule. Further, since G_0 refers to an ideal solution, it follows that

$$\mu_{i(0)} = \mu_i^0 + kT \ln x_i, \tag{45}$$

and hence from equations (43), (44) and (45),

$$\mu_{i(el.)} = \frac{\partial G_{el.}}{\partial N_i} = kT \ln f_i. \tag{46}$$

Introducing the value of $G_{el.}$ as given by equation (41), it is found on differentiating with respect to N_i, remembering that κ involves $\sqrt{n_i}$ and hence $\sqrt{N_i}$, that

$$kT \ln f_i = -\frac{z_i^2\epsilon^2\kappa}{2D},$$

$$\therefore \quad \ln f_i = -\frac{z_i^2\epsilon^2\kappa}{2DkT} = -\frac{Nz_i^2\epsilon^2\kappa}{2DRT}, \tag{47}$$

N being the Avogadro number and R, equal to kN, the gas constant per mole.*

The Debye-Hückel Limiting Law.—The value of κ as given on page 83 is

$$\kappa = \left(\frac{4\pi\epsilon^2}{DkT}\sum_i n_i z_i^2\right)^{\frac{1}{2}}, \tag{48}$$

and if n_i is replaced by $c_i N/1000$, where c_i is the ionic concentration in moles per liter, and R/N is written for k, equation (48) becomes

$$\kappa = \left(\frac{4\pi N^2\epsilon^2}{1000DRT}\sum_i c_i z_i^2\right)^{\frac{1}{2}}. \tag{49}$$

The quantity $\Sigma c_i z_i^2$ is seen to be analogous to twice the ionic strength as defined by Lewis and Randall [equation (34)]; the only difference is that the former involves volume concentrations whereas in the latter molalities are employed. For dilute aqueous solutions, such as were used in the work from which Lewis and Randall made the generalization given on page 140, the two values of the ionic strength are almost identical. It has been stated that if the Debye-Hückel arguments are applied in a rigid manner the expression for κ will actually involve molalities; nevertheless, it is the practice in connection with the application of the equations derived by the method of Debye and Hückel to use an ionic strength defined in terms of molar concentrations, viz.,

$$\mu = \tfrac{1}{2}\sum_i c_i z_i^2, \tag{50}$$

so that equation (49) can be written as

$$\kappa = \left(\frac{8\pi N^2\epsilon^2}{1000DRT}\mu\right)^{\frac{1}{2}}. \tag{51}$$

Introducing this value for κ into equation (47) and at the same time dividing the right-hand side by 2.303 to convert natural to common logarithms, the result is

$$\log f_i = -\frac{N^2\epsilon^3}{2.303R^{\frac{3}{2}}}\left(\frac{2\pi}{1000}\right)^{\frac{1}{2}}\frac{z_i^2}{(DT)^{\frac{3}{2}}}\sqrt{\mu}. \tag{52}$$

* It should be noted that in the differentiation the summation in equation (41) has been reduced to a single term. This is because the numbers of all the ions except of the ith kind remain constant, and so all the terms other than the one involving n_i will be zero.

The universal constants N, ϵ, π and R, as well as the numerical quantities, may be extracted from equation (52), and if the accepted values are employed, this equation becomes

$$\log f_i = -1.823 \times 10^6 \frac{z_i^2}{(DT)^{\frac{3}{2}}} \sqrt{\mu}. \tag{53}$$

For a given solvent and temperature D and T have definite values which may be inserted; equation (53) then takes the general form

$$\log f_i = -Az_i^2\sqrt{\mu}, \tag{54}$$

where A is a constant for the solvent at the specified temperature.

This equation, which represents what has been called the **Debye-Hückel limiting law**, expresses the variation of the activity coefficient of an ion with the ionic strength of the medium. It is called the limiting law because, as seen previously, the approximations made in the derivation of the potential at an ion due to its ionic atmosphere, can be expected to be justifiable only as infinite dilution is approached. The general conclusion may be drawn from equation (53) or (54) that the activity coefficient of an ion should decrease with increasing ionic strength of the solution: the decrease is greater the higher the valence of the ion and the lower the dielectric constant of the solvent.

It will be seen later (p. 230) that there does not appear to be any experimental method of evaluating the activity coefficient of a single ionic species, so that the Debye-Hückel equations cannot be tested in the forms given above. It is possible, however, to derive very readily an expression for the mean activity coefficient, this being the quantity that is obtained experimentally. The mean activity coefficient $f_\pm$ of an electrolyte $M^+_{\nu_+}A^-_{\nu_-}$ is defined by an equation analogous to (30), and upon taking logarithms this becomes

$$\log f_\pm = \frac{\nu_+ \log f_+ + \nu_- \log f_-}{\nu_+ + \nu_-}. \tag{55}$$

The values of $\log f_+$, which is equal to $-Az_+^2\sqrt{\mu}$, and of $\log f_-$, i.e., $-Az_-^2\sqrt{\mu}$, as given by equation (54) can now be inserted in (55); making use of the fact that $z_+\nu_+$ must be equal to $z_-\nu_-$, it is found that

$$\log f_\pm = -Az_+z_-\sqrt{\mu}, \tag{56}$$

which is the statement of the Debye-Hückel limiting law for the mean activity coefficient of an electrolyte whose ions have valences of z_+ and z_-, respectively. The values of the constant A for water at a number of temperatures are given in Table XXXV below.

Attention should be drawn to the fact that the activity coefficients given by the Debye-Hückel treatment are the so-called rational coefficients (p. 135); to express the values in the form of the practical activity coefficients, it is necessary to make use of equation (26). If the solvent is water, so that M_1 is 18, it is seen that

$$\log \gamma = \log f - \log (1 + 0.018\nu m),$$

where γ is the activity coefficient in terms of molalities, f is the value given by the Debye-Hückel equations, and ν is the number of ions produced by one molecule of electrolyte on dissociation. As already seen, however, the difference between the various coefficients is negligible in dilute solutions, and it is in such solutions that the most satisfactory tests of the Debye-Hückel theory can be made.

Debye-Hückel Equation for Appreciable Concentrations.—In the derivation of equation (12), page 83, the approximation was made of regarding the ion as being equivalent to a point charge; this will result in no serious error provided the radius of the ionic atmosphere is large in comparison with that of the ion. An examination of Table XXII, page 85, shows that this condition is satisfied in dilute solutions, but when the concentration approaches a value of about 0.1 molar the radius of the ionic atmosphere is about the same order as that of an ion, i.e., 2×10^{-8} cm. It follows, therefore, that in such solutions the approximation of a point charge is liable to lead to serious errors. A possible method of making the necessary correction has been proposed by Debye and Hückel;[6] it has been found that if a is the mean distance of approach of other ions, e.g., B to the central ion A, as shown in Fig. 47, the potential due to ions of the ith kind is given by the expression

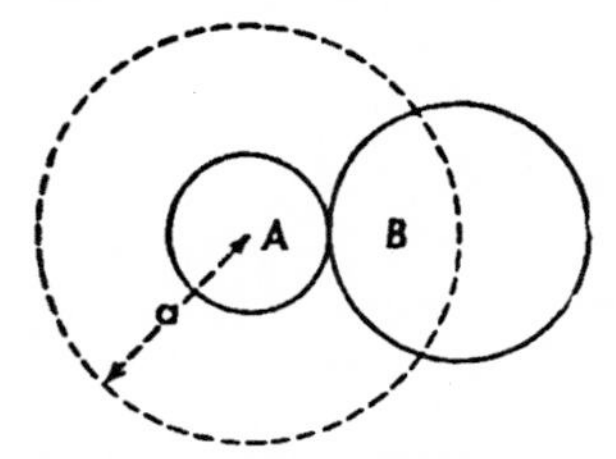

Fig. 47. Mean distance of approach of ions

$$\psi = \frac{z_i \epsilon}{Dr} - \frac{z_i \epsilon \kappa}{D} \cdot \frac{1}{1 + \kappa a}, \tag{57}$$

instead of by equation (35). The mean distance of approach a is often referred to as the "average effective diameter" of the ions, although its exact physical significance probably cannot be expressed precisely. It is seen that the correction term is $(1 + \kappa a)^{-1}$, which approaches unity in dilute solutions when κ is small.

By following through the derivation on page 141, using equation (57) instead of (35), the final result is

$$\ln f_i = - \frac{N z_i^2 \epsilon^2 \kappa}{2DRT} \cdot \frac{1}{1 + \kappa a}, \tag{58}$$

in place of equation (47). It is apparent from equation (51) that, for a given solvent and a definite temperature, κ is equivalent to $B\sqrt{\mathbf{\mu}}$, where B is a constant; hence $1 + \kappa a$ may be replaced by $1 + aB\sqrt{\mathbf{\mu}}$. Making

[6] Debye and Hückel, *Physik. Z.*, **24**, 185 (1923).

this substitution in equation (58),

$$\ln f_i = -\frac{Nz_i^2\epsilon^2\kappa}{2DRT}\cdot\frac{1}{1+aB\sqrt{\mu}}, \tag{59}$$

and hence the Debye-Hückel limiting law, corresponding to equation (54), now becomes

$$\log f_i = -\frac{Az_i^2\sqrt{\mu}}{1+aB\sqrt{\mu}}, \tag{60}$$

where A has the same significance as before. The expression for the mean activity coefficient of an electrolyte is then

$$\log f_\pm = -\frac{Az_+z_-\sqrt{\mu}}{1+aB\sqrt{\mu}}. \tag{61}$$

Both the constants A and B depend on the nature of the solvent and the temperature; the values for water at several temperatures are given in Table XXXV; the corresponding dielectric constants are also recorded.

TABLE XXXV. DEBYE-HÜCKEL CONSTANTS AND DIELECTRIC CONSTANT OF WATER

Temp.	D	A	B
0°	88.15	0.488	0.325×10^8
15°	82.23	0.500	0.328
25°	78.54	0.509	0.330
30°	76.76	0.514	0.331
40°	73.35	0.524	0.333
50°	70.10	0.535	0.335

It will be observed from Table XXXV that at ordinary temperatures the value of B with water as solvent is approximately 0.33×10^8; for most electrolytes the mean ionic diameter a is about 3 to 4×10^{-8} cm. (see Table XXXVI), and hence aB does not differ greatly from unity. A reasonably satisfactory and simple approximation of equation (61) is therefore

$$\log f_\pm = -\frac{Az_+z_-\sqrt{\mu}}{1+\sqrt{\mu}}. \tag{61a}$$

The Hückel and Brønsted Equations.—A further correction to the Debye-Hückel equation has been proposed in order to allow for the polarization of the solvent molecules by the central ion; since these molecules are, in general, more polarizable than the ions themselves, there will be a tendency for the solvent molecules to displace the other ions from the vicinity of a particular ion. The dipolar nature of the solvent molecules will also facilitate the tendency for these molecules to orient themselves about the central ion. It has been suggested that the result of this orientation is equivalent to an increase in the dielectric constant in the immediate vicinity of the ion above that in the bulk of the solvent. By assuming the increase to be proportional to the ionic concentration of the solution, it has been deduced that an additional term $C'\mu$, where C' is an empirical constant, should be added to the right-hand side of equations (60) and (61); hence, the latter now becomes

$$\log f_{\pm} = -\frac{Az_+z_-\sqrt{\mu}}{1 + aB\sqrt{\mu}} + C'\mu. \tag{62}$$

This result has sometimes been called the **Hückel equation.**[7]

It is not certain that the theoretical arguments, which led to the introduction of the term $C'\mu$, are completely satisfactory, but it seems to be established that the experimental data require a term of this type. The aggregation of solvent molecules in the vicinity of an ion is the factor responsible for the so-called "salting-out effect," namely, the decrease in solubility of neutral substances frequently observed in the presence of salts; the constant C' is consequently called the **salting-out constant.** The activity coefficient of a non-electrolyte, as measured by its solubility in the presence of electrolytes, is often given by an expression of the form $\log f = C'\mu$; this is the result to which equation (62) would reduce for the activity of a non-electrolyte, i.e., when z_+ and z_- are zero, in a salt solution of ionic strength μ.

By dividing through the numerator of the fraction on the right-hand side of equation (62) by the denominator, and neglecting all terms in the power series beyond that involving μ, the result is

$$\begin{aligned}\log f_{\pm} &= -Az_+z_-\sqrt{\mu} + (aABz_+z_- + C')\mu \\ &= -Az_+z_-\sqrt{\mu} + C\mu,\end{aligned} \tag{63}$$

where C is a constant for the given electrolyte, equal to $aABz_+z_- + C'$. This relationship is of the same form as an empirical equation proposed by Brønsted,[8] and hence is in general agreement with experiment; it has been called the **Debye-Hückel-Brønsted equation.** In dilute solution, when μ is small, the term $C\mu$ can be neglected, and so this expression then reduces to the Debye-Hückel limiting law.

Qualitative Verification of the Debye-Hückel Equations.—The general agreement of the limiting law equation (54) with experiment is shown by the empirical conclusion of Lewis and Randall (p. 140) that the activity coefficient of an electrolyte is the same in all solutions of a given ionic strength. Apart from the valence of the ions constituting the particular electrolyte under consideration, the Debye-Hückel limiting equation contains no reference to the specific properties of the salts that may be present in the solution. It is of interest to record that the

[7] Hückel, *Physik. Z.*, **26**, 93 (1925); see also, Butler, *J. Phys. Chem.*, **33**, 1015 (1929); Scatchard, *Physik. Z.*, **33**, 22 (1932).

[8] Brønsted, *J. Am. Chem. Soc.*, **44**, 938 (1922); Brønsted and LaMer, *ibid.*, **46**, 555 (1924).

empirical equation proposed by Lewis and Linhart [9] to account for their results on the freezing points of dilute solutions of various electrolytes is of the form $\log f = -\beta c^{\alpha}$, where α was found to be about 0.4 to 0.5 for several salts and β depended on their valence type. Further, as already mentioned, Brønsted's empirical equation for more concentrated solutions is in agreement with the extended equation (62). It can be seen from the Debye-Hückel limiting law equation that at a definite ionic strength the departure of the activity coefficient of a given electrolyte from unity should be greater the higher the valences of the ions constituting the electrolyte; this conclusion is in harmony with the results given in Table XXXIV which have been already discussed.

It was noted on page 139 that although activity coefficients generally decrease with increasing concentration in dilute solutions, in accordance with the requirement of equation (58), the values frequently pass through a minimum at higher concentrations. It is of interest, therefore, to see how far this fact can be explained, at least qualitatively, by means of the Debye-Hückel theory. According to the limiting law equation, the plot of $\log f$ against $\sqrt{\mu}$ should be a straight line of slope $-Az_+z_-$; for a uni-univalent electrolyte in water at 25° this is equal to approximately -0.51, as shown in Fig. 48, I. If the ionic size factor is introduced, as in equation (61), the plot of $\log f$ against $\sqrt{\mu}$ becomes of the form of Fig. 48, II, representing a type of curve which is often obtained experimentally. Finally, the addition of the salting-out factor, as in equation (62), results in a further increase of the activity coefficient by an amount proportional to the ionic strength; the result is that the $\log f$ against $\sqrt{\mu}$ curve becomes similar to Fig. 48, III. It may be mentioned that the latter curve duplicates closely the variation of the activity coefficient of sodium chloride with concentration up to relatively high values of the latter.

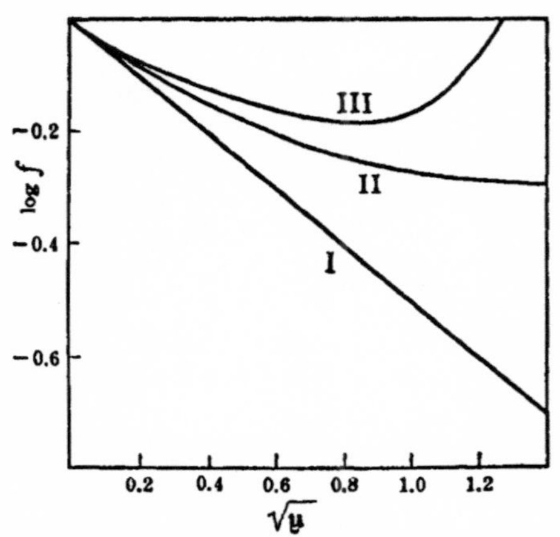

FIG. 48. Simple (I) and extended (II and III) Debye-Huckel equations

Quantitative Tests of the Debye-Hückel Limiting Equation.—Although the Debye-Hückel equations are generally considered as applying to solutions of strong electrolytes, it is important to emphasize that they are by no means restricted to such solutions; they are of general applicability and the only point that must be noted is that in the calculation of the ionic strength the *actual* ionic concentrations must be employed.

[9] Lewis and Linhart, *J. Am. Chem. Soc.*, 41, 1951 (1919).

For incompletely dissociated electrolytes this involves a knowledge of the degree of dissociation, which may not always be available with sufficient accuracy. It is for this reason that the Debye-Hückel equations are generally tested by means of data obtained with strong electrolytes, since they can be assumed to be completely dissociated. It is probable that some of the discrepancies observed with certain electrolytes of high valence types are due to incomplete dissociation for which adequate allowance has not been made.

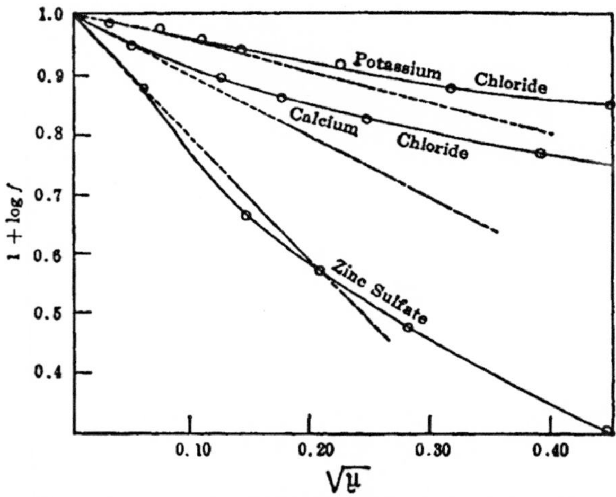

FIG. 49. Test of the limiting Debye-Hückel equation

The experimentally determined activity coefficients, based on vapor pressure, freezing-point and electromotive force measurements, for a number of typical electrolytes of different valence types in aqueous solution at 25°, are represented in Fig. 49, in which the values of $\log f$ are plotted against the square-root of the ionic strength; in these cases the solutions contained no other electrolyte than the one under consideration. Since the Debye-Hückel constant A for water at 25° is seen from Table XXXV to be 0.509, the limiting slopes of the plots in Fig. 49 should be equal to $-0.509\, z_+z_-$; the results to be expected theoretically, calculated in this manner, are shown by the dotted lines. It is evident that the experimental results approach the values required by the Debye-Hückel limiting law as infinite dilution is attained. The influence of valence on the dependence of the activity coefficient on concentration is evidently in agreement with theoretical expectation. Another verification of the valence factor in the Debye-Hückel equation will be given later (p. 177).

A comparison of equations (52) and (53) shows that, for electrolytes of the same valence type, the limiting slope of the plot of $\log f$ against $\sqrt{\mu}$ at constant temperature should be inversely proportional to $D^{\frac{3}{2}}$, where D

is the dielectric constant of the medium. A stringent test of the Debye-Hückel equation is, therefore, to determine the activity coefficients of a given electrolyte in a number of different media of varying dielectric constant; the results are available for hydrochloric acid in methyl and ethyl alcohols, in a number of dioxane-water mixtures, as well as in pure water at 25°. Some of the data are plotted in Fig. 50; the limiting slopes,

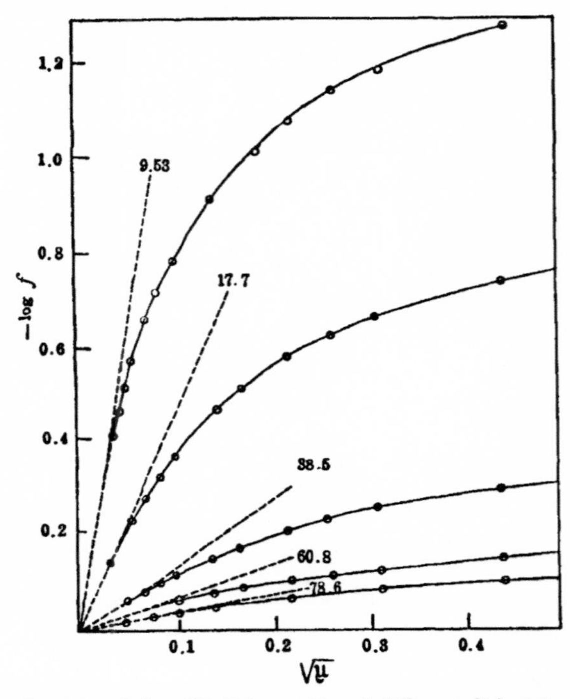

FIG. 50. Limiting Debye-Hückel equation at different dielectric constants (Harned, *et al.*)

marked with the appropriate value of the dielectric constant, are indicated by the dotted lines in each case. The agreement with expectation, over a range of dielectric constant from about 10 to 78.6, is very striking.[10]

The influence of one other variable, namely, the temperature, remains to be considered. It is not an easy matter to vary the temperature without changing the dielectric constant, and so these factors may be considered together. From equations (55) and (56) it is evident that the limiting slope of the plot of $\log f$ against $\sqrt{\mu}$ should vary as $1/(DT)^{\frac{3}{2}}$, where T is the absolute temperature at which the activity coefficients are measured. The experimental results obtained under a wide variety of conditions, e.g., in liquid ammonia at $-75°$ and in water at the boiling

[10] Harned *et al.*, *J. Am. Chem. Soc.*, 61, 49 (1939).

point, are generally in satisfactory agreement with theoretical requirements.[11] Where discrepancies are observed they can probably be explained by incomplete dissociation in media of low dielectric constant.

The Osmotic Coefficient.—Instead of calculating activity coefficients from freezing-point and other so-called osmotic measurements, the data may be used directly to test the validity of the Debye-Hückel treatment. If θ is the depression of the freezing point of a solution of molality m of an electrolyte which dissociates into ν ions, and λ is the molal freezing-point depression, viz., 1.858° for water, a quantity ϕ, called the **osmotic coefficient,** may be defined by the expression

$$\phi = \frac{\theta}{\lambda \nu m}. \tag{64}$$

This coefficient is equivalent to the van't Hoff factor i (see p. 9) divided by ν. It can be shown by means of thermodynamics that if $-\log f$ is proportional to the square-root of the ionic strength, as it undoubtedly is in dilute solutions, then

$$1 - \phi = -\tfrac{1}{3} \ln f. \tag{65}$$

Introducing the Debye-Hückel limiting law for $\log f$, it is seen that

$$1 - \phi = \frac{2.303}{3} A z_+ z_- \sqrt{\mu}, \tag{66}$$

where A has the same significance as before. Since ϕ can be determined directly from freezing-point measurements, by means of equation (64),

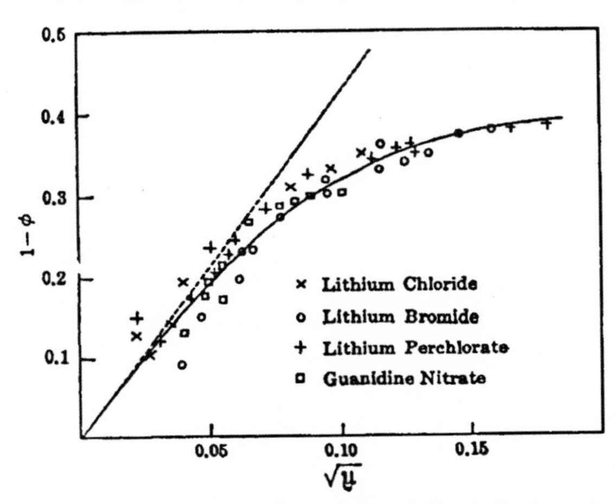

FIG. 51. Test of Debye-Hückel equation by freezing-point measurements in cyclohexanol (Schreiner and Frivold)

[11] Saxton and Smith, *J. Am. Chem. Soc.*, **54**, 2626 (1932); Webb, *J. Am. Chem. Soc.*, **48**, 2263 (1926).

it is possible to test the Debye-Hückel theory in the form of equation (66); the plot of $1 - \phi$ against $\sqrt{\mu}$ should approach a limiting value of $0.768\,Az_+z_-$. The experimental results for electrolytes of different valence types in aqueous solutions are in agreement with expectation; since the data are in principle similar to many that were used in the compilation of Fig. 49, they need not be considered further. It is of interest, however, to examine the values derived from freezing-point measurements in a solvent of low dielectric constant, viz., cyclohexanol, whose dielectric constant is 15.0 and freezing point 23.6°; the full curve in Fig. 51 is drawn through the results for a number of uni-univalent electrolytes, while the dotted curve shows the limiting slope required by equation (66).[12]

Activities at Appreciable Concentrations.—A comparison of the experimental curves in Figs. 49 and 50 with the general form of curve II in Fig. 48 suggests that equation (61) might represent the variation of activity coefficient with concentration in solutions of electrolytes that

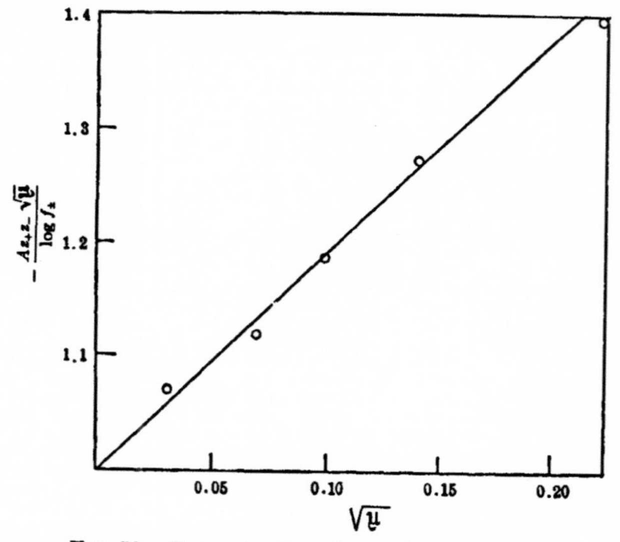

FIG. 52. Determination of mean ionic diameter

were not too concentrated; by a slight rearrangement this equation can be put in the form

$$-\frac{Az_+z_-\sqrt{\mu}}{\log f_\pm} = 1 + aB\sqrt{\mu}, \tag{67}$$

so that if the left-hand side of equation (67) is plotted against $\sqrt{\mu}$ the result should be a straight line of slope aB. Since the value of B is

[12] Schreiner and Frivold, *Z. physik. Chem.*, 124, 1 (1926).

known (cf. Table XXXV), the magnitude of the mean ionic diameter required to satisfy the experimental results can be obtained. The data for aqueous solutions of hydrochloric acid at 25° are shown in Fig. 52; the points are seen to fall approximately on a straight line so that an equation of the form of (61) and (67) is obeyed. The slope of this line is about 1.75 and since B is 0.33×10^8, it follows that for hydrochloric acid a is equal to 5.3×10^{-8}. It has been found in a number of cases that by using values of a that appear to be of a reasonable magnitude it is possible to represent quantitatively the activity coefficients of a number of electrolytes up to ionic strengths of about 0.1. Some of the mean values, collected from those reported in the literature, are given in Table XXXVI. It must be pointed out, however, that such satisfactory results

TABLE XXXVI. MEAN EFFECTIVE IONIC DIAMETERS

Electrolyte	a	Electrolyte	a
HCl	5.3×10^{-8} cm.	$CaCl_2$	5.2×10^{-8} cm.
NaCl	4.4	$MgSO_4$	3.4
KCl	4.1	K_2SO_4	3.0
$CsNO_3$	3.0	$La_2(SO_4)_3$	3.0

are not always obtained; in order to satisfy the experimental data in the case of silver nitrate, for example, a should be 2.3×10^{-8} cm., and for potassium nitrate 0.43×10^{-8} cm., both of which values are lower than would be expected. It is nevertheless of interest that the figures are at least of the correct order of magnitude for an ionic radius, namely about 10^{-8} cm. In some instances, particularly with salts of high valence types, it is found necessary to employ variable or even negative values of a; this may be attributed either to incomplete dissociation or to the approximations made in the Debye-Hückel derivation.

Activities in Concentrated Solutions.—For relatively concentrated solutions it is necessary to use the complete Hückel equation (62); by choosing suitable values for the two adjustable parameters a and C', it has been found possible to represent the variation of activity coefficients with concentration of several electrolytes from 0.001 to 1 molal, and sometimes up to 3 molal. The values of C' seem to lie approximately between 0.05 and 0.15 in aqueous solution. At the higher concentrations it is necessary to make allowance for the difference between the rational and stoichiometric activity coefficients; the latter, which is the experimentally determined quantity, is represented by an extension of equation (62); thus (cf. p. 135),

$$\log \gamma = -\frac{Az_+z_-\sqrt{\mu}}{1 + aB\sqrt{\mu}} + C'\mu - \log(1 + 0.001\,\nu m M_1), \qquad (68)$$

where ν is the number of ions produced by one molecule of electrolyte on dissociation, m is the molality of the solution and M_1 is the molecular weight of the solvent. This equation has been employed for the purpose of extrapolating activity coefficient data to dilute solutions from accurate measurements made at relatively high concentrations. It is not certain that this procedure is altogether justifiable, for the value of a obtained from activity data at high concentrations is often different from that derived from measurements on the same electrolyte in dilute solutions.

Extension of the Debye-Hückel Theory.—In the calculation of the electrical density in the vicinity of an ion (p. 82), it was assumed that $z_i\epsilon\psi/kT$ was negligible in comparison with unity, so that all terms beyond the first in the exponential series could be neglected. According to calculations made by Müller (1927), the neglect of the additional terms is justifiable provided that

$$a > \frac{\epsilon^2 z^2}{4DkT},$$

that is, if the mean ionic diameter a is greater than about $1.4 \times 10^{-6} z^2/D$ at 25°. It follows, therefore, that the additional terms are negligible in aqueous solution if a/z^2 exceeds 1.6×10^{-8}; for a uni-univalent salt, therefore, a should exceed 1.6×10^{-8} cm., but for a bi-bivalent electrolyte a must exceed 6.4×10^{-8} cm. if the Debye-Hückel approximation is to be valid. Since ionic diameters are rarely as high as the latter figure, it is seen that salts of high valence type might be expected to exhibit discrepancies from the simple Debye-Hückel behavior. Since the limiting values of a are larger the smaller the dielectric constant D of the medium, the deviations become more marked and will occur with electrolytes of lower valence type in media of low dielectric constant.

The potential ψ is given approximately by equation (15) on page 83, and hence the assumption, made by Debye and Hückel, that $z_i\epsilon\psi/kT$ is small compared with unity, is equivalent to stating that

$$\frac{z_i^2\epsilon^2}{D} \cdot \frac{e^{-\kappa r}}{r} \ll kT,$$

and this is less likely to be true the higher the valence of the ion and the smaller its radius, and the smaller the dielectric constant of the medium. In order to avoid the approximation involved in neglecting the higher terms in the exponential series, Gronwall, LaMer and Sandved [13] used the complete expansion for the electrical density, and solved the differential equation, following the introduction of the Poisson equation, in the form of a power series. The result obtained for a symmetrical valence type electrolyte, that is one with both ions of the same valence, is given by the following expression, which should be compared with equation

[13] Gronwall, LaMer and Sandved, *Physik. Z.*, **29**, 358 (1928); see also, LaMer, Gronwall and Greiff, *J. Phys. Chem.*, **35**, 2345 (1931).

(58), viz.,

$$\ln f = -\frac{Nz^2\epsilon^2\kappa}{2DRT}\cdot\frac{1}{1+\kappa a} + \sum_{m=1}^{\infty}\left(\frac{Nz^2\epsilon^2}{DRTa}\right)^{2m+1}\left[\tfrac{1}{2}X_{2m+1}(\kappa a) - 2m\,Y_{2m+1}(\kappa a)\right], \quad (69)$$

where $X(\kappa a)$ and $Y(\kappa a)$ are known, but complicated, functions of κa. The summation in equation (69) should be carried over all integral values of m from unity to infinity, but it is found that successive terms in the series decrease rapidly and it is sufficient, in general, to include only two terms.

In the application of equation (69) an arbitrary value of a is chosen so as to give calculated activity coefficients which agree with those derived by direct experiment; the proper choice of a is made by a process of trial and error until a value is found that is satisfactory over a range of concentrations. There is no doubt that the Gronwall-LaMer-Sandved extension represents an important advance over the simple Debye-Hückel treatment, for it frequently leads to more reasonable values of the mean ionic diameter.[14] The validity of equation (69) has been tested by a variety of activity measurements and the results have been found satisfactory; were it not for the tedious nature of the calculations it would probably be more widely used.

It is necessary to call attention to the fact that equation (69) was deduced for symmetrical valence electrolytes; for unsymmetrical types the corresponding equation is of a still more complicated nature.

Ion-Association.—A device, proposed by Bjerrum,[15] for avoiding the difficulty of integrating the Poisson equation when it is not justifiable to assume that $z_i\epsilon\psi/kT$ is much smaller than unity, involves the concept of the association of ions to form **ion-pairs** (cf. p. 96). It may be remarked that, in a sense, a solution, such as that of Gronwall, Sandved and LaMer, of the differential equation resulting from the use of the *complete* expression for the electrical density, makes the Bjerrum treatment unnecessary. The results obtained are, nevertheless, of interest, especially in connection with their application to media of low dielectric constant.

According to the Boltzmann distribution law, the number dn_i of ions of the ith kind in a spherical shell of radius r and thickness dr, surrounding a specified ion, is given by

$$dn_i = n_i 4\pi r^2 e^{-W/kT} dr, \quad (70)$$

[14] LaMer *et al.*, *J. Phys. Chem.*, **35**, 1953 (1931); **40**, 287 (1936); *J. Am. Chem. Soc.*, **53**, 2040, 4333 (1931); **54**, 2763 (1932); **56**, 544 (1934); Partington *et al.*, *Trans. Faraday Soc.*, **30**, 1134 (1934); *Phil. Mag.*, **22**, 857 (1936).

[15] Bjerrum, *K. Danske Vidensk. Selsk. Mat.-fys. Medd.*, **7**, No. 9 (1926); Fuoss and Kraus, *J. Am. Chem. Soc.*, **55**, 1019 (1933); Fuoss, *Trans. Faraday Soc.*, **30**, 967 (1934); *Chem. Revs.*, **17**, 227 (1935).

where n_i is the number of ions of the ith kind in unit volume and W is the work required to separate one of these ions from the central ion; k is the Boltzmann constant and T is the absolute temperature. The central ion, supposed to be positive, carries a charge $z_+\epsilon$ and that of the ith ion, which is of opposite sign, is $z_-\epsilon$; if Coulomb's law is assumed to hold at small interionic distances and the ions are regarded as point charges separated by a medium with an effective dielectric constant (D) equal to that of the solvent, then the work required to separate the ions from a distance r to infinity, and hence the value of W, is given by

$$W = \frac{z_+z_-\epsilon^2}{Dr}. \tag{71}$$

The influence of ions other than the pair under consideration is neglected in this derivation. Substituting this result for W in equation (70), it follows that

$$dn_i = n_i 4\pi r^2 e^{-z_+z_-\epsilon^2/DrkT}dr. \tag{72}$$

The fraction dn_i/dr is a measure of the probability $P(r)$ of finding an ion of charge opposite to that of the central ion at a distance r from the latter; thus

$$P(r) = n_i 4\pi r^2 e^{-z_+z_-\epsilon^2/DrkT}. \tag{73}$$

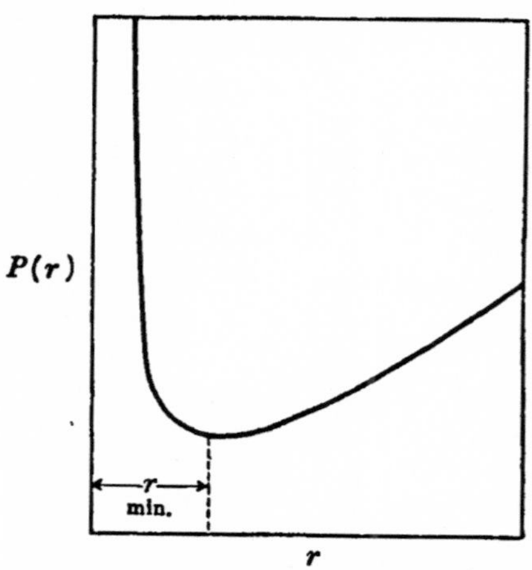

FIG. 53. Distribution of oppositely charged ions about a central ion (Bjerrum)

If the right-hand side of this equation, for various values of r, is plotted against r, the result is a curve of the type shown in Fig. 53, the actual form depending on the valences z_+ and z_- of the oppositely charged ions, and also on the dielectric constant of the medium. It will be observed that at small distances of approach there is a very high probability of finding the two ions together, but this probability falls rapidly, passes through a minimum and then increases somewhat for increasing distances between the ions. The interionic distance $r_{min.}$, for which the probability of finding two oppositely charged ions together is a minimum, can be obtained by differentiating equation (73) with respect to r and setting the result equal to zero; in this way it is found that

$$r_{min.} = \frac{z_+z_-\epsilon^2}{2DkT}. \tag{74}$$

The suggestion was made by Bjerrum that all ions lying within a sphere of radius $r_{min.}$ should be regarded as associated to form ion-pairs, whereas those outside this sphere may be considered to be free. The higher the value of $r_{\text{min.}}$ the greater the volume round a given ion in which the oppositely charged ions can be found, and hence the greater the probability of the occurrence of the ion-pairs. It is evident, therefore, from equation (74) that ion association will take place more readily the higher the valences, z_+ and z_-, of the ions of the electrolyte and the smaller the dielectric constant of the medium. This conclusion is in general agreement with experiment concerning the deviations from the behavior to be expected from the Debye-Hückel treatment based on the assumption of complete dissociation. Attention may be called to the fact, the exact significance of which is not altogether clear, that the value of $r_{\text{min.}}$ given by equation (74) is about twice the mean ionic diameter a which must be exceeded if the additional terms in the Debye-Hückel expansion may be neglected (see p. 154).

The Fraction of Association.—If equation (72) is integrated between $r = a$, where a is the effective mean diameter of the ions, or their distance of closest approach, and $r = r_{\text{min.}}$, the result should give the number, which will be less than unity, of oppositely charged ions that may be regarded as associated with a given ion. In other words, this quantity is equal to the **fraction of association** (θ) of the strong electrolyte into ion-pairs; thus

$$\theta = n_i 4\pi \int_a^{r_{\text{min.}}} r^2 e^{-z_+z_-\epsilon^2/DrkT} dr. \tag{75}$$

If $Nc/1000$, where c is the concentration in moles per liter, is written in place of n_i, and if both ions are assumed to be univalent, equation (75) may be expressed in the form

$$\theta = \frac{4\pi Nc}{1000}\left(\frac{\epsilon^2}{DkT}\right)^3 Q(b), \tag{76}$$

where

$$Q(b) \equiv \int_2^b e^y y^{-4} dy; \qquad y \equiv \frac{\epsilon^2}{DrkT} \qquad \text{and} \qquad b \equiv \frac{\epsilon^2}{DakT}.$$

The values of $Q(b)$ as defined above have been tabulated for various values of b from 1 to 80, and so by means of equation (76) it is possible to estimate the extent of association of a uni-univalent electrolyte consisting of ions of any required mean diameter a, at a concentration c in a medium of dielectric constant D. It will be seen from equation (76) that in general θ increases as b increases, i.e., θ increases as the mean diameter a of the ions and the dielectric constant of the solvent decrease. The values for the fraction of association of a uni-univalent electrolyte in water at 18° have been calculated by Bjerrum for various concentrations for four assumed ionic diameters; the results are recorded in Table XXXVII. The extent of association is seen to increase markedly with decreasing ionic diameter and increasing concentration. The values are appreciably greater in solutions of low dielectric constant, as is apparent from the factor $1/D^3$ in equation (76).

TABLE XXXVII. FRACTION OF ASSOCIATION (θ) OF UNI-UNIVALENT ELECTROLYTE IN WATER AT 18°

	Concentration						
a	0.001	0.005	0.01	0.05	0.1	0.5	1.0 N
2.82Å	0	0.002	0.005	0.017	0.029	0.090	0.138
2.35	0.001	0.004	0.008	0.028	0.048	0.140	0.206
1.76	0.001	0.007	0.012	0.046	0.072	0.204	0.286

The Association Constant.—Suppose that a salt MA is completely ionized in solution and that a certain fraction of the ions are associated as ion-pairs; an equilibrium may be supposed to exist between the free M^+ and A^- ions, on the one hand, and ion-pairs on the other hand. If the law of mass action [cf. equation (20)] is applied to this equilibrium, the result is

$$K = \frac{\text{Activity of M}^+ \times \text{Activity of A}^-}{\text{Activity of ion-pairs}},$$

where K is the dissociation constant (cf. p. 163). If c is the concentration of the salt MA, the concentration of associated ions is θc while that of each of the free ions is $(1 - \theta)c$; further, if f_1 represents the mean activity coefficient of the ions and f_2 is that of the ion-pairs, then

$$K = \frac{(1-\theta)c \times (1-\theta)c}{\theta c} \cdot \frac{f_1^2}{f_2} = \frac{(1-\theta)^2 c}{\theta} \cdot \frac{f_1^2}{f_2}. \tag{77}$$

For very dilute solutions, i.e., when c is small, the activity coefficients are almost unity, while θ is negligible in comparison with unity (see Table XXXVII); equation (77) then reduces to

$$K = \frac{c}{\theta},$$

$$\therefore K^{-1} = \frac{\theta}{c}, \tag{78}$$

where K^{-1}, the reciprocal of the dissociation constant, is called the **association constant** of the completely ionized electrolyte. Introducing the value of θ given by equation (76), the result is

$$K^{-1} = \frac{4\pi N}{1000}\left(\frac{\epsilon^2}{DkT}\right)^3 Q(b), \tag{79}$$

and so the dissociation constant K can be calculated for any assumed value of the distance of closest approach of the ions in a medium of known dielectric constant.

A test of equation (79), based on the theory of ion association, is provided by the measurements of Fuoss and Kraus [16] of the conductance of tetraisoamylammonium nitrate in a series of dioxane-water mixtures of dielectric constant ranging from 2.2 to 78.6 (cf. Fig. 21) at 25°. From the results in dilute solution the dissociation constants were calculated by the method described on page 158. The values of $-\log K$, plotted against $\log D$ of the medium, are indicated by the points in Fig. 54, whereas the full curve is that to be expected from equation (79) if a is taken as 6.4Å. The agreement between the experimental and theoretical results is very striking. It will be observed that as the dielectric constant increases the curve turns sharply downwards and crosses the $\log D$ axis at a value of the dielectric constant of approximately 41. The significance of this result is that for ions of mean effective diameter equal to 6.4Å, the dissociation constant of the electrolyte is very large, and hence the extent of association becomes negligible when the dielectric constant of the solvent exceeds a value of about 41.* For smaller ions or for ions of higher valence, the dielectric constant would have to attain a larger value before an almost completely ionized electrolyte would be also completely dissociated.

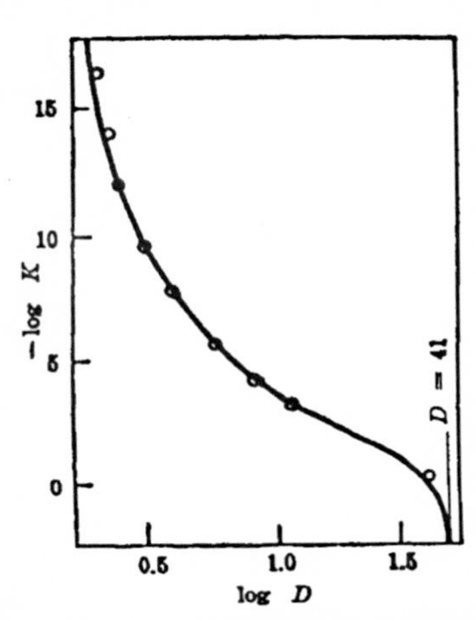

FIG. 54. Association constant and dielectric constant (Fuoss and Kraus)

Triple Ions.—The concept of ion-pairs has been extended to include the possibility of the presence in solution of groups of three ions, viz., $+-+$ or $-+-$, i.e., **triple ions,** held together by electrostatic forces.[17] Such triplets might be expected to form most readily in solvents of low dielectric constant, for it is in such media that the forces of electrostatic attraction would be greatest. Consider an electrolyte MA in a medium of low dielectric constant; there will be an equilibrium between the ions M^+ and A^- and the ion-pairs, as described above. In this case, however, the ion-pair formation will be considerable and θ will approach unity. If $1 - \theta$ is replaced by α, the fraction of the electrolyte present as free ions, and if both θ and the activity coefficient factor are assumed to be

[16] Fuoss and Kraus, *J. Am. Chem. Soc.*, 55, 1019 (1933).

* According to the simple calculations on page 156 the dielectric constant necessary for the solvent in which a uni-univalent electrolyte whose mean ionic diameter is 6.4×10^{-8} cm. should be dissolved in order that there may be no appreciable association is $2.79 \times 10^{-6}/6.4 \times 10^{-8}$, i.e., 42.

[17] Fuoss and Kraus, *J. Am. Chem. Soc.*, 55, 2387 (1933); Fuoss, *ibid.*, 57, 2604 (1935); *Chem. Revs.*, 17, 227 (1935); for reviews, see Kraus, *J. Franklin Inst.*, 225, 687 (1938); *Science*, 90, 281 (1939).

unity, equation (77) can be written as

$$k \approx \alpha^2 c, \tag{80}$$

where k is the approximate dissociation constant and c is the total electrolyte concentration. If in addition to ion-pairs (dual ions) there are present triple ions, viz., MAM^+ and AMA^-, the following equilibria

$$MAM^+ \rightleftharpoons MA + M^+,$$
$$AMA^- \rightleftharpoons MA + A^-,$$

also exist. If the formation of MAM^+ and AMA^- is due to electrical effects only, there will be an equal tendency for both these ions to form; the mass action constant k_3 of the two equilibria may thus be expected to be the same. Hence, neglecting activity coefficients,

$$k_3 = \frac{c_{MA}c_{M^+}}{c_{MAM^+}} = \frac{c_{MA}c_{A^-}}{c_{AMA^-}}, \tag{81}$$

$$\therefore \frac{c_{M^+}}{c_{MAM^+}} = \frac{c_{A^-}}{c_{AMA^-}}. \tag{82}$$

The triple ions should consequently be formed in the same ratio as that in which the simple ions are present in the solution. If α_3 is the fraction of the total electrolyte existing as either of the triple ions, e.g., MAM^+, then c_{MAM^+} is equal to $\alpha_3 c$. Since the amount of these ions will be small, c_{MA} may be taken as approximately equal to the total concentration c, and c_{M^+} can be assumed to remain as αc. Substituting these results in equation (81), it follows that

$$k_3 = \frac{\alpha c}{\alpha_3}, \tag{83}$$

and since k, by equation (80), is equal to $\alpha^2 c$, i.e., α is $\sqrt{k/c}$, it is found that

$$\alpha_3 = \frac{\sqrt{kc}}{k_3}. \tag{84}$$

Although dual ions have no conducting power, since they are electrically neutral, triple ions are able to carry current and contribute to the conductance of the solution. If Λ_0 is the sum of the equivalent conductances of the simple ions at infinite dilution, and λ_0 is the sum of the values for the two kinds of triple ions, then since the latter are formed in the same ratio as the simple ions, it follows that the observed equivalent conductance is given by

$$\Lambda = \alpha\Lambda_0 + \alpha_3\lambda_0,$$

interionic effects being neglected. Substituting $\sqrt{k/c}$ for α, and $\sqrt{kc}/k_3$

for α_3, it is seen that

$$\Lambda = \Lambda_0 \sqrt{\frac{k}{c}} + \lambda_0 \frac{\sqrt{kc}}{k_3}, \tag{85}$$

$$\therefore \quad \Lambda\sqrt{c} = \Lambda_0\sqrt{k} + \frac{\lambda_0\sqrt{k}}{k_3} c. \tag{86}$$

If $\Lambda\sqrt{c}$ is plotted against c for media of low dielectric constant, in which triple ions can form to an appreciable extent, the result should be a straight line; this expectation has been confirmed by experiment, as

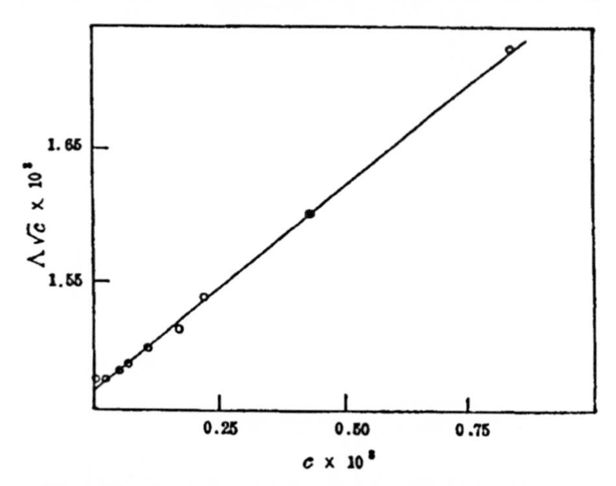

FIG. 55. Test of triple-ion theory (Fuoss and Kraus)

shown by the points in Fig. 55 which are for tetrabutylammonium picrate in anisole. The deviation from the straight line becomes evident only at high concentrations.

Triple Ions and Conductance Minima.—Since equation (85) is of the form

$$\Lambda = \frac{A}{\sqrt{c}} + B\sqrt{c}, \tag{87}$$

where A and B are constants, it is evident that the first term on the right-hand side decreases and the second term increases as the concentration is increased; it is possible, therefore, for a minimum in the equivalent conductance to occur, as has been found experimentally (p. 52). The physical significance of this result is that with increasing concentration the single ions are replaced by electrically neutral ion-pairs, and so the conductance falls; at still higher concentrations, however, the ion-pairs are replaced by triple ions having a relatively high conducting power, and so the equivalent conductance of the solution tends to increase.

The condition for the conductance minimum is found by differentiating equation (87) with respect to c and setting the result equal to zero; this procedure gives

$$c_{\text{min.}} = \frac{A}{B}$$

$$= k_3 \frac{\Lambda_0}{\lambda_0}. \tag{88}$$

By substituting this value in equation (85), and utilizing the relationships given above for α and α_3, it is found that

$$\Lambda_{\text{min.}} = 2(\Lambda_0\alpha)_{\text{min.}} = 2(\lambda_0\alpha_3)_{\text{min.}}. \tag{89}$$

It is seen from equation (88) that the concentration for the minimum conductance is proportional to k_3, and so is inversely proportional to the stability of the triple ions. The minimum occurs when the conductance due to these ions, i.e., $\lambda_0\alpha_3$, is equal to that due to the single ions, i.e., $\Lambda_0\alpha$.

By means of a treatment analogous to that described above for calculating the association constant for the formation of ion-pairs, it is possible to derive an expression for k_3^{-1} which is analogous to equation (79); [18] the result may be put in the form

$$k_3^{-1} = \frac{2\pi N}{1000}\left(\frac{\epsilon^2}{DkT}\right)^3 \frac{1}{b^3} I(b, r), \tag{90}$$

where $I(b, r)$ is a function of b, which has the same significance as before, and of the distance r between the ions. In the region of the minimum conductance, the value of $I(b, r)/b^3$ does not change appreciably, and equation (90) can be written as

$$k_3^{-1} = \frac{A}{D^3},$$

where A is a constant and D is the dielectric constant of the medium; the dissociation constant of the triple ions (k_3) is thus proportional to D^3. Since the concentration $c_{\text{min.}}$ at which the minimum equivalent conductance is observed is proportional to k_3, it follows that

$$\frac{D^3}{c_{\text{min.}}} = \text{constant}; \tag{91}$$

this is the rule derived empirically by Walden (p. 53).[19]

The fact that the concentration at which the conductance minimum occurs decreases with decreasing dielectric constant of the solvent is shown by the results in Fig. 21 (p. 53). In media of very low dielectric

[18] Fuoss and Kraus, *J. Am. Chem. Soc.*, 55, 2387 (1933).

[19] See also, Gross and Halpern, *J. Chem. Phys.*, 2, 188 (1934); Fuoss and Kraus, *ibid.*, 2, 386 (1934).

constant, however, the minimum does not appear, but the conductance curves show inflections; these are attributed to mutual interactions between two dipoles, i.e., ion-pairs, as a result of which quadripoles are formed. The consequence of this is that the normal increase of conductance beyond the minimum, due to the formation of triple ions, is inhibited to some extent. If the dielectric constant of the solvent exceeds a certain value, depending on the mean diameter and valence of the ions, there is no appreciable formation of triple ions at any concentration, and hence there can be no conductance minimum.

Equilibria in Electrolytes: The Dissociation Constant.—When any electrolyte MA is dissolved in a suitable solvent, it yields M^+ and A^- ions in solution to a greater or lesser extent depending on the nature of MA; even if ionization is complete, as is the case with simple salts in aqueous solution, there may still be a tendency for ion-pairs to form in relatively concentrated solution, so that dissociation is not necessarily complete. In general, therefore, there will be set up the equilibrium

$$MA \rightleftharpoons M^+ + A^-,$$

where M^+ and A^- represent the free ions and MA is the undissociated portion of the electrolyte which includes both un-ionized molecules and ion-pairs. Application of the law of mass action, in the form of equation (20), to this equilibrium gives

$$K = \frac{a_{M^+}a_{A^-}}{a_{MA}}, \tag{92}$$

where the a terms are the activities of the indicated species; the equilibrium constant K is called the **dissociation constant** of the electrolyte. The term "ionization constant" is also frequently employed in the literature of electrochemistry, but since the equilibrium is between free ions and *undissociated* molecules, the expression "dissociation constant" is preferred. Writing the activity terms in equation (92) as the product of the concentration and the activity coefficient, it becomes

$$K = \frac{c_{M^+}c_{A^-}}{c_{MA}} \cdot \frac{f_{M^+}f_{A^-}}{f_{MA}}. \tag{93}$$

Further, if α is the degree of dissociation of the electrolyte (cf. p. 96) whose total concentration is c moles per liter, then c_{M^+} and c_{A^-} are each equal to αc, and c_{MA} is equal to $c(1 - \alpha)$; it follows, therefore, that

$$K = \frac{\alpha^2 c}{1 - \alpha} \cdot \frac{f_{M^+}f_{A^-}}{f_{MA}}. \tag{94}$$

If the solution is sufficiently dilute, the activity coefficients are approximately unity, and so equation (94) reduces under these conditions to

$$k = \frac{\alpha^2 c}{1 - \alpha}, \tag{95}$$

which is the form of the so-called **dilution law** as originally deduced by Ostwald (1888). It will be noted that in the approximate equation (95) the symbol k has been used; this quantity is often called the "classical dissociation constant," but as it cannot be a true constant it is preferable to refer to it as the "classical dissociation function" or, in brief, as the "dissociation function."

The relation between the function k and the true or "thermodynamic" dissociation constant K is obtained by combining equations (94) and (95); thus

$$K = k\frac{f_{M^+}f_{A^-}}{f_{MA}}. \tag{96}$$

Provided the ionic strength of the medium is not too high, the activity coefficient of the undissociated molecules never differs greatly from unity; hence, equation (96) may be written as

$$K = k(f_{M^+}f_{A^-}). \tag{97}$$

If the solution is sufficiently dilute for the Debye-Hückel limiting law to be applicable, it follows from equation (54), assuming the ions M^+ and A^- to be univalent, for simplicity, that

$$\log f_{M^+} = \log f_{A^-} = -A\sqrt{\alpha c}, \tag{98}$$

the ionic strength, $\frac{1}{2}\Sigma c_i z_i^2$, being equal to $\frac{1}{2}[(\alpha c \times 1^2) + (\alpha c \times 1^2)]$, i.e., to αc. Upon taking logarithms of equation (97) and substituting the values of $\log f_{M^+}$ and $\log f_{A^-}$ as given by (98), the result is

$$\log K = \log k - 2A\sqrt{\alpha c}. \tag{99}$$

The plot of the values of $\log k$, obtained at various concentrations, against $\sqrt{\alpha c}$ should thus give a straight line of slope $-2A$; for water at 25° the value of A is 0.509 (Table XXXV) and so the slope of the line should be -1.018.

In order to test the reliability of equation (99) it is necessary to know the value of the degree of dissociation at various concentrations of the electrolyte MA; in his classical studies of dissociation constants Ostwald, following Arrhenius, assumed that α at a given concentration was equal to the conductance ratio Λ/Λ_0, where Λ is the equivalent conductance of the electrolyte at that concentration and Λ_0 is the value at infinite dilution. As already seen (p. 95), this is approximately true for weak electrolytes, but it is more correct, for electrolytes of all types, to define α as Λ/Λ' where Λ' is the conductance of 1 equiv. of free ions at the same ionic concentration as in the given solution. It follows therefore, by substituting this value of α in equation (95), that

$$k = \frac{\Lambda^2 c}{\Lambda'(\Lambda' - \Lambda)}. \tag{100}$$

Since Λ for various concentrations can be obtained from conductance data and the Onsager equation, by one of the methods described in Chap. III, it is possible to derive the dissociation function k for the corresponding concentrations. The results obtained for acetic acid in aqueous solution at 25° are given in Table XXXVIII,[20] and the values of

TABLE XXXVIII. DISSOCIATION CONSTANT OF ACETIC ACID AT 25°

$c \times 10^3$	Λ	Λ'	α	$k \times 10^5$	$K \times 10^5$
0.028014	210.38	390.13	0.5393	1.768	1.752
0.11135	127.75	389.79	0.3277	1.779	1.754
0.21844	96.49	389.60	0.2477	1.781	1.751
1.02831	48.15	389.05	0.1238	1.797	1.751
2.41400	32.22	388.63	0.08290	1.809	1.750
5.91153	20.96	388.10	0.05401	1.823	1.749
9.8421	16.37	387.72	0.04222	1.832	1.747
20.000	11.57	387.16	0.02987	1.840	1.737
52.303	7.202	386.18	0.01865	1.854	1.722
119.447	4.760	385.18	0.01236	1.847	1.688
230.785	3.392	384.26	0.008827	1.814	1.632

log k are plotted against $\sqrt{\alpha c}$ in Fig. 56; the dotted line has the theoretical slope required by equation (99). It is clear that in the more dilute solutions the experimental results are in excellent agreement with theory,

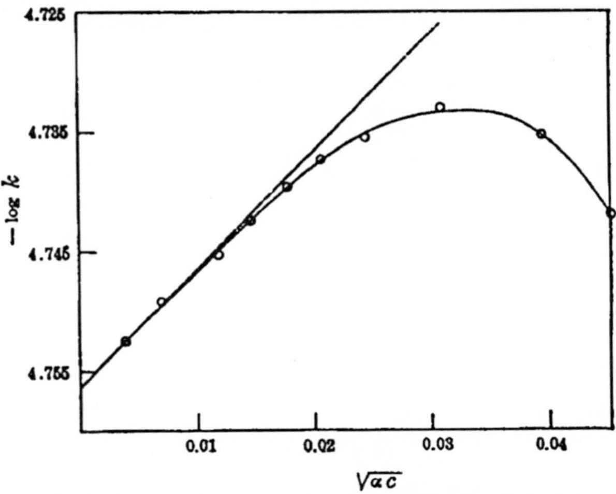

FIG. 56. Dissociation constant of acetic acid (MacInnes and Shedlovsky)

but at higher concentrations deviations become evident. The same conclusion is reached from an examination of the last column in Table

[20] MacInnes and Shedlovsky, *J. Am. Chem. Soc.*, 54, 1429 (1932); MacInnes, *J. Franklin Inst.*, 225, 661 (1938).

XXXVIII which gives the results for K derived from equation (99) using the theoretical value of A, i.e., 0.51. The first figures are seen to be virtually constant, as is to be expected, the mean value of K being 1.752×10^{-5}. At infinite dilution the activity coefficient factor is unity and so the extrapolation of the dissociation functions k to infinite dilution should give the true dissociation constant K; the necessary extrapolation is carried out in Fig. 56, from which it is seen that the limiting value of $\log k$ is -4.7564, so that K is 1.752×10^{-5}, as given above.

Similar results to those described for acetic acid in aqueous solution have been recorded for other weak acids in aqueous solution, and also for several acids in methyl alcohol.[21] In each case the plot of $\log k$ against $\sqrt{\alpha c}$ was found to be a straight line for dilute solutions, the slope being in excellent agreement with that required by the Debye-Hückel limiting law. The deviations observed with relatively concentrated solutions, such as those shown in Fig. 56, are partly due to the failure of the limiting law to apply under these conditions, and partly to the change in the nature, e.g., dielectric constant, of the medium resulting from the presence of appreciable amounts of an organic acid.

Strong Electrolytes.—The arguments presented above are readily applicable to weak electrolytes because the total concentration can be quite appreciable before the ionic strength becomes large enough for the Debye-Hückel limiting law to fail; for example, the results in Table XXXVIII extend up to a concentration of 0.2 N, but the ionic strength is then about 0.04. With relatively strong electrolytes, however, the procedure can be used only for very dilute solutions. In these circumstances it is preferable to return to equation (97), which should hold for all types of electrolytes of the general formula MA, and to employ activity coefficients obtained by direct experimental measurement, instead of the values calculated from the Debye-Hückel equations. The product $f_{M^+}f_{A^-}$ in equation (97) may be replaced by the square of the mean activity coefficient of the electrolyte, i.e., by $f_\pm^2$, in accordance with the definition of equation (30); it follows, therefore, that equation (100) may be modified so as to give

$$K = \frac{\Lambda^2 c}{\Lambda'(\Lambda' - \Lambda)} f_\pm^2. \tag{101}$$

The accuracy of this equation has been confirmed for a number of salts generally regarded as strong electrolytes, as the data in Table XXXIX serve to show.[22] It is evident from these results that the law of mass action holds for strong, as well as for weak electrolytes, provided it is

[21] MacInnes and Shedlovsky, *J. Am. Chem. Soc.*, **57**, 1705 (1935); Saxton *et al.*, *ibid.*, **55**, 3638 (1933); **56**, 1918 (1934); **59**, 1048 (1937); Brockman and Kilpatrick, *ibid.*, **56**, 1483 (1934); Martin and Tartar, *ibid.*, **59**, 2672 (1937); Belcher, *ibid.*, **60**, 2744 (1938); see also, MacInnes, "The Principles of Electrochemistry," 1939, Chap. 19.

[22] Davies *et al.*, *Trans. Faraday Soc.*, **23**, 351 (1927); **26**, 592 (1930); **27**, 621 (1931); **28**, 609 (1932); "The Conductivity of Solutions," 1933, Chap. IX.

applied in the correct manner. The view expressed at one time that the law of mass action was not applicable to strong electrolytes was partly due to the employment of the Arrhenius method of calculating the degree of dissociation, and partly to the failure to make allowance for deviations from ideal behavior.

TABLE XXXIX. APPLICATION OF LAW OF MASS ACTION TO STRONG ELECTROLYTES

Salt	c	Λ/Λ'	$f_\pm$	K
KNO_3	0.01	0.994	0.916	1.40
	0.02	0.989	0.878	1.38
	0.05	0.975	0.806	1.32
	0.10	0.961	0.732	1.37
$AgNO_3$	0.01	0.993	0.902	1.10
	0.02	0.989	0.857	1.31
	0.05	0.973	0.783	1.12
	0.10	0.957	0.723	1.23
	0.50	0.883	0.526	1.18

Intermediate and Weak Electrolytes.—The calculation of the degree of dissociation by the methods given in Chap. III presuppose the availability of suitable conductance data for electrolytes which are virtually completely dissociated at the appropriate concentrations. There is generally no difficulty concerning this matter if the solvent is water, but for non-aqueous media, especially those of low dielectric constant, the proportion of undissociated molecules may be quite large even at small concentrations, and no direct method is available whereby the quantity Λ' can be evaluated from conductance data. For solvents of this type the following method, which can be used for any systems behaving as weak or intermediate electrolytes, may be employed.[23] The Onsager equation for incompletely dissociated electrolytes can be written (cf. p. 95) as

$$\Lambda' = \Lambda_0 - (A + B\Lambda_0)\sqrt{\alpha c},$$

$$\therefore \quad \alpha = \frac{\Lambda}{\Lambda'} = \frac{\Lambda}{\Lambda_0 - (A + B\Lambda_0)\sqrt{\alpha c}}. \tag{102}$$

If a variable x is defined by

$$x \equiv \frac{(A + B\Lambda_0)\sqrt{c\Lambda}}{\Lambda_0^{3/2}}, \tag{103}$$

equation (102) becomes

$$\alpha = \frac{\Lambda}{\Lambda_0 F(x)}, \tag{104}$$

[23] Fuoss and Kraus, *J. Am. Chem. Soc.*, **55**, 476 (1933); Fuoss, *ibid.*, **57**, 488 (1935); *Trans. Faraday Soc.*, **32**, 594 (1936).

where $F(x)$ is a function of x represented by the continued fraction

$$F(x) = 1 - x(1 - x(1 - x(1 - \cdots)^{-\frac{1}{2}})^{-\frac{1}{2}})^{-\frac{1}{2}}$$
$$= \tfrac{4}{3}\cos^2 \tfrac{1}{3}\cos^{-1}(-\tfrac{3}{2}x\sqrt{3}).$$

Values of this function have been worked out and tabulated for values of x from zero to 0.209 in order to facilitate the calculations described below.

Taking the activity coefficient of the undissociated molecules, as usual, to be equal to unity, and replacing $f_{M^+}f_{A^-}$ by $f_\pm^2$, where $f_\pm$ is the mean activity coefficient, equation (94) becomes

$$K = \frac{\alpha^2 c}{1 - \alpha} f_\pm^2, \tag{105}$$

and if the value of α given by equation (104) is inserted, the result is

$$K = \frac{\left[\dfrac{\Lambda}{\Lambda_0 F(x)}\right]^2 c f_\pm^2}{1 - \dfrac{\Lambda}{\Lambda_0 F(x)}},$$

which on multiplying out and rearranging gives

$$\frac{F(x)}{\Lambda} = \frac{1}{K\Lambda_0^2} \cdot \frac{\Lambda c f_\pm^2}{F(x)} + \frac{1}{\Lambda_0}. \tag{106}$$

It is seen from equation (106) that the plot of $F(x)/\Lambda$ against $\Lambda c f_\pm^2/F(x)$ should be a straight line, the slope being equal to $1/K\Lambda_0^2$ and the intercept, for infinite dilution, giving $1/\Lambda_0$. In this manner it should be possible to determine both the dissociation constant K of the electrolyte and the equivalent conductance at infinite dilution (Λ_0) in one operation.

In order to obtain the requisite plot, an approximate estimate of Λ_0 is first made by extrapolating the experimental data of Λ against $\sqrt{c}$, and from this a tentative result for x is derived by means of equation (103), since the Onsager constants A and B are presumably known (see Table XXIII). In this way a preliminary value of $F(x)$ is obtained which is employed in equation (106); the activity coefficients required are calculated from the Debye-Hückel limiting law equation (98), using the value of α given by equation (104) from the rough estimates of Λ_0 and $F(x)$. The results are then plotted as required by equation (106), and the datum for Λ_0 so obtained may be employed to calculate $F(x)$ and α more accurately; the plot whereby Λ_0 and K are obtained may now be repeated. The final results are apparently not greatly affected by a small error in the provisional value of Λ_0 and so it is not often necessary to repeat the calculations. With Λ_0 known accurately, it is possible to determine the degree of dissociation at any concentration, if required, by means of equations (103) and (104), and the tabulated values of $F(x)$.

The work of Fuoss and Kraus and their collaborators and of others has shown that equation (106) is obeyed in a satisfactory manner by a number of electrolytes, both salts and acids, in solvents of low dielectric constant.[24] The results of plotting the values of $F(x)/\Lambda$ against $\Lambda c f_{\pm}^2/F(x)$ for solutions of tetramethyl- and tetrabutyl-ammonium picrates in ethylene chloride are shown in Fig. 57; the intercepts are 0.013549 and 0.17421, and the slopes of the straight lines are 5.638 and 1.3337, re-

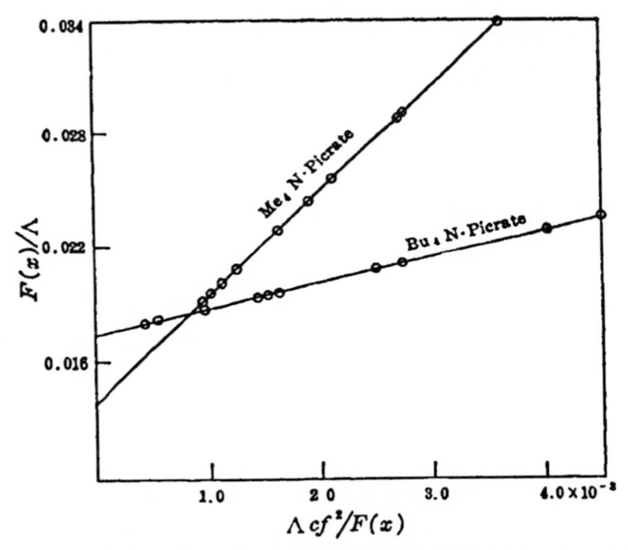

FIG. 57. Salts in media of low dielectric constant (Fuoss and Kraus)

spectively. It follows, therefore, that for tetramethylammonium picrate K is 0.3256×10^{-4} and Λ_0 is 73.81 ohms^{-1} cm.^{2}, whereas the corresponding values for tetrabutylammonium picrate are 2.276×10^{-4} and 57.40 ohms^{-1} cm.^{2}, respectively.

Solubility Equilibria: The Solubility Product Principle.—It was seen on page 133 that the chemical potential of a solid is constant at a definite temperature and pressure; consequently, when a solution is saturated with a given salt $M_{\nu_+}A_{\nu_-}$ the chemical potential of the latter in the solution must also be constant, since the chemical potential of any substance present in two phases at equilibrium must be the same in each phase. It is immaterial whether this conclusion is applied to the undissociated molecules of the salt or to the ions, for the chemical potential is given by

[24] Kraus, Fuoss *et al.*, *Trans. Faraday Soc.*, **31**, 749 (1935); **32**, 594 (1936); *J. Am. Chem. Soc.*, **58**, 255 (1936); **61**, 294 (1939); **62**, 506, 2237 (1940); Owen and Waters, *ibid.*, **60**, 2371 (1938); see also, MacInnes, "The Principles of Electrochemistry," 1939, Chap. 19.

either side of equation (26); thus, taking the left-hand side, it follows that

$$\nu_+(\mu^0_{\mathrm{M}^+} + RT \ln a_+) + \nu_-(\mu^0_{\mathrm{A}^-} + RT \ln a_-) = \text{constant},$$

$$\therefore \quad \nu_+ \ln a_+ + \nu_- \ln a_- = \text{constant},$$

$$\therefore \quad a_+^{\nu_+} a_-^{\nu_-} = \text{constant } (K_s), \tag{107}$$

at a specified temperature and pressure. The constant K_s as defined by equation (107) is the **activity solubility product,** and this equation expresses the **solubility product principle,** first enunciated in a less exact manner by Nernst (1889). If the activity of an ion is written as the product of its concentration, in moles (g.-ions) per liter, and the corresponding activity coefficient, equation (107) becomes

$$c_+^{\nu_+} c_-^{\nu_-} \times f_+^{\nu_+} f_-^{\nu_-} = K_s, \tag{108}$$

and introducing the definition of the mean activity coefficient of the electrolyte $\mathrm{M}_{\nu_+}\mathrm{A}_{\nu_-}$, it follows that

$$c_+^{\nu_+} c_-^{\nu_-} \times f_\pm^{\nu} = K_s, \tag{109}$$

where ν is equal to $\nu_+ + \nu_-$. If the ionic strength of the medium is low, the activity coefficient is approximately unity and equation (109) reduces to the approximate form

$$c_+^{\nu_+} c_-^{\nu_-} = k_s, \tag{110}$$

in which the solubility product principle is frequently employed.

The significance of the solubility product principle is that when a solution is saturated with a given salt the product of the activities, or approximately the concentrations, of its constituent ions must be constant, irrespective of the nature of the other electrolytes present in the solution. If the latter contains an excess of one or other of the ions of the saturating salt, this must be taken into consideration in the activity product. Consider, for example, a solution saturated with silver chloride: then according to the solubility product principle,

$$a_{\mathrm{Ag}^+} a_{\mathrm{Cl}^-} = K_{s(\mathrm{AgCl})}, \tag{111}$$

or, approximately,

$$c_{\mathrm{Ag}^+} c_{\mathrm{Cl}^-} = k_{s(\mathrm{AgCl})}. \tag{112}$$

If the solution which is being saturated with silver chloride already contains one of the ions of this salt, e.g., the chloride ion, then the term a_{Cl^-} will represent the total activity of the chloride ion in the solution; since this is greater than that in a solution containing no excess of chloride ion, the value of a_{Ag^+} required according to equation (111) will be less in the former case. In its simplest terms, based on equation (112), the conclusion is that the silver ion concentration in a saturated solution of silver chloride containing an excess of chloride ions, e.g., due to the presence of potassium chloride in the solution, will be less than in a solution in pure water. Since the silver chloride in solution may be regarded as completely ionized, the silver ion concentration is a measure of the solubility of the salt; it follows, therefore, that silver chloride is less soluble in the presence of excess of chloride ions than in pure water. In general, if there is no formation of complex ions to disturb the equilibrium (cf. p. 172), the solubility of any salt is less in a solution containing a common ion than in water alone; this fact finds frequent application in analytical chemistry.

Solubility in the Presence of a Common Ion.—If S_0 is the solubility of any sparingly soluble salt $M_{\nu_+}A_{\nu_-}$ in moles per liter in pure water, then if the solution is sufficiently dilute for dissociation to be complete, c_+ is equal to ν_+S_0 and c_- is equal to ν_-S_0; hence according to equation (109)

$$\begin{aligned} K_s &= (\nu_+S_0)^{\nu_+}(\nu_-S_0)^{\nu_-} \times f_\pm^\nu \\ &= (\nu_+^{\nu_+}\nu_-^{\nu_-})S_0^\nu f_\pm^\nu. \end{aligned} \tag{113}$$

In the simple case of a uni-univalent sparingly soluble salt, this becomes

$$K_s = S_0^2 f_\pm^2. \tag{114}$$

These equations relate the solubility product to the solubility in pure water and the activity coefficient in the saturated solution; for practical purposes it is convenient to take the activity coefficient to be approximately unity, since the solutions are very dilute, so that equation (114) can be written

$$k_s = S_0^2.$$

For a uni-univalent salt the saturation solubility in pure water is thus equal to the square-root of its solubility product; alternatively, it may be stated that the solubility product is equal to the square of the solubility in water. The solubility of silver chloride in water at 25° is 1.30×10^{-5} mole per liter; the solubility product is consequently 1.69×10^{-10}.

Suppose the addition of x moles per liter of a completely dissociated salt containing a common ion, e.g., the anion, reduces the solubility of the sparingly soluble salt from S_0 to S; for simplicity all the ions present may be assumed to be univalent. The concentrations of cations in the solution, resulting from the complete dissociation of the sparingly soluble salt, is S, while that of the anions is $S + x$; it follows, therefore, by the approximate solubility product principle that

$$\begin{aligned} S(S + x) &= k_s = S_0^2, \\ \therefore \quad S &= -\tfrac{1}{2}x + \sqrt{\tfrac{1}{4}x^2 + S_0^2}. \end{aligned} \tag{115}$$

Using this equation, or forms modified to allow for the valences of the ions which may differ from unity, it is possible to calculate the solubility (S) of a sparingly soluble salt in the presence of a known amount (x) of a common ion, provided the solubility in pure water (S_0) is known. An illustration of the application of equation (115) is provided by the results in Table XL for the solubility of silver nitrite in the presence of silver nitrate (I), on the one hand, and of potassium nitrite (II) on the other hand; the calculated values are given in the last column.[25] The agreement between the observed and calculated results in these dilute solutions is seen to be good, perhaps better than would be expected in view of the neglect of activity coefficients; in the presence of larger amounts of added electrolytes, however, deviations do occur. Much experimental work has been carried out with the object of verifying the solubility product principle in its approximate form, and the general conclusion reached is that it is satisfactory provided the total concentration of the solution is small; at higher concentrations discrepancies are observed, especially if ions of high valence are present. It was found, for example, that in the presence of lanthanum nitrate the solubility of the iodate decreases at first, in agreement with expectation, but as the concentration of the former salt is increased, the solubility of the lanthanum iodate, instead of decreasing steadily, passes through a minimum and then increases. Such deviations from the expected behavior are, of course, due to neglect of the activity coefficients in the application of the simple solubility product principle; the effect of this neglect becomes more evident with increasing concentration, especially if the solution contains ions of high valence. It is evident from the Debye-Hückel limiting law equation that the departure of the activity coefficients from unity is most marked with ions of high valence because the square of the valence appears not only in the factor preceding the square-root of the ionic strength but also in the ionic strength itself. The more exact treatment of solubility, taking the activity coefficients into consideration, is given later.

TABLE XL. SOLUBILITY OF SILVER NITRITE IN THE PRESENCE OF COMMON ION

x moles/liter	S I	S II	S Calculated
0.000	0.0269	0.0269	$(0.0269 = S_0)$
0.00258	0.0260	0.0259	0.0259
0.00588	0.0244	0.0249	0.0247
0.01177	0.0224	0.0232	0.0227

Formation of Complex Ions.—In certain cases the solubility of a sparingly soluble salt is greatly increased, instead of being decreased, by the addition of a common ion; a familiar illustration of this behavior is provided by the high solubility of silver cyanide in a solution of cyanide ions. Similarly, mercuric iodide is soluble in the presence of excess of iodide ions and aluminum hydroxide dissolves in solutions of alkali hydroxides. In cases of this kind it is readily shown by transference measurements that the silver, mercury or other cation is actually present in the solution in the form of a complex ion. The solubility of a sparingly soluble salt can be increased by the addition of any substance, whether it

[25] Creighton and Ward, *J. Am. Chem. Soc.*, 37, 2333 (1915).

contains a common ion or not, which is able to remove the simple ions in the form of complex ions. For example, if either cyanide ion or ammonia is added to a slightly soluble silver compound, such as silver chloride, the silver ions are converted into the complex ions $Ag(CN)_2^-$ or $Ag(NH_3)_2^+$, respectively. In either case the concentration of free silver ions is reduced and the product of the concentrations (activities) of the silver and chloride ions falls below the solubility product value: more silver chloride dissolves, therefore, in order to restore the condition requisite for a saturated solution. If sufficient complex forming material is present the removal of the silver ions will continue until the whole of the silver chloride has dissolved.

Although by far the largest proportion of the silver in a complex cyanide solution is present in the form of argentocyanide ions, $Ag(CN)_2^-$, there is reason for believing that a small concentration of simple silver ions is also present; the addition of hydrogen sulfide, for example, causes the precipitation of silver sulfide which has a very low solubility product. It is probable, therefore, that an equilibrium of the type

$$Ag(CN)_2^- \rightleftharpoons Ag^+ + 2CN^-$$

exists between complex and free ions in an argentocyanide solution and similar equilibria are established in other instances. For the general case of a complex ion $M_qA_r^{\pm}$, the equilibrium is

$$M_qA_r^{\pm} \rightleftharpoons qM^+ + rA^-,$$

and application of the law of mass action gives

$$K_i = \frac{a^q_{M^+}a^r_{A^-}}{a_{M_qA_r^{\pm}}}. \tag{116}$$

or, using concentrations in place of activities,

$$k_i = \frac{c^q_{M^+}c^r_{A^-}}{c_{M_qA_r^{\pm}}}. \tag{117}$$

The constant K_i (or k_i) is called the **instability constant** of the complex ion; it is apparent that the greater its value the greater the tendency of the complex to dissociate into simple ions, and hence the smaller its stability. The reciprocal of the instability constant is sometimes encountered; it is referred to as the **stability constant** of the complex ion.

Determination of Instability Constant.—Two methods have been mainly used for determining the instability constants of complex ions; one involves the measurement of the E.M.F.'s of suitable cells, which will be described in Chap. VII, and the other depends on solubility studies. The latter may be illustrated by reference to the silver-ammonia (argentammine) complex ion.[26] If the formula of the complex is $Ag_m(NH_3)_n^+$, the

[26] See also, Edmonds and Birnbaum, *J. Am. Chem. Soc.*, 62, 2367 (1940); Lanford and Kiehl, *ibid.*, 63, 667 (1941).

instability constant, using concentrations as in equation (117), is given by

$$k_i = \frac{c^m_{Ag^+} c^n_{NH_3}}{c_X}, \quad (118)$$

where, for simplicity of representation, the concentration of the complex ion is given by c_X. If a solution of ammonia is saturated with silver chloride, then by the solubility product principle, $c_{Ag^+}c_{Cl^-}$ gives the solubility product k_s, and hence c_{Ag^+} is equal to k_s/c_{Cl^-}; for such a system equation (118) becomes

$$k_i = \frac{k^m_s c^n_{NH_3}}{c^m_{Cl^-} c_X}. \quad (119)$$

The concentration c of the silver salt in the ammonia solution may be regarded as consisting entirely of the complex ion, since the normal solubility of silver chloride is very small, so that c_X is virtually equal to c; the concentration of the chloride ion may be taken as mc because of the reaction

$$m\,AgCl + n\,NH_3 = Ag_m(NH_3)^+_n + m\,Cl^-,$$

and so equation (119) may be written as

$$k_i = \frac{k^m_s c^n_{NH_3}}{m^m c^{m+1}},$$

$$\therefore \quad \frac{c^n_{NH_3}}{c^{m+1}} = \text{constant}.$$

By means of this equation it is possible to evaluate $n/(m + 1)$ from a number of measurements of the solubility (c) of silver chloride in solutions containing various concentrations (c_{NH_3}) of ammonia.

In order to derive m it is necessary to determine the solubility of silver chloride in ammonia in the presence of an excess of chloride ions; equation (119) then takes the form

$$k_i = \frac{k^m_s c^n_{NH_3}}{c^m_{Cl^-} c}. \quad (120)$$

If in a series of experiments the concentration of ammonia (c_{NH_3}) is kept constant, while the amount of excess chloride (c_{Cl^-}) is varied, equation (120) becomes

$$c^m_{Cl^-} c = \text{constant},$$

so that if the solubility c is measured, the value of m may be determined. Alternatively, solubility measurements may be made in the presence of excess of silver ions; in this case c_{Cl^-} is set equal to k_s/c_{Ag^+} in equation (119), and the subsequent treatment is similar to that given above.*

* For data obtained in an actual experiment, see Problem 11.

Activity Coefficients from Solubility Measurements.—The activity coefficient of a sparingly soluble salt can be determined in the presence of other electrolytes by making use of the solubility product principle.[27] In addition to the equations already given, this principle may be stated in still another form by introducing the definition of the mean ionic concentration, i.e., $c_{\pm}^{\nu}$, which is equal to $c_{+}^{\nu_+}c_{-}^{\nu_-}$, into equation (109); this equation then becomes

$$c_{\pm}^{\nu}f_{\pm}^{\nu} = K_s, \tag{121}$$

$$\therefore \quad f_{\pm} = \frac{K_s^{1/\nu}}{c_{\pm}}. \tag{122}$$

The mean activity coefficient of a sparingly soluble salt in any solution could thus be evaluated provided the solubility product (K_s) and the mean concentration of the ions of the salt in the given solution were known. In order to calculate K_s the value of $c_{\pm}$ is determined in solutions of different ionic strengths and the results are then extrapolated to infinite dilution; under the latter conditions $f_{\pm}$ is, of course, unity and hence $K_s^{1/\nu}$ is equal to the extrapolated value of $c_{\pm}$.

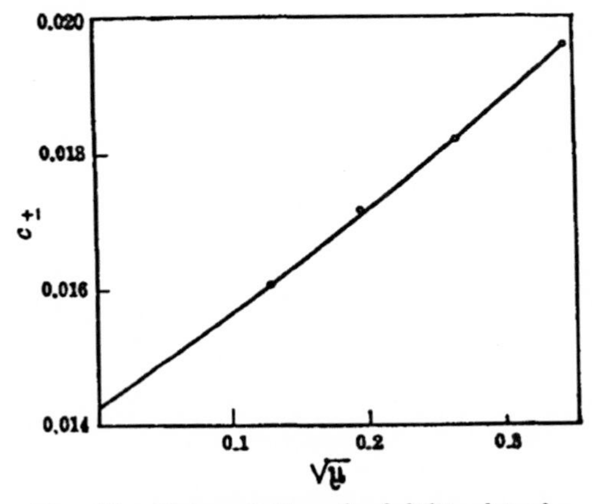

Fig. 58. Extrapolation of solubility data for thallous chloride

The method of calculation will be described with reference to thallous chloride, the solubility of which has been measured in the presence of various amounts of other electrolytes, with and without an ion in common with the saturating salt. By plotting the values of $c_{\pm}$ for the thallium and chloride ions in solutions of different ionic strengths and extrapolating to zero, it is found that $K_s^{1/\nu}$, which in this case is equal to $\sqrt{K_s}$, is 0.01428 at 25° (Fig. 58). It follows, therefore, from equation (122) that the mean activity coefficient of thallous chloride in any saturated solution is given by

$$f_{\pm} = \frac{0.01428}{c_{\pm}}.$$

If the added electrolyte present contains neither thallous nor chloride

[27] Lewis and Randall, *J. Am. Chem. Soc.*, **43**, 1112 (1921); see also, Blagden and Davies, *J. Chem. Soc.*, 949 (1930); Davies, *ibid.*, 2410, 2421 (1930); MacDougall and Hoffman, *J. Phys. Chem.*, **40**, 317 (1936); Pearce and Oelke, *ibid.*, **42**, 95 (1938); Kolthoff and Lingane, *ibid.*, **42**, 133 (1938).

ions, the mean ionic concentration is merely the same as the molar concentration of the thallous chloride in the saturated solution, for then c_{Tl^+} and c_{Cl^-} are both equal to the concentration of the salt. When another thallous salt or a chloride is present, however, appropriate allowance must be made for the ions introduced in this manner. For example, in a solution containing 0.025 mole of thallous sulfate per liter, the saturation solubility of thallous chloride is 0.00677 mole per liter at 25°; assuming both thallium salts to be completely dissociated at this low concentration, the total concentration of thallous ions is $2 \times 0.025 + 0.00677$, i.e., 0.05677 g.-ion per liter. The chloride ion concentration is 0.00677, and so the mean ionic concentration is $(0.05677 \times 0.00677)^{\frac{1}{2}}$, i.e., 0.01961; the mean activity coefficient is then 0.01428/0.01961, that is 0.728. The ionic strength of the solution is

$$\begin{aligned}\mu &= \tfrac{1}{2}[(c_{Tl^+} \times 1^2) + (c_{Cl^-} \times 1^2) + (c_{SO_4^{--}} \times 2^2)] \\ &= \tfrac{1}{2}(0.05677 + 0.00677 + 0.10) \\ &= 0.0817,\end{aligned}$$

so that the mean activity coefficient of a saturated solution of thallous chloride in the presence of thallous sulfate at a total ionic strength of 0.0817 is 0.728 at 25°.

The activity coefficients of thallous chloride at 25°, obtained in the manner described above, in the presence of a number of salts are given in Table XLI; the data are recorded for solutions of various (total) ionic

TABLE XLI. ACTIVITY COEFFICIENTS OF THALLOUS CHLORIDE IN THE PRESENCE OF VARIOUS ELECTROLYTES AT 25°

	Added Electrolyte				
μ	KNO_3	KCl	HCl	$TlNO_3$	Tl_2SO_4
0.02	0.872	0.871	0.871	0.869	0.885
0.05	0.809	0.797	0.798	0.784	0.726
0.10	0.742	0.715	0.718	0.686	0.643
0.20	0.676	0.613	0.630	0.546	—

strengths. It is seen that at low ionic strengths the activity coefficient of the thallous chloride at a given ionic strength is almost independent of the nature of the added electrolyte; it has been claimed that if allowance is made for incomplete dissociation of the latter this independence persists to much higher concentrations.

Solubility and the Debye-Hückel Theory.—The activity coefficients determined by the solubility method apply only to saturated solutions of the given salt in media of different ionic strengths; although their value is therefore limited, in many respects, they are of considerable interest as providing a means of testing the validity of the Debye-Hückel theory of electrolytes. It will be seen from equation (113), if the saturating salt can be assumed to be completely dissociated, that the product $Sf_{\pm}$, where S is the solubility of the given salt in a solution not containing an ion in common with it, must be constant. It follows, therefore, that if S_0 is the solubility of the salt in pure water and S the value in the presence of another electrolyte which has no ion in common with the salt, and f_0 and f are the corresponding mean activity coefficients, then

$$\frac{f_0}{f} = \frac{S}{S_0},$$

$$\therefore \quad \log \frac{S}{S_0} = \log f_0 - \log f.$$

Introducing the values of f and f_0, as given by the Debye-Hückel limiting law equation (54), it follows that

$$\log \frac{S}{S_0} = Az_+z_-(\sqrt{\mu} - \sqrt{\mu_0}), \tag{123}$$

where μ_0 and μ are the ionic strengths of the solutions containing the sparingly soluble salt only and that to which other electrolytes have been added, respectively. Since μ_0 is a constant for a given saturating salt, it follows that the plot of log S/S_0 against $\sqrt{\mu}$ should be a straight line of slope Az_+z_-, where z_+ and z_- are the valences of the two ions of the sparingly soluble substance. The constant A for water at 25° is 0.509, and so the linear slope in aqueous solutions should be 0.509 z_+z_-.

For the purpose of verifying the conclusions derived from the Debye-Hückel theory it is necessary to employ salts which are sufficiently soluble for their concentrations to be determined with accuracy, but not so soluble that the resulting solutions are too concentrated for the limiting law for activity coefficients to be applicable. A number of iodates, e.g., silver, thallous and barium iodates, and especially certain complex cobalt-ammines have been found to be particularly useful in this connection. The results, in general, are in very good agreement with the requirements of equation (123). The solubility measurements with the following four cobaltammines of different valence types, in the presence of such salts as sodium chloride, potassium nitrate, magnesium sulfate, barium

	Salt	Valence Type	Theoretical Slope
I.	$[Co(NH_3)_4(NO_2)(CNS)][Co(NH_3)_2(NO_2)_2(C_2O_4)]$	1 : 1	0.509
II.	$[Co(NH_3)_4(C_2O_4)]_2S_2O_6$	1 : 2	1.018
III.	$[Co(NH_3)_6][Co(NH_3)_2(NO_2)_2(C_2O_4)]_3$	3 : 1	1.527
IV.	$[Co(NH_3)_6][Fe(CN)_6]$	3 : 3	4.581

chloride and potassium cobalticyanide, are of particular interest.[28] The values of log S/S_0 are plotted against the square-root of the ionic strength in Fig. 59; the experimental data are shown by the points and the theoretical slopes are indicated by the full lines in each case. In certain cases the agreement with theory is not as good as depicted in

[28] Brønsted and LaMer, *J. Am. Chem. Soc.*, 46, 555 (1924); LaMer, King and Mason, *ibid.*, 49, 363 (1927).

Fig. 59; this is particularly true if both the saturating salt and the added electrolyte are of high valence types.[29] The deviations are often due to incomplete dissociation, and also to the approximations made in the derivation of the Debye-Hückel equations; as already seen, both these factors become of importance with ions of high vaience.

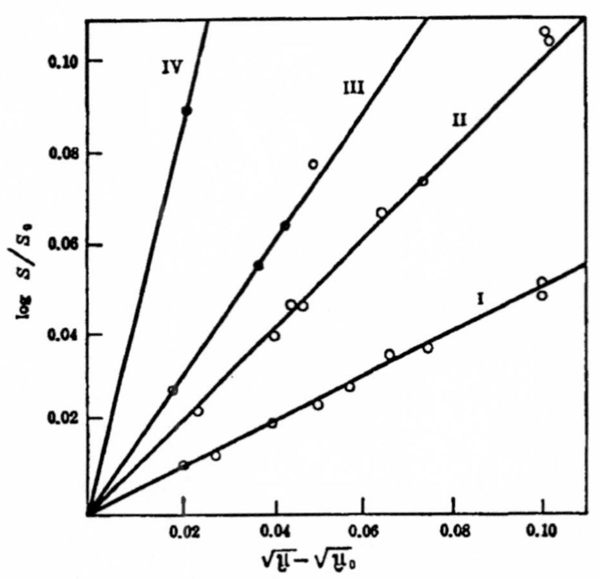

Fig. 59. Dependence of solubility on ionic strength (LaMer, *et al.*)

The factor A in equation (123) is proportional to $1/(DT)^{\frac{3}{2}}$, as shown on page 150; hence, a further test of this equation is to determine the slope of the plot of log S/S_0 against $\sqrt{\mu}$ from solubility data at different temperatures and in media of different dielectric constants. Such measurements have been made in water at 75° (D = 63.7), in mixtures of water and ethyl alcohol (D = 33.8 to 78.6), in methyl alcohol (D = 30), in acetone (D = 21), and in ethylene chloride (D = 10.4). The results have been found in all cases to be in very fair agreement with the requirements of the Debye-Hückel limiting law; as may be expected, appreciable discrepancies occur when the saturating salt is of a high valence type, especially in the presence of added ions of high valence.[30]

[29] LaMer and Cook, *J. Am. Chem. Soc.*, 51, 2622 (1929); LaMer and Goldman, *ibid.*, 51, 2632 (1929); Neuman, *ibid.*, 54, 2195 (1933).

[30] Baxter, *J. Am. Chem. Soc.*, 48, 626 (1926); Williams, *ibid.*, 51, 1112 (1929); Hansen and Williams, *ibid.*, 52, 2759 (1930); Scholl, Hutchison and Chandlee, *ibid.*, 55, 3081 (1933); Seward, *ibid.*, 56, 2610 (1934); see, however, Anhorn and Hunt, *J. Phys. Chem.*, 45, 351 (1941).

Thermal Properties of Strong Electrolytes.—According to equation (42) the free energy of an ionic solution may be expressed in the form

$$G = G_0 + G_{el.}$$

and application of the Gibbs-Helmholtz equation (cf. p. 194) gives

$$H = \left[G_0 - T\left(\frac{\partial G_0}{\partial T}\right)_p \right] + \left[G_{el.} - T\left(\frac{\partial G_{el.}}{\partial T}\right)_p \right], \tag{124}$$

where H is the heat content of a solution of an electrolyte at an appreciable concentration. At infinite dilution the quantity in the second brackets on the right-hand side is zero, since the electrical contribution to the free energy is then zero; the heat content of the solution under these conditions is consequently equal to the quantity in the first brackets. It follows, therefore, that the increase of heat content accompanying the dilution of a solution of an electrolyte from a concentration c to infinite dilution, i.e., $\Delta H_{c\to 0}$, which is the corresponding integral heat of dilution, is given by

$$-\Delta H_{c\to 0} = G_{el.} - T\left(\frac{\partial G_{el.}}{\partial T}\right)_p. \tag{125}$$

Utilizing the value of $G_{el.}$, equal to $W - W_0$ given by equation (41), and remembering that κ involves $T^{-\frac{1}{2}}$, it is found that

$$\Delta H_{c\to 0} = \sum_i \frac{N_i z_i^2 \epsilon^2 \kappa}{2D}\left(1 + \frac{T}{D}\cdot\frac{\partial D}{\partial T} + \frac{T}{3V}\cdot\frac{\partial V}{\partial T}\right), \tag{126}$$

where V is the volume of the system; $\partial D/\partial T$ and $\partial V/\partial T$ refer to constant pressure. Since the heat of dilution is generally recorded for a mole of electrolyte, it follows that N_i is equal to $N\nu_i$ where N is the Avogadro number and ν_i is the number of ions of the ith kind produced by the ionization of a molecule of electrolyte. The expression $\Sigma N_i z_i^2$ in equation (126) may therefore be replaced by $N\Sigma\nu_i z_i^2$, and the result is

$$\Delta H_{c\to 0} = \frac{N\epsilon^2\kappa}{2D}\Sigma\nu_i z_i^2 f(T, D, V), \tag{127}$$

where $f(T, D, V)$ is the function included in the parentheses in equation (126).

The concentration c_i of any ionic species is equal to $\nu_i c$, where c is the concentration of the electrolyte in moles per liter; hence, the ionic strength may be written in an alternative form, thus

$$\mu = \tfrac{1}{2}\Sigma c_i z_i^2 = \tfrac{1}{2}c\Sigma\nu_i z_i^2.$$

It follows, therefore, using equation (51) to define κ, that

$$\Delta H_{c\to 0} = \frac{N^2\epsilon^3\pi^{\frac{1}{2}}}{D^{\frac{3}{2}}(RT)^{\frac{1}{2}}1000^{\frac{1}{2}}}(\Sigma\nu_i z_i^2)^{\frac{3}{2}}\sqrt{c}\, f(T, D, V)$$

$$= \frac{5.91 \times 10^6}{D^{\frac{3}{2}}T^{\frac{1}{2}}}(\Sigma\nu_i z_i^2)^{\frac{3}{2}}\sqrt{c}\, f(T, D, V) \text{ cal. per mole.}$$

For water at 25° this can be written as

$$\Delta H_{c\to 0} = 503(\Sigma\nu_i z_i^2)^{\frac{3}{2}} \sqrt{c}\, f(T, D, V) \text{ cal. per mole.} \qquad (128)$$

The temperature coefficient of the dielectric constant of water is not known with great accuracy, but utilizing the best data to evaluate $f(T, D, V)$, equation (128) becomes, approximately,

$$\Delta H_{c\to 0} = -175(\Sigma\nu_i z_i^2)^{\frac{3}{2}} \sqrt{c},$$

and for a uni-univalent electrolyte at 25°, i.e., $z_+ = z_- = 1$, and $\nu_+ = \nu_- = 1$,

$$\Delta H_{c\to 0} = -495 \sqrt{c} \text{ cal. per mole.}$$

It is seen from these equations that there should be a *negative* increase of heat content when an electrolyte solution is diluted; in other words, the theory of interionic attraction requires that heat should be *evolved* when a solution of an electrolyte is diluted.[31] Further, the integral heat of dilution should be proportional to the square-root of the concentration, the slope of the plot of $\Delta H_{c\to 0}$ against $\sqrt{c}$ should be about -500 for an aqueous solution of a uni-univalent electrolyte at 25°. It must be emphasized that the foregoing treatment presupposes a dilute solution, and in fact the slope mentioned should be the limiting value which is approached at infinite dilution. Accurate measurements of integral heats of dilution are difficult to make, but the careful work of the most recent investigators has given results in general agreement with theoretical expectation. The integral heat of dilution is actually negative for dilute solutions, but at appreciable concentrations it becomes positive, so that heat is then absorbed when the solution is diluted.

The limiting slope of the plot of $\Delta H_{c\to 0}$ against $\sqrt{c}$ has been found to be approximately -500 for a number of uni-univalent electrolytes; the larger the effective size of the ion in solution, the closer the agreement between experiment and the requirements of the interionic attraction theory. By making allowance for the effective ionic diameter, either by the Debye-Hückel method or by utilizing the treatment of Gronwall, LaMer and Sandved, fairly good agreement is obtained at appreciable concentrations.[32]

According to equation (128) the limiting slope of the plot of $\Delta H_{c\to 0}$ against $\sqrt{c}$ for different electrolytes should vary in proportion to the factor $(\Sigma\nu_i z_i^2)^{\frac{3}{2}}$; the results obtained with a number of uni-bivalent and

[31] Bjerrum, *Z. physik. Chem.*, 119, 145 (1926); Gatty, *Phil. Mag.*, 11, 1082 (1931); 18, 46 (1934); Scatchard, *J. Am. Chem. Soc.*, 53, 2037 (1931); Falkenhagen, "Electrolytes" (translated by Bell), 1934.

[32] For summaries, with references, see Lange and Robinson, *Chem. Revs.*, 9, 89 (1931); Falkenhagen, "Electrolytes," 1934; Wolfenden, *Ann. Rep. Chem. Soc.*, 29, 29 (1932); Bell, *ibid.*, 31, 58 (1934); for more recent work, see Robinson *et al.*, *J. Am. Chem. Soc.*, 56, 2312, 2637 (1934); 63, 958 (1941); Sturtevant, *ibid.*, 62, 2171 (1940).

bi-univalent electrolytes are in harmony with this requirement of theory. In spite of the general agreement, the experimental data for integral heats of dilution, especially in non-aqueous solutions, show some discrepancies from the behavior postulated by the interionic attraction theory. It should be noted, however, that heat of dilution measurements provide an exceptionally stringent test of the theory, and the influence of such factors as ionic size, incomplete dissociation and ion-solvent interaction will produce relatively larger effects than is the case with activity coefficients.

PROBLEMS

1. The density of a 0.1 N solution of KI in ethyl alcohol at 17° is 0.8014 while that of the pure solvent is 0.7919; calculate the ratio of the three activity coefficients, f_x, f_c and f_m, in the solution.

2. Compare the molalities and ionic strengths of uni-uni, uni-bi, bi-bi and uni-tervalent electrolytes in solutions of molality m.

3. Use the values of the Debye-Hückel constants A and B at 25°, given in Table XXXV, to plot $-\log f_{\pm}$ for a uni-univalent electrolyte against $\sqrt{\mu}$ for ionic strengths 0.01, 0.1, 0.5 and 1.0, assuming in turn that the mean distance of approach of the ions, a, is either zero, or 1, 2, 4 and 8Å. Investigate, qualitatively, the effect of increasing the valence of the ions.

4. Evaluate the Debye-Hückel constants A and B for ethyl alcohol at 25°, taking the dielectric constant to be 24.3.

5. Utilize the results of the preceding problem, together with the known values of A and B for water, to calculate approximate activity coefficients for uni-uni, uni-bi, and bi-bivalent electrolytes in water and in ethyl alcohol, at ionic strengths 0.1 and 0.01, at 25°. The mean ionic diameter may be taken as 3Å in each case.

6. The following values for the mean activity coefficients of potassium chloride were obtained by MacInnes and Shedlovsky [*J. Am. Chem. Soc.*, **59**, 503 (1937)]:

c	f	c	f
0.005	0.9274	0.04	0.8320
0.01	0.9024	0.06	0.8070
0.02	0.8702	0.08	0.7872
0.03	0.8492	0.10	0.7718

Plot $\sqrt{\mu}/\log f$ against $\sqrt{\mu}$ and determine the value of a which is in satisfactory agreement with these data.

7. Kolthoff and Lingane [*J. Phys. Chem.*, **42**, 133 (1938)] determined the solubility of silver iodate in water and in the presence of various concentrations of potassium nitrate at 25°. The solubility in pure water is 1.771×10^{-4} mole per liter, and the following results were obtained in potassium nitrate solutions:

KNO_3 mole/liter	$AgIO_3$ mole/liter	KNO_3 mole/liter	$AgIO_3$ mole/liter
0.1301×10^{-3}	1.823×10^{-4}	1.410×10^{-3}	1.999×10^{-4}
0.3252	1.870	7.050	2.301
0.6503	1.914	19.98	2.665

Calculate the activity coefficients of the silver iodate in the various solutions; plot the values of $-\log f$ against $\sqrt{\mu}$ to see how far the results agree with the Debye-Hückel limiting law. Determine the mean ionic diameter required to account for the deviations from the law at appreciable concentrations.

8. Utilize the results obtained from the data of Saxton and Waters, given in Problem 7 of Chap. III, together with the activity coefficients derived from the Debye-Hückel limiting equation, to evaluate the dissociation constant of α-crotonic acid.

9. Apply the method of Fuoss and Kraus, described on page 167, to evaluate Λ_0 and K for hydrochloric acid in a dioxane-water mixture, containing 70 per cent of the former, at 25°, utilizing the conductance data obtained by Owen and Waters [*J. Am. Chem. Soc.*, **60**, 2371 (1938)]:

$\sqrt{c} \times 10^2$	1.160	2.037	2.420	2.888	3.919
Λ	89.14	85.20	83.26	81.45	77.20 ohms^{-1} cm.^{2}

The dielectric constant of the solvent is 17.7 and its viscosity is 0.0192 poise. The required values of the function $F(x)$ will be found in the paper by Fuoss, *J. Am. Chem. Soc.*, **57**, 488 (1935).

10. By means of the value of K obtained in the preceding problem, calculate the mean ionic diameter, a, of hydrochloric acid in the given solvent. For this purpose, use equation (79) and the tabulation of $Q(b)$ given by Fuoss and Kraus, *J. Am. Chem. Soc.*, **55**, 1019 (1933).

11. In order to determine the formula of the complex argentammine ion, $Ag_q(NH_3)_r^+$, Bodländer and Fittig [*Z. physik. Chem.*, **39**, 597 (1902)] measured the solubility (S) of silver chloride in ammonia solution at various concentrations (c_{NH_3}) with the following results:

c_{NH_3}	0.1006	0.2084	0.2947	0.4881
$S \times 10^3$	5.164	11.37	15.88	25.58

In the presence of various concentrations (c_{KCl}) of potassium chloride, the solubility (S) of silver chloride in 0.75 molal ammonia was as follows:

c_{KCl}	0.0102	0.0255	0.0511
S	0.0439	0.0387	0.0333

What is the formula of the silver-ammonia ion?

CHAPTER VI

REVERSIBLE CELLS

Chemical Cells and Concentration Cells.—A **voltaic cell,** or element, as it is sometimes called, consists essentially of two electrodes combined in such a manner that when they are connected by a conducting material, e.g., a metallic wire, an electric current will flow. Each electrode, in general, involves an electronic and an electrolytic conductor in contact (cf. p. 6); at the surface of separation between these two phases there exists a potential difference, called the **electrode potential.** If there are no other potential differences in the cell, the E.M.F. of the latter is taken as equal to the *algebraic* sum of the two electrode potentials, allowance being made for the direction of the potential difference when assessing its sign. During the operation of a voltaic cell a chemical reaction takes place at each electrode, and it is the energy of these reactions which provides the electrical energy of the cell. In many cells there is an overall chemical reaction, when all the processes occurring within it are taken into consideration; such a cell is referred to as a **chemical cell,** to distinguish it from a voltaic element in which there is no *resultant* chemical change. In the latter type of cell the reaction occurring at one electrode is exactly reversed at the other; there may, nevertheless, be a net change of energy because of a difference in concentration of one or other of the reactants concerned at the two electrodes. Such a source of E.M.F. is called a **concentration cell,** and the electrical energy arises from the energy change accompanying the transfer of material from one concentration to another.

Irreversible and Reversible Cells.—Apart from the differences mentioned above, voltaic cells may, broadly speaking, be divided into two categories depending on whether a chemical reaction takes place at either electrode even when there is no flow of current, or whether there is no reaction until the electrodes are joined together by a conductor and current flows. An illustration of the former type is the simple cell consisting of zinc and copper electrodes immersed in dilute sulfuric acid, viz.,

$$Zn \mid \text{Dilute } H_2SO_4 \mid Cu;$$

the zinc electrode reacts with the acid spontaneously, even if there is no passage of current. Cells of this type are always **irreversible** in the thermodynamic sense; thermodynamic reversibility implies a state of equilibrium at every stage, and the occurrence of a spontaneous reaction at the electrodes shows that the system is not in equilibrium.

In the Daniell cell, however, which is made up of a zinc electrode in zinc sulfate solution and a copper electrode in copper sulfate solution, viz.,

$$\mathrm{Zn} \mid \mathrm{ZnSO_4}\ \mathrm{soln.} \qquad \mathrm{CuSO_4}\ \mathrm{soln.} \mid \mathrm{Cu},$$

the two solutions being usually separated by means of a porous partition, neither metal is attacked until the electrodes are connected and a current is allowed to flow. The extent of the chemical reaction occurring in such a cell is proportional to the quantity of electricity passing, in accordance with the requirements of Faraday's laws. Many, although not necessarily all, cells in this second category are, however, thermodynamically **reversible cells,** and the test of reversibility is as follows. If the cell under consideration is connected to an external source of E.M.F. which is adjusted so as exactly to balance the E.M.F. of the cell, i.e., so that no current flows, there should be no chemical change in the cell. If the external E.M.F. is decreased by an infinitesimally small amount, current will flow from the cell and a chemical change, proportional in extent to the quantity of electricity passing, should take place. On the other hand, if the external E.M.F. is increased by a small amount, the current should pass in the opposite direction and the cell reaction should be exactly reversed. The Daniell cell, mentioned above, satisfies these requirements and it is consequently a reversible cell. It should be noted that voltaic cells can only be expected to behave reversibly when the currents passing are infinitesimally small and the system is always virtually in equilibrium. If large currents flow, concentration gradients arise on account of diffusion being relatively slow, and the cell can no longer be regarded as being in a state of equilibrium.

Reversible Electrodes.—The electrodes constituting a reversible cell must themselves be reversible, and several types of such electrodes are known. The simplest, sometimes called "electrodes of the first kind," consist of a metal in contact with a solution of its own ions, e.g., zinc in zinc sulfate solution. In this category may be included hydrogen, oxygen and halogen electrodes in contact with solutions of hydrogen, hydroxyl or the appropriate halide ions, respectively; since the electrode material in these latter cases is a non-conductor, and often gaseous, finely divided platinum, or other unattackable metal, which comes rapidly into equilibrium with the hydrogen, oxygen, etc., is employed for the purpose of making electrical contact. Electrodes of the first kind are reversible with respect to the ions of the electrode material, e.g., metal, hydrogen, oxygen or halogen; the reaction occurring if the electrode material is a metal M may be represented by

$$\mathrm{M} \rightleftharpoons \mathrm{M}^+ + \epsilon,$$

the direction of the reaction depending on the direction of the flow of current. If the electrode is that of a non-metal, the corresponding reactions are

$$\mathrm{A} + \epsilon \rightleftharpoons \mathrm{A}^-.$$

With an oxygen electrode, which is theoretically reversible with respect to hydroxyl ions, the reaction may be written

$$\tfrac{1}{2}O_2 + H_2O + 2\epsilon \rightleftharpoons 2OH^-.$$

Electrodes of the "second kind" involve a metal, a sparingly soluble salt of this metal, and a solution of a soluble salt of the same anion; a familiar example is the silver-silver chloride electrode consisting of silver, solid silver chloride and a solution of a soluble chloride, such as hydrochloric acid, viz.,

$$Ag \mid AgCl(s) \;\; HCl \text{ soln.}$$

These electrodes behave as if they were reversible with respect to the common anion, e.g., the chloride ion in the above electrode. The electrode reaction involves the passage of the electrode metal into solution as ions and their combination with the anions of the electrolyte to form the insoluble salt, or the reverse of these stages; thus, for the silver-silver chloride electrode,

$$Ag(s) \rightleftharpoons Ag^+ + \epsilon,$$

followed by

$$Ag^+ + Cl^- \rightleftharpoons AgCl(s),$$

so that the net reaction, writing it for convenience in the reverse order, is

$$AgCl(s) + \epsilon \rightleftharpoons Ag(s) + Cl^-.$$

This is virtually equivalent to the reaction at a chlorine gas electrode, viz.,

$$\tfrac{1}{2}Cl_2 + \epsilon \rightleftharpoons 2Cl^-,$$

except that the silver chloride can be regarded as the source of the chlorine. In fact the silver-silver chloride electrode is thermodynamically equivalent to a chlorine electrode with the chlorine at a pressure equal to the dissociation pressure of the silver chloride, into silver and chlorine, at the experimental temperature. Electrodes of the second kind are of great value in electrochemistry because they permit the ready establishment of an electrode reversible with respect to anions, e.g., sulfate, oxalate, etc., which could not be obtained in a direct manner. Even where it is possible, theoretically, to set up the electrode directly, as in the case of the halogens, it is more convenient, and advantageous in other ways, to employ an electrode of the second kind.

Occasionally electrodes of the "third kind" are encountered;[1] these consist of a metal, one of its insoluble salts, another insoluble salt of the same anion, and a solution of a soluble salt having the same cation as the latter salt, e.g.,

$$Pb \mid PbC_2O_4(s) \;\; CaC_2O_4(s) \;\; CaCl_2 \text{ soln.}$$

[1] Corten and Estermann, *Z. physik. Chem.*, **136**, 228 (1928); LeBlanc and Harnapp, *ibid.*, **166A**, 321 (1933); Joseph, *J. Biol. Chem.*, **130**, 203 (1939).

In this case the lead first dissolves to form lead ions, which combine with $C_2O_4^{--}$ ions to form insoluble lead oxalate, thus

$$Pb \rightleftharpoons Pb^{++} + 2\epsilon$$

and

$$Pb^{++} + C_2O_4^{--} \rightleftharpoons PbC_2O_4(s).$$

The removal of the oxalate ions from the solution causes the calcium oxalate to dissolve and ionize in order that its solubility product may be maintained; thus

$$CaC_2O_4(s) \rightleftharpoons Ca^{++} + C_2O_4^{--},$$

so that the net reaction is

$$Pb(s) + CaC_2O_4(s) \rightleftharpoons PbC_2O_4(s) + Ca^{++} + 2\epsilon.$$

The system thus behaves as an electrode reversible with respect to calcium ions. This result is of great interest since a reversible calcium electrode employing metallic calcium is difficult to realize experimentally.

Another type of reversible electrode involves an unattackable metal, such as gold or platinum, immersed in a solution containing an appropriate oxidized and reduced form of an oxidation-reduction system, e.g., Sn^{++++} and Sn^{++}, or $Fe(CN)_6^{---}$ and $Fe(CN)_6^{----}$; the metal merely acts as a conductor for making electrical contact, just as in the case of a gas electrode. The reaction at an **oxidation-reduction electrode** of this kind is either oxidation of the reduced state or reduction of the oxidized state, e.g.,

$$Sn^{++} \rightleftharpoons Sn^{++++} + 2\epsilon,$$

depending on the direction of the current. In order that it may behave reversibly, the reaction being capable of occurring in either direction, a reversible oxidation-reduction system must contain *both* oxidized and reduced states. It is important to point out that there is no essential difference between an oxidation-reduction electrode and one of the first kind described above; for example, in a system consisting of a metal M and its ions M^+, the former is the reduced state and the latter the oxidized state. Similarly the case of an anion electrode, e.g., chlorine-chloride ions, the anion is the reduced state and the uncharged material, e.g., chlorine, is the oxidized state. In all these instances the electrode process may be written in the general form:

$$\text{Reduced State} \rightleftharpoons \text{Oxidized State} + n\epsilon,$$

where n is the number of electrons by which the oxidized and reduced states differ. It is a matter of convenience, however, to treat separately electrodes involving oxidation-reduction systems in the specialized sense of the terms oxidation and reduction.

Direction of Current Flow and Sign of Reversible Cell.—The combination of two reversible electrodes in a suitable manner will give a reversible cell; in this cell the reaction at one electrode is such that it yields electrons while at the other electrode the reaction removes electrons. The electrons are carried from the former electrode to the latter by the metallic conductor which connects them. The ability to supply or remove electrons is possessed by all reversible electrodes, as is evident from the discussion given above; the particular function which is manifest at any time, i.e., supplying or removing electrons, depends on the direction of the current flow, and this is determined by the nature of the two electrodes combined to form the cell. The electrode Ag, AgCl(*s*) KCl soln., for example, acts as a remover of electrons when combined with Zn, $ZnSO_4$ soln., but it is a source of electrons in the cell obtained by coupling it with the Ag, $AgNO_3$ soln. electrode. Since it is not always possible to say *a priori* in which direction the current in a given cell will flow when the electrodes are connected by an external conductor, it is necessary to adopt a convention for describing the E.M.F. and the reaction occurring in a reversible cell. The convention most frequently employed by physical chemists in America is based on that proposed by Lewis and Randall; it may be stated as follows.

> The E.M.F., including the sign, represents the tendency for positive ions to pass spontaneously through the cell as written from left to right, or of negative ions to pass from right to left.

Since a positive E.M.F. means the passage of positive ions through the cell from left to right, it can be readily seen that electrons must pass through the external conductor in the same direction (cf. Fig. 2). It follows, therefore, that when the E.M.F. of the cell is positive the left-hand electrode acts as a source of electrons while the right-hand electrode removes them; if the E.M.F. is negative, the reverse is true. When expressing the complete chemical reaction occurring in a cell the convention will be adopted of supposing that the condition is the one just derived for a positive E.M.F.*

Reactions in Reversible Cells.—It is of importance in many respects to know what is the reaction occurring in a reversible cell, and some different types of cells will be considered for the purpose of illustrating the procedure adopted in determining the cell reaction. The Daniell cell, for example, is

$$Zn \mid ZnSO_4 \text{ aq.} \vdots CuSO_4 \text{ aq.} \mid Cu,$$

and taking the left-hand electrode as the electron source, i.e., the E.M.F. as stated is positive, the reaction here is

$$Zn = Zn^{++} + 2\epsilon,$$

* Many physical chemists in Europe and practical electrochemists in America use a convention as to the sign of E.M.F. and electrode potential which is the opposite of that employed here.

while at the right-hand electrode the electrons are removed by the process

$$Cu^{++} + 2\epsilon = Cu.$$

The complete reaction is thus

$$Zn + Cu^{++} = Zn^{++} + Cu,$$

and since two electrons are involved in each atomic act, the whole reaction as written, with quantities in gram-atoms or gram-ions, takes place for the passage of two faradays of electricity through the cell (cf. p. 27). Since the cupric ions originate from copper sulfate and the zinc ions form part of zinc sulfate, the reaction is sometimes written as

$$Zn + CuSO_4 = ZnSO_4 + Cu.$$

The E.M.F. of the cell depends on the concentrations of the zinc and cupric ions, respectively, in the two solutions, and so if the cell reaction is to be expressed more precisely, as is frequently necessary, the concentration of the electrolyte should be stated; thus

$$Zn + CuSO_4(m_1) = ZnSO_4(m_2) + Cu,$$

where m_1 and m_2 are the molalities of the copper sulfate and zinc sulfate, respectively, in the Daniell cell.

In the cell

$$Zn \mid ZnSO_4 \text{ aq.} \vdots \text{ KCl aq. } AgCl(s) \mid Ag,$$

the left-hand electrode reaction is the same as above, i.e.,

$$Zn = Zn^{++} + 2\epsilon,$$

while at the right-hand electrode the removal of electrons occurs by means of the process described on page 185, i.e.,

$$AgCl(s) = Ag^+ + Cl^-$$

and

$$Ag^+ + \epsilon = Ag,$$

the net reaction being

$$AgCl(s) + \epsilon = Ag + Cl^-.$$

The complete cell reaction for the passage of two faradays is thus

$$Zn + 2AgCl(s) = Zn^{++} + 2Cl^- + 2Ag,$$

or

$$Zn + 2AgCl(s) = ZnCl_2 + 2Ag.$$

A special case of this type of cell arises when both electrodes are of the same metal, viz.,

$$Ag \mid AgCl(s) \text{ KCl aq.} \vdots AgNO_3 \text{ aq.} \mid Ag.$$

By convention, the reaction at the left-hand electrode is the opposite of that at the right-hand electrode of the previous cell, viz.,

$$Ag + Cl^- = AgCl(s) + \epsilon,$$

and at the right-hand electrode the reaction is

$$Ag^+ + \epsilon = Ag,$$

so that the net reaction in the cell is

$$Ag^+ + Cl^- = AgCl(s)$$

for the passage of one faraday.

Another type of cell in which the two electrodes are constituted of the same material is one involving two hydrogen gas electrodes, viz.,

$$H_2 \mid NaOH\ aq. \vdots HCl\ aq. \mid H_2.$$

If the E.M.F. is positive, the hydrogen passes into solution as ions at the left-hand electrode, i.e.,

$$\tfrac{1}{2}H_2(g) = H^+ + \epsilon,^*$$

but the hydrogen ions react immediately with the hydroxyl ions in the alkaline solution, viz.,

$$H^+ + OH^- = H_2O,$$

to form water. At the right-hand electrode electrons are removed by the discharge of hydrogen ions, thus

$$H^+ \rightleftharpoons \tfrac{1}{2}H_2(g) + \epsilon,$$

so that the net reaction for the passage of one faraday is

$$H^+ + OH^- = H_2O,$$

i.e., the neutralization of hydrogen ions by hydroxyl ions. Since the hydrogen ions are derived from hydrochloric acid and the hydroxyl ions from sodium hydroxide, the reaction can also be written (cf. p. 12) as

$$HCl + NaOH = NaCl + H_2O.$$

Measurement of E.M.F.—The principle generally employed in the measurement of the E.M.F.'s of voltaic cells is that embodied in the **Poggendorff compensation method**; it has the advantage of giving the E.M.F. of the cell on "open circuit," i.e., when it is producing no current. It has been already mentioned that a cell can be expected to behave reversibly only when it is producing an infinitesimally small current, and hence the condition of open circuit is the ideal one for determining the reversible E.M.F.

* The hydrogen ion in aqueous solution is probably $(H_2O)H^+$, i.e., H_3O^+, and not H^+ (cf. p. 308); this does not, however, affect the general nature of the results recorded here.

The **potentiometer**, as the apparatus for measuring E.M.F.'s is called, is shown schematically in Fig. 60; it consists of a working cell C, generally a storage battery, of constant E.M.F. which must be larger than that of the cell to be measured, connected across the ends of a uniform con-

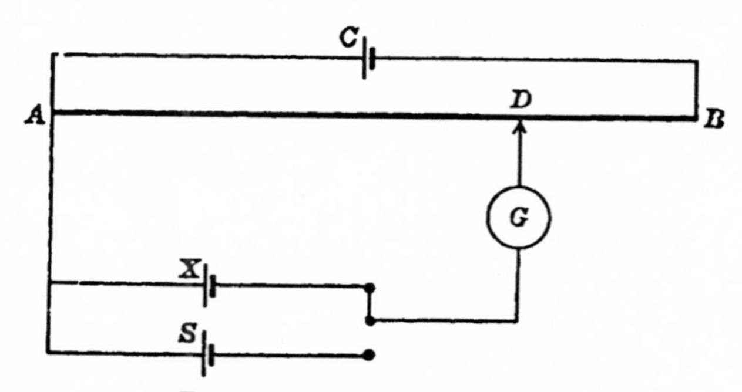

FIG. 60. Measurement of E.M.F.

ductor AB of high resistance. The cell X, which is being studied, is connected to A, with the poles in the same direction as the cell C, and then through a galvanometer G to a sliding contact D which can be moved along AB. The position of D is adjusted until no current flows through the galvanometer; the fall of potential between A and D due to the cell C is exactly compensated by the E.M.F. of X, that is E_X. By means of a suitable switch the cell X is now replaced by a standard cell S, of accurately known E.M.F. equal to E_S, and the sliding contact is re-adjusted until a point of balance is reached at D'. The fall of potential between A and D' is consequently equal to E_S, and since the conductor AB is supposed to be uniform, it follows that

$$\frac{E_X}{E_S} = \frac{AD}{AD'},$$

$$\therefore \quad E_X = \frac{AD}{AD'} \times E_S.$$

Since E_S is known, and AD and AD' can be measured, the E.M.F. of the unknown cell, E_X, can be evaluated.

In its simplest form, the conductor AB may consist of a straight, uniform potentiometer wire of platinum, platinum-iridium, or other resistant metal, stretched tightly along a meter scale; the position of the sliding contact can be read with an accuracy of about 0.5 mm., and if C is 2 volts and AB is 1 meter long, the corresponding error in the evaluation of the E.M.F. is 1 millivolt, i.e., 0.001 volt. Somewhat greater precision can be achieved if the potentiometer wire is several meters in length wound on a slate cylinder. For more accurate work the wire may be replaced by two calibrated resistance boxes; the contact D is fixed where the two boxes are joined, and the potential across AD is varied by changing the resistances in the boxes, keeping the total constant. If R_X is the resistance between A and D with the cell X in circuit, when no current flows through the galvanometer G, then the fall of potential which is equal to E_X must be proportional to R_X; * further, if R is the resistance at the balance point when the standard cell S replaces X, it follows that

$$\frac{E_X}{E_S} = \frac{R_X}{R_S},$$

$$\therefore \quad E_X = \frac{R_X}{R_S} \times E_S.$$

The unknown E.M.F. can thus be calculated from the two resistances. As a general rule the total resistance in the circuit is approximately 11,000 ohms, and hence if the working cell has an E.M.F. of 2 volts, each ohm resistance represents about 0.2 millivolt.

The majority of E.M.F. measurements are made at the present time by means of special potentiometers, operating on the Poggendorff principle, which are purchased from scientific instrument makers. They generally consist of a number of resistance coils with a movable contact, together with a slide wire for fine adjustment. A standard cell is used for calibration purposes, and the E.M.F. of the cell being measured can then be read off directly with an accuracy of 0.1 millivolt, or better.

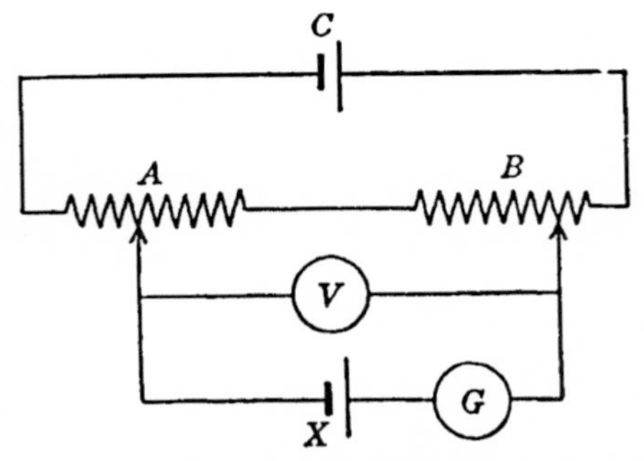

FIG. 61. Potentiometer-voltmeter arrangement

For approximate purposes, as in electroanalytical work or in potentiometric titrations, a simple procedure, known as the potentiometer-voltmeter method, can be employed. The working cell C (Fig. 61) is connected across two continuously variable resistances A and B, as shown; one of these resistances is for coarse and the other for fine adjustment. The experimental cell is placed at X in series with a galvanometer (G), and a millivoltmeter (V) is connected across the variable resistances. The latter are adjusted until no current flows through G; the voltage then indicated on V gives the E.M.F. of the cell.

Current Indicators.—The best form of current detector for accurate work is a suitably damped mirror galvanometer of high megohm sensitivity; for approximate purposes, however, a simple pointer galvanometer

* E_X is actually equal to $E_c \times R_X/R$, where E_c is the E.M.F. of the working cell C, and R is the total resistance of the two boxes in the circuit; since E_c and R are maintained constant, E_X is proportional to R_X.

is generally employed. At one time the capillary electrometer was widely used for the purpose of indicating the attainment of balance in the potentiometer circuit; it has the advantage of being unaffected by electrical and magnetic disturbances, and of not being damaged if large currents are inadvertently passed through it. On the other hand, the capillary electrometer is much less sensitive than most galvanometers and is liable to behave erratically in damp weather; for these and other reasons this form of detector has been discarded in recent years.

An ordinary mirror galvanometer of good quality can detect a current of about 10^{-7} amp., and hence if an accuracy of 0.1 millivolt is desired, as is the case in much work that is not of the highest precision, the resistance of the cell should not exceed 10^3 ohms. Special high-sensitivity galvanometers are available which show an observable deflection with a current of 10^{-11} amp., and so the E.M.F. of cells with resistances up to 10^7 ohms can be measured with their aid; the quadrant electrometer, which detects actually differences of potential rather than current, has also been used for the study of high resistance cells. Another procedure which has been devised is to employ a condenser in series with a ballistic galvanometer to determine the balance point of the potentiometer; the condenser is charged for a definite time by means of the cell being studied and is then discharged through the galvanometer with the aid of a suitable switch. When the potentiometer is balanced the ballistic galvanometer will undergo no deflection when the cell is discharged through it.

For most measurements of E.M.F. of cells of high resistance some form of vacuum-tube potentiometer has been used;[2] this instrument employs the amplifying properties of the vacuum tube, and the principle of operation may be illustrated by means of the simple circuit shown in Fig. 62. The tube is represented by T, and A, B and C indicate the filament, anode and grid batteries, respectively; R_1 and R_2 are variable resistances and G is a galvanometer. The cell X of unknown resistance is connected, as shown, to a potentiometer P from which any desired known voltage can be taken off; by means of the switch S the potentiometer and cell can be included, if required, in the grid circuit of the vacuum tube. The switch is first connected to b and the filament current is adjusted by means of R_1 to provide the optimum sensitivity of the tube; the "compensating current" from A, which passes in the opposite direction to the anode current through

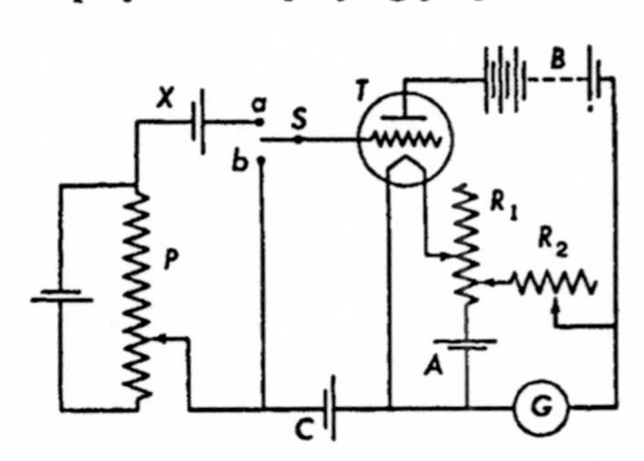

Fig. 62. Vacuum-tube potentiometer for cells of high resistance

[2] See, for example, Garman and Drosz, *Ind. Eng. Chem.* (*Anal. Ed.*), 11, 398 (1939); for review, see Glasstone, *Ann. Rep. Chem. Soc.*, 30, 283 (1933).

the galvanometer G, is then altered by means of the resistance R_2 so as to give a suitable reading on G. The switch S is now turned to a, so that P and X, as well as the battery C, are in the grid circuit; leaving R_1 and R_2 unchanged, the potentiometer is adjusted until the deflection on G is the same as before. The potential on the grid of the tube must, therefore, be the same in both cases: hence the E.M.F. taken from the potentiometer P must be equal and opposite to that of the cell X.

This simple type of vacuum-tube potentiometer is quite satisfactory for cells of not too high resistance, e.g., 10^7 ohms or less, but it is unreliable for still higher resistances. Two sources of error then arise: first, the characteristics of the vacuum tube change as a result of introducing the high resistance, so that a given anode current no longer corresponds to the same grid voltage; second, there is a fall of potential across the high resistance cell due to the flow of current in the grid circuit. With the best ordinary vacuum tubes the grid current may be about 10^{-10} amp., and so with a cell of resistance of 10^8 ohms, the error due to the fall of potential across the cell will be $10^{-10} \times 10^8$, i.e., 10^{-2} volt. Several methods of varying complexity have been devised in order to overcome these sources of error; one of the simplest and most effective, which is employed in commercial potentiometers for the measurement of the E.M.F.'s of cells involving the glass electrode (p. 356), is to use a special type of vacuum tube, known as an "electrometer tube." Although its amplification factor is generally smaller than that of the normal form of tube, the grid-circuit current is very small, 10^{-15} amp. or less, and the characteristics of the tube are not affected by high resistances.

The Standard Cell.—An essential feature of the Poggendorff method of measuring E.M.F.'s, and of all forms of apparatus employing the Poggendorff compensation principle, is a standard cell of accurately known E.M.F. The cell now invariably employed for this purpose is the **Weston standard cell**; it is highly reproducible, its E.M.F. remains constant over long periods of time, and it has a small temperature coefficient. One electrode of the cell is a 12.5 per cent cadmium amalgam in a saturated solution of cadmium sulfate ($3CdSO_4 \cdot 8H_2O$) and the other electrode consists of mercury and solid mercurous sulfate in the same solution, thus

$$12.5\% \text{ Cd in Hg} \mid 3CdSO_4 \cdot 8H_2O \text{ satd. soln. } Hg_2SO_4(s) \mid Hg.$$

The cell is set up in a H-shaped tube as shown in Fig. 63, the left-hand limb containing the cadmium amalgam and the right-hand the mercury; the amalgam is covered with crystals of $3CdSO_4 \cdot 8H_2O$, and the mercury with solid mercurous sulfate, and the whole cell is filled with a saturated solution of cadmium sulfate. The E.M.F. of the Weston cell, in international volts, over a range of temperatures is given by the expression

$$E_t = 1.018300 - 4.06 \times 10^{-5}(t - 20) - 9.5 \times 10^{-7}(t - 20)^2 + 1 \times 10^{-8}(t - 20)^3,$$

so that the value is 1.01830 volt at 20° and decreases about 4×10^{-2} millivolt per degree in this region.*

Although the so-called "saturated" Weston cell, containing a saturated solution of cadmium sulfate, is the ultimate standard for E.M.F. measurement, a secondary standard for general laboratory use has been recommended; this is the **"unsaturated" Weston cell,** which has an even smaller temperature coefficient than the saturated cell. The form of unsaturated cell generally employed contains a solution which has been saturated at 4° C., so that it is unsaturated at room temperatures; its temperature coefficient is so small as to be negligible for all ordinary purposes and its E.M.F. may be taken as 1.0186 volt.[3]

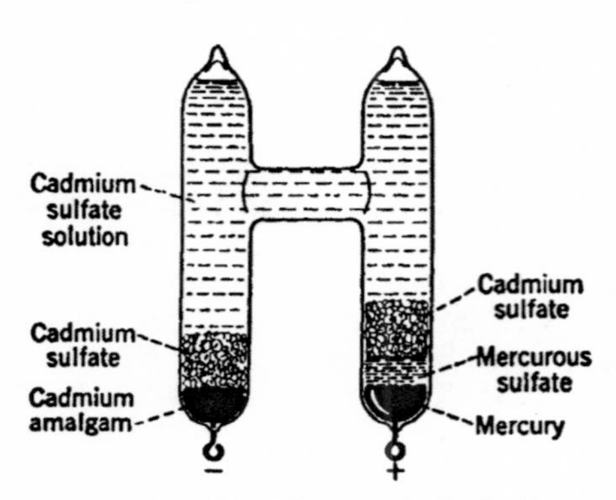

FIG. 63. The Weston standard cell

Free Energy and Heat Changes in Reversible Cells.—Since the quantitative consequences of the second law of thermodynamics are mainly applicable to reversible processes, the study of reversible cells is of particular importance because it is possible to apply thermodynamic methods to the results. If the E.M.F. of a voltaic cell is E volts, and the process taking place in it is accompanied by the passage of n faradays, i.e., nF coulombs, where F represents 96,500 coulombs, the work done by the system in the cell is nFE volt-coulombs or joules (cf. p. 5). If the cell is a reversible one, this work represents "maximum work," and since electrical work does not involve mechanical work resulting from a volume change, it may be taken as equal to the *change of free energy* accompanying the cell reaction. The increase of free energy of a process is equal to the reversible net work, i.e., excluding mechanical work, done *on* the system, and hence it follows that

$$\Delta G = -nFE, \tag{1}$$

where ΔG is the increase of free energy for the process taking place in the cell under consideration. According to the Gibbs-Helmholtz equation, which is derived from the second law of thermodynamics applied to reversible changes,

$$\Delta G = \Delta H + T\left(\frac{\partial(\Delta G)}{\partial T}\right)_p, \tag{2}$$

* It is important to note that the mercury electrode of a commercial Weston cell is always marked *positive*, while the cadmium amalgam electrode is marked *negative*.

[3] See Vinal, *Trans. Electrochem. Soc.*, **68**, 139 (1935).

where ΔH is the increase of heat content * for the cell reaction, and introducing equation (1), the result is

$$-nFE = \Delta H - nFT\left(\frac{\partial E}{\partial T}\right)_p, \tag{3}$$

$$\therefore \quad \Delta H = -nFE + nFT\left(\frac{\partial E}{\partial T}\right)_p$$

$$= -nF\left[E - T\left(\frac{\partial E}{\partial T}\right)_p\right]. \tag{4}$$

It is seen from equation (4) that if the E.M.F. of a reversible cell, i.e., E, and its temperature coefficient, $\partial E/\partial T$, at constant pressure are known, it is possible to evaluate the heat change of the reaction occurring in the cell.

Some of the results obtained in the calculation of heat content changes from E.M.F. measurements are recorded in Table XLII;[4] the values derived from thermochemical measurements are given in the last column for purposes of comparison. The agreement between the results for ΔH derived from E.M.F. measurements and from thermal data is seen to be satisfactory, especially when it is realized that an error of 1×10^{-5} in the temperature coefficient will mean an error of nearly 0.07 kcal. in ΔH at 298° K. It is probable, however, that the temperature coefficients are known with this degree of accuracy, and it is consequently believed that for many reactions the heat changes derived from E.M.F. data are more accurate than those obtained by direct thermal measurement.

TABLE XLII. HEAT CHANGES FROM E.M.F. MEASUREMENTS

Cell Reaction	E	$dE/dT \times 10^4$	ΔH kcal. E.M.F.	ΔH kcal. Thermal
$Zn + 2AgCl = ZnCl_2 + 2Ag$	1.015 (0°)	− 4.02	− 51.99	− 52.05
$Cd + PbCl_2 = CdCl_2 + Pb$	0.1880 (25°)	− 4.80	− 15.25	− 14.65
$Ag + \frac{1}{2}Hg_2Cl_2 = AgCl + Hg$	0.0455 (25°)	+ 3.38	+ 1.275	+ 1.90
$Pb + 2AgCl = PbCl_2 + 2Ag$	0.4900 (25°)	− 1.86	− 25.17	− 24.17

Concentration Cells: Cells without Transference.—In the operation of the cell

$$H_2(1 \text{ atm.}) \mid HCl \text{ aq.}(c) \; AgCl(s) \mid Ag,$$

consisting of a hydrogen and a silver-silver chloride electrode in hydrochloric acid,† the hydrogen at the left-hand electrode dissolves to form hydrogen ions, whereas at the right-hand electrode silver chloride passes into solution and silver is deposited; thus

$$\tfrac{1}{2}H_2(1 \text{ atm.}) = H^+ + \epsilon$$

* The increase of heat content is equal to the heat *absorbed* in the reaction at constant pressure.

[4] Taylor and Perrott, *J. Am. Chem. Soc.*, **43**, 486 (1921); Gerke, *ibid.*, **44**, 1684 (1922).

† The construction of these electrodes is described later (pp. 234, 350).

and

$$AgCl(s) + \epsilon = Ag + Cl^-,$$

so that the net reaction is represented by

$$\tfrac{1}{2}H_2(1 \text{ atm.}) + AgCl(s) = HCl(c) + Ag,$$

since the hydrogen and chloride ions are formed in hydrochloric acid solution of concentration c moles per liter. If two such cells containing hydrochloric acid at concentrations c_1 and c_2, and having E.M.F.'s of E_1 and E_2, respectively, are connected in opposition, the result is the cell

$$H_2(1 \text{ atm.}) \mid HCl(c_1) \; AgCl(s) \mid Ag \mid AgCl(s) \; HCl(c_2) \mid H_2(1 \text{ atm.}),$$

whose E.M.F. is equal to $E_1 - E_2$. The reaction in the left-hand cell for the passage of one faraday, as seen above, is

$$\tfrac{1}{2}H_2(1 \text{ atm.}) + AgCl(s) = HCl(c_1) + Ag,$$

and that in the right-hand cell is the reverse of this, i.e.,

$$HCl(c_2) + Ag = \tfrac{1}{2}H_2(1 \text{ atm.}) + AgCl(s).$$

The net result of the passage of a faraday of electricity through the complete cell is the transfer (i) of hydrogen gas at 1 atm. pressure from the extreme left-hand to the extreme right-hand electrode, (ii) of solid silver chloride from left to right, and (iii) of hydrochloric acid from concentration c_2 to c_1. Since the chemical potentials of the hydrogen gas and solid silver chloride remain unchanged, the free energy change ΔG of the cell reaction is due only to that accompanying the removal of 1 mole of hydrochloric acid, i.e., 1 g.-ion of hydrogen ions and 1 g.-ion of chloride ions, from the solution of concentration c_2 and its addition to c_1. It follows, therefore, that

$$\Delta G = [(\mu_{H^+})_1 - (\mu_{H^+})_2] + [(\mu_{Cl^-})_1 - (\mu_{Cl^-})_2], \tag{5}$$

where μ_{H^+} and μ_{Cl^-} are the chemical potentials of hydrogen and chloride ions, the suffixes 1 and 2 referring to the solutions of concentration c_1 and c_2, respectively. The quantities of solutions in the cells are assumed to be so large that the removal of hydrochloric acid from one and its transfer to the other brings about no appreciable change of concentration; the change of free energy is thus equal to the resultant change in the chemical potentials.

If the chemical potentials are expressed by means of equation (10) on p. 133, the result is

$$\Delta G = RT \ln \frac{(a_{H^+})_1}{(a_{H^+})_2} + RT \ln \frac{(a_{Cl^-})_1}{(a_{Cl^-})_2}$$

$$= RT \ln \frac{(a_{H^+})_1(a_{Cl^-})_1}{(a_{H^+})_2(a_{Cl^-})_2}, \tag{6}$$

where a_{H^+} and a_{Cl^-} refer to the activities of the ions indicated by the subscripts. The electrical energy produced in the cell for the passage of one faraday is EF, where E, as already seen, is equal to $E_1 - E_2$; it follows, therefore, from equation (6), since

$$\Delta G = - EF,$$

that

$$E = \frac{RT}{F} \ln \frac{(a_{H^+})_2(a_{Cl^-})_2}{(a_{H^+})_1(a_{Cl^-})_1} \tag{7}$$

$$= \frac{2RT}{F} \ln \frac{a_2}{a_1}, \tag{8}$$

where a_1 and a_2 are the *mean* activities of the hydrochloric acid in the two solutions (cf. p. 138). The activities may be replaced by the products $m\gamma$ or cf, so that

$$E = \frac{2RT}{F} \ln \frac{c_2 f_2}{c_1 f_1} \tag{9}$$

or

$$E = \frac{2RT}{F} \ln \frac{m_2 \gamma_2}{m_1 \gamma_1}. \tag{10}$$

A cell of the type described above is called a **concentration cell without transference,** for the E.M.F. depends on the relative concentrations, or molalities, of the two solutions concerned, and the operation of the cell is not accompanied by the *direct* transfer of electrolyte from one solution to the other. The transfer occurs indirectly, as shown above, as the result of chemical reactions. In general, a concentration cell without transference results whenever two simple cells whose electrodes are reversible with respect to each of the ions constituting the electrolyte are combined in opposition; in the case considered above, the electrolyte is hydrochloric acid, and one electrode is reversible with respect to hydrogen ions and the other with respect to chloride ions.

If a_1 is the mean ionic activity of the electrolyte in the left-hand side of any concentration cell without transference, and a_2 is the value on the right-hand side, the E.M.F. of the complete cell can be expressed by means of the general equation

$$E = \pm \frac{\nu}{\nu_\pm} \cdot \frac{RT}{z_\pm F} \ln \frac{a_2}{a_1}, \tag{11}$$

where ν is the total number of ions, and ν_+ or ν_- is the number of positive or negative ions produced by the ionization of one molecule of electrolyte; z_+ or z_- is the valence of the ion with respect to which the extreme electrodes are reversible. If this ion is positive, as in the cell already discussed, the positive signs apply throughout, but if it is negative, as in the cell

$$Ag \mid AgCl(s)\ HCl(c_1) \mid H_2(1\ atm.) \mid HCl(c_2)\ AgCl(s) \mid Ag,$$

the negative signs are applicable.

Amalgam Cells.—If the electrolyte in the concentration cell without transference is a salt of an alkali metal, e.g., potassium chloride, it is necessary to set up some form of reversible alkali metal electrode. This is achieved by dissolving the metal in mercury, thus forming a dilute alkali metal amalgam which is attacked much less vigorously by water than is the metal in the pure state. The amalgam nevertheless reacts with water to some extent, and also with traces of oxygen that may be present in the solution: the exposed surface of the amalgam is therefore continuously renewed by maintaining a flow from the end of a tube. For the cell

$$Ag \mid AgCl(s)\ KCl(c_1) \mid KHg_x \mid KCl(c_2)\ AgCl(s) \mid Ag,$$

where KHg_x represents the potassium amalgam, the apparatus is shown in Fig. 64; the reservoir *A* contains the dilute amalgam which flows slowly through the capillary tubes B_1 and B_2, while C_1 and C_2 represent the silver electrodes coated with silver chloride (see p. 234).[5] The potassium chloride solutions of concentrations c_1 and c_2 respectively, from which all dissolved oxygen has been removed, as far as possible, are introduced into the cells by means of the tubes D_1 and D_2. Although reproducible results can be obtained with the exercise of due care, the measurements are not reliable for solutions more dilute than about 0.1 N, because of interaction between the solution and the alkali metal.

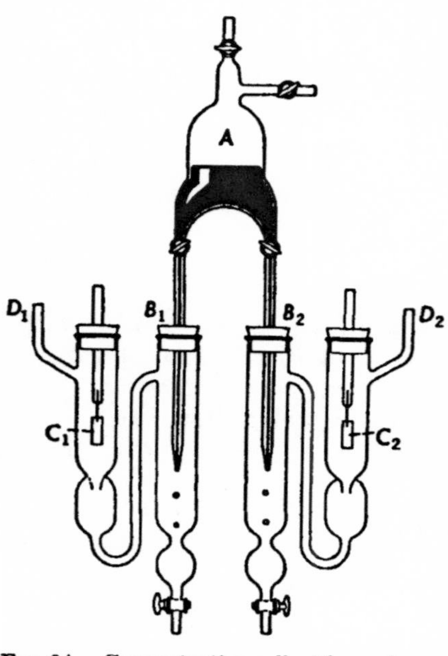

FIG. 64. Concentration cell with amalgam electrodes (MacInnes and Parker)

Amalgam cells are utilized for the study of alkali hydroxides, e.g.,

$$H_2(1\ atm.) \mid NaOH(c_1) \mid NaHg_x \mid NaOH(c_2) \mid H_2(1\ atm.),$$

where the hydrogen electrode is reversible with respect to hydroxyl ions, but equation (11) for the E.M.F. requires some modification in this case, because the cell reaction also involves the transfer of water. The reac-

[5] MacInnes and Parker, *J. Am. Chem. Soc.*, 37, 1445 (1915).

tion in the left-hand cell for the passage of one faraday of electricity is

$$\tfrac{1}{2}H_2(1\text{ atm.}) + NaOH(c_1) = H_2O + Na,$$

and in the right-hand cell it is

$$H_2O + Na = \tfrac{1}{2}H_2(1\text{ atm.}) + NaOH(c_2),$$

and consequently the net process is the transfer of a mole of sodium hydroxide, i.e., one g.-ion each of sodium and hydroxyl ions, from the solution of concentration c_1 to that of concentration c_2, while at the same time a mole of water is transferred in the opposite direction. The increase of free energy accompanying the passage of one faraday is represented by

$$\Delta G = [(\mu_{Na^+})_2 - (\mu_{Na^+})_1] + [(\mu_{OH^-})_2 - (\mu_{OH^-})_1] + [(\mu_{H_2O})_1 - (\mu_{H_2O})_2],$$

and hence, utilizing the equation on page 133 to give the chemical potential of the water in terms of its vapor pressure, it follows that

$$E = -\frac{2RT}{F}\ln\frac{a_2}{a_1} + \frac{RT}{F}\ln\frac{(p_{H_2O})_2}{(p_{H_2O})_1}, \tag{12}$$

where a_1 and a_2 are the mean ionic activities of the sodium hydroxide in the two solutions, and $(p_{H_2O})_1$ and $(p_{H_2O})_2$ are the respective aqueous vapor pressures.*

Determination of Activity Coefficients.—The E.M.F. of a concentration cell without transference is equal to $E_1 - E_2$, where E_1 and E_2 are determined by the concentrations c_1 and c_2, respectively, of the electrolyte; then for a cell to which equation (8) is applicable,

$$E_1 - E_2 = \frac{2RT}{F}\ln\frac{a_2}{a_1}. \tag{13}$$

If in one of the two solutions, e.g., c_2, the activity is unity, and the corresponding E.M.F. of the half-cell containing that solution is E^0, equation (13) reduces to the general form

$$E - E^0 = -\frac{2RT}{F}\ln a. \tag{14}$$

If m is the molality of the electrolyte in the solution of activity a which gives an E.M.F. equal to E in the half-cell, then addition of $(2RT/F)\ln m$ to both sides of equation (14) yields

$$E + \frac{2RT}{F}\ln m - E^0 = -\frac{2RT}{F}\ln\frac{a}{m} \tag{15}$$

$$= -\frac{2RT}{F}\ln\gamma, \tag{16}$$

* It should be noted that the $H_2(g)$, NaOH aq. electrode is to be regarded as reversible with respect to OH^- ions; this accounts for the negative sign in equation (12).

where γ is the mean activity coefficient of the electrolyte in the solution of molality m. In order to convert the Naperian to Briggsian logarithms the corresponding terms are multiplied by 2.3026, and if at the same time the values of R, i.e., 8.313 joules per degree, and of F, i.e., 96,500 coulombs, are inserted, equation (16) can be written as

$$E + 2 \times 1.9835 \times 10^{-4}\, T \log m - E^0 = -2 \times 1.9835 \times 10^{-4}\, T \log \gamma, \quad (17)$$

and, at 25°, this becomes

$$E + 0.1183 \log m - E^0 = -0.1183 \log \gamma. \quad (18)$$

Since E can be measured for any molality m, it would be possible to evaluate the activity coefficient γ if E^0 were known.[6] One method of deriving E^0 makes use of the fact that at infinite dilution, i.e., when m is zero, the activity coefficient γ is unity; under these conditions E^0 will be equal to $E + 0.1183 \log m$ at 25°. If this quantity, for various values of m, is plotted as ordinate against a function of the molality, generally $\sqrt{m}$, as abscissa, and the curve extrapolated to m equal to zero, the limiting value of the ordinate is equal to E^0. To be accurate this extrapolation requires a precise knowledge of the E.M.F.'s of cells containing very dilute solutions, and the necessary data are not easy to obtain. Two alternative methods of extrapolation which avoid this difficulty may be employed; only one of these will, however, be described here.[7]

According to the Debye-Hückel-Brønsted equation (63), p. 147, it is possible to express the variation of the activity coefficient of a uni-univalent electrolyte with molality by the equation

$$\log \gamma = -A\sqrt{m} + Cm, \quad (19)$$

where A is a known constant, equal to 0.509 for water as solvent at 25°. Combination of this with equation (18) then gives

$$E + 0.1183 \log m - 0.0602\sqrt{m} = E^0 - 0.1183\, Cm,$$

$$\therefore \quad E' - 0.0602\sqrt{m} = E^0 - 0.1183\, Cm,$$

where E' is equal to $E + 0.1183 \log m$. According to this result the quantity $E' - 0.0602\sqrt{m}$ should be a linear function of m, and extrapolation of the corresponding plot to m equal zero should give E^0. It is found in practice that the actual plot is not quite linear, as shown by the results in Fig. 65 for the cells

$$H_2(1 \text{ atm.}) \mid HCl(m)\ AgCl(s) \mid Ag,$$

but reasonably accurate extrapolation is nevertheless possible. The

[6] Lewis and Randall, *J. Am. Chem. Soc.*, **43**, 1112 (1921); Randall and Young, *ibid.*, **50**, 989 (1928).

[7] Hitchcock, *J. Am. Chem. Soc.*, **50**, 2076 (1928); Harned *et al.*, *ibid.*, **54**, 1350 (1932); **55**, 2179 (1933); **58**, 989 (1936).

value of E^0 for this cell at 25° is +0.2224 volt, and hence for solutions of hydrochloric acid

$$E + 0.1183 \log m - 0.2224 = -0.1183 \log \gamma.$$

The activity coefficients can thus be determined directly from this equation, using the measured values of the E.M.F. of the cell depicted above,

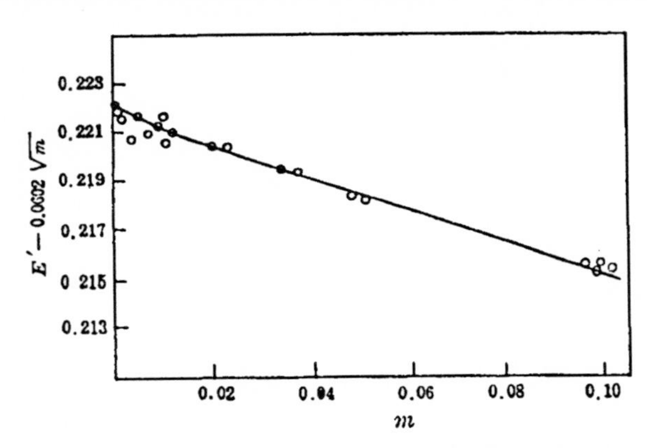

FIG. 65. Extrapolation of E.M.F. to infinite dilution

for various molalities of hydrochloric acid; the results obtained are given in Table XLIII.

TABLE XLIII. MEAN ACTIVITY COEFFICIENTS OF HYDROCHLORIC ACID FROM E.M.F. MEASUREMENTS AT 25°

m	E	$E + 0.1183 \log m$	γ
0.1238	0 34199	0.23466	0.788
0.05391	0.38222	0 23218	0.827
0 02563	0.41824	0 22999	0.863
0 013407	0.44974	0 22820	0.893
0 009138	0.46860	0 22735	0.908
0 005619	0.49257	0.22636	0 926
0.003215	0 52053	0 22562	0.939

Concentration Cells with Transference.—When two solutions of the same electrolyte are brought into actual contact and if identical electrodes, reversible with respect to one or other of the ions of the electrolyte, are placed in each solution, the result is a concentration cell with transference; for example, the removal of the AgCl(s) | Ag | AgCl(s) system from the cell on page 196 gives

$$H_2(1 \text{ atm.}) \mid HCl(c_1) \vdots HCl(c_2) \mid H_2(1 \text{ atm.}),$$

in which the two solutions of hydrochloric acid are in contact, and direct transfer from one to the other is possible. The presence of a **liquid junction,** as the region where the two solutions are brought into contact with one another is called, is represented by the vertical dotted line. When one faraday passes through the cell, 1 g.-atom of hydrogen dissolves at the left-hand electrode to yield 1 g.-ion of hydrogen ions, and the same amount of hydrogen ions will be discharged and 1 g.-atom of hydrogen will be liberated at the right-hand electrode. While the current is passing, t_+ g.-ion of hydrogen ions will migrate across the boundary between the two solutions in the direction of the current, i.e., from left to right, and t_- g.-ion of chloride ions will move in the opposite direction; t_+ and t_- are the transference numbers of the hydrogen and chloride ions, respectively (see Fig. 66). Attention may be drawn to the fact

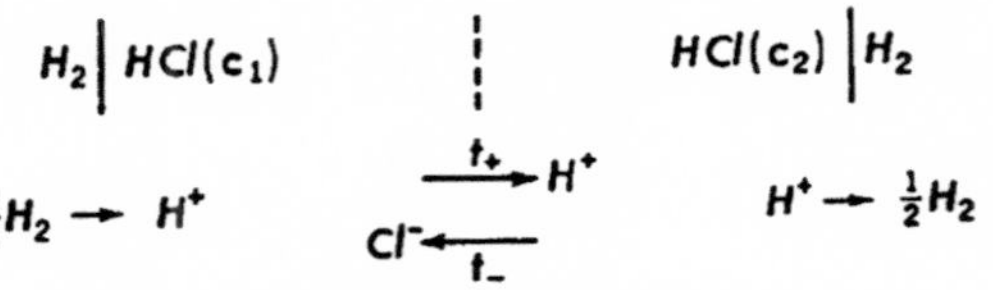

$H_2 | HCl(c_1)$ ⋮ $HCl(c_2) | H_2$

$\frac{1}{2}H_2 \rightarrow H^+$ $\xrightarrow{t_+} H^+$ $Cl^- \xleftarrow{t_-}$ $H^+ \rightarrow \frac{1}{2}H_2$

FIG. 66. Transference at liquid junction

that the transference numbers involved are the Hittorf values, and not the so-called "true" transference numbers (p. 114); this allows for the transfer of water with the ions.

The net result of the passage of one faraday is the transfer of $1 - t_+$, i.e., t_-, g.-ions of hydrogen ions and t_- g.-ions of chloride ions from right to left, so that the increase of free energy is

$$\Delta G = t_-[(\mu_{H^+})_1 - (\mu_{H^+})_2] + t_-[(\mu_{Cl^-})_1 - (\mu_{Cl^-})_2]. \qquad (20)$$

Since the transference numbers vary with concentration, it is convenient to consider two solutions whose concentrations differ by a small amount, viz., c and $c + dc$; under these conditions equation (20) becomes

$$\begin{aligned} \Delta G &= -\, t_-(d\mu_{H^+} + d\mu_{Cl^-}) \\ &= -\, t_-(RT\, d \ln a_{H^+} + RT\, d \ln a_{Cl^-}) \\ &= -\, 2t_- RT\, d \ln a, \end{aligned} \qquad (21)$$

where a is the mean activity of the hydrochloric acid at the concentration c, and t_- is the transference number of the anion at this concentration. The E.M.F. of the cell whose concentrations differ in amount by dc may be represented by dE, and the free energy increase $-F\,dE$ may be equated to the value given by equation (21); hence

$$dE = 2t_- \frac{RT}{F} d \ln a. \qquad (22)$$

For a concentration cell with electrolytes of concentration c_1 and c_2, i.e., mean activities of a_1 and a_2, respectively, the E.M.F. is then obtained by integrating equation (22) between these limits; thus

$$E = \frac{2RT}{F} \int_{a_1}^{a_2} t_- d \ln a. \tag{23}$$

In the general case this becomes

$$E = \pm \frac{\nu}{\nu_\pm} \cdot \frac{RT}{z_\pm F} \int_{a_1}^{a_2} t_\mp d \ln a, \tag{24}$$

where ν_1, $\nu_\pm$ and $z_\pm$ have the same significance as before (p. 197); the transference number $t_\mp$ refers to the ion other than that with respect to which the electrodes are reversible.

If the transference number is taken as constant in the range of concentration c_1 to c_2, equation (24) takes the form

$$E = \pm t_\mp \frac{\nu}{\nu_\pm} \cdot \frac{RT}{z_\pm F} \ln \frac{a_2}{a_1}. \tag{25}$$

In the special case of the hydrogen-hydrochloric acid cell given above, ν is 2, ν_- is 1, and z_+ is 1, and the electrodes are reversible with respect to positive ions; hence

$$E = 2t_- \frac{RT}{F} \ln \frac{a_2}{a_1}. \tag{26}$$

If the concentration cell is one of the type in which water is formed or removed in the cell reaction, e.g.,

$$H_2 \mid NaOH(c_1) \vdots NaOH(c_2) \mid H_2,$$

in which a mole of water is transferred from c_2 to c_1 for the passage of one faraday, due allowance must be made in the manner already described.

Activity Coefficients from Cells With Transference.—In order to set up a cell *without* transference it is necessary to have electrodes reversible with respect to each of the ions of the electrolyte; this is not always possible or convenient, and hence the use of cells *with* transference, which require electrodes reversible with respect to one ion only, has obvious advantages. In order that such cells may be employed for the purpose of determining activity coefficients, however, it is necessary to have accurate transference number data for the electrolyte being studied. Such data have become available in recent years, and in the method described below it will be assumed that the transference numbers are known over a range of concentrations.[8]

The E.M.F. of a cell of the type

$$M \mid MA(c) \vdots MA(c + dc) \mid M.$$

[8] Brown and MacInnes, *J. Am. Chem. Soc.*, 57, 1356 (1935); Shedlovsky and MacInnes, *ibid.*, 58, 1970 (1936); 59, 503 (1937); 61, 200 (1939); MacInnes and Brown, *Chem. Revs.*, 18, 335 (1936).

where M is a metal or hydrogen, yielding cations in solution, is given by equation (22), and since the activity a is equal to cf, this may be written

$$dE = \frac{2RT}{F} t_-(d \ln c + d \ln f),$$

$$\therefore \quad \frac{dE}{t_-} = \frac{2RT}{F}(d \ln c + d \ln f). \tag{27}$$

The activity is expressed in terms of concentrations rather than molalities because the transference numbers are generally known as a function of the former; the procedure described here thus gives the activity coefficient f, but the values can be readily converted into the corresponding γ's by means of the equations on page 135.

The transference number at any concentration can be written as

$$\frac{1}{t_-} = \frac{1}{t_0} + \delta,$$

where t_0 is the value at some reference concentration c_0; if this expression for $1/t_-$ is inserted in equation (27) and the latter multiplied out and rearranged, the result is

$$d \ln f = \frac{F}{2t_0RT} dE - d \ln c + \frac{F}{2RT} \delta \, dE.$$

Integrating between the limits c_0 and c, the corresponding values of the mean activity coefficient of the electrolyte being f_0 and f, it follows, after converting the logarithms, that

$$\log \frac{f}{f_0} = \frac{FE}{2.303 \times 2t_0RT} - \log \frac{c}{c_0} + \frac{F}{2 \times 2.303RT} \int_0^E \delta \, dE. \tag{28}$$

The first two terms on the right-hand side of equation (28) may be evaluated directly from the experimental data, after deciding on the concentration c_0 which is to represent the reference state. The third term is obtained by graphical integration of δ against E, the value of δ being derived from the known variation of the transference number with concentration.

The method just described gives $\log f/f_0$, and hence the activity coefficient f in the solution of concentration c is known in terms of an arbitrary reference scale, i.e., f_0 at concentration c_0; it is necessary now to convert the results to the usual standard state, i.e., the hypothetical ideal solution at unit concentration (see p. 137). For this purpose, use is made of the Debye-Hückel expression for uni-univalent electrolytes,

$$-\log f = \frac{A\sqrt{c}}{1 + B'\sqrt{c}}, \tag{29}$$

where A is the known Debye-Hückel constant for the solvent at the experimental temperature, and B', which is written in place of aB, is a constant for the electrolyte. The term $\log f/f_0$, i.e., $\log f - \log f_0$, may be represented by $\log f + \alpha$, where α is a constant, equal to $-\log f_0$, and hence equation (29) may be rewritten as

$$\log \frac{f}{f_0} + A\sqrt{c} = \alpha + B'\left(\alpha - \log \frac{f}{f_0}\right)\sqrt{c}.$$

For solutions dilute enough for equation (29) to be applicable, the plot of $\log (f/f_0) + A\sqrt{c}$ against $[\alpha - \log (f/f_0)]\sqrt{c}$ should be a straight line with intercept equal to α. The value of α, which is required for the purpose of this plot, is obtained by a short series of approximations. Once α, which is equal to $-\log f_0$, is known, it is possible to derive $\log f$ for any solution from the values of $\log f/f_0$ obtained previously. The activity coefficient of the electrolyte can thus be evaluated from the E.M.F.'s of cells with transference, provided the required transference number information is available.

Determination of Transference Numbers.—Since activity coefficients can be derived from E.M.F. measurements if transference numbers are known, it is apparent that the procedure could be reversed so as to make it possible to calculate transference numbers from E.M.F. data. The method employed is based on measurements of cells containing the same electrolyte, with and without transference. The E.M.F. of a concentration cell without transference (E) is given by equation (11), and if the intermediate electrodes are removed so as to form a concentration cell with transference, the E.M.F., represented by E_t, is now determined by equation (25), provided the transference numbers may be taken as constant within the range of concentrations in the cells. It follows, therefore, on dividing equation (25) by (11), that

$$\frac{E_t}{E} = t_{\mp}, \tag{30}$$

where the transference number $t_{\mp}$ refers to the negative ion if the extreme electrodes are reversible with respect to the positive ion, and *vice versa.*[9]

For example, if the amalgam cell without transference

$$\text{Ag} \mid \text{AgCl}(s)\ \text{LiCl}(c_1) \mid \text{LiHg}_x \mid \text{LiCl}(c_2)\ \text{AgCl}(s) \mid \text{Ag}$$

is under consideration, the corresponding cell with transference is

$$\text{Ag} \mid \text{AgCl}(s)\ \text{LiCl}(c_1) \,\vdots\, \text{LiCl}(c_2)\ \text{AgCl}(s) \mid \text{Ag}.$$

The ratio of the E.M.F.'s of these cells then gives the transference number of the lithium ion, i.e.,

$$\frac{E_t}{E} = t_{\text{Li}^+},$$

[9] The method for determining transference numbers from E.M.F. measurements was first suggested by Helmholtz in 1878.

since the extreme electrodes, i.e., Ag | AgCl(*s*) LiCl aq., are reversible with respect to the chloride ion.

The use of equation (30) gives a mean transference number of the electrolyte within the range of concentrations from c_1 to c_2, but this is of little value because of the variation of transference numbers with concentration; a modified treatment, to give the results at a series of definite concentrations, may, however, be employed. If the concentrations of the solutions are c and $c + dc$, the E.M.F. of the cell with transference is given by the general form of equation (22) as

$$dE_t = \pm t_{\mp} \frac{\nu}{\nu_{\pm}} \cdot \frac{RT}{z_{\pm}F} d \ln a,$$

$$\therefore \quad \frac{dE_t}{d \ln a} = \pm t_{\mp} \frac{\nu}{\nu_{\pm}} \cdot \frac{RT}{z_{\pm}F}, \tag{31}$$

where a is the mean activity of the electrolyte at the concentration c. The corresponding E.M.F. for the cell with transference, derived from equation (11), is

$$dE = \pm \frac{\nu}{\nu_{\pm}} \cdot \frac{RT}{z_{\pm}F} d \ln a,$$

$$\therefore \quad \frac{dE}{d \ln a} = \pm \frac{\nu}{\nu_{\pm}} \cdot \frac{RT}{z_{\pm}F}. \tag{32}$$

It follows, therefore, from equations (31) and (32) that

$$\frac{dE_t/d \ln a}{dE/d \ln a} = t_{\mp},$$

or

$$\frac{dE_t/d \log a}{dE/d \log a} = t_{\mp}. \tag{33}$$

If the E.M.F.'s of the cells, with and without transference, in which the concentration of one of the solutions is varied while the other is kept at a constant low value, e.g., 0.001 molar, are plotted against log a of the variable solution, the slopes of the curves are $dE_t/d \log a$ and $dE/d \log a$, respectively. The transference number of the appropriate ion may thus be determined at any concentration by taking the ratio of the slopes at the value of log a corresponding to this concentration. The activities at the different concentrations, from which the log a data are obtained, must be determined independently by E.M.F. or other methods.

Since the exact measurement of the slopes of the curves is difficult, analytical procedures have been employed. In the simplest one of these,[10] the values of E_t are expressed as a function of the logarithm of the activities of the electrolyte; from this $dE_t/d \log a$ is readily derived by differentiation. Since $dE/d \log a$ is given directly by equation (32),

[10] MacInnes and Beattie, *J. Am. Chem. Soc.*, **42**, 1117 (1920).

t can also be written as a function of log a, and hence it may be evaluated at any desired concentration.

A more rigid but laborious method, for deriving transference numbers from E.M.F. data, makes use of the fact that the activity coefficient of an electrolyte can be expressed, by means of an extended form of the Debye-Hückel equation, as a function of the concentration and of two empirical constants.[11] When applied to the same data, however, this procedure gives results which are somewhat different from those obtained by the method just described. Since the values are in better agreement with the transference data derived from moving boundary and other measurements, they are probably more reliable.

A number of determinations of transference numbers, in both aqueous and non-aqueous solutions, have been made by the E.M.F. method, and the results are in fair agreement with those obtained by other experimental procedures. The results in Table XLIV, for example, are for the

TABLE XLIV. TRANSFERENCE NUMBER OF LITHIUM ION IN LITHIUM CHLORIDE AT 25°

Conc.	E.M.F. Method	Hittorf or Moving Boundary Method
0.005 N	0.3351	0.3303
0.01	0.3333	0.3289
0.02	0.3308	0.3261
0.05	0.3259	0.3211
0.10	0.3203	0.3168
0.20	0.3126	0.3112
0.50	0.3067	0.3079
1.00	0.2809	0.2873

transference number of the lithium ion in lithium chloride at 25°. The discrepancies between the two sets of values are often appreciable, however, and since they are greater than the experimental errors of the best Hittorf or moving boundary measurements, it is probable that the E.M.F. results are in error. It must be concluded, therefore, that the E.M.F.'s of concentration cells cannot yet be obtained with sufficient precision for the transference numbers to be as accurate as the best results obtained by other methods.

Liquid Junction Potentials: Solutions of the Same Electrolyte.—The free energy change occurring in a concentration cell with transference may be divided into two parts; these are (i) the contributions of the reactions at the electrodes, and (ii) that due to the transfer of ions across the boundary between the two solutions. It is evident, therefore, that when two solutions of the same or of different electrolytes are brought into contact, a difference of potential will be set up at the junction between them because of ionic transference. Potentials of this kind are called **liquid junction potentials** or **diffusion potentials.**

[11] Jones and Dole, *J. Am. Chem. Soc.*, 51, 1073 (1929); Jones and Bradshaw, *ibid.*, 54, 138 (1932); see also, Hamer, *ibid.*, 57, 66 (1935); Harned and Dreby, *ibid.*, 61, 3113 (1939).

Consider the simplest case in which the junction is formed between two solutions of the same uni-univalent electrolyte at concentrations c_1 and c_2, e.g.,

$$KCl(c_1) \vdots KCl(c_2).$$

Adopting the usual convention for a positive E.M.F. that the left-hand electrode is the source of electrons, so that positive current flows through the interior of the cell from left to right, it follows that the passage of one faraday of electricity through the cell results in the transfer of t_+ g.-ion of cations, e.g., potassium ions, from left to right, i.e., from solution c_1 to solution c_2, and t_- g.-ion of anions, e.g., chloride ions, in the opposite direction (cf. p. 202). If the approximation is made of taking the transference numbers to be independent of concentration, the free energy change accompanying the passage of one faraday across the liquid junction may be expressed either as $-FE_L$, where E_L is the liquid junction potential, or as

$$\Delta G = t_+ RT \ln \frac{(a_+)_2}{(a_+)_1} + t_- RT \ln \frac{(a_-)_1}{(a_-)_2},$$

$$\therefore \quad E_L = -t_+ \frac{RT}{F} \ln \frac{(a_+)_2}{(a_+)_1} + t_- \frac{RT}{F} \ln \frac{(a_-)_2}{(a_-)_1}. \tag{34}$$

Further, since $t_+ + t_-$ is equal to unity, it follows that

$$E_L = -2t_+ \frac{RT}{F} \ln \frac{a_2}{a_1} + \frac{RT}{F} \ln \frac{(a_-)_2}{(a_-)_1}, \tag{35}$$

where a_1 and a_2 are the mean activities of the electrolyte in the two solutions. By making the further approximation of writing $(a_-)_2/(a_-)_1$ as equal to a_2/a_1, equation (35) reduces to

$$E_L = (1 - 2t_+) \frac{RT}{F} \ln \frac{a_2}{a_1}. \tag{36}$$

Since $1 - t_+$ is equal to t_-, this result may be expressed in the alternative form

$$E_L = (t_- - t_+) \frac{RT}{F} \ln \frac{a_2}{a_1}, \tag{36a}$$

which brings out clearly the dependence of the sign of the liquid junction potential on the relative values of the transference numbers of the anion and cation.

If the liquid junction potential under consideration forms part of the concentration cell

$$Ag \mid AgCl(s) \;\; KCl(c_1) \vdots KCl(c_2) \;\; AgCl(s) \mid Ag,$$

the E.M.F. of the complete cell is given by equation (25) as

$$E_t = -2t_+ \frac{RT}{F} \ln \frac{a_2}{a_1},$$

and hence, from this and equation (36), it is seen that

$$E_L = \frac{2t_+ - 1}{2t_+} E_t. \tag{37}$$

This approximate relationship can be tested by suitable measurements on concentration cells with transference.

As indicated above, the E.M.F. of a cell with transference can be regarded as made up of the potential differences at the two electrodes and the liquid junction potential. It will be seen shortly (p. 229) that each of the former may be regarded as determined by the activity of the reversible ion in the solution contained in the particular electrode. In the cell depicted above, for example, the potential difference at the left-hand electrode is dependent on the activity of the chloride ions in the potassium chloride solution of concentration c_1; similarly the potential difference at the right-hand electrode depends on the chloride ion activity in the solution of concentration c_2. For sufficiently dilute solutions the activity of a given ion, according to the simple Debye-Hückel theory, is determined by the ionic strength of the solution and is independent of the nature of the other ions present. It follows, therefore, that the *electrode potentials* should be the same in all cells of the type

$$\text{Ag} \mid \text{AgCl}(s)\ \ \text{MCl}(c_1) \,\vdots\, \text{MCl}(c_2)\ \ \text{AgCl}(s) \mid \text{Ag},$$

where c_1 and c_2 represent dilute solutions of any uni-univalent chloride MCl, which must be a strong electrolyte. If E is the constant algebraic sum of these potentials, the E.M.F. of the complete cell with transference, which *does* vary with the nature of MCl, will be $E + E_L$, i.e.

$$E_t = E + E_L,$$

$$\therefore\ E = E_t - E_L. \tag{38}$$

The difference between E_t and E_L should thus be constant for given values of c_1 and c_2, irrespective of the nature of the uni-univalent chloride employed in the cell. Inserting the value of E_L given by equation (37) into (38), the result is

$$E = \frac{E_t}{2t_+}.$$

If the right-hand side is constant, for cells with transference containing different chlorides at definite concentrations, it may be concluded that the approximate equation (36) gives a satisfactory measure of the liquid junction potential between two solutions of the same electrolyte. The results in Table XLV provide support for the reliability of this equation, within certain limits; [12] the transference numbers employed are the mean values for the two solutions, the individual figures not differing greatly in the range of concentrations involved.

[12] MacInnes, "The Principles of Electrochemistry," 1939, p. 226; data mainly from MacInnes *et al.*, *J. Am. Chem. Soc.*, 57, 1356 (1935); 59. 503 (1937).

TABLE XLV. TEST OF EQUATION FOR LIQUID JUNCTION POTENTIAL

Electrolyte	c_1	c_2	t_+	E_t	$E_t/2t_+$	E_L
NaCl	0.005	0.01	0.392	13.41 mv.	17.1 mv.	− 3.7 mv.
KCl	0.005	0.01	0.490	16.77	17.1	− 0.3
HCl	0.005	0.01	0.824	28.29	17.2	+ 11.1
NaCl	0.005	0.04	0.391	39.63 mv.	50.7 mv.	− 11.1 mv.
KCl	0.005	0.04	0.490	49.63	50.6	− 1.0
HCl	0.005	0.04	0.826	84.16	50.9	+ 33.3

In order to give some indication of the magnitude of the liquid junction potential, the values of E_L calculated from equation (37) are recorded in the last column. In general, the larger the ratio of the concentrations of the solutions and the more the transference number of either ion departs from 0.5, i.e., the larger the difference between the transference numbers of the two ions, the greater is the liquid junction potential. The sign is determined by the relative magnitudes of the transference numbers of cation and anion of the electrolyte, as seen from equation (36*a*).

General Equation for Liquid Junction Potential.—When the two solutions forming the junction contain different electrolytes, as in many chemical cells, the situation is more complicated; it is convenient, therefore, to consider here the most general case. Suppose a cell contains a solution in which there are several ions of concentration $c_1, c_2, \cdots, c_i, \cdots$ g.-ions per liter, and suppose this forms a junction with another solution in which the corresponding ionic concentrations are $c_1 + dc_1$, $c_2 + dc_2, \cdots$, $c_i + dc_i, \cdots$; the valences of the ions are $z_1, z_2, \cdots, z_i, \cdots$ and their transference numbers are $t_1, t_2, \cdots, t_i, \cdots$, the latter being regarded as constant, since the differences of the ionic concentrations in the two solutions are small. If one faraday of electricity is passed through the cell, t_i/z_i g.-ion of each ionic species will be transferred across the boundary between the two solutions, the positive ions moving in one direction, i.e., left to right according to convention, and the negative ions moving in the opposite direction. The increase of free energy as a result of the transfer of an ion of the ith kind from the solution of concentration c_i to that of concentration $c_i + dc_i$ is given by

$$dG = \frac{t_i}{z_i}[(\mu_i + d\mu_i) - \mu_i]$$

$$= \frac{t_i}{z_i} d\mu_i,$$

where μ_i and $\mu_i + d\mu_i$ are the chemical potentials of the particular ions in the two solutions. For the transfer of all the ions across the boundary when one faraday is passed,

$$\Delta G = \sum_i \frac{t_i}{z_i} d\mu_i,$$

and utilizing the familiar definition of μ_i as $\mu_i^0 + RT \ln a_i$, it follows that

$$\Delta G = \sum_i \frac{t_i}{z_i} RT\, d \ln a_i, \tag{39}$$

where a_i is the activity of the ith ions at the concentration c_i. It should be remembered that in the summation the appropriate signs must be used when considering positive and negative ions, since they move in opposite directions.

Provided the concentrations of any ion do not differ appreciably in the two solutions, the transfer of ions across the boundary when current passes may be regarded as reversible. If dE_L is the potential produced at the junction between the two solutions, then ΔG will also be equal to $-F\,dE_L$ for the passage of one faraday; combination of this result with equation (39) gives

$$dE_L = -\frac{RT}{F} \sum_i \frac{t_i}{z_i} d \ln a_i \tag{40}$$

for the liquid junction potential. Since in actual practice the concentrations of the two solutions differ by appreciable amounts, the liquid junction potential can be regarded as being made up of a series of layers with infinitesimal concentration differences; the resultant potential E_L is obtained by integrating equation (40) between the limits c_1 and c_2, representing the two solutions in the cell; thus

$$E_L = -\frac{RT}{F} \int_{c_1}^{c_2} \sum_i \frac{t_i}{z_i} d \ln a_i. \tag{41}$$

This is the general form of the equation for the liquid junction potential between the two solutions; [13] in order that the integration may be carried out, however, it is necessary to make approximations or to postulate certain properties of the boundary.

For example, if the two solutions contain the same electrolyte, consisting of one cation and one anion, equation (41) becomes

$$E_L = -\frac{RT}{F} \int_{c_1}^{c_2} \frac{t_+}{z_+} d \ln a_+ + \frac{RT}{F} \int_{c_1}^{c_2} \frac{t_-}{z_-} d \ln a_-.$$

If the approximation is made of taking the transference numbers to be independent of concentration, this relationship takes the form

$$E_L = -\frac{t_+}{z_+} \cdot \frac{RT}{F} \ln \frac{(a_+)_2}{(a_+)_1} + \frac{t_-}{z_-} \cdot \frac{RT}{F} \ln \frac{(a_-)_2}{(a_-)_1},$$

which is identical with equation (34) for a uni-univalent electrolyte.

[13] Harned, *J. Phys. Chem.*, **30**, 433 (1926); Taylor, *ibid.*, **31**, 1478 (1927); see also, Guggenheim, *Phil. Mag.*, **22**, 983 (1936).

Type of Boundary and Liquid Junction Potential.—When the two solutions forming the junction contain different electrolytes, the structure of the boundary, and hence the concentrations of the ions at different points, will depend on the method used for bringing the solutions together. It is evident that the transference number of each ionic species, and to some extent its activity, will be greatly dependent on the nature of the boundary; hence the liquid junction potential may vary with the type of junction employed. If the electrolyte is the same in both solutions, however, the potential should be independent of the manner in which the junction is formed. In these circumstances the solution at any point in the boundary layer will consist of only one electrolyte at a definite concentration; hence each ionic species should have a definite transference number and activity. When carrying out the integration of equation (41), the result will, therefore, always be the same no matter what is the type of concentration gradient in the intermediate layer between the two solutions; this theoretical expectation has been verified by experiment.[14] It is the fact that the liquid junction potential is independent of the structure of the boundary, when the electrolyte is the same on both sides, that makes possible accurate measurement of the E.M.F. of concentration cells with liquid junctions. In general, cells of this type are set up with simple "static', junctions, as shown in Fig. 67; the more dilute solution is in the relatively narrow tube which is dipped into the somewhat wider vessel containing the more concentrated solution, so that the boundary is formed at the tip of the narrower tube.

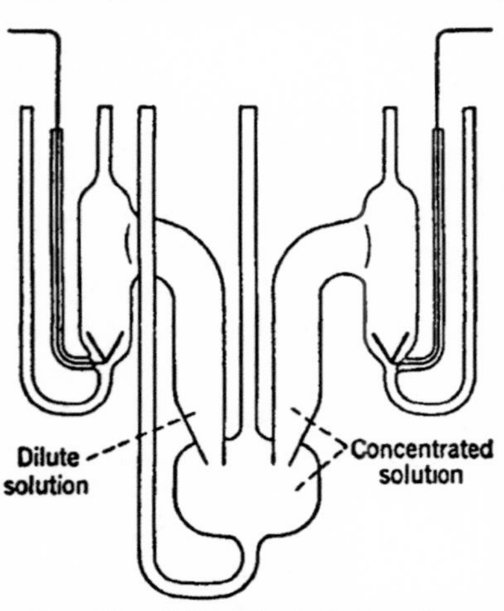

FIG. 67. Cell with static junction (MacInnes)

For solutions of different electrolytes four distinct forms of boundary have been described,[15] but only in two cases is anything like a satisfactory integration of equation (41) possible.

I. **The Continuous Mixture Boundary.**—This type of boundary, which is the one postulated by Henderson,[16] consists of a continuous series of mixtures of the two solutions, *free from the effects of diffusion.* If the two solutions are represented by the suffixes 1 and 2, and $1 - x$ is the frac-

[14] Scatchard and Buehrer, *J. Am. Chem. Soc.*, **53**, 574 (1931); Ferguson *et al.*, *ibid.*, **54**, 1285 (1932); Szábo, *Z. physik. Chem.*, **174A**, 33 (1935).

[15] See Guggenheim, *J. Am. Chem. Soc.*, **52**, 1315 (1930).

[16] Henderson, *Z. physik. Chem.*, **59**, 118 (1907); **63**, 325 (1908); Hermans, *Rec. trav. chim.*, **57**, 1373 (1938); **58**, 99 (1939).

tion of the former solution at a given point in the boundary, the fraction of solution 2 will be x, where x varies continuously from zero to unity; if c_i is the concentration of the ith kind of ion at this point, then

$$c_i = (1 - x)c_{i(1)} + xc_{i(2)},$$

where $c_{i(1)}$ and $c_{i(2)}$ are the concentrations of these ions in the bulk of the solutions 1 and 2, respectively. Making use of this expression, and replacing activities in equation (41) by the corresponding concentrations, as an approximation, it is possible to integrate this equation; the result, known as the **Henderson equation** for liquid junction potentials, is

$$E_L = \frac{RT}{F} \cdot \frac{(U_1 - V_1) - (U_2 - V_2)}{(U_1' + V_1') - (U_2' + V_2')} \ln \frac{U_1' + V_1'}{U_2' + V_2'}, \tag{42}$$

where U_1, V_1, etc., are defined by

$$U_1 \equiv \Sigma(c_+u_+)_1, \qquad V_1 \equiv \Sigma(c_-u_-)_1,$$

$$U_1' \equiv \Sigma(c_+z_+u_+)_1 \quad \text{and} \quad V_1' \equiv \Sigma(c_-z_-u_-)_1,$$

where c_+ and c_- refer to the concentrations of the cations and anions respectively, in g.-ions per liter, u_+ and u_- are the corresponding ionic mobilities, and z_+ and z_- their valences; the suffix 1 refers to the ions in solution 1, and similar expressions hold for U_2, V_2, etc. in which the ions in solution 2 are concerned.

The continuous mixture boundary presupposes the complete absence of diffusion; since diffusion of one solution into the other is inevitable, however, this type of boundary is probably unstable. It is possible that the flowing type of junction considered below may approximate in behavior to the continuous mixture type of boundary.

Two special cases of the Henderson equation are of interest. If the two solutions contain the same uni-univalent electrolyte at different concentrations, then

$$U_1 = U_1' = c_1u_+ \quad \text{and} \quad V_1 = V_1' = c_1u_-,$$

$$U_2 = U_2' = c_2u_+ \quad \text{and} \quad V_2 = V_2' = c_2u_-.$$

Insertion of these values in equation (42) gives

$$E_L = \frac{RT}{F} \cdot \frac{u_+ - u_-}{u_+ + u_-} \ln \frac{c_1}{c_2}. \tag{43}$$

Since $u_+/(u_+ + u_-)$ is equal to the transference number of the cation, i.e., to t_+, this result is equivalent to

$$E_L = (2t_+ - 1)\frac{RT}{F} \ln \frac{c_1}{c_2},$$

which is the same as the approximate equation (36), except that the ratio of the activities has been replaced by the ratio of the concentrations.

Another interesting case is that in which two uni-univalent electrolytes having an ion in common, e.g., sodium and potassium chlorides, are at the same concentration c; in these circumstances, assuming the anion to be common ion,

$$U_1 = U_1' = cu_{+(1)} \quad \text{and} \quad V_1 = V_1' = cu_-,$$
$$U_2 = U_2' = cu_{+(2)} \quad \text{and} \quad V_2 = V_2' = cu_-,$$

and substitution in equation (42) gives

$$E_L = \frac{RT}{F} \ln \frac{u_{+(1)} + u_-}{u_{+(2)} + u_-}$$
$$= \frac{RT}{F} \ln \frac{\Lambda_1}{\Lambda_2}, \qquad (44)$$

where Λ_1 and Λ_2 are the equivalent conductances of the two solutions forming the junction. The resulting relationship is known as the **Lewis and Sargent equation,**[17] tests of which will be described shortly.

II. **The Constrained Diffusion Junction.**—The assumption made by Planck [18] in order to integrate the equation for the liquid junction potential is equivalent to what has been called a "constrained diffusion junction"; this is supposed to consist of two solutions of definite concentration separated by a layer of constant thickness in which a steady state is reached as a result of diffusion of the two solutions from opposite sides. The Planck type of junction could be set up by employing a membrane whose two surfaces are in contact with the two electrolytes which are continuously renewed; in this way the concentrations at the interfaces and the thickness of the intermediate layer are kept constant, and a steady state is maintained within the layer. The mathematical treatment of the constrained diffusion junction is complicated; for electrolytes consisting entirely of univalent ions, the result is the **Planck equation,**

$$E_L = \frac{RT}{F} \ln \xi, \qquad (45)$$

where ξ is defined by the relationship

$$\frac{\xi U_2 - U_1}{V_2 - \xi V_1} = \frac{\ln \frac{c_2}{c_1} - \ln \xi}{\ln \frac{c_2}{c_1} + \ln \xi} \cdot \frac{\xi c_2 - c_1}{c_2 - \xi c_1},$$

U_1, U_2, V_1 and V_2 having the same significance as before.

[17] Lewis and Sargent, *J. Am. Chem. Soc.*, **31**, 363 (1909); see also, MacInnes and Yeh, *ibid.*, **43**, 2563 (1921); Martin and Newton, *J. Phys. Chem.*, **39**, 485 (1935).

[18] Planck, *Ann. Physik*, **40**, 561 (1890); see also, Fales and Vosburgh, *J. Am. Chem. Soc.*, **40**, 1291 (1918); Hermans, *Rec. trav. chim.*, **57**, 1373 (1938).

In the two special cases considered above, first, two solutions of the same electrolyte at different concentrations, and second, two electrolytes with a common ion at the same concentration, the Planck equation reduces to the same form as does the Henderson equation, viz., equations (43) and (44), respectively. It appears, therefore, that in these particular instances the value of the liquid junction potential does not depend on the type of boundary connecting the two solutions.

III. **Free Diffusion Junction.**—The free diffusion type of boundary is the simplest of all in practice, but it has not yet been possible to carry out an exact integration of equation (41) for such a junction.[19] In setting up a free diffusion boundary, an initially sharp junction is formed between the two solutions in a narrow tube and unconstrained diffusion is allowed to take place. The thickness of the transition layer increases steadily, but it appears that the liquid junction potential should be independent of time, within limits, provided that the cylindrical symmetry at the junction is maintained. The so-called "static" junction, formed at the tip of a relatively narrow tube immersed in a wider vessel (cf. p. 212), forms a free diffusion type of boundary, but it cannot retain its cylindrical symmetry for any appreciable time. Unless the two solutions contain the same electrolyte, therefore, the static type of junction gives a variable potential. If the free diffusion junction is formed carefully within a tube, however, it can be made to give reproducible results.[20]

IV. **The Flowing Junction.**—In order to obtain reproducible liquid junctions, in connection with the measurement of the E.M.F.'s of cells involving boundaries between two different electrolytes, Lamb and Larson devised the "flowing junction."[21] In the earlier forms of this type of junction (Fig. 68) an upward current of the more dense solution was allowed to meet a downward flow of the less dense solution at a point where a horizontal tube, leading to an overflow, joined the main tube. The levels of the liquids were so arranged that they flowed at the same slow rate, and a sharp boundary was maintained within the horizontal portion of the overflow tube. Experiments with indicators have

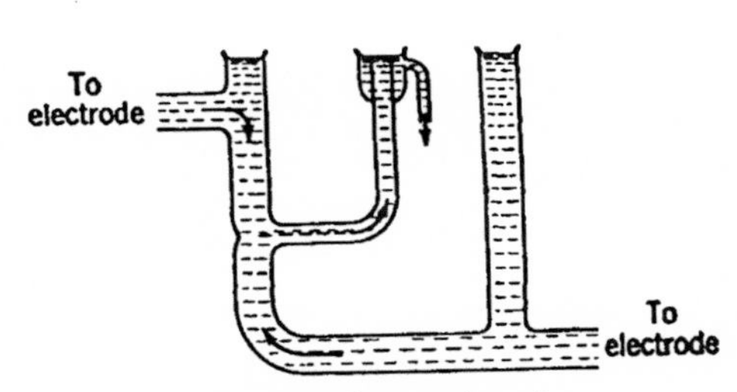

FIG. 68. The flowing junction (Lamb and Larson)

[19] Taylor, *J. Phys. Chem.*, **31**, 1478 (1927).
[20] Guggenheim, *J. Am. Chem. Soc.*, **52**, 1315 (1930).
[21] Lamb and Larson, *J. Am. Chem. Soc.*, **42**, 229 (1920); MacInnes and Yeh, *ibid.*, **43**, 2563 (1921); Scatchard, *ibid.*, **47**, 696 (1925); Scatchard and Buehrer, *ibid.*, **53**, 574 (1931); see also, Roberts and Fenwick, *ibid.*, **49**, 2787 (1927); Lakhani, *J. Chem. Soc.*, 179 (1932); Ghosh, *J. Indian Chem. Soc.*, **12**, 15 (1935).

shown that the boundary between the two solutions in a good flowing junction is extremely thin. With such a junction the potentials between two electrolytes having an ion in common can be reproduced to ± 0.02 millivolt. Simplified forms of flowing junction have been established by allowing the solutions to flow down opposite faces of a thin mica plate having a small hole in which the junction is formed (Fig. 69). The

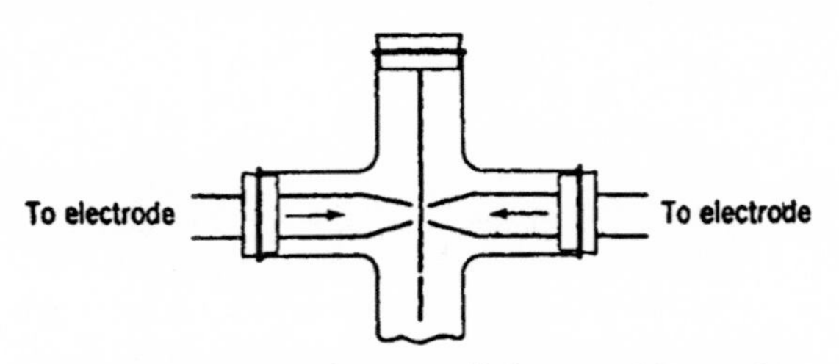

FIG. 69. Flowing junction (Roberts and Fenwick)

mica plate may even be eliminated and fine jets of the two liquids caused to impinge directly on one another.

The problem of the flowing junction is too difficult to be treated theoretically; since the time of contact between the two solutions is so small, the extent of diffusion will probably be negligible, and hence it has been generally assumed that the flowing junction resembles a continuous mixture (Henderson) type of boundary. On the other hand, it has been suggested that since the transition layer between the solutions is extremely thin, diffusion is of importance; the flowing junction would thus resemble the constrained diffusion (Planck) type of boundary. The only reasonably satisfactory experimental determinations of the potential of a flowing junction have been made with solutions of the same concentration and having an ion in common; as already seen, under these conditions the Henderson and Planck junctions lead to the same potentials.

Measurement of Liquid Junction Potentials with Different Electrolytes.—If the same assumption is made as on page 209, that the potential of an electrode reversible with respect to a given ion depends only on the concentration of that ion, then in cells of the type

$$\text{Ag} \mid \text{AgCl}(s)\ \ \text{MCl}(c) \,\vdots\, \text{M}'\text{Cl}(c)\ \ \text{AgCl}(s) \mid \text{Ag},$$

where MCl and M′Cl, the chlorides of two different univalent cations, are present at the same concentration, the total E.M.F. is equal merely to the liquid junction potential. A number of measurements of cells of this form using 0.1 N and 0.01 N solutions of various chlorides have been made with a flowing junction of the type depicted in Fig. 68; the results are in fair agreement with those derived from the Lewis and Sargent equation (44), as shown by the data in Table XLVI.[22] The discrep-

[22] MacInnes and Yeh, *J. Am. Chem. Soc.*, 43, 2563 (1921).

TABLE XLVI. CALCULATED AND OBSERVED FLOWING JUNCTION POTENTIALS AT 25°

Electrolytes		Concentration	Liquid Junction Potential Observed	Liquid Junction Potential Calculated
HCl	KCl	0.1 N	26.78 mv.	28.52 mv.
HCl	NaCl		33.09	33.38
KCl	NaCl		6.42	4.86
KCl	LiCl		8.76	7.62
NaCl	NH_4Cl		− 4.21	− 4.81
HCl	NH_4Cl	0.01 N	27.02 mv.	27.50 mv.
HCl	LiCl		33.75	34.56
KCl	NH_4Cl		1.31	0.02
NaCl	LiCl		2.63	2.53
LiCl	CsCl		− 7.80	− 7.67

ancies are partly due to the assumption that the potentials of the two electrodes in the cell are the same, as well as to the neglect of activity coefficients in the derivation of equation (44). It is possible that the method of producing the flowing junction also has some influence on the observed results; for example, with 0.1 N solutions of hydrochloric acid and potassium chloride, a value of 28.00 mv. was obtained with the type of junction shown in Fig. 69, and 28.27 mv. when jets of the liquids were allowed to impinge on one another directly.

Elimination of Liquid Junction Potentials.—Electromotive force measurements are frequently used to determine thermodynamic quantities of various kinds; in this connection the tendency in recent years has been to employ, as far as possible, cells without transference, so as to avoid liquid junctions, or, in certain cases, cells in which a junction is formed between two solutions of the same electrolyte. As explained above, the potential of the latter type of junction is, within reasonable limits, independent of the method of forming the boundary.

In many instances, however, it has not yet been found possible to avoid a junction involving different electrolytes. If it is required to know the E.M.F. of the cell exclusive of the liquid junction potential, two alternatives are available: either the junction may be set up in a reproducible manner and its potential calculated, approximately, by one of the methods already described, or an attempt may be made to eliminate entirely, or at least to minimize, the liquid junction potential. In order to achieve the latter objective, it is the general practice to place a **salt bridge**, consisting usually of a saturated solution of potassium chloride, between the two solutions that would normally constitute the junction (Fig. 70). An indication of the efficacy of potassium chloride in reducing the magnitude of the liquid junction potential is provided by the data in Table XLVII;[23] the values recorded are the E.M.F.'s of the cell, with "free diffusion" junctions,

$$\mathrm{Hg} \mid \mathrm{Hg_2Cl_2}(s)\ 0.1\ \mathrm{N}\ \mathrm{HCl} \vdots x\ \mathrm{N}\ \mathrm{KCl} \vdots 0.1\ \mathrm{N}\ \mathrm{KCl}\ \mathrm{Hg_2Cl_2}(s) \mid \mathrm{Hg},$$

[23] Guggenheim, *J. Am. Chem. Soc.*, **52**, 1315 (1930); see also, Fales and Vosburgh, *ibid.*, **40**, 1291 (1918); Ferguson *et al.*, *ibid.*, **54**, 1285 (1932).

TABLE XLVII. EFFECT OF SATURATED POTASSIUM CHLORIDE SOLUTION ON LIQUID JUNCTION POTENTIALS

x	E.M.F.	x	E.M.F.
0.2	19.95 mv.	1.75	5.15 mv.
0.5	12.55	2.5	3.4
1.0	8.4	3.5	1.1

where x is varied from 0.2 to 3.5. When x is 0.1 the E.M.F. of the cell is 27.0 mv., and most of this represents the liquid junction potential between 0.1 N hydrochloric acid and 0.1 N potassium chloride. As the concentration of the bridge solution is increased, the E.M.F. falls to a small value, which cannot be very different from that of the cell free from liquid junction potential.

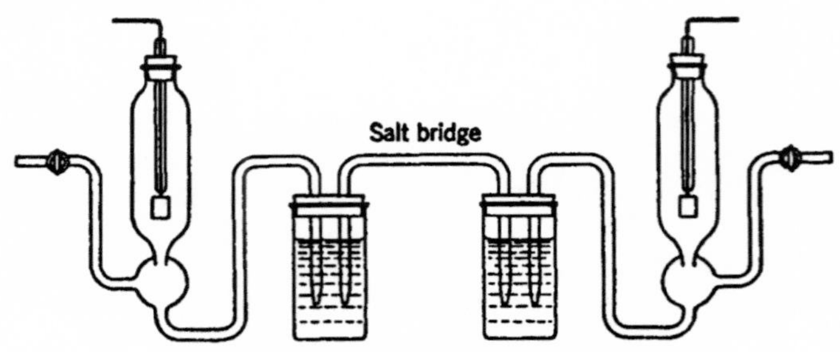

FIG. 70. Cell with salt bridge

When it is not possible to employ potassium chloride solution, e.g., if one of the junction solutions contains a soluble silver, mercurous or thallous salt, satisfactory results can be obtained with a salt bridge containing a saturated solution of ammonium nitrate; the use of solutions of sodium nitrate and of lithium acetate has also been suggested. For non-aqueous solutions, sodium iodide in methyl alcohol and potassium thiocyanate in ethyl alcohol have been employed.

The theoretical basis of the use of a bridge containing a concentrated salt solution to eliminate liquid junction potentials is that the ions of this salt are present in large excess at the junction, and they consequently carry almost the whole of the current across the boundary. The conditions will be somewhat similar to those existing when the electrolyte is the same on both sides of the junction. When the two ions have approximately equal conductances, i.e., when their transference numbers are both about 0.5 in the given solution, the liquid junction potential will then be small [cf. equation (36a)]. The equivalent conductances at infinite dilution of the potassium and chloride ions are 73.5 and 76.3 ohms^{-1} cm.^{2} at 25°, and those of the ammonium and nitrate ions are 73.4 and 71.4 ohms^{-1} cm.^{2} respectively; the approximate equality of the values for the cation and anion in each case accounts for the efficacy of potassium chloride and of ammonium nitrate in reducing liquid junction potentials.

A procedure for the elimination of liquid junction potentials, suggested by Nernst (1897), is the addition of an indifferent electrolyte at the same concentration to both sides of the cell. If the concentration of this added substance is greater than that of any other electrolyte, the former will carry almost the whole of the current across the junction between the two solutions. Since its concentration is the same on both sides of the boundary, the liquid junction potential will be very small. This method of eliminating the potential between two solutions fell into disrepute when it was realized that the excess of the indifferent electrolyte has a marked effect on the activities of the substances involved in the cell reaction. It has been revived, however, in recent years in a modified form: a series of cells are set up, each containing the indifferent electrolyte at a different concentration, and the resulting E.M.F.'s are extrapolated to zero concentration of the added substance.

Concentration Cells with a Single Electrolyte: Amalgam Concentration Cells.—In the concentration cells already described the E.M.F. is a result of the difference of activity or chemical potential, i.e., partial molal free energy, of the electrolyte in the two solutions; it is possible, however, to obtain concentration cells with only one solution, but the activities of the element with respect to which the ions in the solution are reversible are different in the two electrodes. A simple method of realizing such a cell is to employ two amalgams of a base metal at different concentrations as electrodes and a solution of a salt of the metal as electrolyte; thus

$$\text{Zn amalgam } (x_1) \mid ZnSO_4 \text{ soln.} \mid \text{Zn amalgam } (x_2),$$

the mole fractions of zinc in the amalgams being x_1 and x_2, as indicated. The passage of two faradays through this cell is accompanied by the reaction

$$Zn(x_1) = Zn^{++} + 2\epsilon,$$

at the left-hand electrode, and

$$Zn^{++} + 2\epsilon = Zn(x_2)$$

at the right-hand electrode. Since the concentration of zinc ions in the solution remains constant, the net change is the transfer of 1 g.-atom of zinc from the amalgam of concentration x_1 to that of concentration x_2; the increase of free energy is thus

$$\begin{aligned}\Delta G &= \mu_{Zn(2)} - \mu_{Zn(1)} \\ &= RT \ln \frac{a_2}{a_1},\end{aligned}$$

where a_1 and a_2 are the activities of the zinc in the two amalgams. It should be noted that in this derivation it has been assumed that the molecule and atom of zinc are identical.

The free energy change is also given by $-2FE$, where E is the E.M.F. of the cell, so that

$$E = \frac{RT}{2F} \ln \frac{a_1}{a_2}. \tag{46}$$

In the general case of an amalgam concentration cell in which the valence of the metal is z and there are m atoms in the molecule, the equation for the E.M.F. becomes

$$E = \frac{RT}{zmF} \ln \frac{a_1}{a_2}. \tag{47}$$

This result is of particular interest because it can be used to determine the activities of metals in amalgams or other alloys by E.M.F. measurements; such determinations have been carried out in a number of cases.[24]

If the amalgams are sufficiently dilute, the ratio of the activities may be taken as equal to that of their mole fractions, i.e., x_1/x_2, or even to that of their concentrations c_1/c_2; in the latter case equation (47) takes the approximate form

$$E \approx \frac{RT}{zmF} \ln \frac{c_1}{c_2}. \tag{48}$$

Experiments with amalgams of a number of metals, e.g., zinc, lead, tin, copper and cadmium have given results in general agreement with equation (48); the discrepancies observed are due to the approximation of taking the ratio of the concentrations to be equal to that of the activities.

Gas Concentration Cells.—Another form of concentration cell with electrodes of the same material at different activities, employing a single electrolyte, is obtained by using a gas, e.g., hydrogen, for the electrodes at two different pressures; thus

$$H_2(p_1) \mid \text{Solution of hydrogen ions} \mid H_2(p_2),$$

where p_1 and p_2 are the partial pressures of hydrogen in the two electrodes. The passage of two faradays through this cell is accompanied, as may be readily shown, by the transfer of 1 mole of hydrogen gas from pressure p_1 to pressure p_2; if the corresponding activities are a_1 and a_2, it is found, by using the same treatment as for amalgam concentration cells, that the E.M.F. is given by

$$E = \frac{RT}{2F} \ln \frac{a_1}{a_2}. \tag{49}$$

If the gas behaves ideally within the range of pressures employed, the ratio of activities may be replaced by the ratio of the pressures; hence

$$E = \frac{RT}{2F} \ln \frac{p_1}{p_2}. \tag{50}$$

[24] Richards and Daniels, *J. Am. Chem. Soc.*, **41**, 1732 (1919).

If one of the pressures, e.g., p_2, is kept constant while the other is varied, equation (50) takes the general form

$$E = \frac{RT}{2F} \ln p + \text{constant}, \tag{51}$$

where p is the pressure that is varied.

According to equation (51) the plot of the E.M.F. of the cell, in which one hydrogen electrode is kept at constant pressure while the other is changed, against the log p of the variable electrode should give a straight line. It is not convenient to test this equation by actual measurement of cells with two hydrogen electrodes, but an equivalent result should be obtained if the electrode of constant gas pressure is replaced by another not containing a gas, whose potential does not vary appreciably with pressure. Observations have thus been made on cells of the type

$$H_2(p) \mid HCl\ (0.1\ M)\ Hg_2Cl_2(s) \mid Hg,$$

and the results for hydrogen pressures varying from a partial pressure of 0.00517 atm., obtained by admixture with nitrogen, up to 1000 atm. are depicted in Fig. 71, in which the E.M.F.'s of the cells are plotted against

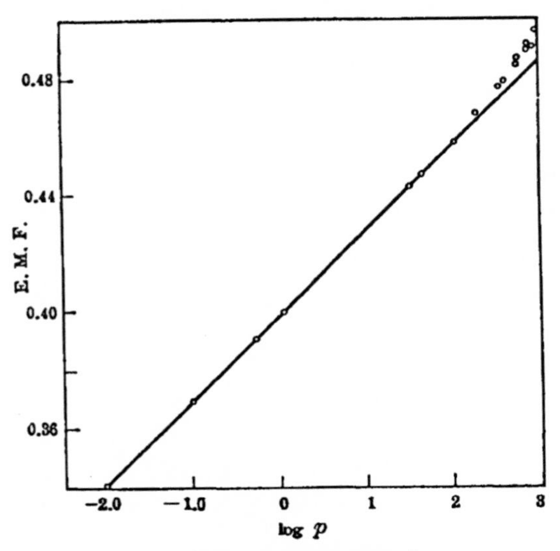

FIG. 71. Hydrogen pressure and E.M.F.

the logarithm of the hydrogen pressure.[25] It is seen that the expected linear relationship holds up to pressures of about 100 atm. The deviations from linearity up to 600 atm. can be accounted for almost exactly

[25] Hainsworth, Rowley and MacInnes, *J. Am. Chem. Soc.*, **46**, 1437 (1924); Romann and Chang, *Bull. Soc. chim.*, **51**, 932 (1932).

by making allowance for departure of the hydrogen gas from ideal behavior. The discrepancies at still higher pressure must be attributed to the neglect of the influence of pressure on the mercury-mercurous chloride electrode.

Since the passage of one mole of chlorine into solution requires two faradays, as is the case for a mole of hydrogen, the E.M.F. of a cell consisting of two chlorine electrodes at different pressures will be given by any of the equations derived above. It follows, therefore, that the E.M.F.'s of cells of the type

$$Cl_2(p) \mid HCl \text{ soln. } Hg_2Cl_2(s) \mid Hg$$

should be represented by equation (51) with the sign preceding the pressure term reversed, because the chlorine yields negative ions; the resulting equation may be put in the alternative form

$$E + \frac{RT}{2F}\ln p = \text{constant.} \tag{52}$$

The data in Table XLVIII were obtained with a cell containing 0.1 N

TABLE XLVIII. ELECTROMOTIVE FORCES OF CHLORINE GAS CELLS AT 25°

p	E	$\frac{RT}{2F}\ln p$	$E + \frac{RT}{2F}\ln p$
0.0492 atm.	− 1.0509	− 0.0387	− 1.0896
0.0247	− 1.0421	− 0.0475	− 1.0896
0.0124	− 1.0330	− 0.0564	− 1.0894
0.0631	− 1.0243	− 0.0650	− 1.0893
0.00293	− 1.0150	− 0.0749	− 1.0899

hydrochloric acid, the pressure of the chlorine gas being reduced by admixture with nitrogen; the constancy of the values in the last column confirm the accuracy of equation (52).[26]

In the case of an oxygen gas cell the electrode reactions may be represented by

$$O_2 + 2H_2O + 4\epsilon \rightleftharpoons 4OH^-,$$

so that the transfer of one mole of oxygen from one electrode to the other requires the passage of four faradays. The E.M.F. of the cell with two oxygen electrodes at different pressures is then

$$E = \frac{RT}{4F}\ln\frac{a_2}{a_1}, \tag{53}$$

or

$$E = \frac{RT}{4F}\ln\frac{p_2}{p_1} \tag{54}$$

if the gas behaves ideally. The sign of the E.M.F. is opposite to that of

[26] Lewis and Rupert, *J. Am. Chem. Soc.*, **33**, 299 (1911); Kameyama *et al.*, *J. Soc. Chem. Ind. (Japan)*, **29**, 679 (1926).

the corresponding hydrogen cell [equations (49) and (50)] because of the opposite charges of the ions. Since the oxygen gas electrode does not normally function in a reversible manner (see p. 353), these equations cannot be tested by direct experiment.

PROBLEMS

1. Determine the reactions taking place at the separate electrodes and in the complete cell in the following reversible cells:

(i) $H_2(g) \mid HCl \mid Cl_2(g)$;

(ii) $Hg \mid HgO(s)NaOH \mid H_2(g)$;

(iii) $Ag \mid AgCl(s)KCl\ Hg_2Cl_2(s) \mid Hg$;

and

(iv) $Pb \mid PbCl_2(s)KCl \,\vdots\, K_2SO_4\ PbSO_4(s) \mid Pb$.

2. Devise reversible cells in which the over-all reactions are:

(i) $Hg + PbO(s) = Pb + HgO(s)$;

(ii) $Zn + Hg_2SO_4(s) = ZnSO_4 + 2Hg$;

(iii) $Pb + 2HCl = PbCl_2(s) + H_2(g)$;

and

(iv) $H_2(g) + \frac{1}{2}O_2(g) = H_2O(l)$.

3. The following values for the E.M.F. of the cell

$$Ag \mid AgBr(s)\ \ KBr\ aq.\ Hg_2Br_2(s) \mid Hg$$

were obtained by Larson [*J. Am. Chem. Soc.*, 62, 764 (1940)] at various temperatures:

20°	25°	30°
0.06630	0.06839	0.07048 volt.

State the reaction occurring in the cell for the passage of one faraday, and evaluate the heat content, free energy and entropy changes at 25°.

4. Harned and Donelson [*J. Am. Chem. Soc.*, 59, 1280 (1937)] report that the variation of the E.M.F. of the cell

$$H_2(1\ atm.) \mid HBr(a = 1)\ \ AgBr(s) \mid Ag$$

with temperature is represented by the equation

$$E = 0.07131 - 4.99 \times 10^{-4}(t - 25) - 3.45 \times 10^{-6}(t - 25)^2.$$

Calculate the change in heat content, in calories, accompanying the reaction

$$H_2(1\ atm.) + 2AgBr(s) = 2Ag + 2HBr(a = 1)$$

at 25°.

5. The reversible cell

$$Zn \mid ZnCl_2(c_1)\ \ Hg_2Cl_2(s) \mid Hg \mid Hg_2Cl_2(s)\ \ ZnCl_2(c_2) \mid Zn$$

was found to have an E.M.F. of 0.09535 volt at 25°. Determine the ratio of the mean ion activities of the zinc chloride in the two solutions.

6. The E.M.F. of the cell

$$H_2(1 \text{ atm.}) \mid HBr(m) \; AgBr(s) \mid Ag$$

with hydrobromic acid at various small molalities (m) was measured at 25° by Keston [*J. Am. Chem. Soc.*, 57, 1671 (1935)] who obtained the results given below:

m	E	m	E
1.262×10^{-4}	0.53300	10.994×10^{-4}	0.42280
1.775	0.51616	18.50	0.39667
4.172	0.47211	37.19	0.36173

Use these data to evaluate E^0 for the cell.

7. The following results were derived from the measurements of Harned, Keston and Donelson [*J. Am. Chem. Soc.*, 58, 989 (1936)] for the cell given in the preceding problem with more concentrated solutions of the acid:

m	E	m	E
0.001	0.42770	0.05	0.23396
0.05	0.34695	0.10	0.20043
0 01	0 31262	0.20	0.16625
0.02	0.27855	0.50	0.11880

Using the value of E^0 obtained above, determine the activity coefficients of hydrobromic acid at the various molalities.

8. The following entropy values at 25° were obtained from thermal measurements: silver, 10.3 cal./deg. per g.-atom; silver chloride, 23.4 per mole; liquid mercury, 17.8 per g.-atom; and mercurous chloride, Hg_2Cl_2, 46.4 per mole. The increase in heat content of the reaction

$$Ag(s) + \tfrac{1}{2}Hg_2Cl_2(s) = AgCl(s) + Hg(l)$$

is 1,900 cal. Calculate the E.M.F. of the cell

$$Ag \mid AgCl(s) \; KCl \text{ aq.} \; Hg_2Cl_2(s) \mid Hg$$

and its temperature coefficient at 25°.

9. Abegg and Cumming [*Z. Elektrochem.*, 13, 18 (1910)] found the E.M.F. of the cell with transference

$$Ag \mid 0.1 \text{ N } AgNO_3 \,\vdots\, 0.01 \text{ N } AgNO_3 \mid Ag$$

to be − 0.0590 volt at 25°. Compare the result with the calculated value using the following data:

0.1 N $AgNO_3$	$f_\pm = 0.733$	$t_+ = 0.468$
0.01 "	0.892	0.465.

10. The E.M.F.'s of the cell with transference

$$Ag \mid AgCl(s) \; 0.1 \text{ N } HCl \,\vdots\, HCl(c) \; AgCl(s) \mid Ag$$

at 25°, and the transference numbers of the hydrogen ion in the hydrochloric acid of concentration c, are from the work of Shedlovsky and MacInnes [*J. Am.*

Chem. Soc., **58**, 1970 (1936)] and of Longsworth [*ibid.*, **54**, 2741 (1932)]:

$c \times 10^3$	E	t_{H^+}
3.4468	0.136264	0.8234
5.259	0.118815	0.8239
10.017	0.092529	0.8251
19.914	0.064730	0.8266
40.492	0.036214	0.8286
59.826	0.020600	0.8297
78.076	0.009948	0.8306
100.000	—	0.8314

Utilize these data to calculate the activity coefficients of hydrochloric acid at the several concentrations.

11. If the E.M.F. of the cell

$$Hg \mid Hg_2Cl_2(s)\ \ 0.01\ N\ KCl \vdots 0.01\ N\ KOH \vdots 0.01\ N\ NaOH\ \ HgO(s) \mid Hg$$

is E, calculate the value of the E.M.F. at 25° free from liquid junction potentials, using the Lewis and Sargent formula.

12. The E.M.F.'s of the cells

$$Zn\ \text{in}\ Hg(c_1) \mid ZnSO_4\ \text{aq.} \mid Zn\ \text{in}\ Hg(c_2)$$

were measured by Meyer [*Z. physik. Chem.*, **7**, 447 (1891)] who obtained the ensuing results:

Temp.	c_1	c_2	E
11.6°	11.30×10^{-5}	3.366×10^{-3}	0.0419 volt
60.0°	6.08×10^{-5}	2.280×10^{-3}	0.0520

Assuming the amalgams are dilute enough to behave ideally, estimate the molecular weight of zinc in the amalgams.

13. The E.M.F. of the cell

$$Cl_2(1\ \text{atm.}) \mid HCl\ \text{aq.}\ \ AgCl(s) \mid Ag$$

is − 1.1364 volt at 25°. The Ag, AgCl(s) electrode may be regarded as a chlorine electrode with the gas at a pressure equal to the dissociation pressure of silver chloride; calculate the value of this pressure at 25°.

CHAPTER VII

ELECTRODE POTENTIALS

Standard Potentials.—When all the substances taking part in a reaction in a reversible cell are in their standard states, i.e., at unit activity, the E.M.F. is the standard value E^0 for the given cell. If the reaction under consideration occurs for the passage of n faradays, then the standard free energy change ΔG^0 is equal to $-nFE^0$; hence by equation (23), page 137, with all the activities equal to unity,

$$-\Delta G^0 = nFE^0 = RT \ln K, \tag{1}$$

where K is the equilibrium constant of the cell reaction. If the reactants and resultants are at any arbitrary concentrations, or activities, the E.M.F. is E and the corresponding free energy change for the reaction ΔG is equal to $-nFE$; it follows, therefore, from equation (22), page 136, that for the reaction

$$a\text{A} + b\text{B} + \cdots = l\text{L} + m\text{M} + \cdots$$

occurring in the cell for the passage of n faradays,

$$-\Delta G = RT \ln K - RT \ln \frac{a_{\text{L}}^l a_{\text{M}}^m \cdots}{a_{\text{A}}^a a_{\text{B}}^b \cdots},$$

$$\therefore \quad nFE = nFE^0 - RT \ln \frac{a_{\text{L}}^l a_{\text{M}}^m \cdots}{a_{\text{A}}^a a_{\text{B}}^b \cdots},$$

$$\therefore \quad E = E^0 - \frac{RT}{nF} \ln \frac{a_{\text{L}}^l a_{\text{M}}^m \cdots}{a_{\text{A}}^a a_{\text{B}}^b \cdots}. \tag{2}$$

This is the general equation for the E.M.F. of any reversible chemical cell in which the reactants and resultants are at any arbitrary activities $a_{\text{A}}, a_{\text{B}}, \cdots$ and $a_{\text{L}}, a_{\text{M}}, \cdots$, respectively.

Since E^0 is related to the equilibrium constant of the reaction, it can clearly be regarded as equal to the difference between two constants E_1^0 and E_2^0 characteristic of the separate electrode reactions which together make up the process occurring in the cell as a whole. Further, the activity fraction may also be separated into two corresponding parts, so that equation (2) can be written as

$$E = \left(E_1^0 - \frac{RT}{nF} \Sigma \ln a_1^{\nu_1} \right) - \left(E_2^0 - \frac{RT}{nF} \Sigma \ln a_2^{\nu_2} \right), \tag{3}$$

where a_1 and a_2 are the activity terms applicable to the two electrodes, and ν_1 and ν_2 are the numbers of molecules or ions of the corresponding species involved in the cell reaction. The actual E.M.F. of the cell can similarly be separated into the separate potentials of the electrodes; if these are represented by E_1 and E_2, it is evident that they may be identified, respectively, with the quantities in the two sets of parentheses in equation (3). In general, therefore, it is possible to write

$$E_i = E_i^0 - \frac{RT}{nF} \Sigma \ln a_i^{\nu_i} \tag{4}$$

for the potential of an electrode in terms of its **standard potential** (E_i^0) and the activities of the species involved in the electrode process. It is evident from equation (4) that the standard potential is the potential of the electrode when all of these substances are at unit activity, i.e., in their standard states.

The application of the procedure outlined above may be illustrated with reference to the reversible cell

$$H_2(1 \text{ atm.}) \mid HCl(c)\ \ AgCl(s) \mid Ag,$$

in which the reaction is

$$\tfrac{1}{2}H_2(1 \text{ atm.}) + AgCl(s) = H^+ + Cl^- + Ag(s)$$

for the passage of one faraday. The appropriate form of equation (2) in this case is

$$E = E^0 - \frac{RT}{F} \ln \frac{a_{H^+}a_{Cl^-}a_{Ag}}{a_{H_2}^{\frac{1}{2}}a_{AgCl}}. \tag{5}$$

The individual electrode reactions (cf. p. 195) are

$$(1) \quad \tfrac{1}{2}H_2(1 \text{ atm.}) = H^+ + \epsilon,$$

and

$$(2) \quad AgCl(s) + \epsilon = Ag(s) + Cl^-,$$

so that equation (5) may be split up as follows

$$E = \left(E_1^0 - \frac{RT}{F} \ln \frac{a_{H^+}}{a_{H_2}^{\frac{1}{2}}}\right) - \left(E_2^0 - \frac{RT}{F} \ln \frac{a_{AgCl}}{a_{Ag}a_{Cl^-}}\right),$$

$$\therefore \quad E_1 = E_1^0 - \frac{RT}{F} \ln \frac{a_{H^+}}{a_{H_2}^{\frac{1}{2}}} \tag{6a}$$

and

$$E_2 = E_2^0 - \frac{RT}{F} \ln \frac{a_{AgCl}}{a_{Ag}a_{Cl^-}}. \tag{6b}$$

The standard state of hydrogen is the ideal gas at 1 atm. pressure, and the standard states of silver and silver chloride are the solids; it follows, therefore, that in this particular case a_{H_2}, a_{AgCl} and a_{Ag} are unity, so that

$$E_{H_2, H^+} = E_{H_2, H^+}^0 - \frac{RT}{F} \ln a_{H^+} \tag{7a}$$

and

$$E_{\text{Ag, AgCl, Cl}^-} = E^0_{\text{Ag, AgCl, Cl}^-} + \frac{RT}{F} \ln a_{\text{Cl}^-}, \tag{7b}$$

where the E^0 terms are the standard potentials of the H_2(1 atm.), H^+ and Ag(s), AgCl(s), Cl^- electrodes. It is seen, therefore, that in the cell under consideration the potential of each electrode depends only on the activity of one ionic species, apart from the standard potential of the system.

The results given by equations (7a) and (7b) may be expressed in a general form applicable to electrodes of all types; using the terms "oxidized" and "reduced" states in their most general sense (cf. p. 186), the potential of the electrode at which the reaction is

$$\text{Reduced State} \rightleftharpoons \text{Oxidized State} + n \text{ Electrons},$$

is given by

$$E = E^0 - \frac{RT}{nF} \ln \frac{(\text{Oxidized State})}{(\text{Reduced State})}.$$

In the electrodes already considered the hydrogen ions and the silver chloride represent the respective oxidized states, whereas hydrogen gas, in the first case, and silver and chloride ions, in the second case, are the corresponding reduced states. For any electrode, therefore, at which the reaction occurring is

$$a\text{A} + b\text{B} + \cdots = x\text{X} + y\text{Y} + \cdots + n\epsilon,$$

the general expression for the electrode potential is

$$E = E^0 - \frac{RT}{nF} \ln \frac{a_{\text{X}}^x a_{\text{Y}}^y \cdots}{a_{\text{A}}^a a_{\text{B}}^b \cdots}.$$

If the electrode is one consisting of a metal M of valence z_+, reversible with respect to M^{z_+} ions, so that the electrode reaction is

$$\text{M} \rightleftharpoons \text{M}^{z_+} + z_+\epsilon,$$

the equation for the potential takes the form

$$E_+ = E^0 - \frac{RT}{z_+F} \ln \frac{a_{\text{M}^+}}{a_{\text{M}}}, \tag{8a}$$

where a_{M} is the activity of the solid metal and a_{M^+} is that of the cations in the solution with which the metal is in equilibrium. By convention, the solid state of the metal is taken as the standard state of unit activity; for an electrode consisting of the pure metal, therefore, a_{M} may be replaced by unity so that equation (8a) becomes

$$E_+ = E^0 - \frac{RT}{z_+F} \ln a_{\text{M}^+}. \tag{8b}$$

For an electrode involving a substance A which is reversible with respect to the anions A^{z-}, the electrode reaction is

$$A^{z-} \rightleftharpoons A + z_{-}\epsilon,$$

the electrode material now being the oxidized state whereas the anions represent the reduced state; the equation for the electrode potential is then

$$E_{-} = E^{0} - \frac{RT}{z_{-}F} \ln \frac{a_{A}}{a_{A^{-}}}. \qquad (9a)$$

As before, the activity a_A of the substance A in the pure state, or if A is a gas then the activity at 1 atm. pressure, is taken as unity so that equation (9a) can be written as

$$E_{-} = E^{0} - \frac{RT}{z_{-}F} \ln \frac{1}{a_{A^{-}}}$$

$$= E^{0} + \frac{RT}{z_{-}F} \ln a_{A^{-}}. \qquad (9b)$$

The general form of equations (9a) and (9b) for any electrode reversible with respect to a single ion of valence $z_{\pm}$ is readily seen to be

$$E_{\pm} = E^{0} \mp \frac{RT}{z_{\pm}F} \ln a_{i}, \qquad (10)$$

where a_i is the activity of the particular ionic species; in this equation the upper signs apply throughout for a positive ion, while the lower signs are used for a negative ion.

For practical purposes the value of R, i.e., 8.313 joules, and F, i.e., 96,500 coulombs, may be inserted in equation (10) and the factor 2.3026 introduced to convert Naperian to Briggsian logarithms; the result is

$$E_{\pm} = E^{0} \mp 1.9835 \times 10^{-4} \frac{T}{z_{\pm}} \log a_{i}. \qquad (10a)^{*}$$

At 25° c., i.e., 298.16° K., which is the temperature most frequently employed for accurate electrochemical measurements, this equation becomes

$$E_{\pm} = E^{0} \mp \frac{0.05915}{z_{\pm}} \log a_{i}.$$

Individual Ion Activities.—The methods described in Chap. V for the determination of the activities or activity coefficients of electrolytes, as well as those depending on vapor pressure, freezing-point or other osmotic measurements, give the *mean* values for both ions into which the solute

* A convenient form of this equation for approximate purposes is

$$E_{\pm} = E^{0} \mp 0.0002 \frac{T}{z_{\pm}} \log a_{i}.$$

dissociates. The question, therefore, arises as to whether it is possible to determine *individual* ion activities experimentally. An examination of the general equation (41), p. 211, or any of the other exact equations, for the liquid junction potential, shows that this potential is apparently determined by the activities of the individual ionic species; hence, if liquid junction potentials could be measured, a possible method would be available for the evaluation of single ion activities. It should be emphasized that the so-called experimental liquid junction potentials recorded in Chap. VI were based on an assumption concerning individual ion activities, e.g., that the activity of the chloride ion is the same in all solutions of univalent chlorides at the same concentration; they cannot, therefore, be used for the present purpose.

The same point can be brought out in another manner. The E.M.F. of the cell with transference

$$\text{Ag} \mid \text{AgCl}(s)\ \ \text{KCl}(c_1) \,\vdots\, \text{KCl}(c_2)\ \ \text{AgCl}(s) \mid \text{Ag}$$

is, according to equation (25), page 203,

$$E = -2t_+ \frac{RT}{F} \ln \frac{a_2}{a_1},$$

whereas the liquid junction potential, as given by equation (35), page 208, is

$$E_L = -2t_+ \frac{RT}{F} \ln \frac{a_2}{a_1} + \frac{RT}{F} \ln \frac{(a_{\text{Cl}^-})_2}{(a_{\text{Cl}^-})_1},$$

$$\therefore \quad E_L = E + \frac{RT}{F} \ln \frac{(a_{\text{Cl}^-})_2}{(a_{\text{Cl}^-})_1}. \tag{11}$$

If the ratio of the activities of the chloride ions were known, the value of the liquid junction potential could be derived precisely from equation (11), provided the E.M.F. of the complete cell, i.e., E, were measured. Although it is true, therefore, that the individual ion activities might be evaluated from a knowledge of the liquid junction potential, the latter can be obtained only if the single ion activities are known.

A further possibility is that by a suitable device the liquid junction potential might be eliminated completely, i.e., E_L might be made equal to zero; under these conditions, therefore, equation (11) would give

$$E = \frac{RT}{F} \ln \frac{(a_{\text{Cl}^-})_1}{(a_{\text{Cl}^-})_2}, \tag{12}$$

and so the individual activities of the chloride ion at different concentrations might be obtained by using an extrapolation procedure similar to that employed in Chap. VI to determine mean activities. It is doubtful, however, whether the results would have any real thermodynamic significance; the apparent individual ion activities obtained in this manner are actually complicated functions of the transference numbers and activities of all the ions present, including those contained in the salt bridge employed to eliminate the liquid junction potential. It is possible that, as a result of a cancellation of various factors, these activities are virtually equal numerically to the individual activities of the ions, but thermodynamically they cannot be the *same* quantities.[1]

Arbitrary Potential Zero: The Hydrogen Scale.—Since the single electrode potential [cf. equation (10)] involves the activity of an individual ionic species, it has no strict thermodynamic significance; the use of such potentials is often convenient, however, and so the difficulty is overcome by defining an arbitrary zero of potential. The definition widely adopted, following on the original proposal by Nernst, is as follows:

The potential of a reversible hydrogen electrode with gas at one atmosphere pressure in equilibrium with a solution of hydrogen ions at unit activity shall be taken as zero at all temperatures.

According to this definition the standard potential of the hydrogen electrode is the arbitrary zero of potential [cf. equation (7a)]: electrode potentials based on this zero are thus said to refer to the **hydrogen scale.** Such a potential is actually the E.M.F. of a cell obtained by combining the given electrode with a standard hydrogen electrode; it has, consequently, a definite thermodynamic value. For example, the potential (E) on the hydrogen scale of the electrode M, $M^{z+}(a_{M^+})$, which is reversible with respect to the z-valent cations M^{z+}, in a solution of activity a_{M^+}, is the E.M.F. of the cell

$$M \mid M^{z+}(a_{M^+}) \;\; H^+(a_{H^+} = 1) \mid H_2(1 \text{ atm.})$$

free from liquid junction, or from which the liquid junction potential has been supposed to be completely eliminated.

The reaction taking place in the cell is

$$M + zH^+(a_{H^+} = 1) = M^{z+}(a_{M^+}) + \tfrac{1}{2}zH_2(1 \text{ atm.}), \tag{13}$$

and the change of free energy is equal to $-zFE$ volt-coulombs. If a_{M^+} is equal to unity, the potential of the electrode is E^0 and the free energy of the reaction is $-zFE^0$; this quantity is called the **standard free energy of formation** of the M^{z+} ions, although it is really the increase of the free energy of the foregoing reaction with all substances in their standard states.

If the electrode is reversible with respect to an anion, e.g., X^{z-}, as in the cell

$$X \mid X^{z-}(a_{X^-}) \;\; H^+(a_{H^+} = 1) \mid H_2(1 \text{ atm.}),$$

the reaction is

$$X^{z-}(a_{X^-}) + zH^+(a_{H^+} = 1) = X + \tfrac{1}{2}zH_2(1 \text{ atm.}), \tag{14}$$

[1] Taylor, *J. Phys. Chem.*, 31, 1478 (1927); Guggenheim, *ibid.*, 33, 842, 1540, 1758 (1929); see also, *Phil. Mag.*, 22, 983 (1936).

and the standard free energy increase is $- zFE^0$. This is the standard free energy of discharge of the X^{z-} ions, and hence the standard free energy of formation of an anion is $+ zFE^0$, where E^0 is its standard potential.

Sign of the Electrode Potential.—The convention concerning the sign of the E.M.F. of a complete cell (p. 187), in conjunction with the interpretation of single electrode potentials just given, fixes the convention as to the sign of electrode potentials. The E.M.F. of the cell

$$M \mid M^+(a_{M^+}) \; H^+(a_{H^+} = 1) \mid H_2(1 \text{ atm.})$$

will clearly be equal and opposite to that of the cell

$$H_2(1 \text{ atm.}) \mid H^+(a_{H^+} = 1) \; M^+(a_{M^+}) \mid M,$$

so that the sign of the potential of the electrode when written M, M^+ must be equal and opposite to that written M^+, M. In accordance with the convention for E.M.F.'s, the positive sign as applied to an electrode potential represents the tendency for positive ions to pass spontaneously from left to right, or of negative ions from right to left, through a cell in which the electrode is combined with a hydrogen electrode. The potential of the electrode M, M^+ represents the tendency for the metal to pass into solution as ions, i.e., for the metal atoms to be oxidized, whereas that of the electrode M^+, M is a measure of the tendency of the ions to be discharged, i.e., for the ions to be reduced.

Subsidiary Reference Electrodes: The Calomel Electrode.—The determination of electrode potentials involves, in principle, the combination of the given electrode with a standard hydrogen electrode and the measurement of the E.M.F. of the resulting cell. For various reasons, such as the difficulty in setting up a hydrogen gas electrode and the desire to avoid liquid junctions, several subsidiary reference electrodes, whose potentials are known on the hydrogen scale, have been devised. The most common of these is the calomel electrode; it consists of mercury in contact with a solution of potassium chloride saturated with mercurous chloride. Three different concentrations of potassium chloride have been employed, viz., 0.1 N, 1.0 N and a saturated solution. By making use of the standard potential of the Ag, AgCl(s), Cl^- electrode described below, the following results have been obtained for the potentials on the hydrogen scale of the three calomel electrodes at temperatures in the vicinity of 25°.[2]

Hg, $Hg_2Cl_2(s)$	0.1 N KCl	$- 0.3338 + 0.00007 (t - 25)$
Hg, $Hg_2Cl_2(s)$	1.0 N KCl	$- 0.2800 + 0.00024 (t - 25)$
Hg, $Hg_2Cl_2(s)$	Saturated KCl	$- 0.2415 + 0.00076 (t - 25)$

These values cannot be regarded as exact, since in their derivation it has been necessary to make allowance for liquid junction potentials or for

[2] Hamer, *Trans. Electrochem. Soc.*, 72, 45 (1937).

single ion activities; the calomel electrodes are, however, useful in connection with various aspects of electrochemical work, as will appear in this and later chapters (see p. 349). The electrode with 0.1 N potassium chloride is preferred for the more precise measurements because of its low temperature coefficient, but the calomel electrode with saturated potassium chloride is often employed because it is easily set up, and when used in conjunction with a saturated potassium chloride salt bridge one liquid junction, at least, is avoided.

Various types of vessels have been described for the purpose of setting up calomel electrodes; the object of the special designs is generally to prevent diffusion of extraneous electrolytes into the potassium chloride solution. In order to obtain reproducible results the mercury and mercurous chloride should be pure; the latter must be free from mercuric compounds and from bromides, and must not be too finely divided. A small quantity of mercury is placed at the bottom of the vessel; it is then covered with a paste of pure mercurous chloride, mercury and potassium chloride solution. The vessel is then completely filled with the appropriate solution of potassium chloride which has been saturated

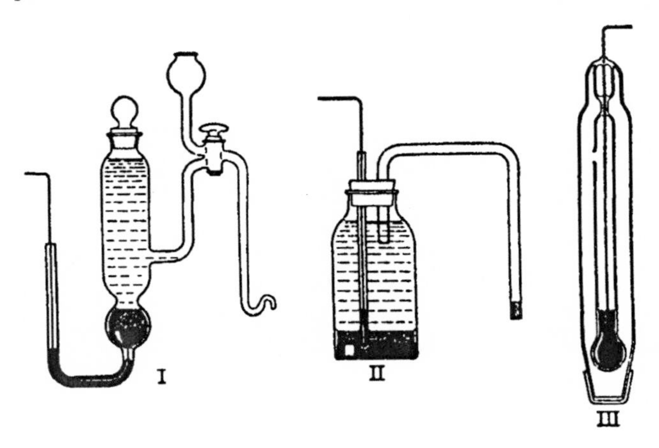

FIG. 72. Forms of calomel electrode

with calomel. Electrical connection is made by means of platinum wire sealed into a glass tube, or through the walls of the vessel. The method employed for connecting the calomel electrode to another electrode so as to make a cell whose E.M.F. can be measured depends on the type of electrode vessel. In the special form used by some workers, Fig. 72, I, this purpose is served by a side tube, sealed into the main vessel, while in the simple apparatus, consisting of a 2 or 4 oz. bottle, often employed for laboratory work (Fig. 72, II), a siphon tube provides the means of connection. The compact calomel electrode of the type used with many commercial potentiometers is dipped directly into the solution of the other electrode system; electrical connection between the two solutions occurs at the relatively loose ground joint (Fig. 72, III).

The Silver-Silver Chloride Electrode.—In recent years the silver-silver chloride electrode has been frequently employed as a reference electrode for accurate work, especially in connection with the determination of standard potentials by the use of cells containing chloride which are thus free from liquid junction potentials. The standard potential of the Ag, AgCl(s), Cl^- electrode is obtained as follows: the E.M.F. of the cell

$$H_2(1 \text{ atm.}) \mid H^+Cl^- \; AgCl(s) \mid Ag,$$

where the activities of the hydrogen and chloride ions in the solution of hydrochloric acid have arbitrary values, is given by equation (5), as

$$E = E^0 - \frac{RT}{F} \ln a_{H^+}a_{Cl^-}, \tag{15}$$

since the hydrogen, silver and silver chloride are in their standard states. Replacing the product $a_{H^+}a_{Cl^-}$ by a^2, where a is the mean activity of the hydrochloric acid, equation (15) becomes

$$E = E^0 - \frac{2RT}{F} \ln a. \tag{16}$$

This equation is seen to be identical with equation (14) of Chap. VI, and in fact the E^0 derived on page 201 by suitable extrapolation of the E.M.F. data of cells of the type shown above, containing hydrochloric acid at different concentrations, is identical with the E^0 of equations (15) and (16). It follows, therefore, that the standard E.M.F. of the cell under consideration is + 0.2224 volt at 25°, and hence the standard E.M.F. of the corresponding cell with the electrodes reversed, i.e.,

$$Ag \mid AgCl(s) \;\; H^+Cl^- \mid H_2(1 \text{ atm.})$$

is − 0.2224 volt.[3] By the convention adopted here, this represents the standard potential of the silver-silver chloride electrode; hence

$$Ag \mid AgCl(s), Cl^-(a_{Cl^-} = 1): \; E^0 = -0.2224 \text{ volt at } 25°.$$

If the potential of this electrode is required in any arbitrary chloride solution, an estimate must be made of the chloride ion activity of the latter; the potential can then be calculated by means of equation (7b).

Several methods have been described for the preparation of silver-silver chloride electrodes: a small sheet or short coil of platinum is first coated with silver by electrolysis of an argentocyanide solution, and this is partly converted into silver chloride by using it as an anode in a chloride solution. Alternatively, a spiral of platinum wire may be covered with a paste of silver oxide which is reduced to finely divided silver by heating

[3] Harned and Ehlers, *J. Am. Chem. Soc.*, 54, 1350 (1932); 55, 2179 (1933).

to about 400°; the silver is then coated with silver chloride by electrolysis in a chloride solution as in the previous case. A third method is to decompose by heat a paste of silver chlorate, silver oxide and water supported on a small spiral of platinum wire; in this way an intimate mixture of silver and silver chloride is obtained. It appears that if sufficient time is permitted for the electrodes to "age," the three methods of preparation give potentials which agree within 0.02 millivolt.[4]

Electrodes similar to that just described, but involving bromide or iodide instead of chloride, have been employed as subsidiary reference electrodes for measurements in bromide and iodide solutions, respectively. They are prepared and their standard potentials (see Table XLIX) are determined by methods precisely analogous to those employed for the silver-silver chloride electrode.[5]

Sulfate Reference Electrodes.—For measurements in sulfate solutions, the electrodes

$$Pb(Hg) \mid PbSO_4(s), SO_4^{--}$$

and

$$Hg \mid Hg_2SO_4(s), SO_4^{--}$$

have been found useful; their standard potentials may be determined by suitable extrapolation, as in the case of the silver-silver chloride electrode, or by measuring one electrode against the other.[6] The best values are

$$Pb(Hg) \mid PbSO_4(s), SO_4^{--}(a_{SO_4^{--}} = 1): \; E^0 = +\,0.3505 \text{ at } 25°$$

and

$$Hg \mid Hg_2SO_4(s), SO_4^{--}(a_{SO_4^{--}} = 1): \; E^0 = -\,0.6141 \text{ at } 25°.$$

If the electrodes are required for use as reference electrodes of known potential in sulfate solutions of arbitrary activity, an estimate of this activity must be made.

Determination of Standard Potentials: Zinc.—The procedure adopted for determining the standard electrode potential of a given metal or non-metal depends on the nature of the substance concerned; a number of examples of different types will be described in order to indicate the different methods that have been employed.

When a metal forms a soluble, highly dissociated chloride, e.g., zinc, the standard potential is best obtained from measurements on cells without liquid junction, viz.,

$$Zn \mid ZnCl_2(m) \;\; AgCl(s) \mid Ag.$$

[4] Smith and Taylor, *J. Res. Nat. Bur. Standards*, 20, 837 (1938); 22, 307 (1939).

[5] Keston, *J. Am. Chem. Soc.*, 57, 1671 (1935); Harned, Keston and Donelson, *ibid.*, 58, 989 (1936); Owen, *ibid.*, 57, 1526 (1935); Cann and Taylor, *ibid.*, 59, 1841 (1937); Gould and Vosburgh, *ibid.*, 62, 2280 (1940).

[6] Shrawder and Cowperthwaite, *J. Am. Chem. Soc.*, 56, 2340 (1934); Harned and Hamer, *ibid.*, 57, 33 (1935).

The cell reaction for the passage of two faradays is

$$Zn(s) + 2AgCl(s) = Ag(s) + Zn^{++} + 2Cl^-,$$

and the E.M.F., according to equation (2), is

$$E = E^0 - \frac{RT}{2F} \ln \frac{a_{Ag} a_{Zn^{++}} a^2_{Cl^-}}{a_{Zn} a^2_{AgCl}}. \tag{17}$$

Since the zinc, silver chloride and silver are present as solids, and hence are in their standard states, their activities are unity; hence, equation (17) becomes

$$E = E^0 - \frac{RT}{2F} \ln a_{Zn^{++}} a^2_{Cl^-}. \tag{18}$$

The standard E.M.F. of the cell, i.e., E^0, is equal to the difference between the standard potentials of the Zn, Zn^{++} and Ag, AgCl(s), Cl^- electrodes; the value of the latter is known, -0.2224 volt at 25°, and hence if E^0 of equation (18) were obtained the standard potential $E^0_{Zn, Zn^{++}}$ would be available. The evaluation of E^0 is carried out by one of the methods described in Chap. VI in connection with determination of activities and activity coefficients; the problem in the latter case is to evaluate E^0 for a particular cell, and this is obviously identical with that involved in the estimation of standard potentials.[7]

Other Bivalent Metals.—The standard potentials of a number of bivalent metals forming highly dissociated soluble sulfates, e.g., cadmium, copper, nickel and cobalt, as well as zinc, have been obtained from cells of the type

$$M \mid M^{++}SO_4^{--}(m) \; PbSO_4(s) \mid Pb(Hg)$$

and

$$M \mid M^{++}SO_4^{--}(m) \quad Hg_2SO_4(s) \mid Hg.$$

The extrapolation procedure is in principle identical with that noted above, and since the standard potentials of the electrodes Pb(Hg), $PbSO_4(s)$, SO_4^{--} and Hg, $Hg_2SO_4(s)$, SO_4^{--} are known, the standard potential of the metal M can be evaluated. In several cases the E.M.F. data for dilute solutions are not easily obtainable and consequently the extrapolation is not reliable. It is apparent, however, from measurements in moderately concentrated solutions that the sulfates of copper, nickel, cobalt and zinc behave in an exactly parallel manner, and hence the mean activity coefficients are probably the same in each case. The values for zinc sulfate are known, since E.M.F. measurements have been made at sufficiently low concentrations for accurate extrapolation and the evaluation of E^0 to be possible. The assumption is then made that the mean activity coefficients are equal in the four sulfate solutions at equal ionic strengths. It is thus possible to derive the appropriate values

[7] Scatchard and Tefft, *J. Am. Chem. Soc.*, 52, 2272 (1930); Getman, *J. Phys. Chem.*, 35, 2749 (1931).

of E^0, for the cells involving copper, nickel or cobalt sulfate, directly from the E.M.F. measurements by means of the equations

$$\begin{aligned} E &= E^0 - \frac{RT}{2F} \ln a_{M^{++}} a_{SO_4^{--}} \\ &= E^0 - \frac{RT}{2F} \ln a_{\pm}^2 \\ &= E^0 - \frac{RT}{F} \ln m_{\pm} - \frac{RT}{F} \ln \gamma_{\pm}, \end{aligned} \tag{19}$$

which are applicable to the sulfate cells; in this instance $m_{\pm}$ is equal to the molality m of the sulfate solution.

The Alkali Metals.—The alkali metals present a special case in the determination of standard potentials since these substances attack water; the difficulty has been overcome by making measurements in aqueous solution with a dilute amalgam which reacts slowly with water (cf. p. 198), and then comparing the potential of the amalgam with that of the pure metal in a non-aqueous medium with which it does not react.[8]

The E.M.F. of the stable and reproducible cell

$$\text{Na (metal)} \mid \text{NaI in ethylamine} \mid 0.206\% \ \text{Na(Hg)}$$

is + 0.8449 volt at 25°, independent of the concentration of the sodium iodide solution; since the process occurring in the cell is merely the transfer of sodium from the pure metal to the dilute amalgam, the potential must also be independent of the nature of the solvent or solute. The E.M.F. of the cell

$$0.206\% \ \text{Na(Hg)} \mid \text{NaCl aq. } 1.022 \ \text{M} \ \ Hg_2Cl_2(s) \mid \text{Hg}$$

is + 2.1582 volts at 25°, and hence that of the combination

$$\text{Na} \mid \text{NaCl aq. } 1.022 \ \text{M} \ \ Hg_2Cl_2(s) \mid \text{Hg}$$

is + 3.0031 volts. The reaction occurring in this cell is

$$Na(s) + \tfrac{1}{2}Hg_2Cl_2(s) = Hg(l) + Na^+ + Cl^-,$$

for the passage of one faraday, and so the E.M.F. is represented by

$$\begin{aligned} E &= E^0 - \frac{RT}{F} \ln a_{Na^+} a_{Cl^-} \\ &= E^0 - \frac{RT}{F} \ln a_{\pm}^2 \\ &= E^0 - \frac{2RT}{F} \ln m_{\pm} - \frac{2RT}{F} \ln \gamma_{\pm}, \end{aligned} \tag{20}$$

[8] Lewis and Kraus, *J. Am. Chem. Soc.*, 32, 1459 (1910); Armbruster and Crenshaw, *ibid.*, 56, 2525 (1934); Bent and Swift, *ibid.*, 58, 2216 (1936); Bent, Forbes and Forziati, *ibid.*, 61, 709 (1939).

where the mean molality $m_{\pm}$ is equal to the molality m of the sodium chloride solution. The molality of the solution is 1.022, and at this concentration the mean activity coefficient of sodium chloride is known from other measurements (Chap. VI) to be 0.655; it is thus readily found from equation (20) that E^0 is + 2.9826 volt at 25°. The standard potential of the electrode Hg, $Hg_2Cl_2(s)$, Cl^- is − 0.2680 volt,* and so the standard potential of the sodium electrode is given by

$$Na \mid Na^+(a_{Na^+} = 1): E^0 = +2.7146 \text{ volts at } 25°.$$

Cells with Liquid Junction.—In the cases described above it has been possible to utilize cells without liquid junctions, but this is not always feasible: the suitable salts may be sparingly soluble, they may hydrolyze in solution, their dissociation may be uncertain, or there may be other reasons which make it impossible, at least for the present, to avoid the use of cells with liquid junctions. In such circumstances it is desirable to choose, as far as possible, relatively simple junctions, e.g., between two electrolytes at the same concentration containing a common ion or between two solutions of the same electrolyte at different concentrations, so that their potentials can be calculated with fair accuracy, as shown in Chap. VI.

The procedure may be illustrated with reference to the determination of the standard potential of silver, of which the only convenient salt for experimental purposes is the nitrate. Since the most reliable reference electrodes contain solutions of halides, it is necessary to interpose a bridge solution between them; the result is

$$Ag \mid AgNO_3(0.1\text{ N}) \vdots KNO_3(0.1\text{ N}) \vdots KCl(0.1\text{ N})\ \ Hg_2Cl_2(s) \mid Hg,$$

in which the liquid junctions, indicated by the dotted lines, are both of the type to which the Lewis and Sargent equation is applicable. The E.M.F. of the complete cell is − 0.3992 volt and the sum of the liquid junction potentials is calculated to be + 0.0007 volt, so that the E.M.F. of the cell

$$Ag \mid AgNO_3(0.1\text{ N}) \parallel KCl(0.1\text{ N})\ \ Hg_2Cl_2(s) \mid Hg,$$

where the double vertical line between the two solutions is used to imply the complete elimination of the liquid junction potential, is − 0.3992 + 0.0007, i.e., − 0.3985 volt at 25°. The potential of the Hg, $Hg_2Cl_2(s)$, KCl(0.1 N) electrode is known to be − 0.3338 volt (p. 232) and so that of the Ag, $AgNO_3$(0.1 N) electrode is − 0.7323 volt. The potential of the silver electrode may be represented by means of equation (9) as

$$E = E^0_{Ag,Ag^+} - \frac{RT}{F} \ln a_{Ag^+}. \tag{21}$$

* This value is obtained by utilizing the observation that the potentials of the Hg, $Hg_2Cl_2(s)$ and Ag, AgCl(s) electrodes in the same chloride solution differ by 0.0456 volt at 25°.

and although E is known, the activity of the silver ions in 0.1 N silver nitrate is, of course, not available. It is necessary, therefore, to make an assumption, and the one commonly employed is to take the activity of the silver ions in the silver nitrate solution as equal to the mean activity of the ions in that solution. The mean activity coefficient of 0.1 N silver nitrate is 0.733, and so the mean activity which is used for a_{Ag^+} in equation (21) is 0.0733. Since E is $-$ 0.7323 volt, it is readily found that E^0_{Ag,Ag^+} is $-$ 0.7994 volt at 25°.

Halogen Electrodes.—The determination of the standard potentials of the halogens is simple in principle; it involves measurement of the potential of a platinum electrode, coated with a thin layer of platinum or iridium black, dipping in a solution of the halogen acid or a halide, and surrounded by the free halogen. The uncertainty due to liquid junction can be avoided by employing the appropriate silver-silver halide or mercury-mercurous halide electrode as reference electrode. In practice, however, difficulties arise because of the possibility of the reactions

$$X_2 + H_2O \rightleftharpoons HXO + H^+ + X^-$$

and

$$X_2 + X^- \rightleftharpoons X_3^-,$$

where X_2 is the halogen molecule; the former reaction occurs to an appreciable extent with chlorine and bromine, and the latter with bromine and iodine. The first of these disturbing effects is largely eliminated by using acid solutions as electrolytes, but due allowance for the removal of halide ions in the form of perhalide must be made from the known equilibrium constants.

The electrode reaction for the system X_2, X^- is

$$X^- = \tfrac{1}{2}X_2 + \epsilon,$$

so that by the arguments on page 228 the electrode potential is given by the equation

$$E = E^0_{X_2,X^-} - \frac{RT}{F}\ln\frac{a^{\frac{1}{2}}_{X_2}}{a_{X^-}}. \qquad (22)$$

For chlorine and bromine the standard states may be chosen as the gas at 1 atm. pressure, and if the gases are assumed to behave ideally, as will be approximately true at low pressures, equation (22) can be written in the form

$$E = E^0_{X_2,X^-} - \frac{RT}{2F}\ln p_{X_2} + \frac{RT}{F}\ln a_{X^-}, \qquad (23)$$

where p_{X_2} is the pressure of the gas in atmospheres.

In the cell

$$Cl_2(p) \mid HCl \text{ soln. } Hg_2Cl_2(s) \mid Hg$$

the reaction for the passage of one faraday is

$$\tfrac{1}{2}Hg_2Cl_2(s) = Hg + \tfrac{1}{2}Cl_2(p)$$

so that the E.M.F., which is independent of the nature of the electrolyte, provided it is a chloride solution, is given by

$$E = E^0 - \frac{RT}{2F}\ln p_{Cl_2}, \tag{24}$$

$$\therefore \quad E^0 = E + \frac{RT}{2F}\ln p_{Cl_2}. \tag{25}$$

The standard E.M.F. of this cell as given by equation (25), with the pressure in atmospheres, is the difference between the standard potentials of the Cl_2(1 atm.), Cl^- and the Hg, $Hg_2Cl_2(s)$, Cl^- electrodes; since the latter is known to be − 0.2680 volt at 25°, the value of the former could be obtained provided E^0 of the cell under consideration were available. This cell is, in fact, identical with the one for which measurements are given on page 222, and the results in the last column of Table XLVIII are actually the values of E^0 required by equation (25) above. It follows, therefore, taking a mean result of − 1.090 volts at 25° for E^0, that the standard potential of the chlorine electrode is − 1.090 − 0.2680, i.e., − 1.358 volts at 25°.

The standard potentials of bromine and iodine have been determined by somewhat similar methods; with bromine the results are expressed in terms of two alternative standard states, viz., the gas at 1 atm. pressure or the pure liquid. The standard state adopted for iodine is the solid state, so that the solution is saturated with respect to the solid phase.[9] The standard potential of fluorine has not been determined by direct experiment, but its value has been calculated from free energies derived from thermal and entropy data.[10]

The Oxygen Electrode.—The standard potential of the oxygen electrode cannot be determined directly from E.M.F. measurements on account of the irreversible behavior of this electrode (cf. p. 353); it is possible, however, to derive the value in an indirect manner. The problem is to determine the E.M.F. of the cell

$$H_2(1\text{ atm.}) \mid H^+(a_{H^+} = 1) \parallel OH^-(a_{OH^-} = 1) \mid O_2(1\text{ atm.}),$$

in which the reaction for the passage of two faradays is essentially

$$H_2(1\text{ atm.}) + \tfrac{1}{2}O_2(1\text{ atm.}) = H_2O(l).$$

The object of the calculations is to evaluate the standard free energy

[9] Lewis and Storch, *J. Am. Chem. Soc.*, 39, 2544 (1917); Jones and Baeckström, *ibid.*, 56, 1524 (1934); Jones and Kaplan, *ibid.*, 50, 2066 (1928).

[10] Latimer, *J. Am. Chem. Soc.*, 43, 2868 (1926); see also, Glasstone, "Text-Book of Physical Chemistry," 1940, p. 993.

(ΔG^0) of this process, for this is equal to $-2FE^0$, where E^0 is the standard E.M.F. of the cell.

According to equation (1),

$$\Delta G^0 = -RT \ln K,$$

where K for the given reaction is defined by

$$K = \frac{a_{H_2O}}{a_{H_2}a_{O_2}^{\frac{1}{2}}}. \tag{26}$$

The activity of liquid water is taken as unity, since this is the usual standard state, and the activities of the hydrogen and oxygen are represented by their respective pressures, since the gases do not depart appreciably from ideal behavior at low pressure; hence, equation (26) may be written as

$$K_p = \frac{1}{p_{H_2}p_{O_2}^{\frac{1}{2}}}. \tag{27}$$

From a study of the dissociation of water vapor into hydrogen and oxygen at high temperatures, it has been found that the variation with temperature of the equilibrium constant K_p', defined by

$$K_p' = \frac{p_{H_2O}}{p_{H_2}p_{O_2}^{\frac{1}{2}}},$$

can be represented, in terms of the free energy change, by

$$\Delta G^{0\prime} = -57{,}410 + 0.94T \ln T + 1.65 \times 10^{-3}T^2 - 3.7 \times 10^{-7}T^3 + 3.92T.$$

If the relationship may be assumed to hold down to ordinary temperatures, then at 25°,

$$-RT \ln K_p' = \Delta G^{0\prime} = -54{,}600 \text{ cal.},$$

and this is the free energy increase accompanying the conversion of one mole of hydrogen gas and one-half mole of oxygen to one mole of water vapor, all at atmospheric pressure. For the present purpose, however, the free energy required is that of the conversion of hydrogen and oxygen at atmospheric pressure to liquid water, i.e., to water vapor at 23.7 mm. pressure at 25°. The difference between these free energy quantities is

$$RT \ln \frac{23.7}{760} = -2{,}050 \text{ cal. at } 25°,$$

and hence the ΔG^0 required is $-54{,}600 - 2{,}050$, i.e., $-56{,}650$ cal.

An entirely different method of arriving at this standard free energy change is based partly on E.M.F. measurements, and partly on equilibrium data. From the dissociation pressure of mercuric oxide at various temperatures it is possible to obtain the standard free energy of the reaction

$$Hg(l) + \tfrac{1}{2}O_2(g) = HgO(s),$$

and when corrected to 25° the result is found to be $-$ 13,940 cal. The E.M.F. of the reversible cell

$$H_2(1 \text{ atm.}) \mid \text{KOH aq. } HgO(s) \mid Hg$$

is $+$ 0.9264 volt at 25°, and so the free energy of the reaction

$$H_2(1 \text{ atm.}) + HgO(s) = H_2O(l) + Hg(l),$$

which occurs in the cell for the passage of two faradays, is $2 \times 96,500 \times 0.9264$ volt-coulombs, i.e., $-$ 42,760 cal. Since all the reactants and resultants in this reaction are in their standard states, this is also the value of the standard free energy change.* Addition of the two results gives the standard free energy of the reaction

$$H_2(g) + \tfrac{1}{2}O_2(g) = H_2O(l)$$

as $-$ 56,700 cal. at 25°.

As a consequence of several different lines of approach, all of which give results in close agreement, it may be concluded that the standard free energy of this reaction is $-$ 56,700 cal. at 25°, and since, as seen above, this is equal to $-2FE^0$, it follows that the standard E.M.F. of the oxygen-hydrogen cell is

$$\frac{56,700 \times 4.185}{2 \times 96,500} = 1.229 \text{ volts}$$

at 25°. It would appear, at first sight, that this is also the standard potential of the oxygen electrode, but such is not the case. The E.M.F. calculated is the standard value for the cell

$$H_2(1 \text{ atm.}) \mid \text{Water} \mid O_2(1 \text{ atm.})$$

in which both oxygen and hydrogen electrodes are in contact with the *same solution*, the latter having the activity of pure water. If the hydrogen ion activity in this solution is unity, the hydrogen electrode potential is zero, by convention, and hence 1.229 volts is the potential of the electrode

$$H_2O(l), H^+(a_{H^+} = 1) \mid O_2(1 \text{ atm.}).$$

The standard potential of oxygen, as usually defined, refers to the electrode

$$O_2(1 \text{ atm.}) \mid OH^-(a_{OH^-} = 1), H_2O(l),$$

that is, in which the hydroxyl ions are at unit activity. It is known from the ionic product of water (see Chap. IX) that in pure water at 25°,

$$a_{H^+}a_{OH^-} = 1.008 \times 10^{-14},$$

and so $-$ 1.229 volts is the potential of the oxygen electrode, at 1 atm.

* A small correction may be necessary because the activity of the water in the KOH solution will be somewhat less than unity.

pressure, when the activity of the hydroxyl ions is 1.008×10^{-14}. The standard potential for unit activity of the hydroxyl ions is then derived from equation (9b) in the form

$$E = E^0_{O_2,OH^-} + \frac{RT}{F} \ln a_{OH^-},$$

which, for a temperature of 25°, becomes in this case

$$-1.229 = E^0_{O_2,OH^-} + 0.05915 \log (1.008 \times 10^{-14}),$$

$$\therefore \quad E^0_{O_2,OH^-} = -0.401 \text{ volt.}$$

Standard Electrode Potentials.—By the use of methods, such as those described above, involving either E.M.F. measurements or free energy and related calculations, the standard potentials of a number of electrodes have been determined; some of the results for a temperature of 25° are recorded in Table XLIX. It should be noted that the signs of the

TABLE XLIX. STANDARD POTENTIALS AT 25°

Electrode	Reaction	Potential	Electrode	Reaction	Potential
Li, Li^+	$Li \rightarrow Li^+ + \epsilon$	+3.024	H_2, OH^-	$\frac{1}{2}H_2 + OH^- \rightarrow H_2O + \epsilon$	+0.828
K, K^+	$K \rightarrow K^+ + \epsilon$	+2.924	O_2, OH^-	$2OH^- \rightarrow \frac{1}{2}O_2 + H_2O + 2\epsilon$	−0.401
Na, Na^+	$Na \rightarrow Na^+ + \epsilon$	+2.714			
Zn, Zn^{++}	$Zn \rightarrow Zn^{++} + 2\epsilon$	+0.761	$Cl_2(g), Cl^-$	$Cl^- \rightarrow \frac{1}{2}Cl_2 + \epsilon$	−1.358
Fe, Fe^{++}	$Fe \rightarrow Fe^{++} + 2\epsilon$	+0.441	$Br_2(l), Br^-$	$Br^- \rightarrow \frac{1}{2}Br_2 + \epsilon$	−1.066
Cd, Cd^{++}	$Cd \rightarrow Cd^{++} + 2\epsilon$	+0.402	$I_2(s), I^-$	$I^- \rightarrow \frac{1}{2}I_2 + \epsilon$	−0.536
Co, Co^{++}	$Co \rightarrow Co^{++} + 2\epsilon$	+0.283			
Ni, Ni^{++}	$Ni \rightarrow Ni^{++} + 2\epsilon$	+0.236	$Ag, AgCl(s), Cl^-$	$Ag + Cl^- \rightarrow AgCl + \epsilon$	−0.2224
Sn, Sn^{++}	$Sn \rightarrow Sn^{++} + 2\epsilon$	+0.140	$Ag, AgBr(s), Br^-$	$Ag + Br^- \rightarrow AgBr + \epsilon$	−0.0711
Pb, Pb^{++}	$Pb \rightarrow Pb^{++} + 2\epsilon$	+0.126	$Ag, AgI(s), I^-$	$Ag + I^- \rightarrow AgI + \epsilon$	+0.1522
H_2, H^+	$\frac{1}{2}H_2 \rightarrow H^+ + \epsilon$	±0.000	$Hg, Hg_2Cl_2(s), Cl^-$	$Hg + Cl^- \rightarrow \frac{1}{2}Hg_2Cl_2 + \epsilon$	−0.2680
Cu, Cu^{++}	$Cu \rightarrow Cu^{++} + 2\epsilon$	−0.340	$Hg, Hg_2SO_4(s), SO_4^{--}$	$2Hg + SO_4^{--} \rightarrow Hg_2SO_4 + 2\epsilon$	−0.6141
Ag, Ag^+	$Ag \rightarrow Ag^+ + \epsilon$	−0.799			
Hg, Hg_2^{++}	$Hg \rightarrow \frac{1}{2}Hg_2^{++} + \epsilon$	−0.799			

potentials correspond to the tendency for positive electricity to pass from left to right, or negative electricity from right to left, in each case; in general, therefore, the potentials in Table XLIX when multiplied by $-nF$ give the *standard free energy increase* for the reaction

$$\text{Reduced State} \rightarrow \text{Oxidized State} + n\epsilon,$$

the corresponding value for hydrogen being taken arbitrarily as zero. For the reverse process, the signs of the potentials would be reversed.

Since the potentials in Table XLIX give the free energies of the oxidation reactions, using the term oxidation in its most general sense, they may be called **oxidation potentials**; the potentials for the reverse processes, i.e., with the signs reversed, are then **reduction potentials** (cf. p. 435).

Potentials in Non-Aqueous Solutions.—Many measurements of varying accuracy have been made of voltaic cells containing solutions in non-aqueous media; in the earlier work efforts were made to correlate the results with the potentials of similar electrodes containing aqueous solutions. Any attempt to combine two electrodes each of which contains a different solvent is doomed to failure because of the large and uncertain potentials which exist at the boundary between the two liquids. It has been realized in recent years that the only satisfactory method of dealing with the situation is to consider each solvent as an entirely independent medium, and not to try to relate the results directly to those obtained in aqueous solutions. Since the various equations derived in this and the previous chapter are independent of the nature of the solvent, they may be applied to voltaic cells containing solutions in substances other than water.

By adopting the convention that the potential of the standard hydrogen electrode, i.e., with ideal gas at 1 atm. pressure in a solution of unit activity of hydrogen ions shall be zero in each solvent, and using methods essentially similar to those described above, the standard potentials of a number of electrodes have been evaluated in methyl alcohol, ethyl alcohol and liquid ammonia. These values represent therefore, in each case, the E.M.F. of the cell

$$M \mid M^+(a_{M^+} = 1) \parallel H^+(a_{H^+} = 1) \mid H_2(1 \text{ atm.}),$$

where M is a metal, or of

$$A \mid A^-(a_{A^-} = 1) \parallel H^+(a_{H^+} = 1) \mid H_2(1 \text{ atm.})$$

if A is a system yielding anions. It would appear at first sight that since the cell reaction, as for example in the former case,

$$M(s) + H^+(a_{H^+} = 1) = M^+(a_{M^+} = 1) + \tfrac{1}{2}H_2(1 \text{ atm.})$$

is the same in all solvents, the E.M.F. should be independent of the nature of the solvent. It must be remembered, however, that both M^+ ions and hydrogen ions are solvated in solution, and since the ions which actually exist in the respective solvents are quite different in each case, the free energy of the reaction will depend on the nature of the solvent. This subject will be considered shortly in further detail.

A number of standard potentials reported for three non-aqueous solvents are compared in Table L with the corresponding values for water as solvent;[11] it should be emphasized that although the standard potential of hydrogen is set arbitrarily at zero for each solvent, the *actual* potentials of these electrodes may be quite different in the various media. The results in each solvent are, however, comparable with one another and it will be observed that there is a distinct parallelism between the

[11] Buckley and Hartley, *Phil. Mag.*, **8**, 320 (1929); Macfarlane and Hartley, *ibid.*, **13**, 425 (1932); **20**, 611 (1935); Pleskow and Monossohn, *Acta Physicochim. U.R.S.S.*, **1**, 871 (1935); **2**, 615, 621, 679 (1935).

TABLE L. STANDARD ELECTRODE POTENTIALS IN DIFFERENT SOLVENTS

Electrode	H_2O (25°)	CH_3OH (25°)	C_2H_5OH (25°)	NH_3 (− 50°)
Li,Li^+	+ 3.024	+ 3.095	+ 3.042	—
K,K^+	+ 2.924	—	—	+ 1.98
Na,Na^+	+ 2.714	+ 2.728	+ 2.657	+ 1.84
Zn,Zn^{++}	+ 0.761	—	—	+ 0.52
Cd,Cd^{++}	+ 0.402	—	—	+ 0.18
Tl,Tl^+	+ 0.338	+ 0.379	+ 0.343	—
Pb,Pb^{++}	+ 0.126	—	—	− 0.33
H_2,H^+	± 0.000	± 0.000	± 0.000	± 0.000
Cu,Cu^{++}	− 0.340	—	—	− 0.43
Ag,Ag^+	− 0.799	− 0.764	− 0.749	− 0.83
Cl_2,Cl^-	− 1.358	− 1.116	− 1.048	− 1.28 *
Br_2,Br^-	− 1.066	− 0.837	− 0.777	− 1.08 *
I_2,I^-	− 0.536	− 0.357	− 0.305	− 0.70 *

* Calculated from free energy data at about 0° c.

standard potentials of the various electrodes in the four solvents. The tendency for the reaction $M \rightarrow M^+ + \epsilon$ to occur, as indicated by a high positive value of the potential, is always greatest with the alkali metals and least with the more noble metals, e.g., copper and silver. The order of the halogens is also the same in each case.

Factors Affecting Electrode Potentials.—If E^0 is the standard potential of a metal in a given solvent, then it is evident from the arguments given above that $-zFE^0$ is equal to the standard free energy of the reaction

$$M + zH^+ = M^{z+} + \tfrac{1}{2}zH_2.$$

This reaction, which is the displacement of hydrogen ions from the solution and their liberation as hydrogen gas, is virtually that occurring when a metal dissolves in a dilute acid solution, provided there are no accompanying complications, e.g., formation of complex ions. It follows, therefore, that $-zFE^0$ may be regarded as the standard free energy of solution of the metal.

According to thermodynamics

$$\Delta G^0 = \Delta H^0 - T\Delta S^0,$$

and experiments have shown that the standard entropy change ΔS^0 resulting from the solution of a metal in dilute acid is relatively small compared with the heat change ΔH^0; it is possible, therefore, to write as a very approximate relationship

$$-zFE^0 \approx \Delta H,$$

where ΔH is heat of solution of the metal. In general, therefore, a parallelism is to be expected between the latter quantity and the standard potential of the metal; hence the factors determining the heat of solution may be regarded as those influencing the standard potential.[12]

In order to obtain some information concerning these factors the reaction involved in the solution of the metal may be imagined to take place in a series of stages, as shown in Fig. 73; the reactants, M and

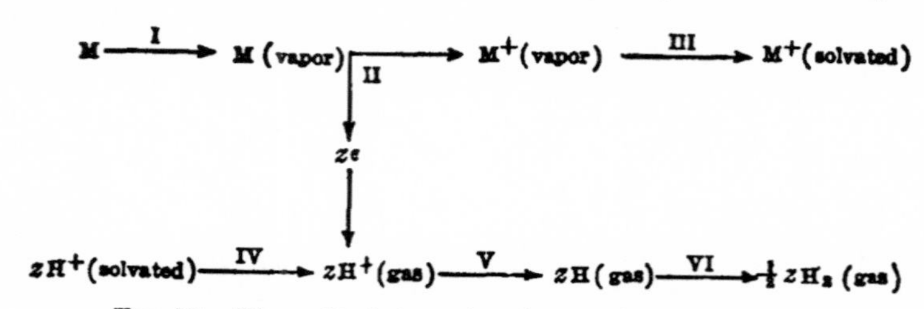

Fig. 73. Theoretical stages in solution of a metal in acid

solvated hydrogen ions, are shown at the left, and the products, hydrogen gas and solvated M^{z+} ions, at the right. The stages are as follows:

I. An atom of the metal is vaporized; the heat supplied is equal to the heat of sublimation, S; hence,

$$\Delta H_{\mathrm{I}} = + S.$$

II. The atom of vapor is ionized to form metal ions M^{z+} and z electrons; the energy which must be supplied is determined by the ionization potential of the metal, the various stages of ionization being taken into consideration if the ion has more than one charge. If I_{M} is the sum of the ionization potentials, the energy of ionization is $I_{\mathrm{M}}\epsilon$, and if it is supposed that this is converted into the standard units of energy used throughout these calculations, then

$$\Delta H_{\mathrm{II}} \approx + I_{\mathrm{M}}\epsilon.$$

III. The gaseous metal ion is dissolved in the solvent, when energy equal to the heat of solvation W_{M^+} is evolved; hence,

$$\Delta H_{\mathrm{III}} = - W_{\mathrm{M}^+}.$$

IV. An equivalent quantity of solvated hydrogen ions (z ions) are removed from the solvent; the energy of solvation W_{H^+} per ion is absorbed, so that

$$\Delta H_{\mathrm{IV}} = + zW_{\mathrm{H}^+}.$$

V. The unsolvated (gaseous) hydrogen ions are combined with the electrons removed from the metal to form atomic hydrogen; if I_{H} is the ionization potential of the hydrogen atom, then

$$\Delta H_{\mathrm{V}} \approx - zI_{\mathrm{H}}\epsilon,$$

since z electrons are added.

[12] Butler, "Electrocapillarity," 1940, Chap. III.

VI. The hydrogen atoms are combined in pairs to form hydrogen molecules; if D_{H_2} is the heat of dissociation of a hydrogen molecule into atoms, then

$$\Delta H_{VI} = -\tfrac{1}{2}zD_{H_2}.$$

The net result of these six stages is the same as the solution of a metal in a dilute acid; hence ΔH for this process is given by the sum of the six heat changes recorded for the separate stages, thus, assuming constant pressure,

$$\Delta H \approx S + I_{M\epsilon} - W_{M^+} - z(\tfrac{1}{2}D_{H_2} + I_{H^+} - W_{H^+}). \quad (28)$$

The quantity in parentheses is characteristic of the hydrogen electrode in the given solvent, and so the factors which determine the heat of solution of a particular metal, and consequently (approximately) its standard potential, may be represented by the expression

$$\Delta H_M \approx S + I_{M\epsilon} - W_{M^+}.$$

The standard potential of a metal in a given solvent thus apparently depends on the sublimation energy of the metal, its ionization potential and the energy of solvation of the ions. Calculations have shown that of these factors the heat of sublimation is much the smallest, but since the other two quantities generally do not differ very greatly, all three factors must play an important part in determining the actual electrode potential.

When comparing the heat changes accompanying the solution of a given metal in different media, it is seen that the factors S, I_M, D_{H_2} and I_H are independent of the nature of the solvent. The standard potential of the metal in different solvents is thus determined by the quantity $W_{H^+} - W_{M^+}$, where W_{H^+} and W_{M^+} are the energies, strictly the *free energies*, of solvation of the hydrogen and M^+ ions, respectively; this result is in agreement with the general conclusion reached previously (p. 244). For a series of similar solvents, such as water and alcohols, the values of $W_{H^+} - W_{M^+}$ for a number of metals will follow much the same order in each solvent; in that case the standard potentials will show the type of parallelism observed in Table L. On the other hand it would not be surprising if for dissimilar solvents. e.g., water and acetonitrile, the order followed by the potentials of a number of electrodes was quite different in the two solvents.

Absolute Single Electrode Potentials.—The electrode potentials discussed hitherto are actually the E.M.F.'s of cells resulting from the combination of the electrode with a standard hydrogen electrode. A single electrode potential, as already seen, involves individual ion activities and hence has no thermodynamic significance;[13] the *absolute* potential difference at an electrode is nevertheless a quantity of theoretical interest. Many attempts have been made to set up so-called "null electrodes"

[13] See, for example, Guggenheim, *J. Phys. Chem.*, 33, 842 (1929).

in which there is actually no difference of potential between the metal and the solution; if such an electrode were available it would be possible by combining it with another electrode to derive the absolute potential of the latter. It appears doubtful, however, whether the "null electrodes" so far prepared actually have the significance attributed to them, since they generally involve relative movement of the metal and the solution (cf. Chap. XVI). A possible approach to the problem is based on a treatment similar to that used in the previous section.

The absolute single potential of a metal is a measure of the standard free energy of the reaction

$$M + \text{solvent} = M^{z+} \text{ (solvated)} + z\epsilon,$$

and this process may be imagined to occur by the series of stages depicted in Fig. 74. These, with the accompanying free energy changes, are vaporization of the metal $(+ S)$; ionization of the atom in the vapor

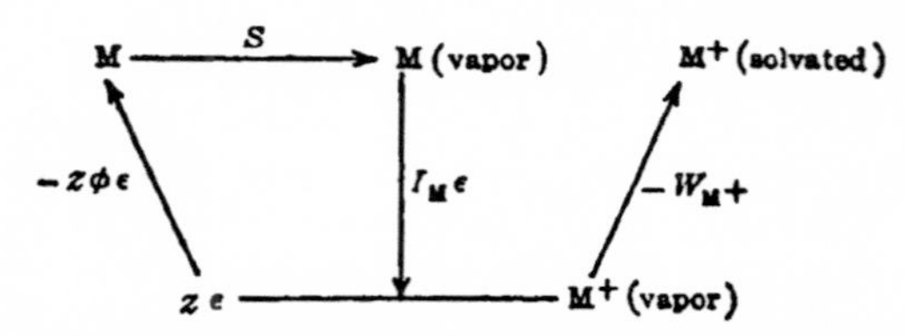

FIG. 74. Theoretical stages in formation of ions in solution

state $(I_M\epsilon)$; solvation of the gaseous ion $(- W_{M^+})$; and finally return of the electrons produced in the ionization stage to the metal $(- z\phi\epsilon)$, where ϕ is the electronic work function of the metal.* It follows, therefore, that

$$\Delta G = S + I_M\epsilon - W_{M^+} - z\phi\epsilon. \qquad (29)$$

Since S, I_M and ϕ, as well as z and ϵ, may be regarded as being known for a given metal, it should be feasible to evaluate ΔG for the ionization process, provided the free energy of solvation of the M^+ ions, i.e., W_{M^+}, were known.

The sum of the energies of solvation of the ions of a salt can be estimated, at least approximately, from the heat of solution of the salt and its lattice energy in the crystalline form. There is, unfortunately, no direct method of dividing this sum into the contributions for the separate ions; it is of interest, however, to consider the theoretical approach to this problem as outlined in the following section.

Free Energy of Solvation of Ions.—If the solvent medium is considered as a continuous dielectric, the free energy of solvation may be

* The electronic work function, or thermionic work function, generally expressed in volts, is a measure of the amount of energy required to remove an electron from the metal; $- z\phi\epsilon$ is, therefore, the free energy change, in electron-volts, accompanying the return of the z electrons to the metal.

regarded as equivalent to the difference in the electrostatic energy of a gaseous ion and that of an ion in the medium of dielectric constant D. In order to evaluate this quantity, use is made of the method proposed by Born:[14] the free energy increase accompanying the charging of a single *gaseous* ion, i.e., in a medium of dielectric constant unity, is $z^2\epsilon^2/2r$, where $z\epsilon$ is the charge carried by the ion and r is its effective radius, the ion being treated as a conducting sphere. If the same ion is charged in a medium of dielectric constant D, the free energy change is $z^2\epsilon^2/2Dr$, and so the increase of free energy accompanying the transfer of the gaseous ion to the particular medium, which may be equated to the **free energy of solvation,** is given by the Born equation as

$$\Delta G = W_{M^+} = -\frac{Nz^2\epsilon^2}{2r}\left(1 - \frac{1}{D}\right),$$

where N, the Avogadro number, is introduced to give the free energy change per mole.

One of the difficulties in applying the Born equation is that the effective radius of the ion is not known; further, the calculations assume the dielectric constant of the solvent to be constant in the neighborhood of the ion. The treatment has been modified by Webb[15] who allowed for the variation of dielectric constant and also for the work required to compress the solvent in the vicinity of the ion; further, by expressing the effective ionic radius as a function of the partial molal volume of the ion, it was possible to derive values of the free energy of solvation without making any other assumptions concerning the effective ionic radius.

Another approach to the problem of ionic solvation has been made by Latimer and his collaborators;[16] by taking the effective radii of negative halogen ions as 0.1Å greater than the corresponding crystal radii and those of positive alkali metal ions as 0.85Å greater than the crystal radii, it has been found possible to divide up the experimental free energies of hydration of alkali halides into the separate values for the individual ions. The results so obtained are in agreement with the requirements of the original form of the Born equation with the dielectric constant equal to the normal value for water.

The free energies of hydration of single ions derived by the different methods of computation show general agreement. For univalent ions the values are approximately 70 to 100 kcal. per g.-ion; the hydrogen ion is exceptional in this respect, its free energy of hydration being about 250 kcal. In any series of ions, e.g., alkali metal ions or halide ions, the hydration free energy usually decreases with increasing mass of the ion.

In spite of the fact that the different treatments yield similar values, it must be emphasized that there is considerable doubt if the results are

[14] Born, *Z. Physik*, **1**, 45 (1920).
[15] Webb, *J. Am. Chem. Soc.*, **48**, 2589 (1926).
[16] Latimer, Pitzer and Slansky, *J. Chem. Phys.*, **7**, 108 (1939).

of sufficient significance to permit of their use in the determination of absolute potentials.[17] The problem of single potentials must, therefore, still be regarded as incompletely solved.

Rates of Electrode Processes.—When a metal M is inserted in a solution of its ions $M(H_2O)_x^+$, the solvent being assumed for simplicity to be water, there will be a tendency for the metal to pass into solution as ions and also for the ions from the solution to discharge on to the metal; in other words the two processes represented by the reversible reaction

$$M(H_2O)_x^+ + \epsilon \underset{k_2}{\overset{k_1}{\rightleftharpoons}} M + xH_2O$$

will occur simultaneously, the ions $M(H_2O)_x^+$ being in solution and the electrons ϵ on the metal. When equilibrium is attained, and the reversible potential of the electrode is established, the two reactions take place at equal rates.

According to modern views,[18] the rate of a process is equal to the specific rate, defined in terms of the accepted standard states, multiplied by the activities of the reacting species;* if k_1 and k_2 are the specific rates of the direct and reverse processes represented above, *in the absence of any potential difference*, then, since a_+ is the activity of the solvated ions in solution and the activity of the solid metal is unity, by convention, the rates of the reactions are k_1a_+ and k_2, respectively. If k_2 is greater than k_1a_+, that is to say, if the reverse reaction in the absence of a potential difference at the electrode, i.e., the passage of ions from the metal into the solution, is more rapid than the direct reaction, i.e., the discharge of ions, the cations will pass into solution from the metal more rapidly than they can return. As a result, therefore, free electrons will be left on the metal and positive ions will accumulate on the solution side of the electrode, thus building up what is known as an **electrical double layer** (see Chap. XVI); the potential difference across this double layer is the single electrode potential. The setting up of the double layer, with its associated potential difference, makes it more difficult for ions to leave the negatively charged metal and enter the solution, while the transfer of ions to the metal, i.e., the direct reaction, is facilitated. When equilibrium is established the two processes are occurring at the same rate and the electrode exhibits its reversible potential.

If E is the actual potential difference across the double layer, formed by the electrons on the metal and the ions in solution, it may be supposed that a fraction α of this potential facilitates the discharge of ions, while the remainder, $1 - \alpha$, hinders the reverse process, i.e., the passage of ions

[17] Frumkin, *J. Chem. Phys.*, 7, 552 (1939).

[18] Glasstone, Laidler and Eyring, "The Theory of Rate Processes," 1941, Chap. X.

* Strictly speaking, the result should be divided by the activity coefficient of the "activated state" for the reaction; in any case this factor cancels out when equilibrium processes are considered.

from the metal into the solution. The actual value of α, which lies between zero and unity, is immaterial for present purposes, since it cancels out at a later stage. In its transfer across the double layer, therefore, the free energy of the discharging ion is increased by αzFE, where z is the valence of the ion, while the free energy of the atom which passes into solution is diminished by an amount $(1 - \alpha)zFE$. The result of these free energy changes is that, in the presence of the double layer potential E, the rates of the forward and reverse reactions under consideration are:

$$\text{Rate of discharge of ions from solution} = k_1 a_+ e^{\alpha zFE/RT}$$

$$\text{Rate of passage of ions into solution} = k_2 e^{-(1-\alpha)zFE/RT},$$

where, as already seen, the corresponding rates in the absence of the potential are $k_1 a_+$ and k_2, respectively. At the equilibrium (reversible) potential the rates of the two processes must be equal; hence

$$k_1 a_+ e^{\alpha zFE/RT} = k_2 e^{-(1-\alpha)zFE/RT},$$

$$\therefore \quad e^{-zFE/RT} = a_+ \frac{k_2}{k_1},$$

$$\therefore \quad E = \frac{RT}{zF} \ln \frac{k_1}{k_2} - \frac{RT}{zF} \ln a_+. \tag{30}$$

Since k_1/k_2 is a constant at definite temperature, this equation is obviously of the same form as the electrode potential equations derived by thermodynamic methods, e.g., equation (8*b*) for an electrode reversible with respect to positive ions. The first term on the right-hand side of equation (30) is clearly the absolute single standard potential of the electrode; it is equal to the standard free energy of the conversion of solid metal to solvated ions in solution divided by zF, and its physical significance has been already discussed.

Electrode Potentials and Equilibrium Constants.—According to equation (1) the standard E.M.F., i.e., E^0, of any reversible cell can be related to the equilibrium constant of the reaction occurring in the cell by the expression

$$E^0 = \frac{RT}{zF} \ln K, \tag{31}$$

and hence a knowledge of the standard E.M.F. permits the equilibrium constant to be calculated, or *vice versa*.

The reaction occurring in the cell

$$Zn \mid ZnSO_4 \text{ aq.} \parallel CuSO_4 \text{ aq.} \mid Zn,$$

for example, for the passage of two faradays is

$$Zn + Cu^{++}\text{aq.} = Zn^{++}\text{aq.} + Cu,$$

and if E^0 is the standard E.M.F., it follows from equation (31) that

$$E^0_{\mathrm{Zn,Cu}} = \frac{RT}{2F} \ln \left(\frac{a_{\mathrm{Zn^{++}}} a_{\mathrm{Cu}}}{a_{\mathrm{Zn}} a_{\mathrm{Cu^{++}}}} \right)_e ,$$

the suffix e indicating that the activities involved are the *equilibrium* values. Since the solid zinc and copper constituting the electrodes are in their standard states, their respective activities are unity; hence,

$$E^0_{\mathrm{Zn,Cu}} = \frac{RT}{2F} \ln \left(\frac{a_{\mathrm{Zn^{++}}}}{a_{\mathrm{Cu^{++}}}} \right)_e . \tag{32}$$

If $E^0_{\mathrm{Zn,Zn^{++}}}$ and $E^0_{\mathrm{Cu,Cu^{++}}}$ represent the standard electrode potentials on the hydrogen scale of the zinc and copper electrodes, as recorded in Table XLIX, then $E^0_{\mathrm{Zn,Zn^{++}}}$ is actually the E.M.F. of the cell

$$\mathrm{Zn} \mid \mathrm{Zn^{++}}(a_{\mathrm{Zn^{++}}} = 1) \parallel \mathrm{H^+}(a_{\mathrm{H^+}} = 1) \mid \mathrm{H_2}(1\ \mathrm{atm.}),$$

while $E^0_{\mathrm{Cu,Cu^{++}}}$ is the E.M.F. of the cell

$$\mathrm{Cu} \mid \mathrm{Cu^{++}}(a_{\mathrm{Cu^{++}}} = 1) \parallel \mathrm{H^+}(a_{\mathrm{H^+}} = 1) \mid \mathrm{H_2}(1\ \mathrm{atm.}).$$

Hence the E.M.F. of the cell

$$\mathrm{Zn} \mid \mathrm{Zn^{++}}(a_{\mathrm{Zn^{++}}} = 1) \parallel \mathrm{Cu^{++}}(a_{\mathrm{Cu^{++}}} = 1) \mid \mathrm{Cu},$$

which has been defined above as $E^0_{\mathrm{Zn,Cu}}$, is also equal to $E^0_{\mathrm{Zn,Zn^{++}}} - E^0_{\mathrm{Cu,Cu^{++}}}$. It follows, therefore, from equation (32) that

$$E^0_{\mathrm{Zn,Zn^{++}}} - E^0_{\mathrm{Cu,Cu^{++}}} = \frac{RT}{2F} \ln \left(\frac{a_{\mathrm{Zn^{++}}}}{a_{\mathrm{Cu^{++}}}} \right)_e , \tag{33}$$

and inserting the standard potentials from Table XLIX, the result is, at 25°,

$$+ 0.761 - (- 0.340) = \frac{0.05915}{2} \log \left(\frac{a_{\mathrm{Zn^{++}}}}{a_{\mathrm{Cu^{++}}}} \right)_e ,$$

$$\therefore \quad \left(\frac{a_{\mathrm{Zn^{++}}}}{a_{\mathrm{Cu^{++}}}} \right)_e = 1.7 \times 10^{37}.$$

The ratio of the activities of the zinc and copper ions at equilibrium will be approximately equal to the ratio of the concentrations under the same conditions; it follows, therefore, that when the system consisting of zinc, copper and their bivalent ions attains equilibrium the ratio of the zinc ion to the copper ion concentration is extremely large. If zinc is placed in contact with a solution of cupric ions, e.g., copper sulfate, the zinc will displace the cupric ions from solution until the $c_{\mathrm{Zn^{++}}}/c_{\mathrm{Cu^{++}}}$ ratio is about 10^{37}; in other words the zinc will replace the copper in solution until the quantity of cupric ions remaining is too small to be detected.

It is thus possible from a knowledge of the standard electrode potentials of two metals to determine the extent to which one metal will replace another, or hydrogen, from a solution of its ions. In the general case of two metals M_1 and M_2, of valence z_1 and z_2, respectively, the reaction which occurs for the passage of z_1z_2 faradays is

$$z_2M_1 + z_1M_2^{z_2+} = z_2M_1^{z_1+} + z_1M_2,$$

and the corresponding general form of equation (33) is

$$E^0_{M_1,M_1^+} - E^0_{M_2,M_2^+} = \frac{RT}{z_1z_2F} \ln \frac{(a_{M_1^+})_e^{z_2}}{(a_{M_2^+})_e^{z_1}}. \quad (34)$$

It can be seen from this equation that the greater the difference between the standard potentials of the two metals M_1 and M_2, the larger will be the equilibrium ratio of activities (or concentrations) of the respective ions. The greater the difference between the standard potentials, therefore, the more completely will one metal displace another from a solution of its ions. The metal with the more positive (oxidation) potential, as recorded in Table XLIX, will, in general, pass into solution and displace the metal with the less positive potential. The series of standard potentials, or **electromotive series,** as it is sometimes called, thus gives the order in which metals are able to displace each other from solution; the further apart the metals are in the series the more completely will the higher one displace the lower one. It is not true, however, to say that a metal lower in the series will not displace one higher in the series; some displacement must always occur until the required equilibrium is established, and the equilibrium amounts of both ions are present in the solution.

By re-arranging equation (34) the result is

$$E^0_{M_1,M_1^+} - \frac{RT}{z_1F} \ln (a_{M_1^+})_e = E^0_{M_2,M_2^+} - \frac{RT}{z_2F} \ln (a_{M_2^+})_e. \quad (35)$$

The left-hand side of this equation clearly represents the reversible potential of the metal M_1 in the equilibrium solution and the right-hand side is that of the metal M_2. It must be concluded, therefore, that when the metal M_1 is placed in contact with a solution of M_2^+ ions, or M_2 is placed in a solution of M_1^+ ions, or in general whenever the conditions are such that the equilibrium

$$M_1 + M_2^+ \rightleftharpoons M_1^+ + M_2$$

is established, the reversible potential of the system M_1, M_1^+ is equal to that of M_2, M_2^+. It is clear from equation (35) that the more positive the standard potential of a given metal, the greater the activity of the corresponding ions which must be present at equilibrium, and hence the more completely will it displace the other metal.

Although equations (34) and (35) are exact, the qualitative conclusions drawn from them are not always strictly correct; for example, since copper has a standard potential of $-$ 0.340 on the hydrogen scale, it would be expected, as is true in the majority of cases, that copper should be unable to displace hydrogen from solution. It must be recorded, however, that copper dissolves in hydrobromic acid, and even in potassium cyanide solution, with the liberation of hydrogen. The reason for this surprising behavior is to be found in the fact that in both instances complex ions are formed whereby the cupric ions are removed from the solution. It is true that when equilibrium is attained the concentration (or activity) of cupric ions is very small in comparison with that of the hydrogen ions, but in order to attain even this small concentration it is necessary for a considerable amount of copper to pass into solution; most of this dissolved copper is present in the form of complex ions, and it is the amount of free cupric ions in equilibrium with these complexes that must be inserted in equation (34) or (35).

It is of interest to note that if the equilibrium constant of the system consisting of two metals and their simple ions could be determined experimentally and the standard potential of one of them were known, the standard potential of the other metal could be evaluated by means of equation (34). This method was actually used to obtain the standard potential of tin recorded in Table XLIX. Finely divided tin and lead were shaken with a solution containing lead and tin perchlorates until equilibrium was attained; the ratio of the concentrations of lead and stannous ions in the solution was then determined by analysis. The standard potential of lead being known, that of tin could be calculated.

Electrode Potentials and Solubility Product.—The solubility product is an equilibrium constant, namely for the equilibrium between the solid salt on the one hand and the ions in solution on the other hand, and methods are available for the evaluation of this property from E.M.F. measurements.

The reaction taking place in the cell

$$Cl_2(1 \text{ atm.}) \mid HCl \;\; AgCl(s) \mid Ag$$

for the passage of one faraday is readily seen to be

$$AgCl(s) = Ag + \tfrac{1}{2}Cl_2(1 \text{ atm.}),$$

but since the solid silver chloride is in equilibrium with silver and chloride ions in the solution, the reaction can be considered to be

$$AgCl(s) \rightleftharpoons Ag^+ + Cl^- = Ag + \tfrac{1}{2}Cl_2(1 \text{ atm.}).$$

The E.M.F. of the cell is then written as

$$E = E^0 - \frac{RT}{F} \ln \frac{a_{Ag}a_{Cl_2}^{\frac{1}{2}}}{a_{Ag^+}a_{Cl^-}}, \tag{36}$$

where a_{Ag^+} and a_{Cl^-} refer to the activities in the saturated solution. The value of E^0 in this equation is the E.M.F. of the cell in which the activity of the chlorine gas and of chloride ions on the one hand, and of solid silver and silver ions on the other hand, are unity; these conditions arise for the standard Cl_2, Cl^- and Ag, Ag^+ electrodes, respectively, so that E^0

in equation (36) is defined by

$$E^0 = E^0_{Cl_2, Cl^-} - E^0_{Ag, Ag^+}.$$

Since solid silver and chlorine gas at atmospheric pressure are the respective standard states, i.e., the activity is unity, equation (36) can be written as

$$E = E^0_{Cl_2, Cl^-} - E^0_{Ag, Ag^+} + \frac{RT}{F} \ln a_{Ag^+}a_{Cl^-}. \qquad (37)$$

The product $a_{Ag^+}a_{Cl^-}$ in the saturated solution may be replaced by the solubility product of silver chloride, i.e., $K_{s(AgCl)}$, and so equation (37) becomes

$$E = E^0_{Cl_2, Cl^-} - E^0_{Ag, Ag^+} + \frac{RT}{F} \ln K_{s(AgCl)}.$$

It is seen from Table XLIX that $E^0_{Cl_2, Cl^-}$ and E^0_{Ag, Ag^+} are respectively -1.358 and -0.799 volt at 25°; hence,

$$E = -1.358 + 0.799 + 0.05915 \log K_{s(AgCl)}.$$

From measurements on the cell depicted at the head of this section, it is found that E is -1.136 volt at 25°, and consequently it follows that

$$K_{s(AgCl)} = 1.78 \times 10^{-10}.$$

The value derived from the solubility of silver chloride obtained by the conductance method is 1.71×10^{-10}.

In general, the above procedure can be applied to any sparingly soluble salt, provided an electrode can be obtainable which is reversible with respect to each ion, viz.,

$$\text{A} \mid \text{Soluble salt of A}^- \text{ ions} \quad \text{MA}(s) \mid \text{M},$$

although for a hydroxide, the oxygen electrode may be replaced by a hydrogen electrode.

A less accurate method for the determination of solubility products, but which is of wider applicability, is the following. If MA is the sparingly soluble salt, and NaA is a soluble salt of the same anion, then the potential of the electrode M, MA(s), NaA aq. may be obtained by combining it with a reference electrode, e.g., a calomel electrode, thus

$$\text{M} \mid \text{MA}(s) \ \text{NaA aq.} \parallel \text{KCl aq.} \ Hg_2Cl_2(s) \mid \text{Hg},$$

with a suitable salt bridge to minimize the liquid junction potential, and measuring the E.M.F. of the resulting cell. Since the potential of the calomel electrode is known, that of the other electrode may be evaluated, on the hydrogen scale. The potential of the M | MA(s) NaA electrode which can be treated as reversible with respect to M^+ ions as well as to

A^- ions, may be written as

$$E = E^0_{M,M^+} - \frac{RT}{zF} \ln a_{M^+},$$

and if E^0_{M,M^+} is known, the activity of the M^+ ions in the solution saturated with MA can be calculated. The activity of the A^- ions may be taken as approximately equal to the mean activity of the salt MA whose concentration is known; the product of a_{M^+} and a_{A^-} in the solution then gives the solubility product of MA.

Electrometric Titration: Precipitation Reactions.—One of the most important practical applications of electrode potentials is to the determination of the end-points of various types of titration;[19] the subject will be treated here from the standpoint of precipitation reactions, while neutralization and oxidation-reduction processes are described more conveniently in later chapters.

Suppose a solution of the soluble salt MX, e.g., silver nitrate, is titrated with a solution of another soluble salt BA, e.g., potassium chloride, with the result that the sparingly soluble salt MA, e.g., silver chloride, is precipitated. Let c moles per liter be the initial concentration of the salt MX, and suppose that at any instant during the titration x moles of BA have been added per liter; further, let y moles per liter be the solubility of the sparingly soluble salt MA at that instant. The value of y will vary throughout the course of the titration since the concentration of M ions is being continuously altered. If the salts are assumed to be completely dissociated, the concentration of M^+ at any instant is given by

$$c_{M^+} = c - x + y,^*$$

where $c - x$ is due to unchanged MX and y to the amount of the sparingly soluble MA remaining in solution. The simultaneous concentration of A^- ions is then

$$c_{A^-} = y,$$

because the A^- ions in solution arise solely from the solubility of MA, the remainder having been removed in the precipitate. Since the solution is saturated with MA, it follows from the approximate solubility product principle, assuming activity coefficients to be unity, that

$$k_s = c_{M^+} \times c_{A^-} = (c - x + y)y, \tag{38}$$

where k_s is the concentration solubility product.

[19] For reviews, see Kolthoff and Furman, "Potentiometric Titrations," 1931; Furman, *Ind. Eng. Chem.* (*Anal. Ed.*), **2**, 213 (1930); *Trans. Electrochem. Soc.*, **76**, 45 (1939); Glasstone, *Ann. Rep. Chem. Soc.*, **30**, 283 (1933); Glasstone, "Sutton's Volumetric Analysis," 1935, Part V.

* The change of volume during titration is neglected since its effect is relatively small.

If an electrode of the metal M, reversible with respect to M^+ ions, were placed in the solution of MX during the titration, its potential would be given by

$$E = E^0_{M,M^+} - \frac{RT}{zF} \ln a_{M^+}$$

$$\approx E^0_{M,M^+} - \frac{RT}{zF} \ln (c - x + y), \qquad (39)$$

where the activity of M^+ ions, i.e., a_{M^+}, has been replaced by the concentration as derived above. If the solubility product k_s is available, then since c and x are known for any point in the titration, it is possible to calculate y by means of equation (38); the values of $c - x + y$ can now be inserted in equation (39) and the variation of electrode potential during the course of titration can be determined. At the equivalence-point, i.e., the ideal end-point of the titration, when the amount of BA added is equivalent to that of MX initially present, c and x are equal; equation (39) then reduces to

$$E \approx E^0_{M,M^+} - \frac{RT}{zF} \ln y$$

$$\approx E^0_{M,M^+} - \frac{RT}{zF} \ln \sqrt{k_s}. \qquad (40)$$

Should the titration be carried beyond the end-point, the value of c_{A^-} now becomes $x - c + y$, while that of c_{M^+} is y, since the solution now contains excess of A^- ions; $x - c$ arises from the excess of BA over MX, and y from the solubility of the sparingly soluble MA. The solubility product is given by

$$k_s = y(x - c + y),$$

and equation (39) becomes

$$E \approx E^0_{M,M^+} - \frac{RT}{zF} \ln y. \qquad (41)$$

The value of y can be calculated as before, if the solubility product is known, and hence the electrode potential of M can be determined.

By means of equations (38), (39) and (41) it is thus possible to calculate the potential of an electrode of the metal M during the course of the whole precipitation titration, from the beginning to beyond the equivalence-point, provided the solubility product of the precipitated salt is known. The calculations show that there is at first a gradual change of potential, but a very rapid increase occurs as the equivalence-point is approached; the change of potential for a given increase in the amount of the titrant added, i.e., dE/dx, is found to be a maximum at the theoretical equivalence-point. This result immediately suggests a method for determining experimentally the end-point of a precipitation titration by E.M.F. measurement. The reversible potential (E) of an M electrode during the course of the titration is plotted against the amount of titrant added (x); the point at which the potential rises most sharply, i.e., the point of inflection where dE/dx is a maximum, is the required end-point. This procedure constitutes the fundamental basis of **potentiometric titration.**

The same general conclusion may be reached without going through the detailed calculations just described. If equation (39) is differentiated twice with respect to x and the resulting expression for d^2E/dx^2 equated to zero, the condition for dE/dx to be a maximum can be obtained. This is found to be that x should be equal to c, which is, of course, the condition for the equivalence-point, in agreement with the conclusion already reached.

By differentiating equation (39) with respect to x it is seen that at the equivalence-point the value of dE/dx is inversely proportional to $\sqrt{k_s}$. The potential jump observed at the end-point is thus greater the smaller the solubility product k_s of the precipitate. The sharpness of a particular titration can thus often be improved by the addition of alcohol to the solution being titrated in order to reduce the solubility of the precipitated salt.

In the treatment given here it has been assumed that the precipitate MA is a salt of symmetrical valence type; if it is an unsymmetrical salt, e.g., M_2A or MA_2, the potential-titration curve, i.e., the plot of the potential (E) against the amount (x) of titrant added, is not symmetrical and the maximum value of dE/dx does not occur exactly at the equivalence-point. The deviations are, however, relatively small if the solubility product of the precipitate is small and the titrated solutions are not too dilute.

Potentiometric Titration: Experimental Methods.—Since the silver electrode generally behaves in a satisfactory manner, the potentiometric method of titration can be applied particularly to the estimation of anions which yield insoluble silver salts, e.g., halides, cyanides, thiocyanates, phosphates, etc. In its simplest form, the experimental procedure is to take a known volume of the solution containing the anion to be titrated and to insert a clean silver sheet or wire, preferably coated with silver by the electrolysis of an argentocyanide solution; this constitutes the "indicator" electrode, and its potential is measured by connecting it, through a salt bridge, with a reference electrode, e.g., a calomel electrode. Since the actual electrode potential is not required, but merely the point at which it undergoes a rapid change, the E.M.F. of the resulting cell is recorded after the addition of known amounts of the silver nitrate solution. The values obtained in the course of the titration of 10 cc. of approximately 0.1 N sodium chloride with 0.1 N silver nitrate, using a silver indicator electrode and a calomel reference

TABLE LI. POTENTIOMETRIC TITRATION OF SODIUM CHLORIDE WITH SILVER NITRATE

$AgNO_3$ (v)	E	ΔE	Δv	$\Delta E/\Delta v$
0.1 cc.	114 mv.			
5.0	130	16	4.9	3.3
8.0	145	15	3.0	5.0
10.0	168	23	2.0	11.5
11.0	202	34	1.0	34
11.10	210	8	0.1	80
11.20	224	14	0.1	140
11.30	250	26	0.1	260
11.35	277	27	0.05	540
11.40	303	26	0.05	520
11.45	318	15	0.05	300
11.50	328	10	0.05	200
12.0	364	36	0.5	72
13.0	389	25	1.0	25
14.0	401	12	1.0	12

electrode, are recorded in Table LI and plotted in Fig. 75; the first column of the table gives the volume v of standard silver nitrate added,

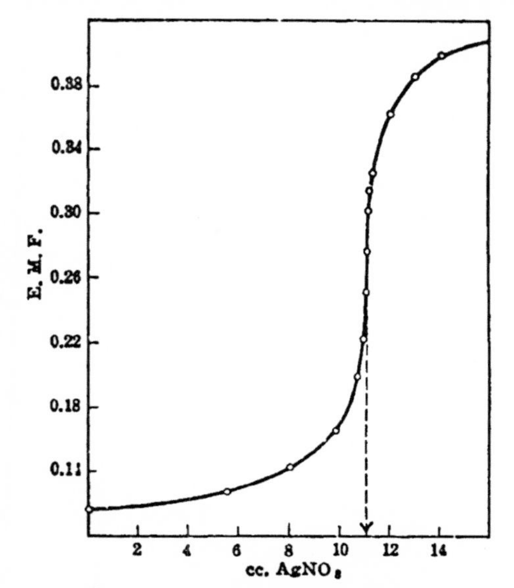

FIG. 75. Potentiometric titration

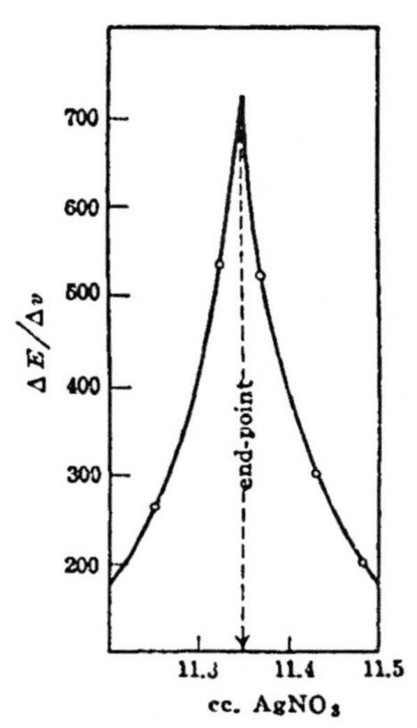

FIG 76. Determination of end-point in potentiometric titration

which is equivalent to x in the treatment given above, and hence $\Delta E/\Delta v$, in the last column, is an approximation to dE/dx. It is clear from the data that $\Delta E/\Delta v$ is a maximum when v is about 11.35 cc., and this must represent the end-point of the titration.

It is not always possible to estimate the end-point directly by inspection of the data and the following method, which is always to be preferred, should be used. The values of $\Delta E/\Delta v$ in the vicinity of the end-point are plotted against $v + \frac{1}{2}\Delta v$, i.e., the volume of titrant corresponding to the middle of each titration interval, as in Fig. 76; the volume of titrant corresponding to the maximum value of $\Delta E/\Delta v$ can now be determined very precisely. This graphical method is particularly useful when the inflection in the potential-titration curve at the end-point is relatively small.

Differential Titration.—The object of potentiometric titration is to determine the point at which $\Delta E/\Delta v$ is a maximum, and this can be achieved directly, without the use of graphical methods, by utilizing the principle of **differential titration.** If to two identical solutions, e.g., of sodium chloride, are added v and $v + 0.1$ cc. respectively of titrant, e.g., silver nitrate, the difference of potential between similar electrodes placed in the two solutions gives a direct measure of $\Delta E/\Delta v$, where Δv is 0.1 cc., at the point in the titration corresponding to the addition of $v + 0.05$ cc. of silver nitrate. The E.M.F. of the cell made up of these two electrodes will thus be a maximum at the end-point.

In the earliest applications of the method of differential titration the solution to be titrated was divided into two equal parts; similar electrodes were placed in each and electrical connection between the two solutions was made with wet filter-paper. The electrodes were connected through a suitable high resistance to a galvanometer. Titrant was then added to the two solutions from two separate burettes, one being always kept a small amount, e.g., 0.1 cc., in advance of the other. The point of maximum potential difference, and hence that at which $\Delta E/\Delta v$ was a maximum, was indicated by the largest deflection of the galvanometer; the total titrant added at this point was then equivalent to the total solution titrated. By this means the end-point of the titration was obtained without the use of a reference electrode or a potentiometer, and the necessity for graphical estimation of the titration corresponding to the maximum $\Delta E/\Delta v$ was avoided.[20]

The method of differential titration has been modified so that the process can be carried out in one vessel with one burette; by means of special devices, a small quantity of the titrated solution surrounding one of the two identical electrodes is kept temporarily from mixing with the bulk of the solution before each addition of titrant. The difference of potential between the two electrodes after the addition of an amount Δv of titrant gives a measure of $\Delta E/\Delta v$. The form of apparatus devised by MacInnes and Dole,[21] which is capable of giving results of great accuracy, is depicted in Fig. 77. One of the two identical indicator electrodes,

[20] Cox, *J. Am. Chem. Soc.*, **47**, 2138 (1925).

[21] MacInnes *et al.*, *J. Am. Chem. Soc.*, **48**, 2831 (1926); **51**, 1119 (1929); **53**, 555 (1931); *Z. physik. Chem.*, **130**, 217 (1927).

viz., E_1, is placed directly in the titration vessel, and the other, E_2, is inserted in the tube A, which should be as small as convenient; at the bottom of this tube there is a small hole B, and a "gas-lift" C is sealed into its side. The hole D in the tube A permits the overflow of liquid when the gas-lift is in operation. In order to carry out a titration, a known volume of solution is placed in a beaker and the two electrodes are inserted; the liquid is allowed to enter A, but the gas-stream is turned off. Titrant is added from the burette, with constant stirring, until there is a large increase in the E.M.F. of the cell formed by the two electrodes; this may be indicated by a potentiometer, for precision work, or by means of a galvanometer with a resistance in series. The solution in the beaker is actually somewhat over-titrated, but when the gas-stream is started the reserve solution in the tube A, which normally mixes only slowly with the bulk of the liquid, because of the smallness of the hole B, is forced out; in this way the titration is brought back, although the end-point is near. The difference of potential between the two electrodes is now zero, since the same solution surrounds both of them. The gas-stream is stopped, and a drop (Δv) of titrant is added to the bulk of the solution in the beaker; the galvanometer deflection, or potential difference, is then a measure of $\Delta E/\Delta v$, since one electrode, E_2, is immersed in a solution to which v cc. of titrant have been added, while the other, E_1, is surrounded by one to which $v + \Delta v$ cc. have been added. The gas-stream is started once more so as to obtain complete mixing of the solutions; it is then stopped, another drop of titrant added, and the potential reading again noted. This procedure is continued until the end-point is passed, the end-point itself being characterized by the maximum potential difference between the two electrodes.

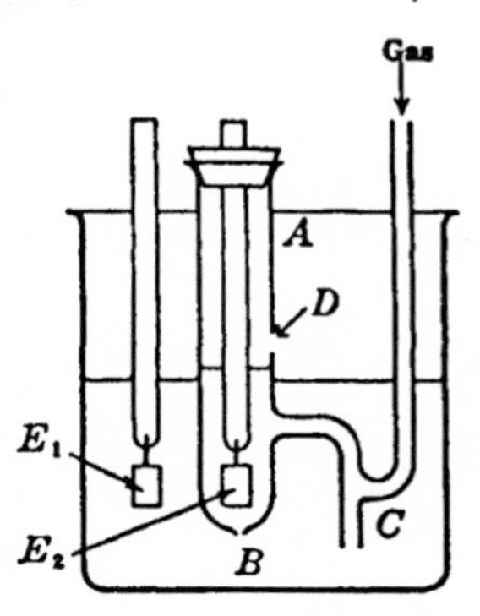

Fig. 77. Apparatus for differential titration (MacInnes and Dole)

Many simplified potentiometric titration methods have been described from time to time, and various forms of apparatus have been devised to facilitate the performance of these titrations; for reference to these matters the more specialized literature should be consulted.[22]

Complex Ions.—The formula of a relatively stable complex ion can be determined by means of E.M.F. measurements; in the general case already considered on page 173, viz.,

$$M_qA_r^{\pm} \rightleftharpoons qM^+ + rA^-,$$

[22] See the books and review articles to which reference is made on page 256; the subject of potentiometric titrations, among others, is also treated in Kolthoff and Laitinen, "pH and Electro-Titrations," 1941.

it was seen that the instability constant K_i can be represented by

$$K_i = \frac{a^q_{M^+} a^r_{A^-}}{a_{M_qA_r^\pm}},$$

$$\therefore \quad a^q_{M^+} = K_i \frac{a_{M_qA_r^\pm}}{a^r_{A^-}}.$$

If an electrode of the metal M is inserted in the solution of the complex ion, the reversible potential should be given by

$$E = E^0_{M,M^+} - \frac{RT}{zF} \ln a_{M^+}$$

$$= E^0_{M,M^+} - \frac{RT}{qzF} \ln K_i - \frac{RT}{qzF} \ln \frac{a_{M_qA_r^\pm}}{a^r_{A^-}}. \tag{42}$$

For two solutions containing different total amounts of the complex ion, but the same relatively large excess of the anion A^-, it follows from equation (42) that

$$E_1 - E_2 = \frac{RT}{qzF} \ln \frac{(a_{M_qA_r^\pm})_2}{(a_{M_qA_r^\pm})_1}, \tag{43}$$

where the suffixes 1 and 2 refer to the two solutions; the value of a_{A^-} is assumed to be the same for the two cases. If the complex ion $M_qA_r^\pm$ is relatively stable, then in the presence of excess of A^- ions, virtually the whole of the M present in solution will be in the form of complex ions. As an approximation, therefore, the ratio of the activities of the $M_qA_r^\pm$ ions in the two solutions in equation (43) may be replaced by the ratio of the total concentrations of M; hence

$$E_1 - E_2 = \frac{RT}{qzF} \ln \frac{(c_M)_2}{(c_M)_1}. \tag{44}$$

If $(c_M)_1$ and $(c_M)_2$, the total concentrations of the species M in the respective solutions, are known, and the potentials E_1 and E_2 are measured, it is possible to evaluate q by means of equation (44).

If the solutions are made up with same concentration of M, i.e., approximately the same concentration, or activity, of the complex ions $M_qA_r^\pm$, but with different amounts of the anion A^-, it follows from equation (42) that in this case

$$E_1 - E_2 = \frac{RT}{qzF} \ln \frac{(a_{A^-})^r_1}{(a_{A^-})^r_2}. \tag{45}$$

The ratio of the activities of the A^- ions may be replaced, as an approximation, by the ratio of the concentrations; hence, from equation (45),

$$E_1 - E_2 = \frac{rRT}{qzF} \ln \frac{(c_{A^-})_1}{(c_{A^-})_2}. \tag{46}$$

Since q has been already determined, the value of r can be derived from equation (46) so that the formula of the complex ion has been found.

Another method for deriving the ratio r/q involves the same principle as is used in potentiometric titration; for simplicity of explanation a definite case, namely the formation of the argentocyanide ion, $Ag(CN)_2^-$, will be considered. If a solution of potassium cyanide is titrated with silver nitrate, the potential of a silver electrode in the titrated solution will be found to undergo a sudden change of potential when the whole of the cyanide has been converted into argentocyanide ions. From the relative amounts of silver and cyanide ions at the point where dE/dx is a maximum the formula of the complex ion can be calculated. An analogous titration method can be used to determine the formula of any stable complex ion; the procedure actually gives the ratio of M^+ to A^- in the complex ion $M_qA_r^{\pm}$, but if this ratio is known there is generally no difficulty, from valence and other chemical considerations, in deriving the molecular formula.

By expressing the concentration, or activity, of the M^+ ions in the titrated solution, and hence the potential of an M electrode, in terms of c, the initial concentration of the solution, x, the amount of titrant added, and k_i, the instability constant of the complex ion, it is possible, utilizing the method of differentiation described in connection with precipitation titrations (page 258), to show that dE/dx is a maximum at the point corresponding to complete formation of the complex ion. Further, the value of dE/dx at this point, and hence the sharpness of the inflection in the titration curve, can be shown to be greater the smaller the instability constant.

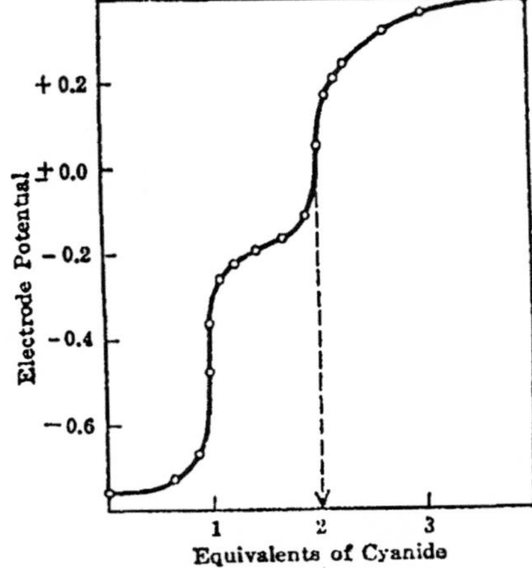

Fig. 78. Formula of complex argentocyanide ion

The potential of a silver electrode during the course of the titration of silver nitrate with potassium cyanide is shown in Fig. 78; the first marked change of potential occurs when one equivalent of cyanide has been added to one of silver, so that the whole of the silver cyanide is precipitated, and the second, when two equivalents of cyanide have been added, corresponds to the complete formation of the $Ag(CN)_2^-$ ion. It will be seen that the changes of potential occur very sharply in each case; this means that the silver cyanide is very slightly soluble and that the complex ion is very stable.

Electrode Potential and Valence.—The equation (8*b*) for the potential of an electrode reversible with respect to positive ions may be written in the approximate form

$$E \approx E^0 - 0.0002 \frac{T}{z_i} \log c_i,$$

where the activity of the ionic species is replaced by the concentration. At ordinary laboratory temperatures, about 20°c., i.e., T is 293° K., this equation becomes

$$E \approx E^0 - \frac{0.058}{z_i} \log c_i. \qquad (47)$$

It follows, therefore, that a ten-fold change of concentration of the ions will produce a change of $0.058/z_i$ volt in the electrode potential, where z_i is the valence of the ions with respect to which the electrode is reversible. It is possible, therefore, to utilize equation (47) to determine the valence of an ion.[23] For example, the result of a ten-fold change in the concentration of a mercurous nitrate solution was found to cause a change of 0.029 volt in the potential of a mercury electrode at 17°; it is evident, therefore, that z must be 2, so that the mercurous ions are *bivalent*. These ions are therefore written as Hg_2^{++} and mercurous chloride and nitrate are represented by Hg_2Cl_2 and $Hg_2(NO_3)_2$.

PROBLEMS

1. Work out the expressions for the E.M.F.'s and single potentials of the cells and electrodes given in Problem 1 of Chap. VI in terms of the variable activities.

2. From the standard potential data in Table XLIX determine (i) the standard free energies at 25° of the reactions

$$Ag^+aq. + Cl^-aq. = AgCl(s)$$

and

$$Ag + \tfrac{1}{2}Cl_2(1 \text{ atm.}) = AgCl(s),$$

and (ii) the solubility product of silver chloride.

3. It is known from thermal measurements that the entropy of aluminum at 25° is 6.7 cal./deg. per g.-atom, and that of hydrogen gas at 1 atm. pressure is 31.2 per mole. The heat of solution of aluminum in dilute acid shows that ΔH for the reaction

$$Al + 3H^+aq. = Al^{+++}aq. + \tfrac{3}{2}H_2(1 \text{ atm.})$$

is − 127,000 cal. From measurements on the entropy of solid cesium alum and its solubility, etc., Latimer and his collaborators [*J. Am. Chem. Soc.*, **60**, 1829 (1938)] have estimated the entropy of the Al^{+++}aq. ion to be − 76 cal./deg. per g.-ion. Calculate the standard potential of aluminum on the usual hydrogen scale.

[23] Ogg, *Z. physik. Chem.*, **27**, 285 (1898); see also, Reichinstein, *ibid.*, **97**, 257 (1921); Kasarnowsky, *Z. anorg. Chem.*, **128**, 117 (1923).

4. Jones and Baeckström [*J. Am. Chem. Soc.*, 56, 1524 (1934)] found the E.M.F. of the cell

$$Pt \mid Br_2(l)\ \ KBr\ aq.\ \ AgBr(s) \mid Ag$$

to be − 0.9940 volt at 25°. The vapor pressure of the saturated solution of bromine in the potassium bromide solution is 159.45 mm. of mercury; calculate the standard potential of the $Br_2(g$, 1 atm.), Br^- electrode.

5. The standard free energy of the process

$$\tfrac{1}{2}H_2(1\ atm.) + \tfrac{1}{2}Cl_2(1\ atm.) = HCl(1\ atm.),$$

is given in International Critical Tables, VII, 233, by the expression

$$\Delta G^0 = -21{,}870 + 0.45T \ln T - 0.25 \times 10^{-5}T^2 - 5.31T.$$

The partial pressure of hydrogen chloride over 1.11 N hydrochloric acid solution is 4.03×10^{-4} mm. at 25°. Calculate the E.M.F. of the cell

$$H_2(1\ atm.) \mid 1.11\ \text{N}\ HCl\ aq. \mid Cl_2(1\ atm.)$$

at this temperature. Use the result to determine the standard potential of the chlorine electrode, the mean activity coefficient of the hydrochloric acid being estimated from the data in Table XXXIV.

6. Calculate from the standard potentials of cadmium and thallium the ratio of the activities of Cd^{++} and Tl^- ions when metallic cadmium is shaken with thallous perchlorate solution until equilibrium is attained.

7. Knüppfer [*Z. physik. Chem.*, 26, 255 (1898)] found the E.M.F. of the cell

$$Tl\ (Hg) \mid TlCl(s)\ \ KCl(c_1) \vdots KCNS(c_2)\ \ TlCNS(s) \mid Tl\ (Hg)$$

to be − 0.0175 volt at 0.8° and − 0.0105 volt at 20° with c_1/c_2 equal to 0.84. Assuming the solutions to behave ideally, calculate the equilibrium ratios of c_1/c_2 at the two temperatures and estimate the temperature at which the arbitrary ratio, i.e., 0.84, will become the equilibrium value.

8. The E.M.F. of the cell

$$Pb \mid Pb(OH)_2(s)\ \ \text{N}\ NaOH\ \ HgO(s) \mid Hg$$

is 0.554 volt at 20°; the potential of the Hg, $HgO(s)$ N NaOH electrode is − 0.114 volt. Calculate the approximate solubility product of lead hydroxide.

9. In the potentiometric titration of 25 cc. of a potassium cyanide solution with 0.1 N silver nitrate, using a silver indicator electrode and a calomel reference electrode, the following results were obtained:

cc. $AgNO_3$ (v)	2.20	11.70	15.50	18.00	19.60	20.90
E.M.F. (E)	0.550	0.481	0.445	0.422	0.392	0.363
cc. $AgNO_3$ (v)	21.50	21.75	21.95	22.15	22.35	22.55
E.M.F. (E)	0.343	0.309	0.259	0.187	− 0.255	− 0.319

Plot E against v, and ΔE against Δv in the vicinity of the end-point; from the results determine the concentration of the potassium cyanide solution.

10. When studying the behavior of a tin anode in potassium oxalate solution, Jeffery [*Trans. Faraday Soc.*, 20, 390 (1924)] noted that a complex anion, having the general formula $Sn_q(C_2O_4)_r^=$, was formed. In order to deter-

mine its constitution, measurements of the cell

$$Sn \mid Sn_q(C_2O_4)_r^- \; K_2C_2O_4 \text{ aq.} \vdots KCl \text{ (satd.) } Hg_2Cl_2(s) \mid Hg$$

were made: in one series of experiments (A) the concentration of potassium oxalate was large and approximately constant while the total amount of tin in solution (c_{Sn}) was varied; in the second series (B), c_{Sn} was kept constant at 0.01 g.-atom per liter, while the concentration of potassium oxalate ($c_{ox.}$) was varied. The results were as follows:

A		B	
c_{Sn}	E	$c_{ox.}$	E
1.00×10^{-2}	0.7798	2.0	0.7866
0.833	0.7823	2.5	0.7937
0.714	0 7842	3.0	0.7990
0.625	0.7859	3.5	0.8002
0.556	0.7877	4.0	0.8052

Devise a graphical method, based on equations (44) and (46), to evaluate q and r; activity corrections may be neglected, and the whole of the tin present in solution may be assumed to be in the form of the complex anion.

CHAPTER VIII

OXIDATION-REDUCTION SYSTEMS

Oxidation-Reduction Potentials.—It was seen on page 186 that a reversible electrode can be obtained by inserting an inert electrode in a solution containing the oxidized and reduced forms of a given system; such electrodes are called **oxidation-reduction electrodes.** It has been pointed out, and it should be emphasized strongly, that there is no essential difference between electrodes of this type and those already considered involving a metal and its cations, or a non-metal and its corresponding anions. This lack of distinction is brought out by the fact that the iodine-iodide ion system is frequently considered from the oxidation-reduction standpoint. Nevertheless, certain oxidation-reduction systems, using the expression in its specialized meaning, have interesting features and they possess properties in common which make it desirable to consider them separately.

According to the general arguments at the beginning of Chap. VI, which are applicable to reactions of all types, including those involving oxidation and reduction, the potential of an electrode containing the system

$$\text{Reduced State} \rightleftharpoons \text{Oxidized State} + n\ \text{Electrons}$$

is given by the general equation

$$E = E^0 - \frac{RT}{nF} \ln \frac{(\text{Oxidized State})}{(\text{Reduced State})}, \tag{1}$$

where n is the number of electrons difference between the two states, and the parentheses represent activities.

Oxidation-reduction potentials, like the other types discussed in the preceding chapter, are generally expressed on the hydrogen scale, so that for the system

$$Fe^{++} \rightleftharpoons Fe^{+++} + \epsilon,$$

for example, the electrode potential as usually recorded is really the E.M.F. of the cell

$$Pt \mid Fe^{++}, Fe^{+++} \parallel H^+(a_{H^+} = 1) \mid H_2(1\ \text{atm.}).$$

Using the familiar convention that a positive E.M.F. represents the tendency of positive current to flow from left to right through the cell, the reaction at the left-hand electrode may evidently be written as

$$Fe^{++} = Fe^{+++} + \epsilon,$$

for the passage of one faraday. This result may be obtained directly by analogy with the process occurring at the electrode M, M^+, namely $M = M^+ + \epsilon$. At the right-hand electrode, the reaction is

$$H^+ + \epsilon = \tfrac{1}{2}H_2,$$

so that the net cell reaction, for one faraday of electricity, is

$$Fe^{++} + H^+ = Fe^{+++} + \tfrac{1}{2}H_2.$$

The E.M.F. of the complete cell is then given in the usual manner by

$$E = E^0 - \frac{RT}{F} \ln \frac{a_{Fe^{+++}} a_{H_2}^{\frac{1}{2}}}{a_{Fe^{++}} a_{H^+}},$$

and since, by convention, the activities of the hydrogen gas and the hydrogen ions are taken as unity, it follows that

$$E = E^0_{Fe^{++}, Fe^{+++}} - \frac{RT}{F} \ln \frac{a_{Fe^{+++}}}{a_{Fe^{++}}}. \tag{2}$$

The oxidation-reduction potential is thus seen to be determined by the ratio of the activities of the oxidized and reduced states, in agreement with the general equation (1). The standard potential E^0 is evidently that for a system in which both states are at unit activity.

In the most general case of an oxidation-reduction system represented by

$$aA + bB + \cdots \rightleftharpoons xX + yY + \cdots + n\epsilon,$$

for which there is a difference of n electrons between the reduced state, involving A, B, etc., and the oxidized state, involving X, Y, etc., the potential is given by (cf. page 228)

$$E = E^0 - \frac{RT}{nF} \ln \frac{a_X^x a_Y^y \cdots}{a_A^a a_B^b \cdots}. \tag{3}$$

When all the species concerned, viz., A, B, $\cdots$, X, Y, etc., are in their standard states, i.e., at unit activity, the potential is equal to E^0, the standard oxidation-reduction potential of the system. It is important to remember that in order that a stable reversible potential may be obtained, *all* the substances involved in the system must be present; the actual potential will, according to equation (3), depend on their respective activities.

Types of Reversible Oxidation-Reduction Systems.—Various types of reversible oxidation-reduction systems have been studied: the simplest consist of ions of the same metal in two stages of valence, e.g., ferrous and ferric ions. If M^{z_1+} and M^{z_2+} are two cations of the metal M, carrying charges z_1 and z_2, respectively, where z_2 is greater than z_1, the electrode reaction is

$$M^{z_1+} \rightleftharpoons M^{z_2+} + (z_2 - z_1)\epsilon,$$

and the potential is given by

$$E = E^0 - \frac{RT}{(z_2 - z_1)F} \ln \frac{a_2}{a_1},$$

where a_2 and a_1 are the activities of the oxidized and reduced forms, respectively.

Another type of system consists of two anions carrying different charges, e.g., ferro- and ferri-cyanide, i.e.,

$$Fe(CN)_6^{----} \rightleftharpoons Fe(CN)_6^{---} + \epsilon,$$

and the electrode potential for this system is

$$E = E^0 - \frac{RT}{F} \ln \frac{a_{Fe(CN)_6^{---}}}{a_{Fe(CN)_6^{----}}}.$$

In certain cases both anions and cations of the same metal are concerned; for such systems the equilibria, and hence the equations for the electrode potential, involve hydrogen ions. An instance of this kind is the permanganate-manganous ion system, viz.,

$$Mn^{++} + 4H_2O \rightleftharpoons MnO_4^- + 8H^+ + 5\epsilon,$$

for which the electrode potential is

$$E = E^0 - \frac{RT}{5F} \ln \frac{a_{MnO_4^-} a_{H^+}^8}{a_{Mn^{++}}},$$

the activity of the water being unity provided the solutions are relatively dilute.

In some important oxidation-reduction systems one or more solids are concerned; for example, in the case of the equilibrium

$$Mn^{++} + 2H_2O \rightleftharpoons MnO_2(s) + 4H^+ + 2\epsilon,$$

the potential is

$$E = E^0 - \frac{RT}{2F} \ln \frac{a_{H^+}^4}{a_{Mn^{++}}},$$

since the activity of the solid manganese dioxide is taken as unity, in accordance with the usual convention as to standard states.

In the equilibrium

$$PbSO_4(s) + 2H_2O \rightleftharpoons PbO_2(s) + 4H^+ + SO_4^{--} + 2\epsilon,$$

which is of importance in connection with the lead storage battery, two solids are involved, namely lead sulfate and lead dioxide, and hence

$$E = E^0 - \frac{RT}{2F} \ln a_{H^+}^4 a_{SO_4^{--}}.$$

The potential thus depends on the fourth power of the activity of the hydrogen ions and also on that of the sulfate ions in the solution.

A large number of reversible oxidation-reduction systems involving organic compounds are known; most of these, although not all, are of the quinone-hydroquinone type. The simplest example is

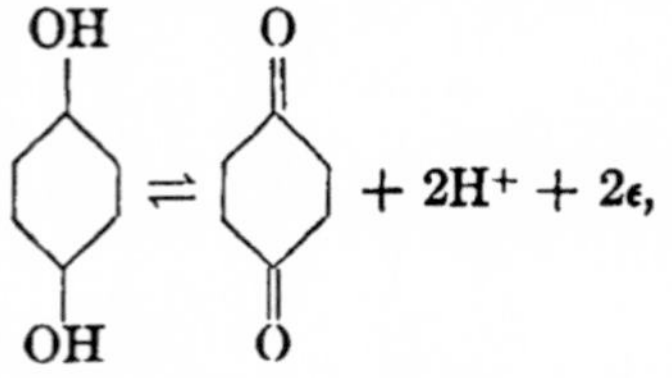

OH O

$$\rightleftharpoons \quad + 2H^+ + 2\epsilon,$$

OH O

and such systems may be represented by the general equation

$$H_2Q \rightleftharpoons Q + 2H^+ + 2\epsilon,$$

where H_2Q is the reduced, i.e., hydroquinone, form and Q is the oxidized, i.e., quinone, form. The potential of such a system is given by

$$E = E^0 - \frac{RT}{2F} \ln \frac{a_Q a_{H^+}^2}{a_{H_2Q}}. \tag{4}$$

For many purposes it is convenient to maintain the hydrogen ion activity constant and to include the corresponding term in the standard potential; equation (4) then becomes

$$E = E^{0\prime} - \frac{RT}{2F} \ln \frac{a_Q}{a_{H_2Q}}, \tag{5}$$

where $E^{0\prime}$ is a subsidiary standard potential applicable to the system at the specified hydrogen ion activity.

Determination of Standard Oxidation-Reduction Potentials.—In principle, the determination of the standard potential of an oxidation-reduction system involves setting up electrodes containing the oxidized and reduced states at known activities and measuring the potential E by combination with a suitable reference electrode; insertion of the value of E in the appropriate form of equation (3) then permits E^0 to be calculated. The inert metal employed in the oxidation-reduction electrode is frequently of smooth platinum, although platinized platinum, mercury and particularly gold are often used.

In the actual evaluation of the standard potential from the experimental data a number of difficulties arise, and, as a result of the failure to overcome or to make adequate allowance for them, most of the measurements of oxidation-reduction potentials carried out prior to about 1925 must be regarded as lacking in accuracy. In the first case, it is rarely possible to avoid a liquid junction potential in setting up the cell for measuring the oxidation-reduction potential; secondly, there is often uncertainty concerning the actual concentrations of the various species, because of complex ion formation and because of incomplete dissociation and hydrolysis of the salts present; finally, activity coefficients, which were neglected in the earlier work, have an important influence, as will be apparent from the following considerations.

In the simple case of a system consisting of two ions carrying different charges, e.g., Fe^{++}, Fe^{+++} or $Fe(CN)_6^{----}$, $Fe(CN)_6^{---}$, designated by the suffixes 1 and 2, respectively, the equation for the potential is

$$E = E^0 - \frac{RT}{nF} \ln \frac{a_2}{a_1}$$

$$= E^0 - \frac{RT}{nF} \ln \frac{c_2}{c_1} - \frac{RT}{nF} \ln \frac{f_2}{f_1}, \tag{6}$$

where the activity has been replaced by the product of the concentration and the activity coefficient. Utilizing the Debye-Hückel limiting equation (p. 144), viz.,

$$\log f = - Az_i^2\sqrt{\mu}$$

it follows that

$$\log \frac{f_2}{f_1} = A(z_1^2 - z_2^2)\sqrt{\mu},$$

and insertion in equation (6) gives

$$E = E^0 - \frac{RT}{nF} \ln \frac{c_2}{c_1} - 2.303 \frac{RT}{nF} A(z_1^2 - z_2^2)\sqrt{\mu}.$$

If water is the solvent, then at 25° the constant A is 0.509; hence, this equation becomes

$$E = E^0 - \frac{0.05915}{n} \log \frac{c_2}{c_1} - \frac{0.0301}{n} (z_1^2 - z_2^2)\sqrt{\mu}. \tag{7}$$

For most oxidation-reduction systems $z_1^2 - z_2^2$ is relatively high, e.g., 7 for the $Fe(CN)_6^{----}$, $Fe(CN)_6^{---}$ system, and so the last term in equation (7), which represents the activity coefficient factor, may be quite considerable; further, the terms in the ionic strength involve the square of the valence and hence μ will be large even for relatively dilute solutions.[1] In any case, the presence of neutral salts, which were frequently added to the solution in the earlier studies of oxidation-reduction potentials, increases the ionic strength; they will consequently have an appreciable influence on the potential, although the ratio of the amounts of oxidized to reduced forms remains constant.

A striking illustration of the effect of neglecting the activity coefficient is provided by the results obtained by Peters (1898) in one of the

[1] Kolthoff and Tomsicek, *J. Phys. Chem.*, 39, 945 (1935); Glasstone, "The Electrochemistry of Solutions," 1937, p. 346.

earliest quantitative studies of reversible oxidation-reduction electrodes. From measurements made in solutions containing various proportions of ferrous and ferric chloride chloride in 0.1 N hydrochloric acid, an approximately constant value of − 0.713 volt at 17° was calculated for the standard potential of the ferric-ferrous system, using the ratio of concentrations instead of activities. This result was accepted as correct for some years, but it differs from the most recent values by about 0.07 volt; the discrepancy is close to that estimated from equation (5) on the basis of an ionic strength of 0.25, which is approximately that existing in the experimental solutions. Actually, of course, the Debye-Hückel limiting equation would not hold with any degree of exactness at such a high ionic strength, but it is of interest to observe that it gives an activity correction of the right order.

In recent years care has been taken to eliminate, or reduce, as far as possible the sources of error in the evaluation of standard oxidation-reduction potentials; highly dissociated salts, such as perchlorates, are employed wherever possible, and corrections are applied for hydrolysis if it occurs. The cells are made up so as to have liquid junction potentials whose values are small and which can be determined if necessary, and the results are extrapolated to infinite dilution to avoid activity corrections. One type of procedure adopted is illustrated by the case described below.*

In order to determine the oxidation-reduction potential of the system involving penta- (VO_2^+) and tetra-valent (VO^{++}) vanadium, viz.,

$$VO^{++} + H_2O = VO_2^+ + 2H^+ + \epsilon,$$

measurements were made with cells of the form

$$\text{Pt} \mid VO_2Cl,\ VOCl_2,\ HCl \ \vdots\ HCl\ \ Hg_2Cl_2(s) \mid Hg$$

containing the three constituents, VO_2Cl, $VOCl_2$ and hydrochloric acid at various concentrations.[2] By employing acid of the same concentration in both parts of the cell, the liquid junction potential was reduced to a negligible amount. The reaction taking place in the cell for the passage of one faraday is

$$VO^{++} + H_2O + \tfrac{1}{2}Hg_2Cl_2(s) = VO_2^+ + 2H^+ + Cl^- + Hg(l),$$

so that the E.M.F. is given by

$$E = E^0 - \frac{RT}{F} \ln \frac{a_{VO_2^+} a_{H^+}^2 a_{Cl^-}}{a_{VO^{++}}}, \tag{8}$$

where the standard potential for the cell (E^0) is equal to the difference between the standard potentials of the V^5, V^4 system and that of the

* See also, Problem 4, page 304.

[2] Carpenter, *J. Am. Chem. Soc.*, 56, 1847 (1934); Hart and Partington, *J. Chem. Soc.*, 1532 (1940).

Hg, $Hg_2Cl_2(s)$, Cl^- electrode, the latter being − 0.2680 volt at 25°. Replacing the activities of the VO^{++} and VO_2^+ ions by the products of their respective concentrations and activity coefficients, represented by f_2 and f_1, respectively, equation (8) becomes, after rearrangement,

$$E + \frac{RT}{F} \ln a_{H^+}^2 a_{Cl^-} + \frac{RT}{F} \ln \frac{c_{VO^{++}}}{c_{VO_2^+}} = E^0 - \frac{RT}{F} \ln \frac{f_1}{f_2}. \quad (9)$$

Since the hydrochloric acid may be regarded as being completely ionized, c_{H^+} and c_{Cl^-} may each be taken as equal to c_{HCl}, the concentration of this acid in the cell; further, the product of f_{H^+} and f_{Cl^-} is equal to f_{HCl}^2, where f_{HCl} is the mean activity coefficient of the hydrochloric acid. It follows, therefore, that the quantity $a_{H^+}^2 a_{Cl^-}$, which is equal to $(c_{H^+}^2 c_{Cl^-}) f_{H^+}^2 f_{Cl^-}$, may be replaced by $c_{HCl}^3 f_{HCl}^2 f_{H^+}$; upon inserting this result in equation (9) and rearranging, it is found that

$$E + \frac{3RT}{F} \ln c_{HCl} + \frac{RT}{F} \ln \frac{c_{VO^{++}}}{c_{VO_2^+}} = E^0 - \frac{RT}{F} \ln \frac{f_1 f_{HCl}^2 f_{H^+}}{f_2}. \quad (10)$$

The activity coefficient term in this equation becomes zero at infinite dilution; it follows, therefore, that extrapolation of the left-hand side to zero concentration, using the results obtained with cells containing various concentrations of the three constituents, should give E^0 for the cell. The value obtained in this manner, by plotting the left-hand side of equation (10) against a suitable function of the ionic strength, was − 0.7303 volt; it follows, therefore, that the standard potential of the VO^{++}, $VO_2^+ + 2H^+$ system is − 0.730 + (− 0.268), i.e., − 0.998 volt.

An alternative extrapolation procedure is based on the approximation of taking f_{H^+} to be equal to f_{HCl}; equation (9) can then be written as

$$E + \frac{3RT}{F} \ln c_{HCl} + \frac{3RT}{F} \ln f_{HCl} + \frac{RT}{F} \ln \frac{c_{VO^{++}}}{c_{VO_2^+}} = E^0 - \frac{RT}{F} \ln \frac{f_1}{f_2}. \quad (11)$$

The values of the activity coefficients of hydrochloric acid at the ionic strengths existing in the cell are obtained from tabulated data, and hence the left-hand side of this equation, for various concentrations, may be extrapolated to zero ionic strength, thus giving E^0. A further possibility is to replace $\log f_{HCl}$ by the Debye-Hückel expression $- A\sqrt{\mu}$, and to extrapolate, as before, by plotting against a suitable function of the ionic strength. As a general rule, several methods of extrapolation are possible; the procedure preferred is the one giving an approximate straight line plot, for this will probably give the most reliable result when extrapolating to infinite dilution.

Another method of evaluating standard oxidation-reduction potentials is to make use of chemical determinations of equilibrium constants.[3]

[3] Schumb and Sweetser, *J. Am. Chem. Soc.*, 57, 871 (1935).

The chemical reaction occurring in the hypothetical cell, free from liquid junction,

$$Ag \mid Ag^+ \parallel Fe^{++}, Fe^{+++} \mid Pt,$$

for the passage of one faraday is

$$Ag + Fe^{+++} = Ag^+ + Fe^{++}.$$

The standard E.M.F. of this cell (E^0) with all reactants at unit activity is given by (cf. p. 251)

$$E^0 = \frac{RT}{F} \ln K = \frac{RT}{F} \ln \left(\frac{a_{Ag^+} a_{Fe^{++}}}{a_{Fe^{+++}}} \right)_e, \tag{12}$$

where the activities are those at equilibrium, indicated by the suffix e; the activity of the solid silver is equal to unity, and so is omitted from the equilibrium constant. The standard E.M.F. is also equal to the difference of the standard potentials of the silver and ferrous-ferric electrodes, thus

$$E^0 = E^0_{Ag,\,Ag^+} - E^0_{Fe^{++},\,Fe^{+++}}, \tag{13}$$

and hence if the equilibrium constant of the cell reaction could be determined by chemical analysis, the value of $E^0_{Fe^{++},\,Fe^{+++}}$ could be calculated, since the standard potential of silver is known (Table XLIX).

A solution of ferric perchlorate, containing free perchloric acid in order to repress hydrolysis, was shaken with finely divided silver until equilibrium of the system

$$Ag + Fe(ClO_4)_3 \rightleftharpoons AgClO_4 + Fe(ClO_4)_2$$

was attained. Since perchlorates are very strong electrolytes, they are generally regarded as being completely dissociated at not too high concentrations; this reaction is, therefore, equivalent to that of the hypothetical cell considered above. By analyzing the solution at equilibrium, a concentration equilibrium "constant" (k), for various total ionic strengths, was calculated; this function k is related to the true equilibrium constant in the following manner:

$$K = \frac{a_{Ag^+} a_{Fe^{++}}}{a_{Fe^{+++}}} = \frac{c_{Ag^+} c_{Fe^{++}}}{c_{Fe^{+++}}} \cdot \frac{f_{Ag^+} f_{Fe^{++}}}{f_{Fe^{+++}}}$$

$$= k \frac{f_{Ag^+} f_{Fe^{++}}}{f_{Fe^{+++}}},$$

and if the activity coefficients are expressed in terms of the ionic strength by means of the extended form of the Debye-Hückel equation (p. 147), it is found that

$$\log K = \log k + \log f_{Ag^+} + \log f_{Fe^{++}} - \log f_{Fe^{+++}}$$

$$= \log k - (z^2_{Ag^+} + z^2_{Fe^{++}} - z^2_{Fe^{+++}}) A \sqrt{\mu} + C \sqrt{\mu}$$

$$= \log k + 4A \sqrt{\mu} + C\mu.$$

The value of A is known to be 0.509 for water at 25°, and that of C is found empirically; another term, $D\mu^2$, with an empirical value of D, may be added if necessary, and the true dissociation constant K can then be calculated from the experimental data. In this manner, it was found that K is 0.531 at 25°, and hence from equations (12) and (13), making use of the fact that the standard potential of silver is − 0.799, it follows that at 25°

$$-0.799 - E^0_{Fe^{++}, Fe^{+++}} = 0.05915 \log 0.531$$
$$= -0.016,$$
$$\therefore \quad E^0_{Fe^{++}, Fe^{+++}} = -0.783 \text{ volt.}$$

Direct measurements of the potential of the ferric-ferrous system have also been made; after allowing for hydrolysis and activity effects, the standard potential at 25° was found to be − 0.772 volt, but so many corrections were involved in arriving at this result that the value based on equilibrium measurements is probably more accurate.[4]

Approximate Determination of Standard Potentials.—Many studies have been made of oxidation-reduction systems with which, for one reason or another, it is not possible to obtain accurate results: this may be due to the difficulty of applying activity corrections, uncertainty as to the exact concentrations of the substances involved, or to the slowness of the establishment of equilibrium with the inert metal of the electrode. It is probable that whenever the difference in the number of electrons between the oxidized and reduced states, i.e., the value of n for the oxidation-reduction system, is relatively large the processes of oxidation and reduction occur in stages, one or more of which may be slow. In that event equilibrium between the system in the solution and the electrode will be established slowly, and the measured potential may be in error. To expedite the attainment of the equilibrium **a potential mediator** may be employed;[5] this is a substance that undergoes reversible oxidation-reduction and rapidly reaches equilibrium with the electrode.

Consider, for example, a system of two ions M^+ and M^{++} which is slow in the attainment of equilibrium with the electrode, and suppose a very small amount of a ceric salt (Ce^{++++}) is added to act as potential mediator; the reaction

$$M^+ + Ce^{++++} \rightleftharpoons M^{++} + Ce^{+++}$$

takes place until equilibrium is attained. At this point the potential of the M^+, M^{++} system must be identical with that of the Ce^{+++}, Ce^{++++} system (cf. p. 284). The ceric-cerous system comes to equilibrium rapidly with the inert metal, e.g., platinum, electrode and the potential registered is consequently both that of the Ce^{+++}, Ce^{++++} and M^+, M^{++}

[4] Popoff and Kunz, *J. Am. Chem. Soc.*, **51**, 382 (1929); Bray and Hershey, *ibid.*, **56**, 1889 (1934).

[5] Loimaranta, *Z. Elektrochem.*, **13**, 33 (1907); Foerster and Pressprich, *ibid.*, **33**, 176 (1927); Goard and Rideal, *Trans. Faraday Soc.*, **19**, 740 (1924).

systems in the experimental solution. If the potential mediator is added in very small amount, a negligible quantity of M^+ is used up and M^{++} formed in the establishment of the chemical equilibrium represented above: the measured potential in the presence of the mediator may thus be regarded as the value for the original system. In addition to ceric salts, iodine has been used as a potential mediator; the platinum electrode then measures the potential of the iodine-iodide ion system. If the results obtained in the presence of a mediator are to have definite thermodynamic significance they should be independent of the nature of the mediator and of the electrode material, provided the latter is not attacked in any way.

Standard Potentials from Titration Curves.—A method of studying oxidation-reduction systems involving the determination of potentials during the course of titration with a suitable substance, which frequently acts as a potential mediator, has been employed to a considerable extent in work on systems containing organic compounds. The pure oxidized form of the system, e.g., a quinone or related substance, is dissolved in a solution of definite hydrogen ion concentration, viz., a buffer solution (see Chap. XI); known amounts of a reducing solution, e.g., titanous chloride or sodium hydrosulfite, are added, in the absence of air, and the solution is kept agitated by means of a current of nitrogen. The potential of an inert electrode, e.g., platinum, gold or mercury, immersed in the reacting solution is measured after each addition of the titrant, by combination with a reference electrode such as a form of calomel electrode. The results obtained are of the type shown in Fig. 79, in which the electrode potentials observed during the course of the addition of various amounts of titanous chloride to a buffered (pH 6.98) solution of 1-naphthol-2-sulfonate indophenol at 30° are plotted as ordinates against the volumes of added reagent as abscissae.[6] The point at which the potential undergoes a rapid change is that corresponding to complete reduction (cf. p. 286), and the quantity of reducing solution then added is equivalent to the whole of the oxidized organic compound originally present. From the amounts of reducing agent added at various

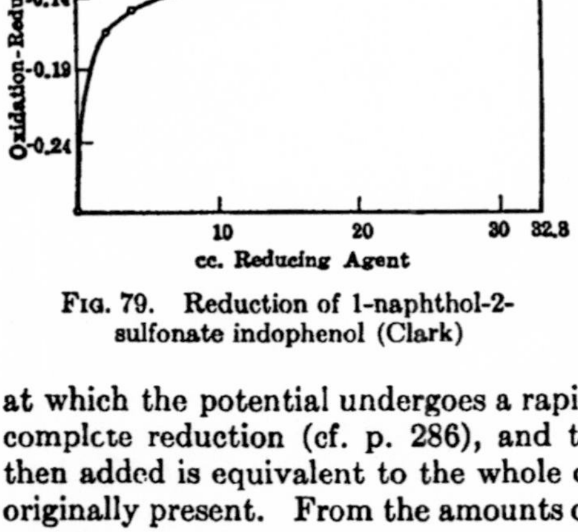

Fig. 79. Reduction of 1-naphthol-2-sulfonate indophenol (Clark)

[6] Clark *et al.*, "Studies on Oxidation-Reduction," Hygienic Laboratory Bulletin, No. 151, 1928; see also, Conant *et al.*, *J. Am. Chem. Soc.*, **44**, 1382, 2480 (1922); LaMer and Baker, *ibid.*, **44**, 1954 (1922).

stages the corresponding ratios of the concentrations of the oxidized form (o) to the reduced form (r) may be calculated without any knowledge of the initial amount of the former or of the concentration of the titrating agent. If t_c is the volume of titrant added when the sudden change of potential occurs, i.e., when the reduction is complete, and t is the amount of titrant added at any point in the titration, then at this point o is equivalent to $t_c - t$, and r is equivalent to t, provided the titrant employed is a powerful reducing agent.* According to equation (1), replacing the ratio of the activities by the ratio of concentrations, it follows that

$$E = E^{0\prime} - \frac{RT}{nF} \ln \frac{o}{r}$$

$$= E^{0\prime} - \frac{RT}{nF} \ln \frac{t_c - t}{t}, \tag{14}$$

where $E^{0\prime}$ is the standard potential of the system for the hydrogen ion concentration employed in the experiment. Values of $E^{0\prime}$ can thus be obtained for a series of points on the titration curve; if the system is behaving in a satisfactory manner these values should be approximately constant. The results obtained by applying equation (14) to the data in Fig. 79 are recorded in Table LII.

TABLE LII. EVALUATION OF APPROXIMATE STANDARD POTENTIAL AT 30° OF 1-NAPHTHOL-2-SULFONATE INDOPHENOL AT pH 6.98

t	Per cent Reduction	E	$\frac{RT}{2F} \ln \frac{t_c - t}{t}$	$E^{0\prime}$
4.0	12.2	− 0.1479	− 0.0258	− 0.1221
8.0	24.4	− 0.1368	− 0.0148	− 0.1220
12.0	36.6	− 0.1292	− 0.0072	− 0.1220
16.0	48.8	− 0.1224	− 0.0006	− 0.1218
20.0	61.0	− 0.1159	+ 0.0058	− 0.1217
24.0	73.2	− 0.1085	+ 0.0131	− 0.1216
28.0	85.4	− 0.0985	+ 0.0230	− 0.1215
32.8 (t_c)	100.0	− 0.036	—	—

The experiment described above can also be carried out by starting with the reduced form of the system and titrating it with an oxidizing agent, e.g., potassium dichromate. The standard potentials obtained in this manner agree with those derived from the titration of the oxidized form with a reducing agent, and also with the potentials measured in mixtures made up from known amounts of oxidized and reduced forms. The presence of the inorganic oxidizing or reducing system, which often has the advantage of serving as a potential mediator, does not affect the results to any appreciable extent.

It will be seen shortly that the value of n, the number of electrons involved in the oxidation-reduction system, is of some interest; if this is

* The precise conditions for efficient reduction are discussed on page 286.

not known, it can be evaluated from the slope of the flat portion of the titration curve such as that in Fig. 79. This slope is determined by the value of n only, and is independent of the chemical nature of the system; the larger is n the flatter is the curve. An exact estimate of n may be made by plotting the measured potential E against $\log o/r$, or its equivalent $\log (t_e - t)/t$; the plot, according to equation (14), should be a straight line of slope $- 2.303RT/nF$, i.e., $- 0.059/n$ at 25° or $- 0.060/n$ at 30°. The results derived from Fig. 79 are plotted in this manner in Fig. 80; the points are seen to fall approximately on a straight line, in

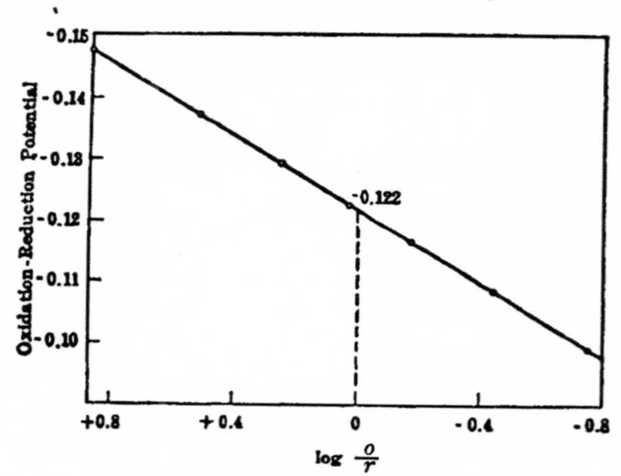

FIG. 80. Determination of n and $E^{0\prime}$

agreement with expectation, and the slope is $- 0.03$ at 30°, so that n is equal to 2. The standard potential of the system at the given hydrogen ion concentration, i.e., $E^{0\prime}$, is given by the point at which the ratio o/r is unity, i.e., $\log o/r$ is zero; this is seen to be $- 0.122$ volt, in agreement with the values in Table LII.

Standard Oxidation-Reduction Potentials.—Some values of standard oxidation-reduction potentials at 25° are given in Table LIII.[7] The sign of the potential is based on the usual convention (p. 187), and the assumption that an inert material precedes the system mentioned in each case; for example, for Pt | Fe^{++}, Fe^{+++} the standard potential is $- 0.783$ volt. A positive sign would indicate the tendency for negative electricity, e.g., electrons, to pass from solution to the metal, i.e.,

$$Fe^{++}(+ \text{Pt}) = Fe^{+++} + \epsilon(\text{Pt}),$$

so that in this particular case the standard free energy change of the process

$$Fe^{++} = Fe^{+++} + \epsilon$$

[7] For further data, see International Critical Tables, Vol. VI, and Latimer, "The Oxidation States of the Elements and their Potentials in Aqueous Solutions," 1938.

is given by

$$\Delta G^0 = -nFE^0 = +0.783F.$$

If the electrode had been represented by Fe^{++}, Fe^{+++} | Pt, i.e., with the inert metal succeeding the system, the sign of the potential would be reversed, i.e., + 0.783 volt. A positive potential in this case means a tendency for the process

$$Fe^{+++} + \epsilon(Pt) = Fe^{++} + (Pt)$$

to occur, which is the reverse of that just given. The order of writing the components present in the solution, viz., Fe^{++}, Fe^{+++} or Fe^{+++}, Fe^{++} is immaterial, although the usual convention is to employ the former method of representation.

TABLE LIII. STANDARD OXIDATION-REDUCTION POTENTIALS AT 25°

Electrode	Reaction	Potential
Co^{++}, Co^{+++}	$Co^{++} \rightarrow Co^{+++} + \epsilon$	− 1.82
Pb^{++}, Pb^{++++}	$Pb^{++} \rightarrow Pb^{++++} + 2\epsilon$	− 1.75
$PbSO_4(s)$, $PbO_2(s)$, SO_4^{--}	$PbSO_4 + 2H_2O \rightarrow PbO_2 + 4H^+ + SO_4^{--} + 2\epsilon$	− 1.685
Ce^{+++}, Ce^{++++}	$Ce^{+++} \rightarrow Ce^{++++} + \epsilon$	− 1.61
Mn^{++}, MnO_4^-, H^+	$Mn^{++} + 4H_2O \rightarrow MnO_4^- + 8H^+ + 5\epsilon$	− 1.52
Tl^+, Tl^{+++}	$Tl^+ \rightarrow Tl^{+++} + 2\epsilon$	− 1.22
Hg_2^{++}, Hg^{++}	$Hg_2^{++} \rightarrow 2Hg^{++} + 2\epsilon$	− 0.906
Fe^{++}, Fe^{+++}	$Fe^{++} \rightarrow Fe^{+++} + \epsilon$	− 0.783
MnO_4^{--}, MnO_4^-	$MnO_4^{--} \rightarrow MnO_4^- + \epsilon$	− 0.54
$Fe(CN)_6^{----}$, $Fe(CN)_6^{---}$	$Fe(CN)_6^{----} \rightarrow Fe(CN)_6^{---} + \epsilon$	− 0.356
Cu^+, Cu^{++}	$Cu^+ \rightarrow Cu^{++} + \epsilon$	− 0.16
Sn^{++}, Sn^{++++}	$Sn^{++} \rightarrow Sn^{++++} + 2\epsilon$	− 0.15
Ti^{+++}, Ti^{++++}	$Ti^{+++} \rightarrow Ti^{++++} + \epsilon$	− 0.06
Cr^{++}, Cr^{+++}	$Cr^{++} \rightarrow Cr^{+++} + \epsilon$	+ 0.41

The potentials recorded in Table LIII may be called "oxidation potentials" (cf. p. 243) since they give a measure of the free energies of the oxidation processes; for the reverse reactions, the potentials, with the signs reversed, are the corresponding "reduction potentials."

Variation of Oxidation-Reduction Potential.—From a knowledge of the standard oxidation-reduction potential of a given system it is possible to calculate, with the aid of the appropriate form of equation (3), the potential of any mixture of oxidized and reduced forms. For approximate purposes it is sufficient to substitute concentrations for activities; the results are then more strictly applicable to dilute solutions, but they serve to illustrate certain general points. A number of curves, obtained in this manner, for the dependence of the oxidation-reduction potential on the proportion of the system present in the oxidized form, are depicted in Fig. 81; these curves are obviously of the same form as the experimental curve in Fig. 79. The position of the curve on the oxidation-reduction scale depends on the standard potential of the system, which corresponds approximately to 50 per cent oxidation, while its slope is determined by the number of electrons by which the oxidized and reduced states differ. The influence of hydrogen ion concentration in the case of the permanganate-manganous ion system is shown by the curves for a_{H^+} equal to 1 and 0.1, respectively.

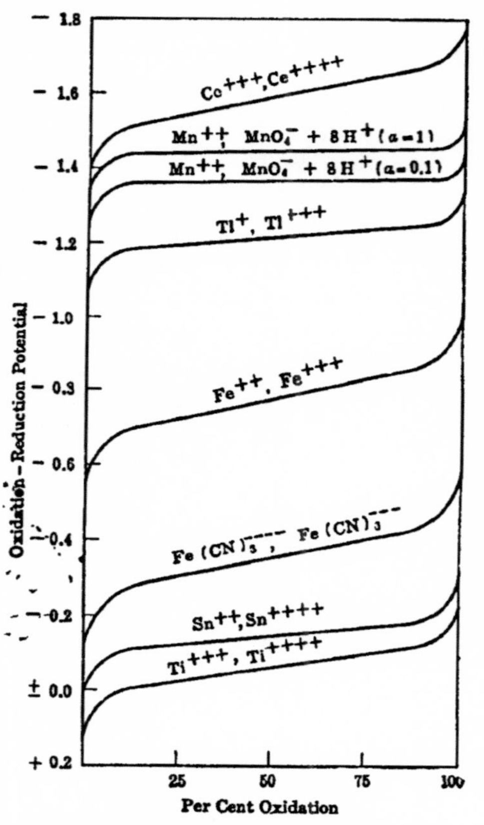

FIG. 81. Oxidation-reduction potentials at 25°.

It is seen from the curves in Fig. 81 that the potential rises rapidly at first as the amount of oxidized form is increased: this is due to the fact that when the proportion of the latter is small a relatively small *actual* increase in its amount brings about a large *relative* change. For example, if the solution contained 0.1 per cent of oxidized form and 99.9 per cent of reduced form, the potential would be

$$E = E^0 - \frac{RT}{nF} \ln \frac{o}{r}$$

$$= E^0 - \frac{0.059}{n} \log \frac{1}{999}$$

$$\approx E^0 + \frac{0.177}{n}$$

A change of 1 per cent in the proportion of oxidized form in the system would make the actual proportion 1.1 per cent, while there would be 98.9 per cent of oxidized form: the oxidation-reduction potential would then be given by

$$E = E^0 - \frac{0.059}{n} \log \frac{1}{98.9}$$

$$\approx E^0 + \frac{0.118}{n},$$

indicating a change of potential of about $0.059/n$ volt. As the amounts of oxidized and reduced states become of the same order, the potential changes only slowly, since an increase or decrease in either brings about little change in the ratio which determines the oxidation-reduction potential. Thus a change of 1 per cent in the amount of the oxidized form from 49 to 50 per cent, for example, alters the ratio of oxidized to reduced forms from 49/51 to 50/50; this will correspond to a change of $0.0052/n$ volt in potential. Solutions in this latter condition are said to be "poised":* the addition of appreciable amounts of an oxidizing or reducing agent to such a solution produces relatively little change in the oxidation-reduction potential. Finally, when the system consists almost exclusively of the oxidized form, i.e., at the right-hand side of Fig. 81, the potential again changes rapidly; the amount of reduced form is now very small, and consequently a small actual change means a large change in the ratio of oxidized to reduced forms in the solution.

Ionization in Stages.—When a metal yields two positive ions, M^{z_1+} and M^{z_2+}, there are three standard potentials of the system; these are the potentials of the electrodes M, M^{z_1+} and M, M^{z_2+} in addition to the oxidation-reduction potential M^{z_1+}, M^{z_2+}. If the values of these standard potentials are E_1^0, E_2^0 and $E_{1,2}^0$, respectively, then the free energy changes for the following process are as indicated below:

$$M = M^{z_1+} + z_1\epsilon, \qquad \Delta G_1^0 = -z_1FE_1^0;$$

$$M = M^{z_2+} + z_2\epsilon, \qquad \Delta G_2^0 = -z_2FE_2^0;$$

and

$$M^{z_1+} = M^{z_2+} + (z_2 - z_1)\epsilon, \qquad \Delta G_{1,2}^0 = -(z_2 - z_1)FE_{1,2}^0.$$

It follows from these three equations that

$$z_2E_2^0 - z_1E_1^0 = (z_2 - z_1)E_{1,2}^0,$$

so that the three potentials are not independent. If any two of the three potentials are known, the third can be evaluated directly. For example, the standard potentials for Cu, Cu^{++} and Cu^+, Cu^{++}, which are equivalent to E_2^0 and $E_{1,2}^0$, respectively, are -0.340 and -0.160 volt at 25°. It follows, therefore, since z_1 is equal to 1 and z_2 to 2, that

$$-2 \times 0.340 - E_1^0 = -0.160,$$

$$\therefore \quad E_1^0 = -0.520 \text{ volt}.$$

When a metal M is placed in contact with a solution containing either M^{z_1+} or M^{z_2+} ions, or both, reaction will occur until the equilibrium

$$(z_2 - z_1)M + z_1M^{z_2+} \rightleftharpoons z_2M^{z_1+}$$

is established; in this condition, it follows from the law of mass action

* This is the equivalent of the term "buffered" as applied to hydrogen ion potentials (cf. p. 410).

that

$$\frac{a_1^{z_2}}{a_2^{z_1}} = K,$$

where a_1 and a_2 are the activities of the M^{z_1+} and M^{z_2+} ions, respectively, *at equilibrium*. The activity of the solid metal M is taken as unity.

The value of this equilibrium constant can be calculated from the standard potentials derived above. It can be deduced, although it is obvious from general considerations, that when equilibrium is attained the potential of the metal M must be the same with respect to both M^{z_1+} and M^{z_2+} ions; hence,

$$E_1^0 - \frac{RT}{z_1F}\ln a_1 = E_2^0 - \frac{RT}{z_2F}\ln a_2,$$

$$\therefore \quad \ln K = (E_1^0 - E_2^0)\frac{z_1z_2F}{RT}. \tag{15}$$

It has been seen above that for the copper-copper ion system, E_1^0 is $-$ 0.520 and E_2^0 is $-$ 0.340, and so at 25°,

$$\log K = -\frac{0.180 \times 2}{0.05915} = -6.085,$$

$$\therefore \quad K = \frac{a_{Cu^+}^2}{a_{Cu^{++}}} = 8.22 \times 10^{-7}.$$

When metallic copper comes to equilibrium with a solution containing its ions, therefore, the concentration of cuprous ions will be very much smaller than that of cupric ions. For mercury on the other hand, E_1^0 for Hg, Hg_2^{++} is $-$ 0.799 volt, while $E_{1,2}^0$ is $-$ 0.906; from these data it is found that at equilibrium $a_{Hg_2^{++}}/a_{Hg^{++}}$ is 91. The ratio of the activity of the mercurous ions to that of the mercuric ions is thus 91, and hence the system in equilibrium with metallic mercury consists mainly of mercurous ions, although mercuric ions are also present to an appreciable extent. It can be seen from equation (15) that the equilibrium constant between the two ions of a given metal in the presence of that metal is greater the larger the difference of the standard potentials with respect to the two ions; the ions giving the less negative standard potential are present in excess at equilibrium (cf. p. 253).

Attention may be called to the fact that if the equilibrium constant could be determined by chemical methods, and if one of the three standard potentials of a particular metal-ion system is known, the other two could be evaluated. This procedure was actually used for copper, the calculations given above being carried out in the reverse direction.[8]

Oxidation-Reduction Equilibria.—When two reversible oxidation-reduction systems are mixed a definite equilibrium is attained which is

[8] Heinerth, *Z. Elektrochem.*, 37, 61 (1931).

determined largely by the standard potentials of the systems. For example, for the reaction between the ferrous-ferric and stannous-stannic systems, the equilibrium can be represented by

$$2Fe^{++} + Sn^{++++} \rightleftharpoons 2Fe^{+++} + Sn^{++},$$

and when equilibrium is attained, the law of mass action gives

$$K = \left(\frac{a_{Sn^{++}} a^2_{Fe^{+++}}}{a_{Sn^{++++}} a^2_{Fe^{++}}} \right)_e . \tag{16}$$

The reaction takes place in the cell

$$Pt \mid Fe^{++}, Fe^{+++} \parallel Sn^{++}, Sn^{++++} \mid Pt,$$

for the passage of two faradays, and so it follows that the standard E.M.F. is given by

$$E^0 = \frac{RT}{2F} \ln K = \frac{RT}{2F} \ln \left(\frac{a_{Sn^{++}} a^2_{Fe^{+++}}}{a_{Sn^{++++}} a^2_{Fe^{++}}} \right)_e , \tag{17}$$

where E^0 is equal to the difference in the standard potentials of the ferrous-ferric and stannous-stannic systems, i.e.,

$$E^0 = E^0_{Fe^{++}, Fe^{+++}} - E^0_{Sn^{++}, Sn^{++++}}. \tag{18}$$

It is evident, therefore, from equations (17) and (18), that the equilibrium constant depends on the difference of the standard potentials of the interacting systems; if the equilibrium constant were determined experimentally it would be possible to calculate the difference of standard potentials, exactly as in the case of the replacement of one metal by another (cf. p. 254). Alternatively, if the difference in standard potentials is known, the equilibrium constant can be evaluated.

The value of $E^0_{Fe^{++}, Fe^{+++}}$ is $-$ 0.783 and that of $E_{Sn^{++}, Sn^{++++}}$ is $-$ 0.15 at 25°; hence making use of the relationship, from equations (17) and (18),

$$E^0_{Fe^{++}, Fe^{+++}} - E^0_{Sn^{++}, Sn^{++++}} = \frac{RT}{2F} \ln K, \tag{19}$$

it is readily found that

$$K = \left(\frac{a_{Sn^{++}} a^2_{Fe^{+++}}}{a_{Sn^{++++}} a^2_{Fe^{++}}} \right)_e = 4.1 \times 10^{-22}.$$

This low value of the equilibrium constant means that when equilibrium is attained in the ferrous-ferric and stannous-stannic mixture, the concentrations (activities) of ferric and stannous ions must be negligibly small in comparison with those of the ferrous and stannic ions. In other words, when these two systems are mixed, reaction occurs so that the ferric ions are virtually completely reduced to ferrous ions while the stannous are oxidized to stannic ions. This fact is utilized in analytical work for the reduction of ferric to ferrous ions prior to the estimation of the latter by means of dichromate.

Inserting the expression for K, given by equation (16), into equation (19) and rearranging, the result is

$$E^0_{Fe^{++}, Fe^{+++}} - \frac{RT}{F} \ln \frac{(a_{Fe^{+++}})_e}{(a_{Fe^{++}})_e} = E^0_{Sn^{++}, Sn^{++++}} - \frac{RT}{2F} \ln \frac{(a_{Sn^{++++}})_e}{(a_{Sn^{++}})_e}, \quad (20)$$

the left-hand side of this equation being the potential of the ferrous-ferric system and the right-hand side that of the stannous-stannic system at equilibrium. When this condition is attained, therefore, both systems must exhibit the same oxidation-reduction potential; this fact has been already utilized in connection with the employment of potential mediators.

Oxidation-Reduction Systems in Analytical Chemistry.—An examination of the calculation just made shows that the very small equilibrium constant,* and hence the virtually complete interaction of one system with the other, is due to the large difference in the standard potentials of the two systems. The system with the more negative standard potential as recorded in Table LIII, e.g., Pt | Fe^{++}, Fe^{+++} in the case considered above, always oxidizes the system with the less negative standard potential, e.g., Pt | Sn^{++}, Sn^{++++}, the extent of the oxidation being greater the larger the difference between the standard potentials. The same conclusion may be stated in the alternative manner: the system with the less negative potential reduces the one with the more negative potential, the extent being greater the farther the systems are apart in the table of standard potentials. It is of interest to call attention to the fact that as a consequence of these arguments the terms "oxidizing agent" and "reducing agent" are to be regarded as purely relative. A given system, e.g., ferrous-ferric, will reduce a system above it in Table LIII, e.g., cerous-ceric, but it will oxidize one below it, e.g., stannous-stannic.

The question of the extent to which one system oxidizes or reduces another is of importance in connection with oxidation-reduction titrations in analytical chemistry. The reason why ceric sulfate and acidified potassium permanganate are such useful reagents in volumetric analysis is because they have large negative standard potentials and are consequently able to bring about virtually complete oxidation of many other systems. If the permanganate system had a standard potential which did not differ greatly from that of the system being titrated, the equilibrium constant might be of the order of unity; free permanganate, indicated by its pink color, would then be present in visible amount long before oxidation of the other system was complete. The titration values would thus have no analytical validity. In order that oxidation or reduction of a system should be "complete," within the limits of accuracy of ordinary volumetric analysis, it is necessary that the concentration of one form at the end-point should be at least 10^3 times that of the other; that is to say, oxidation or reduction is complete within 0.1 per cent or

* If the reaction were considered in the opposite direction the equilibrium constant would be the reciprocal of the value given, and hence would be very large.

better. The equilibrium constant should thus be smaller than 10^{-6} if n is the same for both interacting oxidation-reduction systems, or 10^{-9} if n is unity for one system and two for the other. By making use of equations similar to (19), it can be readily shown that if two oxidation-reduction systems are to react completely in the ordinary analytical sense, the standard potentials should differ by at least 0.35 volt if n is unity for both systems, 0.26 volt if n is unity for one and two for the other, or 0.18 volt if n is two for both.

Potentiometric Oxidation-Reduction Titrations.—The variation of potential during the course of the conversion of the completely reduced state of any system to the completely oxidized state is represented by a curve of the type shown in Figs. 79 and 81; these curves are, therefore, equivalent to potential-titration curves, the end-point of the titration in each case being marked by a relatively rapid change of potential. The question arises as to whether this end-point could be estimated with sufficient accuracy in any given case by measuring the potential of an inert electrode, e.g., platinum, inserted in the titration system. An answer can be obtained by considering the further change in potential after the end-point has been passed; before the equivalence-point the potentials are determined by the titrated system, since this is present in

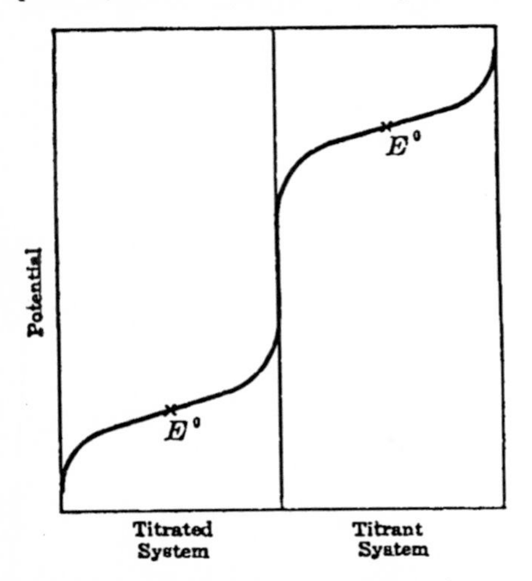

Fig. 82. Potential-titration curve; determination of end-point is possible

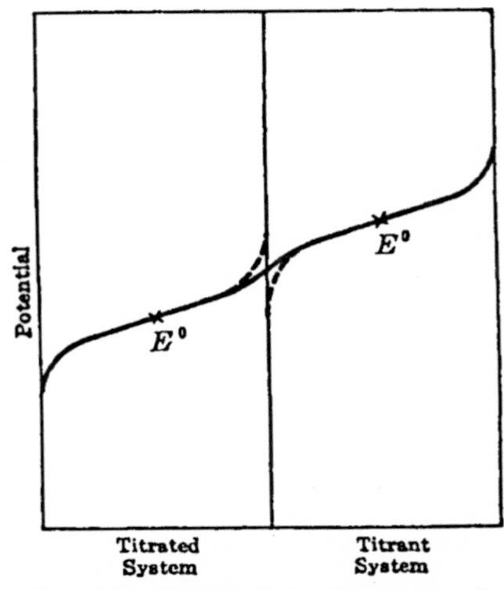

Fig. 83. Potential-titration curve; determination of the end-point is not satisfactory

excess, while after the equivalent point they are determined by the titrant system. The potential-titration curve from one extreme to the other can then be derived by placing side by side the curves for the two separate systems and joining them by a tangent. Two examples are shown in Figs. 82 and 83; in the former the standard potentials, represented by the respective mid-points, are reasonably far apart, but in the latter they are close together. In Fig. 82 there is a rapid increase of potential at the titration end-point, and so its position can be determined accurately; systems of this type, therefore, lend themselves to potentiometric titration. When the standard potentials of the titrated and titrant systems are close together, however, the change of potential at the equivalence-point is not marked to any appreciable extent; satisfactory potentiometric detection of the end-point in such a titration is therefore not possible.

It will be recalled that the condition for reliable potentiometric titration is just that required for one system to reduce or oxidize another completely within the normal limits of analytical accuracy. It follows, therefore, that when the standard potentials of the two interacting systems are such as to make them suitable for analytical work, the reaction is also one whose end-point can be derived reasonably accurately potentiometrically. The minimum differences between the standard potentials given on page 285 for an analytical accuracy of about 0.1 per cent, with systems of different types, may also be taken as those requisite for satisfactory potentiometric titration. The greater the actual difference, of course, the more precisely can the end-point be estimated.

The method of carrying out oxidation-reduction titrations potentiometrically is essentially similar to that for precipitation reactions, except that the indicator electrode now consists merely of an inert metal. The determination of the end-point graphically or by some form of differential titration procedure is carried out in a manner exactly analogous to that described in Chap. VII; various forms of simplified methods of oxidation-reduction titration have also been described.[9]

Potential at the Equivalence-Point.—Since the potentials of the two oxidation-reduction systems, represented by the subscripts I and II, involved in a titration must be the same, it follows that

$$E = E_{\mathrm{I}}^0 - \frac{RT}{nF}\ln\frac{\mathrm{O_I}}{\mathrm{R_I}} = E_{\mathrm{II}}^0 - \frac{RT}{nF}\ln\frac{\mathrm{O_{II}}}{\mathrm{R_{II}}}, \qquad (21)$$

where E is the actual potential and E_{I}^0 and E_{II}^0 are the respective standard potentials. Consider the case in which the reduced form of the system I, i.e., $\mathrm{R_I}$, is titrated with the oxidized form of the system II, i.e., $\mathrm{O_{II}}$, so that the reaction

$$\mathrm{R_I} + \mathrm{O_{II}} = \mathrm{O_I} + \mathrm{R_{II}}$$

occurs during the titration. At the equivalence-point, not only are the concentrations of $\mathrm{O_I}$ and $\mathrm{R_{II}}$ equal, as at any point in the titration, but $\mathrm{R_I}$ and $\mathrm{O_{II}}$ are also equal to each other; hence $\mathrm{O_I/R_I}$ is then equal to $\mathrm{R_{II}/O_{II}}$. Substitution of this result into equation (21) immediately gives for $E_{\text{equiv.}}$, the potential at the equivalence-point,

$$E_{\text{equiv.}} = \frac{E_{\mathrm{I}}^0 + E_{\mathrm{II}}^0}{2}.$$

[9] See general references to potentiometric titration on page 256.

This result holds for the special case in which each oxidation-reduction system involves the transfer of the same number of electrons, i.e., the value of n is the same in each case. If they are different, however, the equation for the reaction between the two systems becomes

$$n_{II}R_I + n_I O_{II} = n_{II}O_I + n_I R_{II},$$

where n_I and n_{II} refer to the systems I and II, respectively. By using the same general arguments as were employed above, it is found that the potential at the equivalence-point is given by

$$E_{\text{equiv.}} = \frac{n_{II}E_I^0 + n_I E_{II}^0}{n_I + n_{II}}.$$

Oxidation-Reduction Indicators.—A reversible oxidation-reduction indicator is a substance or, more correctly, an oxidation-reduction system, exhibiting different colors in the oxidized and reduced states, generally colored and colorless, respectively. Mixtures of the two states in different proportions, and hence corresponding to different oxidation-reduction potentials, will have different colors, or depths of color; every color thus corresponds to a definite potential which depends on the standard potential of the system, and frequently on the hydrogen ion concentration of the solution. If a small amount of an indicator is placed in another oxidation-reduction system, the former, acting as a potential mediator, will come to an equilibrium in which its oxidation-reduction potential is the same as that of the system under examination. The potential of the given indicator can be estimated from its color in the solution, and hence the potential of the system under examination will have the same value.

Since the eye, or even mechanical devices, are capable of detecting color variations within certain limits only, any given oxidation-reduction indicator can be effectively employed only in a certain range of potential. Consider, for example, the simple case of an indicator system for which n is unity; the oxidation-reduction potential at constant hydrogen ion concentration is given approximately by

$$E = E^0 - \frac{RT}{F}\ln\frac{o}{r}.$$

Suppose the limits within which color changes can be detected are 9 per cent of oxidized form, i.e., o/r is $9/91 \approx 1/10$, at one extreme, to 91 per cent of oxidized form, i.e., o/r is $91/9 \approx 10$; the corresponding potential limits at ordinary temperatures are then given by the foregoing equation as $E^0 + 0.058$, and $E^0 - 0.058$, respectively. If n for the indicator system had been 2, the limits of potential would have been $E^0 + 0.029$ and $E^0 - 0.029$. It is seen, therefore, that an oxidation-reduction indicator can be used for determining the potentials of unknown systems only if the values lie relatively close to the standard potential E^0 of the indicator. In other words, it is only in the vicinity of its standard potential, at the particular hydrogen ion concentration of the medium, that an oxidation-reduction indicator undergoes detectable color changes. In order to cover an appreciable range of potentials, it is clearly necessary to have a range of indicators with different standard potentials.

Indicators for Biological Systems.[10]—Many investigations have been carried out of substances which have the properties necessary for a suitable oxidation-reduction indicator. As a result of this work it is convenient for practical purposes to divide such indicators into two categories: there are those of relatively low potential, viz., − 0.3 to + 0.5 volt in neutral solution, which are especially useful for the study of biological systems, and those of more negative standard potentials that are employed in volumetric analysis. The majority of substances proposed as oxidation-reduction indicators for biological purposes are also acid-base indicators, exhibiting different colors in acid and alkaline solutions. They are frequently reddish-brown in acid media, i.e., at high hydrogen ion concentrations, and blue in alkaline solutions, i.e., at low hydrogen ion concentrations, and since the former color is less intense than the latter it is desirable to use the indicator in its blue form. In biological systems it is generally not possible to alter the hydrogen ion concentration from the vicinity of the neutral point, i.e., pH 7,* and so indicators are required with relatively strong acidic, or weakly basic, groups so that they exhibit their alkaline colors at relatively high hydrogen ion concentrations (cf. Chap. X). A number of such indicators have been synthesized by Clark and his co-workers, by introducing halogen atoms into one of the phenolic groups of phenol-indophenol, e.g., 2 : 6-dichlorophenol-indophenol. In addition to the members of this series, other indicators of biological interest are indamines, e.g., Bindschedler's green and toluylene blue; thiazines, e.g., Lauth's violet and methylene blue; oxazines, e.g., cresyl blue and ethyl Capri blue; and certain indigo-sulfonates, safranines and rosindulines. A group of oxidation-reduction indicators of special interest are the so-called "viologens," introduced by Michaelis; they are NN′-di-substituted-4 : 4-dipyridilium chlorides which are deeply colored in the reduced state, and have the most positive standard potentials of any known indicators. A few typical oxidation-reduction indicators used in biological work, together with their standard potentials ($E^{0\prime}$) at pH 7, determined by direct measurement, are given in Table LIV; it will be observed that these cover almost the whole range of potentials from − 0.3 to + 0.45 volt, with but few gaps.

It is rarely feasible in biological investigations to determine the actual potential from the color of the added indicator, although this should be possible theoretically, because the indicators are virtually of the one

[10] Clark *et al.*, "Studies on Oxidation-Reduction," 1928 *et seq.*; Michaelis, "Oxydations-Reductions Potentiale," 1933; for review, see Glasstone, *Ann. Rep. Chem. Soc.*, 31, 305 (1934).

* For a discussion of pH and its significance, see Chap. X; see also, page 292.

TABLE LIV. OXIDATION-REDUCTION INDICATORS FOR BIOLOGICAL WORK

Indicator	$E^{0\prime}$	Indicator	$E^{0\prime}$
Phenol-*m*-sulfonate indo-2 : 6-dibromophenol	− 0.273	Ethyl Capri blue	+ 0.072
m-Bromophenol indophenol	− 0.248	Indigo trisulfonate	+ 0.081
2 : 6-Dichlorophenol indophenol	− 0.217	Indigo disulfonate	+ 0.125
2 : 6-Dichlorophenol indo-*o*-cresol	− 0.181	Cresyl violet	+ 0.173
2 : 6-Dibromophenol indoguaiacol	− 0.159	Phenosafranine	+ 0.252
Toluylene blue	− 0.115	Tetramethyl phenosafranine	+ 0.273
Cresyl blue	− 0.047	Rosinduline scarlet	+ 0.296
Methylene blue	− 0.011	Neutral red	+ 0.325
Indigo tetrasulfonate	+ 0.046	Sulfonated rosindone	+ 0.380
		Methyl viologen	+ 0.445

color type. For most purposes, therefore, it is the practice to take a number of samples of the solution under examination, to add different indicators to each and to observe which are reduced; if one indicator is decolorized and the other not, the potential must lie between the standard potentials of these two indicators at the hydrogen ion concentration (pH) of the solution. Similarly, indicators may be used in the reduced state and their oxidation observed. Indicators are also often employed as potential mediators in solutions for which equilibrium with the electrode is established slowly; the potential is then measured electrometrically. When employing an oxidation-reduction indicator it is essential that the solution to which it is added should be well poised (p. 281), so that in oxidizing or reducing the indicator the ratio of oxidized to reduced states of the experimental system should not be appreciably altered. The amount of indicator added must, of course, be relatively small.

Indicators for Volumetric Analysis.—The indicators described above are frequently too unstable for use in volumetric analysis and, in addition, they show only feeble color changes in acid solution. The problem of suitable indicators for detecting the end-points of oxidation-reduction titrations is, however, in some senses, simpler than that of finding a series of indicators for use over a wide range of potentials. It has been seen that if two oxidation-reduction systems interact sufficiently completely to be of value for analytical purposes, there is a marked change of potential of the system at the equivalence-point (cf. Fig. 82). Ideally, the standard potential of the indicator should coincide with the equivalence-point potential of the titration; actually it is sufficient, however, for the former to lie somewhere in the region of the rapidly changing potential of the titration system. When the end-point is reached, therefore, and the oxidation-reduction potential undergoes a rapid alteration, the color of the indicator system will change sharply from one extreme to the other. If the standard potential of the indicator is either below or above the region in which the potential inflection occurs, the color change will take place either before or after the equivalence-point, and in any case will be gradual rather than sharp. Such indicators would be of no value for the particular titration under consideration. It has been found (p. 285) that if two systems are to interact sufficiently for analytical purposes their standard potentials must differ by about 0.3 volt, and hence the standard potential of a suitable oxidation-reduction indicator must be about 0.15 volt below that of one system and 0.15 volt above that of the other. Since the most important volumetric oxidizing agents have high negative potentials, however, a large number of indicators is not necessary for most purposes.

The interest in the application of indicators in oxidation-reduction titrations has followed on the discovery that the familiar color change undergone by diphenylamine on oxidation could be used to determine the end-point of the titration of ferrous ion by dichromate in acid solution. Diphenylamine, preferably in the form of its soluble sulfonic acid, at first undergoes irreversible oxidation to diphenylbenzidine, and it is this substance, with its oxidation product diphenylamine violet, that constitutes the real indicator.[11]

The standard potential of the indicator system is not known exactly, but experiments have shown that in not too strongly acid solutions the sharp color change from colorless to violet, with green as a possible intermediate, occurs at a potential of about − 0.75 volt. The standard potential of the ferrous-ferric system is − 0.78 whereas that of the dichromate-chromic ion system in an acid medium is approximately − 1.2 volt; hence a suitable oxidation-reduction indicator might be expected to have a standard potential of about − 0.95 volt. It would thus appear that diphenylamine would not be satisfactory for the titration of ferrous ions by acid dichromate, and this is actually true if a simple ferrous salt is employed. In actual practice, for titration purposes, phosphoric acid or a fluoride is added to the solution; these substances form complex ions with the ferric ions with the result that the effective standard potential of the ferrous-ferric system is lowered (numerically) to about − 0.5 volt. The change of potential at the end-point of the titration is thus from about − 0.6 to − 1.1 volt, and hence diphenylamine, changing color in the vicinity of − 0.75 volt, is a satisfactory indicator.

Ceric sulfate is a valuable oxidizing agent, the employment of which in volumetric work was limited by the difficulty of detecting the end-point unless a potentiometric method was used. A number of indicators are now available, however, which permit direct titration with ceric sulfate solution to be carried out. One of the most interesting and useful of these is *o*-phenanthroline ferrous sulfate, the cations of which, viz., $Fe(C_{12}H_8N_2)_3^{++}$, with the corresponding ferric ions, viz., $Fe(C_{12}H_8N_2)_3^{+++}$, form a reversible oxidation-reduction system; the reduced state has an intense red color and the oxidized state a relatively feeble blue color, so that there is a marked change in the vicinity of the standard potential which is about − 1.1 volt.[12] The high potential of the phenanthroline-

[11] Kolthoff and Sarver, *J. Am. Chem. Soc.*, 52, 4179 (1930); 53, 2902 (1931); 59, 23 (1937); for review, see Glasstone, *Ann. Rep. Chem. Soc.*, 31, 309 (1934); also, Whitehead and Wills, *Chem. Revs.*, 29, 69 (1941).

[12] Walden, Hammett and Chapman, *J. Am. Chem. Soc.*, 53, 3908 (1931); 55, 2649 (1933); Walden and Edmonds, *Chem. Revs.*, 16, 81 (1935).

ferrous ion indicator permits it to be used in connection with the titration of ferrous ions without the addition of phosphoric acid or fluoride ions. The indicator has been employed for a number of titrations with ceric sulfate and also with acid dichromate, and even with very dilute solutions of permanganate when the color of the latter was too feeble to be of any value for indicator purposes. Another indicator having a high standard potential is phenylanthranilic acid; this is a diphenylamine derivative which changes color in the vicinity of $-$ 1.08 volt. It has been recommended for use with ceric sulfate as the oxidizing titrant.[13]

Although there are now several useful indicators for titrations involving strongly oxidizing reactants, the situation is not so satisfactory in connection with reducing reagents, e.g., titanous salts. The standard potential of the titanous-titanic system is approximately $-$ 0.05 volt, and hence a useful indicator should show a color change at a potential of about $-$ 0.2 volt or somewhat more negative. The only substance that is reasonably satisfactory for this purpose, as far as is known at present, is methylene blue which changes color at about $-$ 0.3 volt in acid solution.

Quinone-Hydroquinone Systems.—In the brief treatment of the quinone-hydroquinone system on page 270 no allowance was made for the possibility of the hydroquinone ionizing as an acid; actually such ionization occurs in alkaline solutions and has an important effect on the oxidation-reduction potential of the system. Hydroquinone, or any of its substituted derivatives, can function as a dibasic acid. It ionizes in two stages, viz.,

$$H_2Q \rightleftharpoons H^+ + HQ^-$$

and

$$HQ^- \rightleftharpoons H^+ + Q^{--},$$

and the dissociation constants corresponding to these two equilibria (cf. p. 318) are given by

$$K_1 = \frac{a_{H^+}a_{HQ^-}}{a_{H_2Q}} \qquad \text{and} \qquad K_2 = \frac{a_{H^+}a_{Q^{--}}}{a_{HQ^-}}.$$

The hydroquinone in solution thus exists partly as undissociated H_2Q, and also as HQ^- and Q^{--} ions formed in the two stages of ionization; the total stoichiometric concentration h of the hydroquinone is equal to the sum of the concentrations of these three species, i.e.,

$$h = c_{H_2Q} + c_{HQ^-} + c_{Q^{--}},$$

and if the values of c_{HQ^-} and $c_{Q^{--}}$ derived from the expressions for K_1 and K_2 are inserted in this equation, the approximation being made of taking the activity coefficients of H_2Q, HQ^- and Q^{--} to be equal to

[13] Syrokomsky and Stiepin, *J. Am. Chem. Soc.*, 58, 928 (1936).

unity, the result is

$$h = c_{H_2Q} + \frac{c_{H_2Q}}{a_{H^+}} k_1 + \frac{c_{H_2Q}}{a_{H^+}^2} k_1 k_2,$$

$$c_{H_2Q} = \frac{h a_{H^+}^2}{a_{H^+}^2 + k_1 a_{H^+} + k_1 k_2}. \tag{22}$$

In view of the neglect of the activity coefficients, the constants K_1 and K_2 have been replaced by k_1 and k_2 which become identical with the former at infinite dilution. If q is the concentration of the quinone form, which is supposed to be a neutral substance exhibiting neither acidic nor basic properties, the oxidation-reduction potential, which according to equation (4) may be written as

$$E = E^0 - \frac{RT}{2F} \ln \frac{a_Q}{a_{H_2Q}} - \frac{RT}{2F} \ln a_{H^+}^2, \tag{23}$$

is given by

$$E = E^0 - \frac{RT}{2F} \ln \frac{q}{c_{H_2Q}} - \frac{RT}{2F} \ln a_{H^+}^2, \tag{24}$$

the ratio of the activities of Q and H_2Q being taken as equal to the ratio of their concentrations. Introduction of the value of c_{H_2Q} from equation (22) into (24) now gives

$$E = E^0 - \frac{RT}{2F} \ln \frac{q}{h} - \frac{RT}{2F} \ln (a_{H^+}^2 + k_1 a_{H^+} + k_1 k_2). \tag{25}$$

If k_1 and k_2 are small, the terms $k_1 a_{H^+}$ and $k_1 k_2$ may be neglected in comparison with $a_{H^+}^2$, and equation (25) then reduces to

$$E = E^0 - \frac{RT}{2F} \ln \frac{q}{h} - \frac{RT}{F} \ln a_{H^+}, \tag{26}$$

which is the conventional form for the quinone-hydroquinone system, q and h representing the total concentrations of the two constituents.

According to equation (26) the variation of the oxidation-reduction potential with hydrogen ion concentration is relatively simple, but if the acidic dissociation functions k_1 and k_2 of the hydroquinone are appreciable, equation (25) must be employed, and the situation becomes somewhat more complicated. The method of studying this problem is to maintain the ratio q/h constant, i.e., the stoichiometric composition of the quinone-hydroquinone mixture is unchanged, but to suppose the hydrogen ion concentration is altered. For this purpose the equations for the electrode potential are differentiated with respect to $-\log a_{H^+}$; this quantity is a very useful function of the hydrogen ion concentration, designated by the symbol pH and referred to as the **hydrogen ion exponent.** Differentiation of equation (25) thus gives

$$-\frac{dE}{d \log a_{H^+}} = \frac{dE}{d(\text{pH})} = 2.303 \frac{RT}{2F} \cdot \frac{2a_{H^+}^2 + k_1 a_{H^+}}{a_{H^+}^2 + k_1 a_{H^+} + k_1 k_2} \tag{27}$$

as applicable over the whole pH range. If a_{H^+} is large in comparison with k_1 and k_2, i.e., in relatively acid solutions, this equation reduces to

$$\frac{dE}{d(\text{pH})} = 2.303\,\frac{RT}{F} \qquad \text{for} \qquad a_{H^+} \gg k_1 > k_2, \tag{28}$$

which can also be derived directly from equation (26).

If, however, k_1 is much greater than a_{H^+} and this is much greater than k_2, the terms $2a^2_{H^+}$ in the numerator and $a^2_{H^+}$ and k_1k_2 in the denominator of equation (27) may be neglected; the result is

$$\frac{dE}{d(\text{pH})} = 2.303\,\frac{RT}{2F} \qquad \text{for} \qquad k_1 \gg a_{H^+} > k_2. \tag{29}$$

Finally, when a_{H^+} becomes very small, i.e., in alkaline solutions, both terms in the numerator of equation (27) may be disregarded, and so

$$\frac{dE}{d(\text{pH})} = 0 \qquad \text{for} \qquad k_1 > k_2 \gg a_{H^+}. \tag{30}$$

The slope of the plot of the oxidation-reduction potential, for constant quinone-hydroquinone ratio, against the pH, i.e., against $-\log a_{H^+}$, thus undergoes changes, as shown in Fig. 84; the temperature is 30°, so

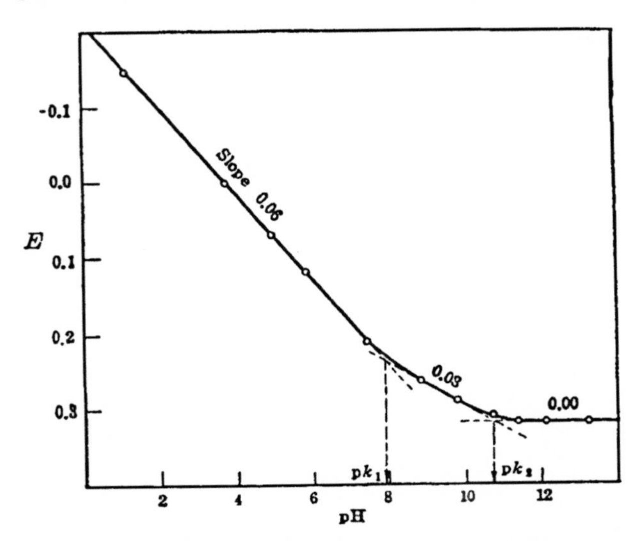

FIG. 84. Variation of a quinone-hydroquinone (anthraquinone sulfonate) potential with pH

that the slopes corresponding to equations (28), (29) and (30) are 0.060, 0.030 and zero, respectively. The position and length of the intermediate portion of slope 0.030 depend on the actual values and ratio of the acidic dissociation functions k_1 and k_2; this may be seen by investigating the conditions for which a_{H^+} is equal to k_1 and k_2 respectively.

If a_{H^+} in equation (27) is set equal to k_1, the result is

$$\frac{dE}{d(\text{pH})} = 2.303\,\frac{RT}{2F}\cdot\frac{3k_1}{2k_1 + k_2} \qquad \text{for} \qquad a_{H^+} = k_1,$$

and since k_2 is generally much smaller than k_1, this becomes

$$\frac{dE}{d(\text{pH})} = 2.303\,\frac{RT}{2F}\cdot\frac{3}{2} = 0.045 \text{ at } 30°.$$

When the hydrogen ion activity a_{H^+} is equal to the acidic function k_1, i.e., when the pH is equal to $-\log k_1$, the latter quantity being represented by pk_1, the slope of the pH-potential curve is thus seen to be intermediate between 0.060 and 0.030. Such a slope corresponds, in general, to a point on the first bend of the curve in Fig. 84; the exact position for a slope of 0.045 is obtained by finding the point of intersection of the two lines of slope 0.060 and 0.030, as shown. At this point, therefore, the pH is equal to pk_1.

To find the slope of the pH potential curve when a_{H^+} is equal to k_2, i.e., when the pH is equal to pk_2, the values of a_{H^+} in equation (27) are replaced by k_2; hence

$$\frac{dE}{d(\text{pH})} = 2.303\,\frac{RT}{2F}\cdot\frac{2k_2 + k_1}{k_2 + 2k_1} \qquad \text{for} \qquad a_{H^+} = k_2,$$

and since, as before, k_2 may be regarded as being much smaller than k_1,

$$\frac{dE}{d(\text{pH})} = 2.303\,\frac{RT}{2F}\cdot\frac{1}{2} = 0.015 \text{ at } 30°.$$

The pH is thus equal to pk_2 when the slope of the pH-potential curve is midway between 0.030 and zero; the value of pk_2 can be found by extending the lines of slopes 0.030 and zero until they intersect, as shown in Fig. 84. An examination of the pH-potential curve thus gives the values of the acidic dissociation functions for the particular hydroquinone as 7.9 and 10.6 for pk_1 and pk_2, respectively, at 30°.

The case considered here is relatively simple, but more complex behavior is frequently encountered: the reduced form may have more than two stages of acidic dissociation and in addition the oxidized form may exhibit one or more acidic dissociations. There is also the possibility of basic dissociation occurring, but this can be readily treated as equivalent to an acidic ionization (cf. p. 362). The method of treatment given above can, however, be applied to any case, no matter how complex, and the following general rules have been derived which facilitate the analysis of pH-potential curves for oxidation-reduction systems of constant stoichiometric composition.[14]

[14] Clark, "Studies on Oxidation-Reduction," Hygienic Laboratory Bulletin, 1928.

(1) Each bend in the curve may be correlated with an acidic dissociation constant; if the curve becomes steeper with increasing pH, i.e., as the solution is made more alkaline, the dissociation has occurred in the oxidized form, but if it becomes flatter it has occurred in the reduced form (cf. Fig. 84).

(2) The intersection of the extensions of adjacent linear parts of the curve occurs at the pH equal to pk for the particular dissociation function responsible for the bend.

(3) Each dissociation constant changes the slope by $2.303RT/nF$ volt per pH unit, where n is the number of electrons difference between oxidized and reduced states.

Two Stage Oxidation-Reduction.—The completely oxidized, i.e., holoquinone, form of a quinone differs from the completely reduced, i.e., hydroquinone, form by two hydrogen atoms, involving the addition or removal, respectively, of two electrons and two protons in one stage, viz.,

$$H_2Q \rightleftharpoons Q + 2H^+ + 2\epsilon.$$

It is known from chemical studies, however, that in many cases there is an intermediate stage between the hydroquinone (H_2Q) and the quinone (Q); this may be a meriquinone, which may be regarded as a molecular compound ($Q \cdot H_2Q$), or it may be a semiquinone (HQ). The latter is a true intermediate with a molecular weight of the same order as that of the quinone, instead of double, as it is for the meriquinone. The possibility that oxidation and reduction of quinonoid compounds might take place in two stages, each involving one electron, i.e., n is unity, with the intermediate formation of a semiquinone was considered independently by Michaelis and by Elema.[15] If the two stages of oxidation-reduction do not interfere, a ready distinction between meriquinone and semiquinone formation as intermediate is possible by means of E.M.F. measurements.

For meriquinone formation the stages of oxidation-reduction may be written

$$(1) \quad 2H_2Q \rightleftharpoons H_2Q \cdot Q + 2H^+ + 2\epsilon,$$

and

$$(2) \quad H_2Q \cdot Q \rightleftharpoons 2Q + 2H^+ + 2\epsilon,$$

so that if E_1 represents the *standard* potential of the first stage at a definite hydrogen ion concentration,

$$E = E_1 - \frac{RT}{2F} \ln \frac{(H_2Q \cdot Q)}{(H_2Q)^2}, \tag{31}$$

where the parentheses represent activities. If the original amount of the reduced form (H_2Q) in a given solution is a, and x equiv. of a strong

[15] Friedheim and Michaelis, *J. Biol. Chem.*, **91**, 355 (1931); Michaelis, *ibid.*, **92**, 211 (1931); **96**, 703 (1932); Elema, *Rec. trav. chim.*, **50**, 807 (1931); **52**, 569 (1933); *J. Biol. Chem.*, **100**, 149 (1933).

oxidizing agent are added, $\frac{1}{2}x$ moles of Q are formed, and these combine with an equivalent amount of H_2Q to form $\frac{1}{2}x$ moles of meriquinone, $H_2Q \cdot Q$; an amount $a - x$ moles of H_2Q remains unchanged. It follows, therefore, neglecting activity coefficients, that in a solution of volume v, equation (31) becomes

$$E = E_1 - \frac{RT}{2F} \ln \frac{\frac{1}{2}x/v}{(a - x)^2/v^2}$$

$$= E_1 - \frac{RT}{2F} \ln \frac{x}{(a - x)^2} - \frac{RT}{2F} \ln \frac{v}{2}. \tag{32}$$

The potential thus depends on the volume of the solution, and hence the position of the curve showing the variation of the oxidation-reduction potential during the course of the titration of H_2Q by a strong oxidizing agent varies with the concentration of the solution. At constant volume equation (32) becomes

$$E = E_1' - \frac{RT}{2F} \ln x + \frac{RT}{F} \ln (a - x),$$

so that in the early stages of oxidation, i.e., when x is small, the last term on the right-hand side may be regarded as constant, and the slope of the titration curve will correspond to a process in which two electrons are involved, i.e., n is 2. In the later stages, however, the change of potential is determined mainly by the last term, and the slope of the curve will change to that of a one-electron system, i.e., n is effectively unity.

When a true semiquinone is formed, the two stages of oxidation-reduction are

$$(1) \quad H_2Q \rightleftharpoons HQ + H^+ + \epsilon,$$

and

$$(2) \quad HQ \rightleftharpoons Q + H^+ + \epsilon,$$

so that

$$E = E_1 - \frac{RT}{F} \ln \frac{(HQ)}{(H_2Q)}$$

$$= E_1 - \frac{RT}{F} \ln \frac{x}{a - x}, \tag{33}$$

for a definite hydrogen ion concentration. The value of the potential is seen to depend on the ratio of x to $a - x$, and not on the actual concentration of the solution; the position of the titration curve is thus independent of the volume. Further, it is evident from equation (33) that the type of slope is the same throughout the curve, and corresponds to a one-electron process, i.e., n is unity.

If the two stages of oxidation are fairly distinct, it is thus possible to distinguish between meriquinone and semiquinone formation. In the former case the position of the titration curve will depend on the volume of the solution and it will be unsymmetrical, the earlier part corresponding to an n value of 2, and the later part to one of unity. If semiquinone formation occurs, however, the curve will be symmetrical, with n equal to unity over the whole range, and its position will not be altered by changes in the total volume of the solution. A careful investigation along these lines has shown that many oxidation-reduction systems satisfy the conditions for semiquinone formation; in one way or another, this has been found to be true for α-oxyphenazine and some of its derivatives, e.g., Wurster's red, and for a number of anthraquinones.

Semiquinone Formation Constant.—It was assumed in the foregoing treatment that the two stages of oxidation are fairly distinct, but when this is not the case the whole system behaves as a single two-electron process, as in Fig. 79. In view of the interest associated with the formation of semiquinone intermediates in oxidation-reduction reactions, methods have been developed for the study of systems in which the two stages may or may not overlap. The treatment is somewhat complicated, and so the outlines only will be given here.[16]

If R represents the completely reduced form (H_2Q), S the semiquinone (HQ), and T the totally oxidized form (Q), the electrical equilibria, assuming a constant hydrogen ion concentration, are

$$(1)\quad R \rightleftharpoons S + \epsilon \qquad \text{and} \qquad (2)\quad S \rightleftharpoons T + \epsilon,$$

so that if r, s and t are the concentrations of the three forms,

$$E = E_1 - \frac{RT}{F} \ln \frac{s}{r}, \tag{34}$$

and

$$E = E_2 - \frac{RT}{F} \ln \frac{t}{s}, \tag{35}$$

during the first and second stages, respectively; E_1 and E_2 are the standard potentials of these stages. The potential can also be formulated in terms of the equilibrium between initial and final states, viz.,

$$R \rightleftharpoons T + 2\epsilon,$$

so that

$$E = E_m - \frac{RT}{2F} \ln \frac{t}{r}, \tag{36}$$

where E_m is the usual standard potential for the system as a whole at some definite hydrogen ion concentration. It can be seen from equations (34), (35) and (36) that

$$E_m = \tfrac{1}{2}(E_1 + E_2),$$

[16] For reviews, see Michaelis, "Oxydations-Reductions Potentiale," 1933; *Trans. Electrochem. Soc.*, 71, 107 (1937); *Chem. Revs.*, 16, 243 (1935); Michaelis and Schubert, *ibid.*, 22, 437 (1938); Michaelis, *Ann. New York Acad. Sci.*, 40, 39 (1940); Müller, *ibid.*, 40, 91 (1940).

and since E_1 and E_2 will be in the centers of the first and second parts of the titration curves, i.e., when s/r and t/s are unity, respectively, it follows that E_m will be the potential in the middle of the whole curve. In addition to the electrical equilibria, there will be a chemical equilibrium between R, S and T, viz.,

$$R + T \rightleftharpoons 2S,$$

so that by the approximate form of the law of mass action

$$k = \frac{s^2}{rt}$$

where k is known as the **semiquinone formation constant.**

If a is the initial amount of reduced form H_2Q which is being titrated, and x equiv. of strong oxidizing agent are added, then x/a is equal to 1 in the middle of the complete titration curve and to 2 at the end. By making use of the relationships given above, it is possible to derive an equation of some complexity giving the variation of $E - E_m$ with x/a,

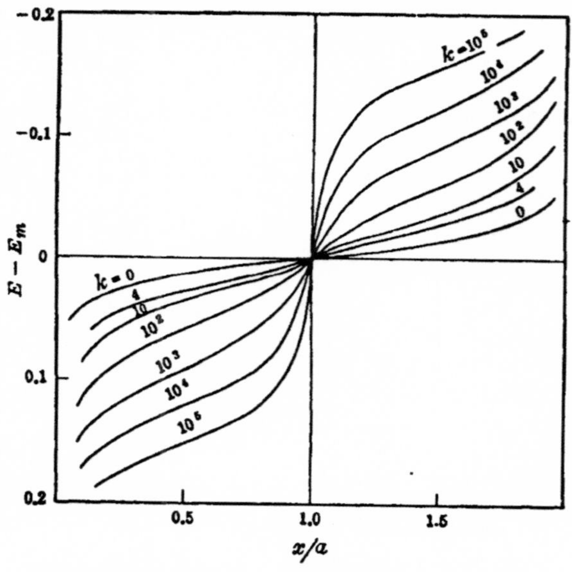

Fig. 85. Titration curves for semiquinone formation

i.e., during the course of the titration, for any value of k, the semiquinone formation constant. Some of the results obtained in this manner are shown in Fig. 85; as long as k is small, the titration curve throughout has the shape of a normal two-electron oxidation-reduction system, there being no break at the midpoint where x/a is unity. As k increases, the slope changes until it corresponds to that of a one-electron process; in fact when the value of k lies between 4 and 16, the slope is that for a system with n equal to unity, but there is no break at the midpoint. The presence of a semiquinone is often indicated in these cases, however, by the appearance of a color which differs from that of either the completely oxidized or the completely reduced forms. When the semiquinone formation constant k exceeds 16, a break appears at the midpoint, and the extent of this break becomes more marked as k increases. The detection of semiquinone formation by the shape of the titration curve is only possible, therefore, when the semiquinone formation constant is large.

If actual oxidation-reduction measurements are made on a particular system during the course of a titration, it is possible, by utilizing the equation from which the data in Fig. 85 were calculated, to evaluate the semiquinone formation constant for that system. The standard potentials E_1, E_2 and E_m can also be obtained for the hydrogen ion concentration existing in the experimental solution.

Influence of Hydrogen Ion Concentration.—The values of E_1, E_2 and E_m will depend on the pH of the solution, and since the forms R, S and T may possess acidic or basic functions, the slopes of the curves of these three standard potentials against pH may change direction at various points and crossings may occur. A system for which E_2 is above E_1 at one pH, i.e., the semiquinone formation constant is large, may thus behave in a reverse manner, i.e., E_1 is above E_2, and the semiquinone formation is very small, at another pH. It is apparent, therefore, that although a given system may show distinct semiquinone formation at one hydrogen ion concentration, there may be no definite indication of such formation at another hydrogen ion concentration. If the oxidized form of the system consists of a positive ion, e.g., anthraquinone sulfonic acid, semiquinone formation is readily observed in alkaline solutions only, but if it is a negative ion, e.g., α-oxyphenazine, the situation is reversed and the semiquinone formation can be detected most easily in acid solution.

Storage Batteries (Secondary Cells).[17]—When an electric current is passed through an electrolytic cell chemical changes are produced and electrical energy is converted into chemical energy. If the cell is reversible, then on removing the source of current and connecting the electrodes of the cell by means of a conductor, electrical energy will be produced at the expense of the stored chemical energy and current will flow through the conductor. Such a device is a form of **storage battery, or secondary cell**; * certain chemical changes occur when the cell is "charged" with electricity, and these changes are reversed during dis-

[17] Vinal, "Storage Batteries," 1940.

* A primary cell is one which acts as a source of electricity without being previously charged up by an electric current from an external source; in the most general sense, every voltaic cell is a primary cell, although the latter term is usually restricted to cells which can function as practical sources of current, e.g., the Leclanché cell.

charge. Theoretically, any reversible cell should be able to store electrical energy, but for practical purposes most of them are unsuitable because of low electrical capacity, incomplete reversibility as to the physical form of the substances involved, chemical action or other changes when idle, etc. Only two types of storage battery have hitherto found any wide application, and since they both involve oxidation-reduction systems their theoretical aspects will be considered here.

The Acid Storage Cell.—The so-called "acid" or "lead" storage cell consists essentially of two lead electrodes, one of which is covered with lead dioxide, with approximately 20 per cent sulfuric acid, i.e., with a specific gravity of about 1.15 at 25°, as the electrolyte. The charged cell is generally represented simply as Pb, H_2SO_4, PbO_2, but it is more correct to consider it as

$$Pb \mid PbSO_4(s)\ H_2SO_4aq.\ PbSO_4(s), PbO_2(s) \mid Pb,$$

the right-hand lead electrode acting as an inert electrode for an oxidation-reduction system. The reactions occurring in the cell when it produces current, i.e., on discharge, are as follows.

Left-hand electrode:

$$Pb = Pb^{++} + 2\epsilon$$
$$Pb^{++} + SO_4^{--} = PbSO_4(s).$$

∴ Net reaction for two faradays is

$$Pb + SO_4^{--} = PbSO_4(s) + 2\epsilon.$$

Right-hand electrode:

$$PbO_2(s) + 2H_2O \rightleftharpoons Pb^{++++} + 4OH^-,$$
$$Pb^{++++} + 2\epsilon = Pb^{++},$$
$$Pb^{++} + SO_4^{--} = PbSO_4(s),$$
$$4OH^- + 4H^+ = 4H_2O,$$

∴ Net reaction for two faradays is

$$PbO_2(s) + 4H^+ + SO_4^{--} + 2\epsilon = PbSO_4(s) + 2H_2O.$$

Since both electrodes are reversible, the processes occurring when electricity is passed through the cell, i.e., on charge, are the reverse of those given above; it follows, therefore, that the complete cell reaction in both directions may be written as the sum of the individual electrode processes, thus

$$Pb + PbO_2 + 2H_2SO_4 \underset{\text{charge}}{\overset{\text{discharge}}{\rightleftharpoons}} 2PbSO_4 + 2H_2O$$

for two faradays. The mechanism of the operation of the lead storage battery as represented by this equation was first proposed by Gladstone and Tribe (1883) before the theory of electrode processes in general was well understood; it is known as the "double sulfation" theory, because it postulates the formation of lead sulfate at both electrodes. Various alternative theories concerning the lead cell have been proposed from time to time but these appear to have little to recommend them; apart from certain processes which occur to a minor extent, e.g., formation of oxides higher than PbO_2, there is no doubt that the reactions given here represent essentially the processes occurring at the electrodes of an acid storage battery.

It will be observed that according to the suggested cell reaction, two molecules of sulfuric acid should be removed from the electrolyte and two molecules of water formed for the discharge of two faradays of electricity from the charged cell. This expectation has been confirmed experimentally. Further, it is possible to calculate the free energy of this change thermodynamically in terms of the aqueous vapor pressure of sulfuric acid solutions; the values should be equal to $-2FE$, where E is the E.M.F. of the cell and this has been found to be the case.

A striking confirmation of the validity of the double sulfation theory is provided by thermal measurements; since the E.M.F. of the storage cell and its temperature coefficient are known, it is possible to calculate the heat change of the reaction taking place in the cell by means of the Gibbs-Helmholtz equation (p. 194). The value of the heat of the reaction believed to occur can be derived from direct thermochemical measurements, and the results can be compared. The data obtained in this manner for lead storage cells containing sulfuric acid at various concentrations, given in the first column with the density in the second, are quoted in Table LV; [18] the agreement between the values in the last two

TABLE LV. HEAT CHANGE OF REACTION IN LEAD STORAGE BATTERY

H_2SO_4 per cent	$d^{25°}_{25°}$	$E_{25°}$ volts	$dE/dT \times 10^4$	ΔH E.M.F.	ΔH Thermal
4.55	1.030	1.876	—	—	—
7.44	1.050	1.905	+ 1.5	− 85.83	− 86.53
14.72	1.100	1.962	+ 2.9	− 86.54	− 87.44
21.38	1.150	2.005	+ 3.3	− 87.97	− 87.37
27.68	1.200	2.050	+ 3.0	− 90.46	− 90.32
33.80	1.250	2.098	+ 2.2	− 93.77	− 93.08
39.70	1.300	2.148	+ 1.8	− 96.63	− 96.22

columns is very striking, and appears to provide conclusive proof of the suggested mechanism.

It is evident from the data in Table LV that the E.M.F. of the lead storage cell increases with increasing concentration of sulfuric acid; this result is, of course, to be expected from the cell reactions. According to the reaction occurring at the Pb, $PbSO_4$ electrode, generally referred to as the **negative electrode** of the battery, its potential (E_-) is given by

$$E_- = E^0_{\text{Pb, PbSO}_4\text{, SO}_4^{--}} + \frac{RT}{2F} \ln a_{\text{SO}_4^{--}}. \tag{37}$$

Since the activity, or concentration, of sulfate ions depends on the con-

[18] Craig and Vinal, *J. Res. Nat. Bur. Standards*, **24**, 475 (1940).

centration of sulfuric acid, it is clear that the potential of this electrode will vary accordingly. The standard potential in equation (37) is + 0.350 volt at 25°, and if the activity of the sulfate ion is taken as equal to the mean activity of sulfuric acid, it is readily calculated that for a storage battery containing acid of the usual concentration, i.e., 4 to 5 N, in which the mean activity coefficient is about 0.18 to 0.2, the actual potential of the negative electrode is about + 0.33 volt. The so-called negative electrode potential may also be represented by

$$E_- = E^0_{\mathrm{Pb,Pb^{++}}} - \frac{RT}{2F} \ln a_{\mathrm{Pb^{++}}}, \tag{38}$$

but since the solution is saturated with lead sulfate, $a_{\mathrm{Pb^{++}}}$ will be inversely proportional to $a_{\mathrm{SO_4^{--}}}$; equations (37) and (38) are thus consistent.

The potential of the $PbSO_4$, PbO_2 electrode, usually called the **positive electrode,** can be represented by (cf. p. 269)

$$-E_+{}^* = -E^0_{\mathrm{PbSO_4,PbO_2,SO_4^{--}}} + \frac{RT}{2F} \ln \frac{a^4_{\mathrm{H^+}} a_{\mathrm{SO_4^{--}}}}{a^2_{\mathrm{H_2O}}} \tag{39}$$

and hence will be very markedly dependent on the concentration of sulfuric acid, since this affects $a_{\mathrm{H^+}}$, $a_{\mathrm{SO_4^{--}}}$ and $a_{\mathrm{H_2O}}$. The standard potential required for equation (39) is − 1.68 volts at 25° (see Table LIII); making the assumption that the activities of the hydrogen and sulfate ions are equal to the mean activity of sulfuric acid in which the activity of water from vapor pressure data is 0.3, it is found that, for 4 to 5 N acid, the potential $-E_+$ of the positive electrode is about 1.70 volts.

The positive electrode may also be regarded as a simple oxidation-reduction electrode involving the plumbous-plumbic system; thus

$$-E_+ = -E^0_{\mathrm{Pb^{++},Pb^{++++}}} + \frac{RT}{2F} \ln \frac{a_{\mathrm{Pb^{++++}}}}{a_{\mathrm{Pb^{++}}}}. \tag{40}$$

The activity of plumbic ions in a solution saturated with lead dioxide (or plumbic hydroxide) will be inversely proportional to the fourth power of the hydroxyl ion activity, and hence it is directly proportional to the fourth power of the hydrogen ion activity (cf. p. 339), in agreement with the requirements of equation (39).

The Alkaline Storage Battery.—The alkaline or Edison battery is made up of an iron (negative) and a nickel sesquioxide (positive) electrode in potassium hydroxide solution; it may be represented as

$$\mathrm{Fe} \mid \mathrm{FeO}(s) \;\; \mathrm{KOH\ aq.} \;\; \mathrm{NiO}(s), \mathrm{Ni_2O_3}(s) \mid \mathrm{Ni},$$

the nickel acting virtually as an inert electrode material. The reactions taking place in the charged cell during discharge are as follows.

* The negative sign is used because the potential of the electrode as written, viz., $PbSO_4(s)$, $PbO_2(s)$, Pb, is opposite in direction to that corresponding to the convention on which the standard potentials in Tables XLIX and LIII are based.

Left-hand electrode:

$$Fe = Fe^{++} + 2\epsilon,$$
$$Fe^{++} + 2OH^- = FeO(s) + H_2O,$$

$\therefore$ Net reaction for two faradays is

$$Fe + 2OH^- = FeO(s) + H_2O + 2\epsilon.$$

Right-hand electrode:

$$Ni_2O_3(s) + 3H_2O \rightleftharpoons 2Ni^{+++} + 6OH^-,$$
$$2Ni^{+++} + 2\epsilon = 2Ni^{++},$$
$$2Ni^{++} + 4OH^- = 2NiO(s) + 2H_2O,$$

$\therefore$ Net reaction for two faradays is

$$Ni_2O_3(s) + H_2O + 2\epsilon = 2NiO(s) + 2OH^-.$$

The complete cell reaction during charge and discharge, respectively, may be represented by

$$Fe + Ni_2O_3 \underset{\text{charge}}{\overset{\text{discharge}}{\rightleftharpoons}} FeO + 2NiO.$$

The potential of the iron ("negative") electrode, which is about + 0.8 volt in practice, is given by the expression

$$E_- = E^0_{Fe,\,FeO,\,OH^-} + \frac{RT}{2F} \ln \frac{a^2_{OH^-}}{a_{H_2O}},$$

and similarly that of the nickel sesquioxide ("positive") electrode, which is approximately + 0.55 volt, is represented by

$$-E_+ = -E^0_{NiO,\,Ni_2O_3} - \frac{RT}{2F} \ln \frac{a^2_{OH^-}}{a_{H_2O}}.$$

The potentials of both individual electrodes are dependent on the hydroxyl ion activity (or concentration) of the potassium hydroxide solution employed as electrolyte. It is evident, however, that in theory the E.M.F. of the complete cell, which is equal to $E_- - E_+$, should be independent of the concentration of the hydroxide solution. In practice a small variation is observed, viz., 1.35 to 1.33 volts for N to 5 N potassium hydroxide; this is attributed to the fact that the oxides involved in the cell reactions are all in a "hydrous" or "hydrated" form, with the result that a number of molecules of water are transferred in the reaction. The equations for the potentials of the separate electrodes should then contain different terms for the activity of the water in each case: the E.M.F. of the complete cell thus depends on the activity of the water in the electrolyte, and hence on the concentration of the potassium hydroxide.

PROBLEMS

1. Write down the electrochemical equations for the oxidation-reduction systems involving (i) ClO_3^- and Cl_2, and (ii) $Cr_2O_7^{--}$ and Cr^{+++}. Use the results to derive the complete equations for the reactions of each of these with the Sn^{++++}, Sn^{++} system.

2. According to Brønsted and Pedersen [*Z. physik. Chem.*, **103**, 307 (1924)] the equilibrium constant of the reaction

$$Fe^{+++} + I^- = Fe^{++} + \tfrac{1}{2}I_2$$

at 25° is approximately 21, after allowing for the tri-iodide equilibrium. The standard potential of the $I_2(s)$, I^- electrode is $-$ 0.535 volt and the solubility of iodine in water is 0.00132 mole per liter; calculate the approximate standard potential of the (Pt)Fe^{++}, Fe^{+++} system.

3. From the measurements of Sammet [*Z. physik. Chem.*, **53**, 678 (1905)] the standard potential of the system (Pt)$IO_3^- + 6H^+$, $\tfrac{1}{2}I_2$ has been estimated as $-$ 1.197 volt. Determine the theoretical equilibrium constant of the reaction

$$IO_3^- + 5I^- + 6H^+ = 3I_2 + 3H_2O.$$

What conclusion may be drawn concerning the quantitative determination of iodate by the addition of acidified potassium iodide followed by titration with thiosulfate?

4. Kolthoff and Tomsicek [*J. Phys. Chem.*, **39**, 945 (1935)] measured the potentials of the electrode (Pt)$Fe(CN)_6^{----}$, $Fe(CN)_6^{---}$ at 25°; the concentrations of potassium ferro- and ferri-cyanide were varied, but the ratio was unity in every case. The concentrations (c) of each of the salts, in moles per liter, and the corresponding electrode potentials (E_0'), on the hydrogen scale, are given below:

c	E_0'	c	E_0'
0.04	$-$ 0.4402	0.0004	$-$ 0.3754
0.02	$-$ 0.4276	0.0002	$-$ 0.3714
0.01	$-$ 0.4154	0.0001	$-$ 0.3664
0.004	$-$ 0.4011	0.00008	$-$ 0.3652
0.002	$-$ 0.3908	0.00006	$-$ 0.3642
0.001	$-$ 0.3834	0.00004	$-$ 0.3619

Plot the values of E_0' against $\sqrt{\mu}$ and extrapolate the results to infinite dilution to obtain the standard potential of the ferrocyanide-ferricyanide system. Alternatively, derive the value of E^0 from each E_0' by applying the activity correction given by the Debye-Hückel limiting law.

5. The oxidation-reduction system involving 5- and 4-valent vanadium may be represented by the general equation

$$\frac{x}{z}V_zO_z^{2z+} + (y - x)H_2O = V_xO_y^{(5x-2y)+} + 2(y - x)H^+ + x\epsilon.$$

Using the symbol V^5 to represent the oxidized form V_xO_y and V^4 for the reduced form V_zO_z, write the equation for the E.M.F. of the cell consisting of the V^4, V^5 and H^+, H_2 electrodes. Derive the expressions to which this equation reduces (i) when V^4 and H^+ are kept constant, (ii) when V^5 and V^4 are constant, and (iii) when V^5 and H^+ are constant. The experimental results of Carpenter [*J. Am. Chem. Soc.*, **56**, 1847 (1934)] are as follows:

(i)		(ii)		(iii)	
V^5	E	H^+	E	V^4	E
0.529×10^{-3}	-0.9031	0.0240	-0.9098	4.42×10^{-3}	-0.9554
2.489	-0.9395	0.1077	-0.9554	35.11	-0.9048
9.855	-0.9723	0.4442	-0.9974		
19.67	-0.9875	0.9000	-1.0198		

Using the expressions already derived, show that the values of x, $(2y - 3x)/x$ and z can be obtained by plotting E against log V^5, log H^+ and log V^4, respectively. Insert the values of x, y and z in the expression given above and so derive the actual equation for the oxidation-reduction system.

6. In an investigation of the oxidation-reduction potentials of the system in which the oxidized form was anthraquinone 2 : 6-disulfonate, Conant and his collaborators [*J. Am. Chem. Soc.*, **44**, 1382 (1922)] obtained the following values for $E^{0\prime}$ at various pH's:

pH	6.90	7.64	9.02	9.63	10.49	11.27	11.88	12.20
$E^{0\prime}$	0.181	0.220	0.275	0.292	0.311	0.324	0.326	0.326

Plot $E^{0\prime}$ against the pH and interpret the results.

7. By extrapolating the E.M.F.'s to infinite dilution, Andrews and Brown [*J. Am. Chem. Soc.*, **57**, 254 (1935)] found E^0 for the cell

$$\text{Pt} \mid KMnO_4, MnO_2(s) \ \ KOH \text{ aq. } HgO(s) \mid Hg$$

to be -0.489 at 25°. The standard potential of the Hg, HgO(s), OH^- electrode is -0.098, and the equilibrium constant of the system

$$3MnO_4^{--} + 2H_2O = 2MnO_4^- + MnO_2(s) + 4OH^-$$

is 16 at this temperature. Calculate the standard potential of the (Pt)MnO_4^-, MnO_4^{--} electrode.

8. The standard potential of the (Pt) | $PbSO_4(s)$, $PbO_2(s)$, SO_4^{--} electrode is -1.685 volts at 25°; calculate the E.M.F.'s of the cell

$$\text{Pt} \mid PbSO_4(s), PbO_2(s) \ \ H_2SO_4(c) \mid H_2(1 \text{ atm.})$$

for 1.097 and 6.83 molal sulfuric acid solutions. The mean activity coefficients (γ) and aqueous vapor pressures (p) of the solutions are:

m	γ	p
1.097	0.146	22.76 mm.
6.83	0.386	12.95

The vapor pressure of water at 25° is 23.76 mm. of mercury.

9. From the standard potentials of the systems (Pt)Cu^+, Cu^{++} and I_2, I^- evaluate the equilibrium constant of the reaction

$$Cu^{++} + I^- = Cu^+ + \tfrac{1}{2}I_2,$$

and show that it is entirely owing to the low solubility product of cuprous iodide, CuI, i.e., approximately 10^{-12}, that this reaction can be used for the analytical determination of cupric ions.

10. The solubility products of cupric and cuprous hydroxides, $Cu(OH)_2$ and CuOH, respectively, are approximately 10^{-19} and 10^{-14} at ordinary temperatures [Allmand, *J. Chem. Soc.*, **95**, 2151 (1909)]; show that the solid cupric hydroxide is unstable in contact with metallic copper and tends to be reduced to cuprous hydroxide.

CHAPTER IX

ACIDS AND BASES

Definition of Acids and Bases.*—The old definitions of an acid as a substance which yields hydrogen ions, of a base as one giving hydroxyl ions, and of neutralization as the formation of a salt and water from an acid and a base, are reasonably satisfactory for aqueous solutions, but there are serious limitations when non-aqueous media, such as ethers, nitro-compounds, ketones, etc., are involved. As a result of various studies, particularly those on the catalytic influence of un-ionized molecules of acids and bases and of certain ions, a new concept of acids and bases, generally associated with the names of Brønsted and of Lowry, has been developed in recent years.[1] According to this point of view an acid is defined as a substance with a tendency to lose a proton, while a base is any substance with a tendency to gain a proton; the relationship between an acid and a base may then be written in the form

$$\underset{\text{acid}}{A} \rightleftharpoons \underset{\text{proton}}{H^+} + \underset{\text{base}}{B}. \tag{1}$$

The acid and base which differ by a proton according to this relationship are said to be **conjugate** to one another; every acid must, in fact, have its conjugate base, and every base its conjugate acid. It is unlikely that free protons exist to any extent in solution, and so the acidic or basic properties of any species cannot become manifest unless the solvent molecules are themselves able to act as proton acceptors or donors, respectively: that is to say, the medium must itself have basic or acidic properties. The interaction between an acid or base and the solvent, and in fact almost all types of acid-base reactions, may be represented as an equilibrium between two acid-base systems, viz.,

$$\underset{\text{acid}_1}{A_1} + \underset{\text{base}_2}{B_2} \rightleftharpoons \underset{\text{base}_1}{B_1} + \underset{\text{acid}_2}{A_2}, \tag{2}$$

where A_1 and B_1 are the conjugate acid and base of one system, and

* G. N. Lewis [*J. Franklin Inst.*, **226**, 293 (1938); see also, *J. Am. Chem. Soc.*, **61**, 1886, 1894 (1939); **62**, 2122 (1940)] proposes to define a base as a substance capable of furnishing a pair of electrons to a bond, i.e., an electron donor, whereas an acid is able to accept a pair of electrons, i.e., an electron acceptor. The somewhat restricted definitions employed in this book are, however, more convenient from the electrochemical standpoint.

[1] Lowry, *Chem. and Ind.*, **42**, 43 (1923); Brønsted, *Rec. trav. chim.*, **42**, 718 (1923); *J. Phys. Chem.*, **30**, 377 (1926); for reviews, see Brønsted, *Chem. Revs.*, **3**, 231 (1928); Hall, *ibid.*, **8**, 191 (1931); Bjerrum, *ibid.*, **16**, 287 (1935); Bell, *Ann. Rep. Chem. Soc.*, **31**, 71 (1934).

A_2 and B_2 are those of the other system, e.g., the solvent. Actually A_1 possesses a proton in excess of B_1, while A_2 has a proton more than B_2; the reaction, therefore, involves the transfer of a proton from A_1 to B_2 in one direction, or from A_2 to B_1 in the other direction.

Types of Solvent.—In order that a particular solvent may permit a substance dissolved in it to behave as an acid, the solvent itself must be a base, or proton acceptor. A solvent of this kind is said to be **protophilic** in character; instances of protophilic solvents are water and alcohols, acetone, ether, liquid ammonia, amines and, to some extent, formic and acetic acids. On the other hand, solvents which permit the manifestation of basic properties by a dissolved substance must be proton donors, or acidic; such solvents are **protogenic** in nature. Water and alcohols are examples of such solvents, but the most marked protogenic solvents are those of a strongly acidic character, e.g., pure acetic, formic and sulfuric acids, and liquid hydrogen chloride and fluoride. Certain solvents, water and alcohols, in particular, are **amphiprotic,** for they can act both as proton donors and acceptors; these solvents permit substances to show both acidic and basic properties, whereas a purely protophilic solvent, e.g., ether, or a completely protogenic one, e.g., hydrogen fluoride, would permit the manifestation of either acidic or basic functions only. In addition to the types of solvent already considered, there is another class which can neither supply nor take up protons: these are called **aprotic** solvents, and their neutral character makes them especially useful when it is desired to study the interaction of an acidic and a basic substance without interference by the solvent.

Acids.—Since an acid must possess a labile proton it can be represented by HA, and if S is a protophilic, i.e., basic, solvent, the equilibrium existing in the solution, which is of the type represented by equation (2), may be written as

$$\underset{\text{acid}_1}{HA} + \underset{\text{base}_2}{S} \rightleftharpoons \underset{\text{acid}_2}{HS^+} + \underset{\text{base}_1}{A^-}, \tag{3}$$

where HS^+ is the form of the hydrogen ion in the particular solvent and A^- is the conjugate base of the acid HA. There are a number of important consequences of this representation which must be considered. In the first place, it is seen that the anion A^- of every acid HA must be regarded as the conjugate base of the latter. If the acid is a strong one, it will tend to give up its proton very readily; this is, in fact, what is meant by a "strong acid." For such an acid, e.g., hydrochloric acid, the equilibrium between acid and solvent, represented by equation (3), lies considerably to the right; that is to say, the reverse process occurs to a small extent only. This means that the anion of a strong acid, e.g., the chloride ion, will not have a great affinity for a proton, and hence it must be regarded as a "weak base." On the other hand, if HA is a very weak acid, e.g., phenol, the equilibrium of equation (3) lies well to the left, so that the process

$$A^- + HS^+ \rightleftharpoons HA + S$$

will take place to an appreciable extent; the anion A^-, e.g., the phenoxide ion, will be a moderately strong base.

Another consequence of the interaction between the acid and the solvent is that the hydrogen ion in solution is not to be regarded as a bare proton, but as a combination of a proton with, at least, one molecule of solvent; the hydrogen ion thus depends on the nature of the solvent. In water, for example, there are good reasons for believing that the hydrogen ion is actually H_3O^+, sometimes called the "oxonium" or "hydronium" ion: the free energy of hydration of the proton is so high, approximately 250 kcal. (see p. 249), that the concentration of free protons in water must be quite negligible, and hence almost all the protons must have united with water molecules to form H_3O^+ ions. Further hydration of the H_3O^+ ions probably occurs in aqueous solution, but this is immaterial for present purposes.

Striking evidence of the part played by the water in connection with the manifestation of acidic properties is provided by observations on the properties of hydrogen bromide solutions in liquid sulfur dioxide.[2] The latter is only feebly basic and, although it dissolves hydrogen bromide, the solution is a poor conductor; there is consequently little or no ionization under these conditions. The solution of hydrogen bromide in sulfur dioxide is able, however, to dissolve a mole of water for every mole of hydrogen bromide present, and the resulting solution is an excellent conductor. Since water is sparingly soluble in sulfur dioxide alone, it is clear that the reaction

$$HBr + H_2O = H_3O^+ + Br^-$$

must take place between the hydrogen bromide and water. Confirmation of this view is to be found in the observation that on electrolysis of the solution one mole of water is liberated at the cathode for each faraday passing; the discharge of the H_3O^+ ion clearly results in the formation of an atom, or half a molecule, of hydrogen and a molecule of water.

It is of interest to note in connection with the question of the nature of the hydrogen ion in solution that the crystalline hydrate of perchloric acid, $HClO_4 \cdot H_2O$, has been shown by X-ray diffraction methods to have the same fundamental structure as ammonium perchlorate. Since the latter consists of interpenetrating lattices of NH_4^+ and ClO_4^- ions, it is probable that the former is built up of H_3O^+ and ClO_4^- ions.

A third conclusion to be drawn from the equilibrium represented by equation (3) is that since the solvent S is to be regarded as a base, the corresponding hydrogen ion SH^+ is an acid. The hydronium ion H_3O^+ is thus an acid, and in fact the acidity of the strong acids, e.g., perchloric,

[2] Bagster and Cooling, *J. Chem. Soc.*, 117, 693 (1920).

hydrobromic, sulfuric, hydrochloric and nitric acids, in water is due almost exclusively to the H_3O^+ ion. It is because the process

$$\underset{\text{acid}_1}{HA} + \underset{\text{base}_2}{H_2O} \rightleftharpoons \underset{\text{acid}_2}{H_3O^+} + \underset{\text{base}_1}{A^-},$$

where HA is a strong acid, goes almost completely to the right, that the aforementioned acids appear to be equally strong in aqueous solution, provided the latter is not too concentrated. In solutions more concentrated than about 2 N, however, these acids do show differences in catalytic behavior for the inversion of sucrose; the results indicate that the strengths decrease in the order given (Fig. 86).

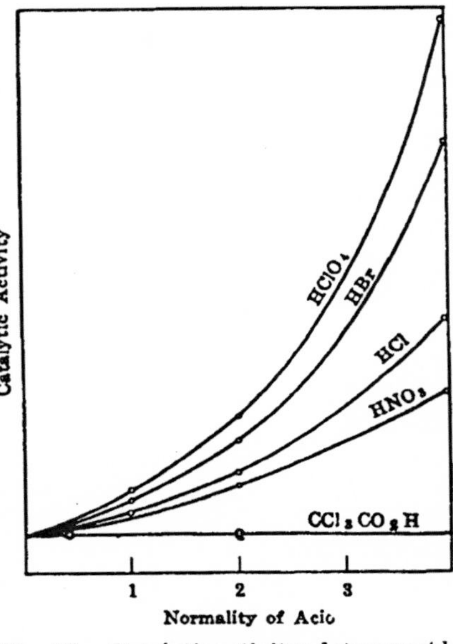

Fig. 86. Catalytic activity of strong acids

In order that it may be possible to distinguish in strength between the so-called strong acids, it is evidently necessary to employ a solvent which is less strongly protophilic than water; the equilibrium of equation (3) will then not lie completely to the right, but its position will be determined by the relative proton-donating tendencies, i.e., strengths, of the various acids. A useful solvent for this purpose is pure acetic acid; this is primarily a protogenic (acidic) solvent, but it has slight basic properties, so that the reaction

$$HA + CH_3CO_2H \rightleftharpoons CH_3CO_2H_2^+ + A^-$$

occurs to some extent, although the equilibrium cannot lie far to the right. Even acids, such as perchloric and hydrochloric, which are regarded as strong acids, will interact to a small extent only with the solvent, and the number of ions in solution will be relatively small; the extent of ionization will, therefore, depend on the strength of the acid in a manner not observed in aqueous solution. The curves in Fig. 87 show the variation of the conductance of a number of acids in pure acetic acid at 25°; the very low equivalent conductances recorded are due to the very small degrees of ionization. It is seen, therefore, that acids which appear to be equally strong in aqueous solution behave as weak acids when dissolved in acetic acid; moreover, it is possible to distinguish between their relative strengths, the order being as follows:

$$HClO_4 > HBr > H_2SO_4 > HCl > HNO_3.$$

This order agrees with that found by catalytic methods and also by potentiometric titration.[3]

In spite of the small extent of ionization of acids in a strongly protogenic medium such as acetic acid, the activity of the resulting hydrogen ions is very high; this may be attributed to the strong tendency of the $CH_3CO_2H_2^+$ ion to lose a proton, so that the ion will behave as an acid of exceptional strength. The intense acidity of these solutions, as shown by hydrogen electrode measurements, by their catalytic activity, and in other ways, has led to them being called **superacid solutions.**[4] The property of superacidity can, of course, be observed only with solvents which are strongly protogenic, but which still possess some protophilic nature. Hydrogen fluoride, for example, has no protophilic properties, and so it cannot be used to exhibit superacidity; in fact no known substance exhibits acidic behavior in this solvent, as explained below.

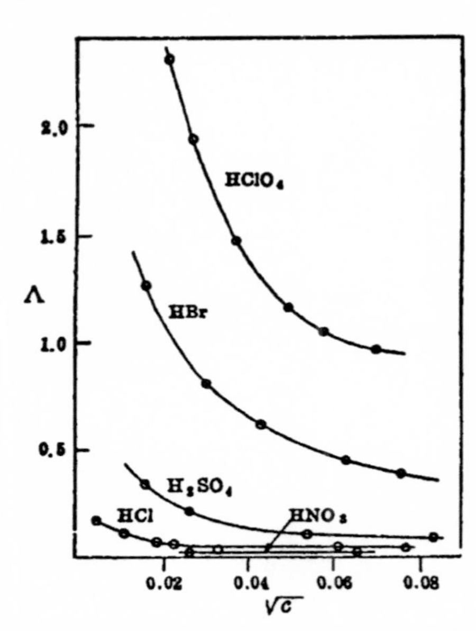

Fig. 87. Conductance of acids in glacial acetic acid (Kolthoff and Willman)

It is an obvious corollary, from the discussiom given here concerning the influence of the solvent, that in a highly basic, i.e., protophilic, medium, even acids that are normally regarded as weak would be highly ionized. It is probable that in liquid ammonia interaction with a weak acid, such as acetic acid, would occur to such an extent that it would appear to be as strong as hydrochloric acid.

Bases.—The equilibrium between an acidic, i.e., protogenic, solvent and a base may be represented by another form of the general equation (2), viz.,

$$\underset{\text{base}}{B} + \underset{\text{acid}}{SH} \rightleftharpoons \underset{\text{acid}}{BH^+} + \underset{\text{base}}{S^-}, \tag{4}$$

where the solvent is designated by SH to indicate its acidic property. It is seen from this equilibrium that the cation BH^+ corresponding to the base B is to be regarded as an acid; for example, if the base is NH_3,

[3] Hall and Conant, *J. Am. Chem. Soc.*, **49**, 3047, 3062 (1927); Hall and Werner, *ibid.*, **50**, 2367 (1928); Hantzsch and Langbein, *Z. anorg. Chem.*, **204**, 193 (1932); Kolthoff and Willman, *J. Am. Chem. Soc.*, **56**, 1007 (1934); Weidner, Hutchison and Chandlee, *ibid.*, **56**, 1285 (1934).

[4] Hall and Conant, *J. Am. Chem. Soc.*, **49**, 3047, 3062 (1927); Hall and Werner, *ibid.*, **50**, 2367 (1928); Conant and Werner, *ibid.*, **52**, 4436 (1930).

the corresponding cation is NH_4^+, and so the ammonium ion and, in fact, all mono-, di-, and tri-substituted ammonium ions are to be regarded as the conjugate acids of the corresponding amine (anhydro-) bases. It can be readily shown, by arguments analogous to those used in connection with acids, that when the base is a strong one, e.g., hydroxyl ions, its conjugate acid, i.e., water, will be a weak acid; similarly, the conjugate acid to a very weak base will be moderately strong.

The strength of a base like that of an acid must depend on the nature of the solvent: in a strongly protogenic medium, such as acetic acid or other acid, the ionization process

$$\underset{\text{base}}{B} + \underset{\text{acid}}{CH_3CO_2H} = \underset{\text{acid}}{BH^+} + \underset{\text{base}}{CH_3CO_2^-}$$

will take place to a very considerable extent even with bases which are weak in aqueous solution. Just as it is impossible to distinguish between the strengths of weak acids in liquid ammonia, weak bases are indistinguishable in strength when dissolved in acetic acid; it has been found experimentally, by measurement of dissociation constants, that all bases stronger than aniline, which is a very weak base in water, are equally strong in acetic acid solution.[5] To arrange a series of weak bases in the order of their strengths, it would be necessary to use a protophilic solvent, such as liquid ammonia: water is obviously better than acetic acid for this purpose, but it is not possible to distinguish between the strong bases in the former medium, since they all produce OH^- ions almost completely.

Substances which are normally weak bases in water exhibit considerable basicity in strongly acid media; the results in Table LVI, for

TABLE LVI. EQUIVALENT CONDUCTANCES IN HYDROGEN FLUORIDE SOLUTIONS AT $-15°$ IN OHMS^{-1} CM.[3]

Concentration	Methyl alcohol	Acetone	Glucose
0.026 N	243	244	279
0.115	200	190	208
0.24	164	181	165
0.50	139	176	114

example, show that methyl alcohol, acetone and glucose, which are non-conductors in aqueous solution, are excellent conductors when dissolved in hydrogen fluoride.[6] These, and other oxygen compounds, behave as bases and ionize in the following manner:

$$\underset{\text{base}}{>O} + \underset{\text{acid}}{HF} = \underset{\text{acid}}{>\overset{+}{O}H} + \underset{\text{base}}{F^-}.$$

A number of substances which are acids in aqueous solution function as

[5] Hall, *J. Am. Chem. Soc.*, **52**, 5115 (1930); *Chem. Revs.*, **8**, 191 (1931).
[6] Fredenhagen *et al.*, *Z. physik. Chem.*, **146A**, 245 (1930); **164A**, 176 (1933); Simons, *Chem. Revs.*, **8**, 213 (1931).

bases in hydrogen fluoride, e.g.,

$$\underset{\text{base}}{CH_3CO_2H} + \underset{\text{acid}}{HF} = \underset{\text{acid}}{CH_3CO_2H_2^+} + \underset{\text{base}}{F^-}.$$

This reaction occurs because the acid possesses some protophilic properties, and these become manifest in the presence of the very strongly protogenic solvent. As may be expected, the stronger the acid is in water, the weaker does it behave as a base in hydrogen fluoride.

Dissociation Constants of Acids and Bases.—If the law of mass action is applied to the equilibrium between an acid HA and the basic solvent S, i.e., to the equilibrium

$$HA + S \rightleftharpoons HS^+ + A^-,$$

the result is

$$K = \frac{a_{HS^+}a_{A^-}}{a_{HA}a_S}. \tag{5}$$

If the concentration of dissolved substances in the solvent is not large, the activity of the latter, i.e., a_S, may be regarded as unity, as for the pure solvent; equation (5) then becomes

$$K_a = \frac{a_{HS^+}a_{A^-}}{a_{HA}}. \tag{6}$$

The substance HS^+ is the effective hydrogen ion in the solvent S, so that a_{HS^+} is equivalent to the quantity conventionally written in previous chapters as a_{H^+}, it being understood that the symbol H^+ does not refer to a proton but to the appropriate hydrogen ion in the given solvent; it follows, therefore, that equation (6) may be written in the form

$$K_a = \frac{a_{H^+}a_{A^-}}{a_{HA}}, \tag{7}$$

which is identical with that obtained by regarding the acid as HA ionizing into H^+ and A^-, in accordance with the general treatment on page 163. The constant as defined by equation (6) or (7) is thus identical with the familiar dissociation constant of the acid HA in the given solvent as obtained by the methods described in Chap. V; further reference to the determination of dissociation constants is made below.

Application of the law of mass action to the general base-solvent equilibrium

$$B + SH \rightleftharpoons BH^+ + S^-,$$

gives

$$K_b = \frac{a_{BH^+}a_{S^-}}{a_B}, \tag{8}$$

the activity of the solvent SH being regarded as constant; the quantity K_b is the dissociation constant of the base. If the base is an amine,

RNH_2, in aqueous solution, then the equilibrium

$$RNH_2 + H_2O \rightleftharpoons RNH_3^+ + OH^-$$

is established, and the dissociation constant is given by

$$K_b = \frac{a_{RNH_3^+}a_{OH^-}}{a_{RNH_2}}.$$

The result is therefore the same as would be obtained by means of the general treatment given in Chap. V for an electrolyte MA, if the undissociated base were regarded as having the formula RNH_3OH in aqueous solution.

The dissociation constants of acids, and bases, are of importance as giving a measure of the **relative strengths** of the acids, and bases, in the given medium. The strength of an acid is measured by its tendency to give up a proton, and hence the position of the equilibrium with a given solvent, as determined by the dissociation constant, is an indication of the strength of the acid. Similarly, the strength of a base, which depends on its ability to take up a proton, is also measured by its dissociation constant, since this is the equilibrium constant for the reaction in which the solvent molecule transfers a proton to the base.

Determination of Dissociation Constants: The Conductance Method. —As seen in Chap. V, equation (7) may be written in the form

$$K_a = \frac{c_{H^+}c_{A^-}}{c_{HA}} \cdot \frac{f_{H^+}f_{A^-}}{f_{HA}}$$

and if α is the true degree of dissociation of the solution of acid whose stoichiometric concentration is c, then

$$K_a = \frac{\alpha^2 c}{1 - \alpha} \cdot \frac{f_{H^+}f_{A^-}}{f_{HA}}. \quad (9)$$

Accurate methods for evaluating K_a based on this equation, involving the use of conductance measurements, have been already described in Chap. V; these require a lengthy experimental procedure, but if carried out carefully the results are of high precision. For solvents of high dielectric constant the calculation based on the Onsager equation may be employed (p. 165), but for low dielectric constant media the method of Fuoss and Kraus (p. 167) should be used.

Many of the dissociation constants in the older literature have been determined by the procedure originally employed by Ostwald (1888), which is now known to be approximate in nature; if the activity coefficient factor in equation (9) is neglected, and the degree of dissociation α is set equal to the conductance ratio (Λ/Λ_0), the result is

$$k \approx \frac{\Lambda^2 c}{\Lambda_0(\Lambda_0 - \Lambda)}. \quad (10)$$

An approximate dissociation function k was thus calculated from the measured equivalent conductance of the solution of weak acid, or weak base, at the concentration c, and the known value at infinite dilution. For moderately weak acids, of dissociation constant of 10^{-5} or less, the degree of dissociation is not greatly different from the conductance ratio, provided the solutions are relatively dilute; under these conditions, too, the activity coefficient factor will be approximately unity. If the acid solutions are sufficiently dilute, therefore, the dissociation constants given by equation (10) are not seriously in error. For example, if the data in Table XXXVIII on page 165 for acetic acid solutions are treated by the Ostwald method they give k_a values varying from about 1.74×10^{-5} in the most dilute solutions to 1.82×10^{-5} in the more concentrated. The results in dilute solution do not differ appreciably from those obtained by the more complicated but more accurate method of treating the data. It may be mentioned, however, that the earlier determinations of dissociation constants were generally based on conductance measurements with solutions which were rarely more dilute than 0.001 N, whereas those in Table XXXVIII refer to much less concentrated solutions. For acids whose dissociation constants are greater than about 10^{-5} the Ostwald method would give reasonably accurate results for the dissociation constant only at dilutions which are probably too great to yield reliable conductance measurements.

Electromotive Force Method.—An alternative procedure for the evaluation of dissociation constants, which also leads to very accurate results, involves the study of cells without liquid junction.[7] The chemical reaction occurring in the cell

$$H_2(1 \text{ atm.}) \mid HA(m_1) \;\; NaA(m_2) \;\; NaCl(m_3) \;\; AgCl(s) \mid Ag,$$

where HA is an acid, whose molality is m_1 in the solution, and NaA is its sodium salt, of molality m_2, is

$$\tfrac{1}{2}H_2(1 \text{ atm.}) + AgCl(s) = Ag + H^+ + Cl^-$$

for the passage of one faraday. The E.M.F. of the cell is therefore given by (cf. p. 226)

$$E = E^0 - \frac{RT}{F} \ln a_{H^+}a_{Cl^-}, \qquad (11)$$

where E^0 is the standard E.M.F. of the hydrogen-silver chloride cell, i.e., of the hypothetical cell

$$H_2(1 \text{ atm.}) \mid H^+(a_{H^+} = 1) \parallel Cl^-(a_{Cl^-} = 1) \;\; AgCl(s) \mid Ag.$$

The E.M.F. of this cell is clearly equal in magnitude but opposite in sign to the standard potential of the Ag, AgCl(s) Cl^- electrode, and hence E^0 in equation (11) is + 0.2224 volt at 25°. The subscripts H^+ and Cl^-

[7] Harned and Ehlers, *J. Am. Chem. Soc.*, **54**, 1350 (1932); for reviews, see Harned, *J. Franklin Inst.*, **225**, 623 (1938); Harned and Owen, *Chem. Revs.*, **25**, 31 (1939).

in equation (11), etc., refer to the hydrogen and chloride ions, respectively, it being understood that the former is really H_3O^+ in aqueous solution, and the corresponding oxonium ion in other solvents. The activities in equation (11) may be replaced by the product of the respective molalities (m) and the stoichiometric activity coefficients (γ), so that

$$E = E^0 - \frac{RT}{F} \ln m_{H^+}m_{Cl^-} - \frac{RT}{F} \ln \gamma_{H^+}\gamma_{Cl^-}. \tag{12}$$

The activities in equation (7) for the dissociation constant may be expressed in a similar manner, so that

$$K = \frac{a_{H^+}a_{A^-}}{a_{HA}} = \frac{m_{H^+}m_{A^-}}{m_{HA}} \cdot \frac{\gamma_{H^+}\gamma_{A^-}}{\gamma_{HA}},$$

and combination of this expression with equation (12) gives

$$E = E^0 - \frac{RT}{F} \ln \frac{m_{HA}m_{Cl^-}}{m_{A^-}} - \frac{RT}{F} \ln \frac{\gamma_{HA}\gamma_{Cl^-}}{\gamma_{A^-}} - \frac{RT}{F} \ln K, \tag{13}$$

$$\therefore \quad \frac{F(E - E^0)}{2.303RT} + \log \frac{m_{HA}m_{Cl^-}}{m_{A^-}} = - \log \frac{\gamma_{HA}\gamma_{Cl^-}}{\gamma_{A^-}} - \log K, \tag{14}$$

or, at 25°,

$$\frac{(E - E^0)}{0.05915} + \log \frac{m_{HA}m_{Cl^-}}{m_{A^-}} = - \log \frac{\gamma_{HA}\gamma_{Cl^-}}{\gamma_{A^-}} - \log K. \tag{15}$$

The right-hand side of equation (14) may be set equal to $-\log K'$, where K' becomes identical with K at infinite dilution, for then the activity coefficient factor $\gamma_{HA}\gamma_{Cl^-}/\gamma_{A^-}$ becomes unity and the term $\log \gamma_{HA}\gamma_{Cl^-}/\gamma_{A^-}$ in equation (15) is zero.

Since E^0 is known, and the E.M.F. of the cell (E) can be measured with various concentrations of acid, sodium salt and sodium chloride, i.e., for various values of m_1, m_2 and m_3 in the cell depicted above, it is possible to evaluate the left-hand side of equation (14) or (15). In dilute solution, the sodium chloride may be assumed to be completely dissociated so that the molality of the chloride ion can be taken as equal to that of the sodium chloride, i.e., m_{Cl^-} is equal to m_3. The acid HA will be partly in the undissociated form and partly dissociated into hydrogen and A^- ions; the stoichiometric molality of HA is m_1, and if m_{H^+} is the molality of the hydrogen ions resulting from dissociation, the molality of undissociated HA molecules, i.e., m_{HA} in equation (15), is equal to $m_1 - m_{H^+}$. Finally, it is required to known m_{A^-}: the A^- ions are produced by the dissociation of NaA, which may be assumed to be complete, and also by the small dissociation of the acid HA; it follows, therefore, that m_{A^-} is equal to $m_2 + m_{H^+}$. Since m_{H^+}, the hydrogen ion concentration, is required for these calculations, a sufficiently accurate value is estimated from the approximate dissociation constant (cf. p. 390); this procedure is satisfactory provided the dissociation constant of the acid is about 10^{-4} or less, as is generally the case. If the values of the left-hand side of equation (14) or (15) are plotted against the ionic strength of the solution and extrapolated to infinite dilution, the intercept gives $-\log K$, from which the dissociation constant K can be readily obtained. The general practice is to keep the ratio of acid to salt, i.e., m_1 to m_2, constant, approximately unity, in a series of experiments, and to vary the ionic strength by using different concentrations of sodium chloride. The results obtained for acetic acid are shown in Fig. 88; the value of $\log K_a$ is seen to be -4.756, so that K_a is 1.754×10^{-5} at 25°.

When comparing the dissociation constant obtained by the conductance method with that derived from E.M.F. measurements, it must be remembered that the former is based on volume concentrations, i.e., g.-ions or moles per liter, while the latter involves molalities. This difference arises because it is more convenient to treat conductance data in terms of volume concentrations, whereas the standard states for E.M.F. studies are preferably chosen in terms of molalities. If K_c and K_m are the dissociation constants based on volume concentrations and molalities, respectively, then it can be readily seen that K_c is equal to $K_m\rho$, where ρ is the density of the solvent at the experimental temperature. For water at 25°, ρ is 0.9971, and hence K_c for acetic acid, calculated from E.M.F. measurements, is 1.749×10^{-5}, compared with 1.753×10^{-5} from conductance data. Considering the difference in principle involved in the two methods, the agreement is very striking. Almost as good correspondence has been found for other acids with which accurate conductance and E.M.F. studies have been made; this may be regarded as providing strong support for the theoretical treatments involved, especially in the case of the conductance method.

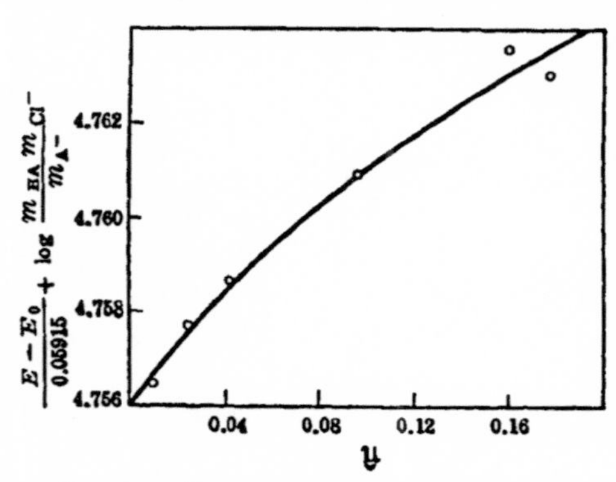

FIG. 88. Dissociation constant of acetic acid (Harned and Ehlers)

The procedure described here may be regarded as typical of that adopted for any moderately weak acid, i.e., of dissociation constant 10^{-3} to 10^{-5}; for weaker acids, however, some modification is necessary. In addition to the acid dissociation

$$HA + H_2O \rightleftharpoons H_3O^+ + A^-,$$

allowance must be made for the equilibrium

$$A^- + H_2O \rightleftharpoons OH^- + HA,$$

which is due to the water, i.e., the solvent, functioning as an acid to some extent; this corresponds to the phenomenon of hydrolysis to be discussed in Chap. XI. It follows, therefore, that if the stoichiometric molality of HA is m_1, then

$$m_{HA} = m_1 - m_{H^+} + m_{OH^-},$$

since HA is used up in the dissociation process while it is formed in the hydrolysis reaction, in amounts equivalent to the hydrogen and hydroxyl ions, respectively. Further, if the molality of the salt NaA is m_2, then

$$m_{A^-} = m_2 + m_{H^+} - m_{OH^-},$$

since A^- ions are formed in the dissociation process but are used up in the hydrolysis. If the dissociation constant is greater than 10^{-5} and the ratio of acid to salt, i.e., m_1/m_2, is approximately unity, m_{OH^-} is found by calculation to be less than 10^{-9}, and so this term can be neglected in the expressions for m_{HA} and m_{A^-}, as was done above. If the dissociation constant lies between 10^{-5} and 10^{-9}, and m_1/m_2 is about unity, $m_{H^+} - m_{OH^-}$ is negligibly small, so that m_{HA} and m_{A^-} may be taken as equal to m_1 and m_2, respectively. For still weaker acids, m_{H^+} is so small that it may be ignored in comparison with m_{OH^-}; m_{HA} is now equal to $m_1 + m_{OH^-}$, and m_{A^-} is $m_2 - m_{OH^-}$. The values of m_{OH^-} required for determining m_{HA} and m_{A^-} are obtained by utilizing the fact that $m_{H^+}m_{OH^-}$ is equal to 10^{-14} at 25°.

Dissociation Constants of Bases.—The dissociation constants of bases can be determined, in principle, by methods which are essentially similar to those employed for acids. Replacing activities in equation (8) by the product of molalities and activity coefficients, it is seen that for a base

$$K_b = \frac{m_{BH^+}m_{S^-}}{m_B} \cdot \frac{\gamma_{BH^+}\gamma_{S^-}}{\gamma_B}, \tag{16}$$

and this may be replaced by

$$K_b = \frac{\alpha^2 c}{1 - \alpha} \cdot \frac{\gamma_{BH^+}\gamma_{S^-}}{\gamma_B}, \tag{17}$$

where α represents the degree of dissociation of the hypothetical solvated base, e.g., $BH \cdot OH$ in water. By neglecting the activity coefficient factor in equation (17) and replacing α by the conductance ratio, an approximate equation identical in form with (10) is obtained; the value of Λ_0 in this equation is the sum of the equivalent conductances of the BH^+ and OH^- ions, e.g., of NH_4^+ and OH^- if the base is ammonia.

Very little accurate E.M.F. work has been done on the dissociation constants of bases, chiefly because moderately weak bases are very volatile, while the non-volatile bases, e.g., anilines, are usually very weak. An exception to this generalization is to be found in the aliphatic amino-acids which will be considered in connection with the subject of amphoteric electrolytes. Since silver chloride is soluble in aqueous solutions of ammonia and of many amines, it is not possible to use silver-silver chloride electrodes with such bases; the employment of sodium amalgam has been proposed, but it is probable that the silver-silver iodide electrode will prove most useful for the purpose of the accurate determination of the dissociation constants of bases by the E.M.F. method.

Apart from determinations of dissociation constants made from conductance data, most values derived from E.M.F. measurements have been obtained by an approximate procedure which will be described later.

Dissociation Constants of Polybasic Acids: Conductance Method.— A polybasic acid ionizes in stages, each stage having its own characteristic dissociation constant: for example, the ionization of a tribasic acid H_3A, such as phosphoric acid, may be represented by:

$$1. \quad H_3A + H_2O \rightleftharpoons H_3O^+ + H_2A^-, \qquad K_1 = \frac{a_{H^+}a_{H_2A^-}}{a_{H_3A}}. \tag{18a}$$

$$2. \quad H_2A^- + H_2O \rightleftharpoons H_3O^+ + HA^{--}, \qquad K_2 = \frac{a_{H^+}a_{HA^{--}}}{a_{H_2A^-}}. \tag{18b}$$

$$3. \quad HA^{--} + H_2O \rightleftharpoons H_3O^+ + A^{---}, \qquad K_3 = \frac{a_{H^+}a_{A^{---}}}{a_{HA^{--}}}. \tag{18c}$$

The fact that ionization occurs in these three stages successively with increasing dilution shows that $K_1 > K_2 > K_3$; this is always true, because the presence of a negative charge on H_2A^- and of two such charges on HA^{--} makes it increasingly difficult for a proton to be lost.

If the dissociation constants for any two successive stages are sufficiently different it is sometimes feasible to apply the methods employed for monobasic acids; the conditions under which this is possible will be considered with reference to a dibasic acid, but the general conclusions can be extended to more complex cases. If H_2A is a dibasic acid for which K_1, the dissociation constant of the first stage,*

$$H_2A + H_2O \rightleftharpoons H_3O^+ + HA^-,$$

is of the order of 10^{-3} to 10^{-5}, while the constant K_2 of the second stage of dissociation,

$$HA^- + H_2O \rightleftharpoons H_3O^+ + A^{--},$$

* The first stage dissociation constant of a dibasic acid is actually the sum of two constants; consider, for example, the unsymmetrical dibasic acid $HX \cdot X'H$, where X and X′ are different. This acid can dissociate in two ways, viz.,

$$HX \cdot X'H + H_2O \rightleftharpoons H_3O^+ + {}^-X \cdot X'H,$$

and

$$HX \cdot X'H + H_2O \rightleftharpoons H_3O^+ + HX \cdot X'^-,$$

and if K_1' and K_1'' are the corresponding dissociation constants, the experimental first stage dissociation constant K_1 is actually equal to $K_1' + K_1''$. If the acid is a symmetrical one, e.g., of the type $CO_2H(CH_2)_n CO_2H$, the constants K_1' and K_1'' are identical, so that K_1 is equal to $2K_1'$. Similar considerations apply to all polybasic acids.

is very small, i.e., the acid is moderately weak in the first stage and very weak in the second stage, then it may be treated virtually as a monobasic acid. The value of K_1 may be determined in the usual manner from conductance measurements on the acid H_2A and its salt NaHA at various concentrations, together with the known values for hydrochloric acid and sodium chloride (cf. p. 164). Provided the dissociation constant K_2 of the acid HA^- is very small, the extent of the second stage dissociation will be negligible in the solutions of both H_2A and NaHA. This method has been applied to the determination of the first dissociation constant of phosphoric acid;[8] for this acid K_1 is 7.5×10^{-3} at 25°, whereas K_2 is 6.2×10^{-8}.

If the dissociation constant of the second stage is relatively large, e.g., about 10^{-5} or more, it is not possible to carry out the normal conductance procedure for evaluating K_1; this is because the HA^- ion in the solution of the completely ionized salt NaHA dissociates to an appreciable extent to form H_3O^+ and A^{--} ions, and the measured conductance is much too large. As a result of this further dissociation, it is not possible to derive the equivalent conductances of NaHA required for the calculation of the dissociation constant. An attempt has been made to overcome this difficulty by estimating the equivalent conductance of the ion HA^- in an indirect manner, so that the value for the salt NaHA may be calculated. By assuming that the intermediate ion of an organic dibasic acid, viz., $OH \cdot CORCO_2^-$, has the same equivalent conductance at infinite dilution as the anion of the corresponding amic acid, viz., $NH_2 \cdot CORCO_2^-$, which can be obtained by direct measurement, it has been concluded that the equivalent conductance λ_{HA^-} of the intermediate ion is equal to $0.53\lambda_{A^{--}}$, where $\lambda_{A^{--}}$ is the conductance of the A^{--} ion, i.e., of $^-CO_2RCO_2^-$ in the case under consideration. Since the latter quantity can be determined without great difficulty by conductance measurements with the salt Na_2A, the value of λ_{HA^-} for the given acid at infinite dilution can be obtained. The known equivalent conductance of sodium is now added to that of the HA^-, thus giving the value of Λ_0 for the salt NaHA; the variation of the equivalent conductance with concentration can now be expressed by assuming the Onsager equation to be applicable. Since the conductance of the acid H_2A at various concentrations is known, as well as that of HCl and NaCl, all the information is available for calculating the dissociation constant of H_2A as a monobasic acid. This method cannot be regarded as accurate, however, for the identification of λ_{HA^-} with $0.53\lambda_{A^{--}}$ is known to be an approximation.[9]

The determination of the second dissociation constant (K_2) of a dibasic acid also requires a knowledge of the equivalent conductance of the intermediate ion HA^-, and if the value of K_2 is large enough to be determined from conductance measurements, the further dissociation of HA^-

[8] Sherrill and Noyes, *J. Am. Chem. Soc.*, **48**, 1861 (1926).

[9] Jeffery and Vogel, *J. Chem. Soc.*, 21 (1935); 1756 (1936); Davies, *ibid.*, 1850 (1939).

is too great for the equivalent conductance to be derived accurately from the experimental data for the salt NaHA. In the earlier attempts to evaluate K_2 the assumption was made of a constant ratio of λ_{HA^-} to $\lambda_{A^{--}}$, as described above; this, however, leads to results that are too uncertain to have any serious worth. If transference data are available, it is possible in certain cases to determine the required value of λ_{HA^-} and hence to calculate the second dissociation constant of the acid. The method has been used to evaluate K_2 for sulfuric acid: in its first stage of dissociation this is a very strong acid, but the second stage dissociation, although very considerable, is much smaller.[10]

Dissociation Constants of Dibasic Acids by E.M.F. Measurement.—If the ratio of the dissociation constants of a dibasic acid, or of any two successive stages of ionization of a polybasic acid, is greater than about 10^2 or 10^3, it is possible to treat each stage as a separate acid and to determine its dissociation constant by means of cells without liquid junction in the manner already described. In a mixture of the free dibasic acid H_2A with its salt NaHA, the essential equilibria are

$$1. \quad H_2A + H_2O \rightleftharpoons H_3O^+ + HA^-,$$

and

$$2. \quad HA^- + H_2O \rightleftharpoons H_3O^+ + A^{--},$$

and from these, by subtraction, may be obtained the equilibrium

$$3. \quad 2HA^- \rightleftharpoons H_2A + A^{--}.$$

If K_1 and K_2 are the dissociation constants for the stages 1 and 2, it can be readily shown that the equilibrium constant for the process 3 is equal to K_2/K_1.

If the stoichiometric molality of H_2A is m_1 in a given solution and that of the salt NaHA, assumed to be completely dissociated into HA^- ions, is m_2, then

$$m_{H_2A} = m_1 - m_{H^+} + m_{A^{--}}, \tag{19}$$

since H_2A is removed to form hydrogen ions in process 1, while it is formed in process 3 in an amount equivalent to A^{--}; further,

$$m_{HA^-} = m_2 + m_{H^+} - 2m_{A^{--}}, \tag{20}$$

since HA^- is formed in reaction 1 and removed in 3, in amounts equivalent to H_3O^+ and $2A^{--}$ respectively. It has been seen that the equilibrium constant of process 3 is equal to K_2/K_1, and the smaller this ratio the less will be the tendency of the reaction to take place from left to right; if K_2/K_1 is smaller than about 10^{-3}, i.e., $K_1/K_2 > 10^3$, the extent of the reaction will be negligible, and then the $m_{A^{--}}$ terms in equations (19) and (20) can be ignored. The expressions for m_{H_2A} and m_{HA^-} then reduce to the same form as do the corresponding ones for m_{HA} and m_{A^-},

[10] Sherrill and Noyes, *J. Am. Chem. Soc.*, **48**, 1861 (1926).

respectively, for a monobasic acid. If K_1 lies between 10^{-5} and 10^{-9} and m_1/m_2 is approximately unity, m_{H^+} may be neglected, as explained on page 317; for weaker acids, however, the term m_{OH^-}, arising on account of hydrolysis, must be included.

It follows, therefore, that when K_2/K_1 is small, or K_1/K_2 is large, the value of K_1 can be readily determined by measurements on cells of the type

$$H_2(1\text{ atm.}) \mid H_2A(m_1)\ NaHA(m_2)\ NaCl(m_3)\ AgCl(s) \mid Ag,$$

the equation for the E.M.F. being, by analogy with equation (14),

$$\frac{F(E - E^0)}{2.303RT} + \log \frac{m_{H_2A}m_{Cl^-}}{m_{HA^-}} = -\log \frac{\gamma_{H_2A}\gamma_{Cl^-}}{\gamma_{HA^-}} - \log K_1. \qquad (21)$$

The values of m_{H_2A} and m_{HA^-} are derived as explained above, and m_{Cl^-} is taken as equal to m_3; the method of extrapolation, which yields $\log K_1$, is the same as described for a monobasic acid.

In order to investigate the second stage dissociation constant, the system studied consists of a mixture of highly ionized NaHA, which is equivalent to the acid HA^-, of molality m_1, and its salt Na_2A, of molality m_2. In this case, it follows from the three processes given above, that

$$m_{HA^-} = m_1 - m_{H^+} - 2m_{H_2A}$$

and

$$m_{A^{--}} = m_2 + m_{H^+} + m_{H_2A}.$$

If K_2/K_1 is small the m_{H_2A} terms may be neglected, just as the $m_{A^{--}}$ terms were neglected in the previous case, since process 3 occurs to a small extent only; under these conditions the expressions for m_{HA^-} and $m_{A^{--}}$ are equivalent to those applicable to a monobasic acid. The determination of K_2 can then be carried out by means of the cell

$$H_2(1\text{ atm.}) \mid NaHA(m_1)\ Na_2A(m_2)\ NaCl(m_3)\ AgCl(s) \mid Ag,$$

the E.M.F. of which is given by the expression

$$\frac{F(E - E^0)}{2.303RT} + \log \frac{m_{HA^-}m_{Cl^-}}{m_{A^{--}}} = -\log \frac{\gamma_{HA^-}\gamma_{Cl^-}}{\gamma_{A^{--}}} - \log K_2. \qquad (22)$$

The values of m_{HA^-}, $m_{A^{--}}$ and m_{Cl^-} are determined in the usual manner, but since the activity coefficient factor $\gamma_{HA^-}\gamma_{Cl^-}/\gamma_{A^{--}}$ involves two univalent ions in the numerator with a bivalent ion in the denominator, it will differ more from unity than does the corresponding factor in equations (14) and (21); the usual extrapolation procedure is consequently liable to be less accurate. Utilizing the form

$$\log \gamma_i = -Az_i^2\sqrt{\mu} + C\mu$$

of the extended Debye-Hückel equation, however, it is seen that equation

(22) may be written as

$$\frac{F(E - E^0)}{2.303RT} + \log \frac{m_{HA^-} m_{Cl^-}}{m_{A^{--}}} + 2A\sqrt{\mu} = C\mu - \log K_2. \quad (23)$$

The plot of the left-hand side of this expression, where A is 0.509 at 25°, against the ionic strength μ should thus be a straight line, at least approximately; the intercept for zero ionic strength gives the value of $-\log K_2$. The results obtained in the determination of the second dissociation constant of phosphoric acid are shown in Fig. 89; the upper curve is for cells containing the salts KH_2PO_4 and Na_2HPO_4, and the lower for the two corresponding sodium salts in a different proportion. In this case the acid is $H_2PO_4^-$, and its dissociation constant is seen to be $-$ antilog 7.206, i.e., K_2 is 6.223×10^{-8} at 25°.[11]

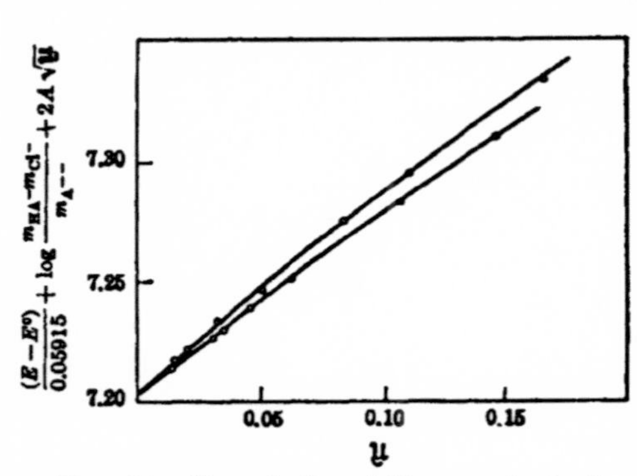

FIG. 89. Second dissociation constant of phosphoric acid (Nims)

If the ratio K_1/K_2 for two successive stages is smaller than 10^3, it would be necessary to include the $m_{A^{--}}$ and m_{H_2A} terms, which were neglected previously, in the determination of K_1 and K_2, respectively. The evaluation of these quantities, as well as of m_{H^+} or m_{OH^-}, would require preliminary values of K_1 and K_2, and the calculations, although feasible, would be tedious. No complete determination by means of cells without liquid junction appears yet to have been made of the dissociation constants of a dibasic acid for which K_1/K_2 is less than 10^3.

Dissociation Constants by Approximate E.M.F. Methods.—When, for various reasons, it is not convenient or desirable to carry out the lengthy series of measurements required for the determination of accurate dissociation constants by the conductance method or by means of cells without liquid junction, approximate E.M.F. methods, utilizing cells with liquid junctions, can be applied. These methods involve the determination of the hydrogen ion concentration, or activity, in solutions containing a series of mixtures of the acid and its salt with a strong base, generally obtained by adding definite quantities of the latter to a known amount of acid. The procedures used for actual measurement of hydrogen ion activities are described in Chap. X, but the theoretical basis of the evaluation of dissociation constants will be considered here.

[11] Nims, *J. Am. Chem. Soc.*, **55**, 1946 (1933); for application to malonic acid, see Hamer, Burton and Acree, *J. Res. Nat. Bur. Standards*, **24**, 269 (1940).

If a is the initial concentration (molality) of the weak or moderately weak acid HA, and b is the amount of strong, monoacid base MOH added at any instant, then b is also equal to m_{M^+}, the molality of M^+ ions at that instant, since the salt MA produced on neutralization may be taken as being completely dissociated. The acid HA is only partially neutralized to form A^- ions, and so

$$a = m_{HA} + m_{A^-}. \tag{24}$$

Further, as the solution must be electrically neutral, the sum of all the positive charges will be equal to the sum of the negative charges; hence

$$m_{M^+} + m_{H^+} = m_{A^-} + m_{OH^-},$$

or

$$b + m_{H^+} = m_{A^-} + m_{OH^-}. \tag{25}$$

The dissociation constant K_a of the acid HA may be expressed in the form

$$K_a = \frac{a_{H^+}a_{A^-}}{a_{HA}}$$

$$= a_{H^+}\frac{m_{A^-}}{m_{HA}} \cdot \frac{\gamma_{A^-}}{\gamma_{HA}},$$

and if m_{A^-} and m_{HA} are eliminated by means of equations (24) and (25), it is found that

$$a_{H^+} = K_a \frac{a - b - m_{H^+} + m_{OH^-}}{b + m_{H^+} - m_{OH^-}} \cdot \frac{\gamma_{HA}}{\gamma_{A^-}}. \tag{26}$$

If the quantity B is defined by

$$B \equiv b + m_{H^+} - m_{OH^-},$$

then equation (26) may be written as

$$a_{H^+} = K_a \frac{a - B}{B} \cdot \frac{\gamma_{HA}}{\gamma_{A^-}},$$

or, taking logarithms,

$$\log a_{H^+} = \log K_a + \log \frac{a - B}{B} + \log \frac{\gamma_{HA}}{\gamma_{A^-}}.$$

It was seen on page 292 that the pH, or hydrogen ion exponent, of a solution may be defined as $-\log a_{H^+}$; in an analogous manner the symbol pK_a, called the **dissociation exponent,** may be substituted for $-\log K_a$; hence

$$\text{pH} = pK_a + \log \frac{B}{a - B} + \log \frac{\gamma_{A^-}}{\gamma_{HA}}. \tag{27}$$

According to the extended Debye-Hückel theory, it is possible to write

$$\log \frac{\gamma_{A^-}}{\gamma_{HA}} = -A\sqrt{\mu} + C\mu, \tag{28}$$

remembering that A^- is a univalent ion and HA an undissociated molecule, and so equation (27) becomes

$$pH = pK_a + \log \frac{B}{a - B} - A\sqrt{\mu} + C\mu, \tag{29}$$

$$\therefore \quad pH - \log \frac{B}{a - B} + A\sqrt{\mu} = pK_a + C\mu.$$

If the left-hand side of this equation for a series of acid-base mixtures is plotted against the ionic strength of the solution, the intercept for μ equal to zero would give the value of pK_a, i.e., $-\log K_a$.

The methods used for the determination of the pH of the solution will be described in the following chapter, but in the meantime the evaluation of B and μ will be considered. If the hydrogen ion concentration of the solution is greater than 10^{-4} g.-ion per liter, i.e., for an acid of medium strength, the hydroxyl ion concentration m_{OH^-} will be less than 10^{-10} and so can be neglected in comparison with m_{H^+}; B then becomes equal to $b + m_{H^+}$. On the other hand, for a very weak acid, when the hydrogen ion concentration is less than 10^{-10} g.-ion per liter, the quantity m_{H^+} may be ignored, so that B is equal to $b + m_{OH^-}$. For solutions of intermediate hydrogen ion concentration, i.e., between 10^{-4} and 10^{-10} g.-ion per liter, $m_{H^+} - m_{OH^-}$ is negligibly small and so B may be taken as equal to b. The values of a and b are known from the amounts of acid and base, respectively, employed to make up the given mixture, and m_{H^+} and m_{OH^-} are readily determined by the aid of the relationships $m_{H^+} = a_{H^+}/\gamma_{H^+}$ and $m_{H^+}m_{OH^-} = k_w$, which is 10^{-14} at 25° (cf. p. 339). The quantity a_{H^+} is derived from the measured pH, and γ_{H^+} is calculated with sufficient accuracy by means of the simple Debye-Hückel equation. The ionic strength μ of the solution is given by $b + m_{H^+} - m_{OH^-}$; except at the beginning of the neutralization, however, when b is small, the value of μ may be taken as equal to b.

The data obtained for acetic acid at 25° are plotted in Fig. 90; [12] the results are seen to fall approximately on a straight line, and from the intercept at zero ionic strength pK_a is seen to be 4.72. The difference between this value and that given previously is to be attributed to an incorrect standardization of the pH scale (cf. footnote, p. 349).

Instead of employing the graphical method described above, the general practice is to make use of equation (27); the quantities pH and B are obtained for each solution and the corresponding pk_a evaluated. The activity correction may be applied by means of equation (28) since

[12] Walpole, *J. Chem. Soc.*, **105**, 2501 (1914).

A is known, and C can be guessed approximately or neglected as being small; alternatively, the tentative pk_a values obtained by neglecting the activity coefficients may be plotted against a function of the ionic strength and extrapolated to infinite dilution.

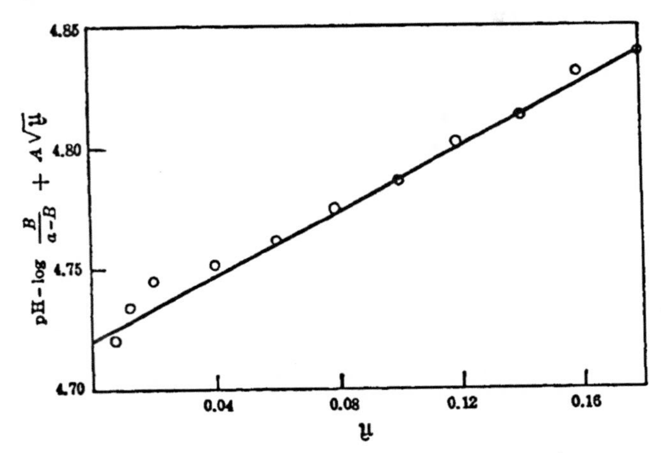

FIG. 90. Dissociation constant of acetic acid

If B is equal to $\frac{1}{2}a$, and the solution is relatively dilute, so that the terms involving the ionic strength are small, equation (29) reduces to

$$\text{pH} = \text{p}k_a.$$

Provided the pH of the system lies between 4 and 10, the quantity B is virtually equal to b, and hence it follows that when b is equal to $\frac{1}{2}a$ the pH of the solution is (approximately) equal to the $\text{p}k_a$ of the acid. In other words, the pH of a half-neutralized solution of an acid, i.e., of a solution containing equivalent amounts of the acid and its salt, is equal to $\text{p}k_a$. This fact is frequently utilized for the approximate determination of dissociation functions.

Dibasic Acids.—The treatment given above is applicable to any stage of ionization of a polybasic acid, provided its dissociation constant differs by a factor of at least 10^3 from those of the stages immediately preceding and following it: the activity correction, equivalent to equation (28), will however depend on the charges carried by the undissociated acid and the corresponding anion. If these are $r - 1$ and r, respectively, then according to the extended Debye-Hückel equation

$$\log \frac{\gamma_r}{\gamma_{r-1}} = -A[r^2 - (r-1)^2]\sqrt{\mu} + C\mu$$
$$= -A(2r-1)\sqrt{\mu} + C\mu,$$

so that equation (29) for the rth dissociation constant of a polybasic acid becomes

$$\mathrm{pH} = \mathrm{p}K_r + \log \frac{B}{a - B} - A(2r - 1)\sqrt{\mu} + C\mu. \tag{30}$$

When the dissociation constants of successive stages are relatively close together, a more complicated treatment becomes necessary.[13] The dissociation constants of the first and second stages of a dibasic acid H_2A may be written in a form analogous to that given above, viz.,

$$K_1 = a_{H^+} \frac{m_{HA^-}}{m_{H_2A}} \cdot \frac{\gamma_{HA^-}}{\gamma_{H_2A}} \quad \text{and} \quad K_2 = a_{H^+} \frac{m_{A^{--}}}{m_{HA^-}} \cdot \frac{\gamma_{A^{--}}}{\gamma_{HA^-}}. \tag{31}$$

If to a solution containing the acid H_2A at molality a there are added b *equivalents* of a strong monoacid base, MOH, the solution will contain H^+, M^+, HA^-, A^{--} and OH^- ions; for electrical neutrality therefore,

$$m_{M^+} + m_{H^+} = m_{HA^-} + 2m_{A^{--}} + m_{OH^-},$$

the term $2m_{A^{--}}$ arising because the A^{--} ions carry two negative charges. Replacing the concentration of M^+ ions, i.e., m_{M^+}, by b, as in the previous case, this equation becomes

$$b + m_{H^+} = m_{HA^-} + 2m_{A^{--}} + m_{OH^-}. \tag{32}$$

Further, the initial amount of the acid a will be equivalent to the total quantity of un-neutralized H_2A and of HA^- and A^{--} ions present at any instant; that is

$$a = m_{H_2A} + m_{HA^-} + m_{A^{--}}. \tag{33}$$

If a quantity B is defined, as before, by

$$B \equiv b + m_{H^+} - m_{OH^-},$$

it can be shown that equations (31), (32) and (33) lead to the result

$$a_{H^+}^2 \frac{B}{2a - B} \cdot \frac{\gamma_{A^{--}}}{\gamma_{H_2A}} = a_{H^+} \frac{a - B}{2a - B} \cdot \frac{\gamma_{A^{--}}}{\gamma_{HA^-}} K_1 + K_1K_2. \tag{34}$$

It follows, therefore, that if the left-hand side of this expression (X) is plotted against the coefficient of K_1 in the first term on the right-hand side (Y), a straight line of slope K_1 and intercept K_1K_2 should result. The evaluation of B involves the same principles as described in connection with monobasic acids. In the first stage of neutralization, i.e., when $a > b$, the ionic strength may be taken as $b + m_{H^+}$, as before, but in the

[13] Auerbach and Smolczyk, *Z. physik. Chem.*, **110**, 83 (1924); Britton, *J. Chem. Soc.*, **125**, 423 (1924); **127**, 1896 (1925); Morton, *Trans. Faraday Soc.*, **24**, 14 (1928); Partington *et al.*, *ibid.*, **30**, 598 (1934); **31**, 922 (1935); Gane and Ingold, *J. Chem. Soc.*, 2151 (1931); German, Jeffery and Vogel, *ibid.*, 1624 (1935); German and Vogel, *J. Am. Chem Soc.*, **58**, 1546 (1936); Jones and Soper, *J. Chem. Soc.*, 133 (1936); see also, Simms, *J. Am. Chem. Soc.*, **48**, 1239 (1926); Muralt, *ibid.*, **52**, 3518 (1930).

second stage, i.e., when $b > a$, a sufficient approximation is $2b - a$. Provided the solutions are reasonably dilute the limiting law of Debye and Hückel may be used to derive $\gamma_{A^{--}}$ and the ratio $\gamma_{A^{--}}/\gamma_{HA^-}$, the activity coefficient of the undissociated molecules γ_{H_2A} being taken as

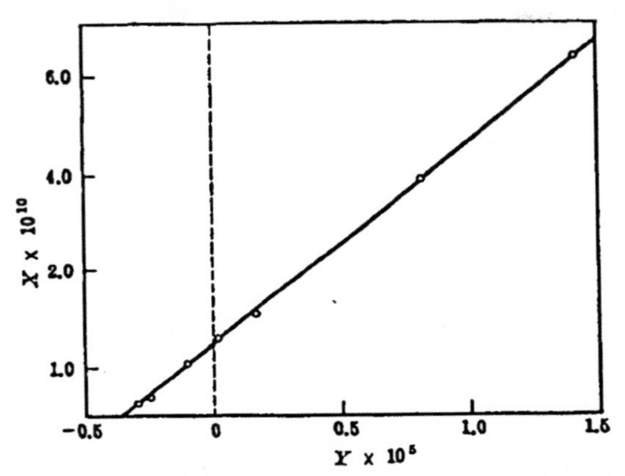

FIG. 91. Dissociation constants of adipic acid (Speakman)

unity. The experimental results obtained in this manner for adipic acid are shown in Fig. 91;[14] the plot is seen to approximate very closely to a straight line, the values of K_1 and K_1K_2 being 3.80×10^{-5} and 1.43×10^{-10} respectively, so that K_2 is 3.76×10^{-6}.

An alternative treatment of equation (34) is to write it in the form

$$X = K_1Y + K_1K_2, \tag{35}$$

where X and Y are defined by

$$X \equiv a_{H^+}^2 \frac{B}{2a - B} \cdot \frac{\gamma_{A^{--}}}{\gamma_{H_2A}}$$

and

$$Y \equiv a_{H^+} \frac{a - B}{2a - B} \cdot \frac{\gamma_{A^{--}}}{\gamma_{HA^-}}.$$

The solutions of equation (35) are

$$K_1 = \frac{X}{Y + K_2} \quad \text{and} \quad K_2 = \frac{X - K_1Y}{K_1}.$$

If two points during the neutralization are chosen, such that the quantities X and Y have the values X' and Y' and X'' and Y'', respectively, then it is readily found that

$$K_1 = \frac{X' - X''}{Y' - Y''} \quad \text{and} \quad K_2 = \frac{X'Y'' - X''Y'}{X'' - X'}.$$

[14] Speakman, *J. Chem. Soc.*, 855 (1940).

Since the X's and Y's can be evaluated, as already described, the two dissociation constants of a dibasic acid can be determined from pairs of pH measurements.

The methods just described can be extended so as to be applicable to acids of higher basicity, irrespective of the ratio of successive dissociation constants.

Colorimetric Determination of Dissociation Constants.—The colorimetric method for determining or comparing dissociation constants has been chiefly applied in connection with non-aqueous solvents, but it has also been used to study certain acids in aqueous solution. It can be employed, in general, whenever the ionized and non-ionized forms of an acid, or base, have different absorption spectra in the visible, i.e., they have different visible colors, or in the near ultra-violet regions of the spectrum. If the acid is a moderately strong one, e.g., picric acid, it will dissociate to a considerable extent when dissolved in water, and the amounts of un-ionized form HA and of ions A^- will be of the same order; under these conditions an accurate determination of the dissociation constant is possible. By means of preliminary studies on solutions which have been made either definitely acid, so as to suppress the ionization entirely, or definitely alkaline, so that the salt only is present and ionization is complete, the "extinction coefficient" for light of a given wave length of the form HA or A^- can be determined. As a general rule the ions A^- have a more intense color and it is the extinction coefficient of this species which is actually measured. Once this quantity is known, the amount of A^- in any system, such as the solution of the acid in water, can be found, provided Beer's law is applicable.* In a solution of the pure acid of concentration a in pure water, c_{H^+} is equal to c_{A^-}, while the concentration of undissociated acid c_{HA} is equal to $a - c_{H^+}$ or to $a - c_{A^-}$; hence if c_{A^-} is determined colorimetrically, it is possible to evaluate directly the concentration dissociation function $c_{H^+}c_{A^-}/c_{HA}$. This function, as already seen, depends on the ionic strength of the medium, but extrapolation to infinite dilution should give the true dissociation constant.

If the acid is too weak to yield an appreciable amount of A^- ions when dissolved in pure water, e.g., p-nitrophenol, it is necessary to employ a modified procedure which is probably less accurate. A definite quantity of the acid being studied is added to excess of a "buffer solution" (see Chap. XI) of known pH; the pH chosen should be close to the expected pK_a of the acid, for under these conditions the resulting solution will contain approximately equal amounts of the undissociated acid HA and of A^- ions. The amount of either HA or A^-, whichever is the more convenient, is then determined by studying the absorption of light of

* According to Beer's law, $\log I_0/I = \epsilon cd$, where I_0 is the intensity of the incident light and I is that of the emergent light for a given wave length, for which the extinction coefficient is ϵ, d is the thickness of the layer of solution, and c is its concentration. If ϵ is known, the value of c can be estimated from the experimental value of I_0/I.

suitable wave length, the corresponding extinction coefficient having been obtained from separate experiments, as explained previously. If c_{A^-} is determined in this manner, c_{HA} is known, since it is equal to $c - c_{A^-}$, where c is the stoichiometric concentration of the acid. In this way it is possible to calculate the ratio c_{A^-}/c_{HA}, and since a_{H^+} is known from the pH of the solution, the function $a_{H^+}c_{A^-}/c_{HA}$ can be evaluated. For many purposes this is sufficiently close to the dissociation constant to be employed where great accuracy is not required. Alternatively, the values of the function in different solutions may be extrapolated to zero ionic strength. This method has been used to study acids which exhibit visible color changes in alkaline solutions, e.g., nitrophenols,[15] as well as for substances that are colorless in both acid and alkaline media but have definite absorption spectra in the ultra-violet region of the spectrum, e.g., benzoic and phenylacetic acids.[16]

Approximate Methods for Bases.—The procedures described for determining the dissociation constants of acids can also be applied, in principle, to bases; analogous equations are applicable except that hydroxyl ions replace hydrogen ions, and *vice versa*, in all the expressions. Since the value of the product of a_{H^+} and a_{OH^-} is known to have a definite value at every temperature (cf. Table LXI), it is possible to derive a_{OH^-} from a_{H^+} obtained experimentally.

Dissociation Constant Data.—The dissociation constants at 25° of a number of acids and bases obtained by the methods described above are recorded in Table LVII; the varying accuracy of the results is indicated, to some extent, by the number of significant figures quoted.[17] The pK_a and pK_b values are given in each case, since these are more frequently employed in calculations than are the dissociation constants themselves. Acids and bases having dissociation constants of about 10^{-5}, i.e., pK is in the vicinity of 5, are generally regarded as "weak," but if the values are in the region of 10^{-9}, i.e., pK is about 9, they are referred to as "very weak." If the dissociation constant is about 10^{-2} or 10^{-3}, the acid or base is said to be "moderately strong," and at the other extreme, when the dissociation constant is 10^{-12} or less, the term "extremely weak" is employed.

Aprotic Solvents.—The colorimetric method of studying dissociation constants has found a special application in aprotic solvents such as benzene; these solvents exhibit neither acidic nor basic properties, and so they do not have the levelling effects observed with acids in proto-

[15] von Halban and Kortüm, *Z. physik. Chem.*, **170A**, 351 (1934); **173A**, 449 (1935); Kilpatrick *et al.*, *J. Am. Chem. Soc.*, **59**, 572 (1937); **62**, 3047 (1940); *J. Phys. Chem.*, **43**, 259 (1939).

[16] Flexser, Hammett and Dingwall, *J. Am. Chem. Soc.*, **57**, 2103 (1935); Martin and Butler, *J. Chem. Soc.*, 1366 (1939).

[17] For further data, see Harned and Owen, *Chem. Revs.*, **25**, 31 (1939); Dippy, *ibid.*, **25**, 151 (1939); Gane and Ingold, *J. Chem. Soc.*, 2153 (1931); Jeffery and Vogel, *ibid.*, 21 (1935); 1756 (1936); German, Jeffery and Vogel, *ibid.*, 1624 (1935); 1604 (1937).

TABLE LVII. DISSOCIATION CONSTANT EXPONENTS OF ACIDS AND BASES AT 25°

Monobasic Organic Acids

Acid	pK_a	Acid	pK_a
Formic	3.751	Benzoic	4.20
Acetic	4.756	*o*-Chlorobenzoic	2.92
Propionic	4.874	*m*-Chlorobenzoic	3.82
n-Butyric	4.820	*p*-Chlorobenzoic	3.98
iso-Butyric	4.821	*p*-Bromobenzoic	3.97
n-Valeric	4.86	*p*-Hydroxybenzoic	4.52
Trimethylacetic	5.05	*p*-Nitrobenzoic	3.42
Diethylacetic	4.75	*p*-Toluic	4.37
Chloroacetic	2.870	Phenylacetic	4.31
Lactic	3.862	Cinnamic (cis)	3.88
Gycolic	3.831	Cinnamic (trans)	4.44
Acrylic	4.25	Phenol	9.92

Dibasic Organic Acids

Acid	pK_1	pK_2	Acid	pK_1	pK_2
Oxalic	1.30	4.286	Pimelic	4.51	5.42
Malonic	2.84	5.695	Suberic	4.53	5.40
Succinic	4.20	5.60	Maleic	2.00	6.27
Glutaric	4.35	5.42	Fumaric	3.03	4.48
Adipic	4.42	5.41	Phthalic	2.89	5.42

Bases

Base	pK_b	Base	pK_b
Ammonia	4.76	Triethylamine	3.20
Methylamine	3.30	Aniline	9.39
Dimethylamine	3.13	Benzylamine	4.63
Trimethylamine	4.13	Diphenylamine	13.16
Ethylamine	3.25	Pyridine	8.80
Diethylamine	2.90	Piperidine	2.88

Inorganic Acids

Acid				Acid		
Sulfuric (2nd stage)			1.02	Hydrogen sulfide	7.2,	11.9
Phosphoric	2.124,	7.206,	12 32	Hydrogen cyanide		9.14
Carbonic		6.35,	10.25	Boric		9.24

philic and with bases in protogenic media (cf. pp. 309, 311). It is thus possible to make a comparison of the strengths of acids and bases without any interfering influence of the solvent. Suppose a certain amount of an acid HA is dissolved in an aprotic solvent and a known quantity of a base B is added; although neither acid nor base can function alone, they can exercise their respective functions when present together, so that an acid-base equilibrium of the familiar form

$$\underset{\text{acid}}{HA} + \underset{\text{base}}{B} \rightleftharpoons \underset{\text{acid}}{BH^+} + \underset{\text{base}}{A^-}$$

is established. Application of the law of mass action to the equilibrium then gives

$$K = \frac{c_{BH^+}c_{A^-}}{c_{HA}c_B} \cdot \frac{f_{BH^+}f_{A^-}}{f_{HA}f_B}. \tag{36}$$

If the essential dissociations of the acids HA and BH^+, to yield protons, i.e.,

$$HA \rightleftharpoons H^+ + A^- \quad \text{and} \quad BH^+ \rightleftharpoons H^+ + B,$$

where H^+ represents a proton, are considered, the *fundamental* dissociation constants are

$$K_{HA} = \frac{a_{H^+}a_{A^-}}{a_{HA}} \quad \text{and} \quad K_{BH^+} = \frac{a_{H^+}a_B}{a_{BH^+}}, \tag{37}^*$$

respectively; comparison of these quantities with the equilibrium constant of equation (36) shows that

$$K = \frac{K_{HA}}{K_{BH^+}},$$

and hence is equal to the ratio of the fundamental dissociation constants of the acids HA and BH^+, the latter being the conjugate acid of the added base B.

If the color of the base B differs from that of its conjugate acid BH^+, it is possible by light absorption experiments to estimate the value of either c_B or c_{BH^+}; since the stoichiometric composition of the solution is known, the concentrations of all the four species c_{HA}, c_{A^-}, c_B and c_{BH^+} can be thus estimated, and value of K in equation (36), apart from the activity coefficient factor, can be calculated. In this way the approximate ratio of the dissociation constant of the acid HA to that of BH^+ is obtained. The procedure is now repeated with an acid HA′ using the same base B, and from the two values of K the ratio of the dissociation constants of HA and HA′ can be found. This method can be carried through for a number of acids, new bases being used as the series is extended.[18]

On account of the low dielectric constants of aprotic solvents, considerable proportions of ion-pairs and triple ions are present, but spectrometric methods are unable to distinguish between these and single ions; the determinations of the amounts of *free ions*, which are required by the calculations, will thus be in error. The activity coefficient factor, neglected in the above treatment, will also be of appreciable magnitude, but this can be diminished if the base is a negatively charged ion B^-;

* In these expressions a_{H^+} stands for the activity of *protons*.

[18] LaMer and Downes, *J. Am. Chem. Soc.*, 53, 888 (1931); 55, 1840 (1933); *Chem. Revs.*, 13, 47 (1933).

the activity factor will then be $f_{HA}f_{B^-}/f_{A^-}f_{BH}$ which involves a neutral molecule and a singly charged ion in both numerator and denominator, and hence will not differ greatly from unity.

The Acidity Function.—A property of highly acid solutions, which is of some interest in connection with catalysis, is the **acidity function** H_0: it is defined with reference to an added electrically neutral base B, and measures the tendency of the solution to transfer a proton to the base;[19] thus

$$H_0 = -\log \frac{a_{H^+}f_B}{f_{BH^+}}. \tag{38}$$

There are reasons for believing that the fraction f_B/f_{BH^+} is practically constant for all bases of the same electrical type, and so the acidity function may be regarded as being independent of the nature of the base B. Combination of equation (38) with the usual definition of K_a, the conventional dissociation constant of the acid BH^+, gives

$$H_0 = pK_a + \log \frac{c_B}{c_{BH^+}}. \tag{39}$$

This equation provides a method for evaluating the acidity function of any acid solution; a small amount of a base B, for which pK_{BH^+} is known, is added to the given solution and the ratio c_B/c_{BH^+} is estimated colorimetrically. The acidity functions of a number of mixtures of perchloric, sulfuric and formic acids with water have been determined in this manner.

By reversing the procedure, equation (39) may be used, in conjunction with the known acidity functions of strongly acid media, to determine the dissociation constants of the conjugate acids BH^+ of a series of extremely weak bases. The relative amounts of B and BH^+ can be determined by suitable light-absorption measurements. The method has been applied to the study of a number of bases which are much too weak to exhibit basic properties in water. The results obtained in certain cases are given in Table LVIII; the figures in parentheses are the reference points for each solvent medium.[20] It is seen, therefore, that all the dissociation constants recorded are based on the pK_a value of 2.80 for the acid conjugate to aminoazobenzene, this being the normal result in aqueous solution. The results in Table LVIII, which are seen to be independent of the acidic medium used as the solvent, thus refer to dissociation constants of the various conjugate acids BH^+ in aqueous solutions. The dissociation exponents pK_b of the bases (B) themselves can be derived by subtracting the corresponding pK_a values, for BH^+, from pK_w, i.e., from 14. It is evident that many of the bases included in

[19] Hammett and Deyrup, *J. Am. Chem. Soc.*, **54**, 2721, 4239 (1932); Hammett and Paul, *ibid.*, **56**, 827 (1934); Hall *et al.*, *ibid.*, **62**, 2487, 2493 (1940).

[20] Hammett and Paul, *J. Am. Chem. Soc.*, **56**, 827 (1934); Hammett, *Chem. Revs.*, **16**, 67 (1935).

TABLE LVIII. DISSOCIATION CONSTANTS (pK_a) OF CONJUGATE ACIDS

Base	Solvent Medium			
	$HCl - H_2O$	$H_2SO_4 - H_2O$	$HClO_4 - H_2O$	HCO_2H
Aminoazobenzene	(2.80)	—	—	—
Benzeneazodiphenylamine	1.52	—	—	—
p-Nitroaniline	1.11	(1.11)	(1.11)	—
o-Nitroaniline	− 0.17	− 0.13	− 0.19	(− 0.17)
p-Chloronitroaniline	− 0.91	− 0.85	− 0.91	− 0.94
p-Nitrodiphenylamine	—	− 2.38	—	− 2.51
2 : 4-Dichloro-6-nitroaniline	—	− 3.22	− 3.18	− 3.31
p-Nitroazobenzene	—	− 3.35	− 3.35	− 3.29
2 : 4-Dinitroaniline	—	− 4.38	− 4.43	—
Benzalacetophenone	—	− 5.61	—	—
Anthraquinone	—	− 8.15	—	—
2 : 4 : 6-Trinitroaniline	—	− 9.29	—	—

Table LVIII are extremely weak; the dissociation constant of 2 : 4 : 6-trinitroaniline, for example, is as low as 5×10^{-24}.

Effect of Solvent on Dissociation Constants.—The dissociation equilibrium of an uncharged acid HA in the solvent S can be represented as

$$HA + S \rightleftharpoons SH^+ + A^-;$$

the dissociation process consequently involves the formation of a positive and a negative ion from two uncharged molecules. Since the electrostatic attraction between two oppositely charged particles decreases with increasing dielectric constant of the medium, it is to be expected that, other factors being more or less equal, an increase of the dielectric constant of the solvent will result in an increase in the dissociation constant of an electrically neutral acid. It has been found experimentally, in agreement with expectation, that the dissociation constant of an uncharged carboxylic acid decreases by a factor of about 10^5 or 10^6 on passing from water to ethyl alcohol as solvent. In the same way, the dissociation constant of an uncharged base is diminished by a factor of approximately 10^3 to 10^4 for the same change of solvent.

If the acid is a positive ion, e.g., NH_4^+, or the base is a negative ion, e.g., $CH_3CO_2^-$, the process of dissociation does not involve the separation of charges, viz.,

$$NH_4^+ + S = SH^+ + NH_3,$$

or

$$CH_3CO_2^- + HS = CH_3CO_2H + S^-.$$

The effect of changing the dielectric constant of the medium would thus be expected to be small, and in fact the dissociation constants do not differ very greatly in water and in ethyl alcohol. The value of pK_a for the ammonium ion acid, for example, is about 9.3 in water and 11.0 in methyl alcohol. It should be noted that the foregoing arguments do not take into consideration the different tendencies of the solvent molecule to take up a proton; the conclusions arrived at are consequently more likely to be applicable to a series of similar solvents, e.g., hydroxylic substances.

A quantitative approach to the problem of the influence of the medium on the dissociation constants of acids, which eliminates the proton accepting tendency of the solvent, involves a comparison of the dissociation constants of a series of acids with the value for a reference acid. Consider the acid HA in the solvent S; the dissociation constant is given by

$$K_a = \frac{a_{SH^+}a_{A^-}}{a_{HA}},$$

whereas that for the reference acid HA_0 in the same solvent is

$$K_0 = \frac{a_{SH^+}a_{A_0^-}}{a_{HA_0}},$$

so that, since SH^+ is the same in both cases,

$$\frac{K_a}{K_0} = \frac{a_{A^-}a_{HA_0}}{a_{A_0^-}a_{HA}} = K,$$

where K is the equilibrium constant of the reaction between the two acid-base systems, viz.,

$$HA + A_0^- \rightleftharpoons A^- + HA_0.$$

The standard free energy change of this process is then given by

$$\begin{aligned} -\Delta G^0 &= RT \ln K \\ &= 2.30\, RT \log K, \end{aligned}$$

where $\log K$, equal to $\log (K_a/K_0)$, is equivalent to $pK_0 - pK_a$.

This free energy change may be regarded as consisting of a non-electrostatic term ΔG_n and an electrostatic term $\Delta G_{el.}$ equivalent to the gain in electrostatic free energy resulting from the charging up of the ion A^- and the discharge of A_0^- in the medium of dielectric constant D. According to the Born equation (see p. 249), the electrostatic free energy increase per mole accompanying the charging of a *spherical* univalent ion is given by

$$\Delta G_{el.} = \frac{N\epsilon^2}{2Dr},$$

and so in the case under consideration, for charge and discharge of the ions A^- and A_0^-, respectively,

$$\Delta G_{el.} = \frac{N\epsilon^2}{2D}\left(\frac{1}{r_{A^-}} - \frac{1}{r_{A_0^-}}\right),$$

where r_{A^-} and $r_{A_0^-}$ are the radii of the corresponding spherical ions. It follows, therefore, that

$$-\ln K = \frac{\Delta G_n}{RT} + \frac{N\epsilon^2}{2DRT}\left(\frac{1}{r_{A^-}} - \frac{1}{r_{A_0^-}}\right).$$

If the effective radii of the two ions remain approximately constant in a series of solvents, it follows that

$$-\log K = \frac{\Delta G_n}{2.3RT} + \frac{a}{D},$$

where a is a constant. The plot of $-\log K$, that is, of $-\log(K_a/K_0)$, against $1/D$, i.e., the reciprocal of the dielectric constant of the solvent, should thus be a straight line; the intercept for $1/D$ equal to zero, i.e., for infinite dielectric constant, should give a measure of the dissociation constant of the acid HA free from electrostatic effects.

Measurements of dissociation constants of carboxylic acids, e.g., of substituted acetic and benzoic acids, using either acetic or benzoic acid as the reference substance HA, made in water, methyl and ethyl alcohols and ethylene glycol, are in good agreement with expectation.[21] The plot of the values of $-\log(K_a/K_0)$ against $1/D$ is very close to a straight line for each acid, provided D is greater than about 25. The slope of the line, however, varies with the nature of the acid, so that an acid which is stronger than another in one solvent may be weaker in a second solvent. The comparison of the dissociation constants of a series of acids in a given solvent may consequently be misleading, since a different order of strengths would be obtained in another solvent. It has been suggested, therefore, that when comparing the dissociation constants of acids the values employed should be those extrapolated to infinite dielectric constant; in this way the electrostatic effect, at least, of the solvent would be eliminated.

Attempts to verify the linear relationship between $-\log K$ and $1/D$ by means of a series of dioxane-water mixtures have brought to light considerable discrepancies.[22] The addition of dioxane to water results in a much greater decrease in the dissociation constant than would be expected from the change in the dielectric constant of the medium. Since the organic acids studied are more soluble in dioxane than in water, it is probable that molecules of the former solvent are preferentially oriented about the acid anion; the effective dielectric constant would then be less than in the bulk of the solution. It is thus possible to

[21] Wynne-Jones, *Proc. Roy. Soc.*, **140A**, 440 (1933); Kilpatrick *et al.*, *J. Am. Chem. Soc.*, **59**, 572 (1937); **62**, 3051 (1940); *J. Phys. Chem.*, **43**, 259 (1939); **45**, 454, 466, 472 (1941); Lynch and LaMer, *J. Am. Chem. Soc.*, **60**, 1252 (1938); see also, Hammett, *ibid.*, **59**, 96 (1937); *J. Chem. Phys.*, **4**, 613 (1936).

[22] Elliott and Kilpatrick, *J. Phys. Chem.*, **45**, 472 (1941); see also, Harned, *ibid.*, **43**, 275 (1939).

account for the unexpectedly low dissociation constants in the dioxane-water mixtures.

Dissociation Constant and Temperature.—The dissociation constants of uncharged acids do not vary greatly with temperature, as may be seen from the results recorded in Table LIX for a number of simple fatty

TABLE LIX. INFLUENCE OF TEMPERATURE ON DISSOCIATION CONSTANT

Acid	0°	10°	20°	30°	40°	50°	60°
Formic acid	1.638	1.728	1.765	1.768	1.716	1.650	1.551×10^{-4}
Acetic acid	1.657	1.729	1.753	1.750	1.703	1.633	1.542×10^{-5}
Propionic acid	1.274	1.326	1.338	1.326	1.280	1.229	1.160×10^{-5}
n-Butyric acid	1.563	1.576	1.542	1.484	1.395	1.302	1.199×10^{-5}

acids. A closer examination of the figures, however, reveals the fact that in each case the dissociation constant at first increases and then decreases as the temperature is raised; this type of behavior has been found to be quite general, and Harned and Embree [23] showed that the temperature variation of dissociation constants could be represented by the general equation

$$\log K_a = \log K_\theta - p(t - \theta)^2,$$

where K_a is the dissociation constant of the acid at the temperature t, K_θ is the maximum value, attained at the temperature θ, and p is a constant. It is an interesting fact that for a number of acids p has the same value, viz., 5×10^{-5}; this means that if $\log K_a - \log K_\theta$ for a number of acids is plotted against the corresponding value of $t - \theta$, the results all fall on a single parabolic curve. The actual temperature at which the maximum value of the dissociation constant is attained depends on the nature of the acid; for acetic acid it is 22.6°, but higher and lower values have been found for other acids. For some acids, e.g., chloroacetic acid and the first stage of phosphoric acid, the maximum dissociation constant would be reached only at temperatures below the freezing point of water.

An alternative relationship [24]

$$\log K = A + \frac{B}{T} - 20 \log T,$$

where A and B are constants, has been proposed by Pitzer to represent the dependence of dissociation constant on the absolute temperature T. This equation has a semi-theoretical basis, involving the empirical facts that the entropy change and the change in heat capacity accompanying the dissociation of a monobasic acid are approximately constant.

Some attempts have been made to account for the observed maximum in the dissociation constant. It was seen on page 334 that the division

[23] Harned and Embree, *J. Am. Chem. Soc.*, **56**, 1050, 2797 (1934); see also, Harned, *J. Franklin Inst.*, **225**, 623 (1938); Harned and Owen, *Chem. Revs.*, **25**, 131 (1939).

[24] Pitzer, *J. Am. Chem. Soc.*, **59**, 2365 (1937); see also, Walde, *J. Phys. Chem.*, **39**, 477 (1935); Wynne-Jones and Everett, *Trans. Faraday Soc.*, **35**, 1380 (1939).

of the free energy of dissociation of an acid into non-electrostatic and electrostatic terms leads to the expectation that log K_a is related to the reciprocal of the dielectric constant of the solvent. Since $1/D$ for water increases with increasing temperature, the value of log K_a should decrease; in addition to this effect there is the normal tendency for the dissociation constant, regarded as the equilibrium constant of an endothermic reaction, to increase with increasing temperature. The simultaneous operation of these two factors will lead to a maximum dissociation constant at a particular temperature.[25]

Amphiprotic Solvents: The Ionic Product.—In an amphiprotic solvent both an acid and its conjugate base can function independently; for example, if the acid is HA the conjugate base is A^-, and if the amphiprotic solvent is SH, the acidic and basic equilibria are

$$HA + SH \rightleftharpoons SH_2^+ + A^-$$

and

$$\underset{\text{acid}}{SH} + \underset{\text{base}}{A^-} \rightleftharpoons \underset{\text{acid}}{HA} + \underset{\text{base}}{S^-},$$

respectively. The ion SH_2^+ is the hydrogen ion, sometimes called the **lyonium ion,** in the given medium, and S^- is the anion, or **lyate ion,** of the solvent. The conventional dissociation constants of the acid HA and of its conjugate base A^- are then written as

$$K_a = \frac{a_{SH_2^+}a_{A^-}}{a_{HA}} \quad \text{and} \quad K_b = \frac{a_{HA}a_{S^-}}{a_{A^-}},$$

and the product is thus

$$K_aK_b = a_{SH_2^+}a_{S^-}, \tag{40}$$

which is evidently a specific property of the solvent. Since the solvent is amphiprotic and can itself function as either an acid or a base, the equilibrium

$$\underset{\text{acid}}{SH} + \underset{\text{base}}{SH} \rightleftharpoons \underset{\text{acid}}{SH_2^+} + \underset{\text{base}}{S^-}$$

must always exist, and if the activity of the undissociated molecules of solvent is taken as unity, it follows that the equilibrium constant K_S of this process is given by

$$K_S = a_{SH_2^+}a_{S^-}, \tag{41}$$

the constant K_S defined in this manner being called the **ionic product** or **ionization constant** of the solvent. It is sometimes referred to as the **autoprotolysis constant,** since it is a measure of the spontaneous tendency for the transfer of a proton from one molecule of solvent to another to

[25] Gurney, *J. Chem. Phys.*, **6**, 499 (1938); Baughan, *ibid.*, **7**, 951 (1939); see also, Magee, Ri and Eyring, *ibid.*, **9**, 419 (1941); LaMer and Brescia, *J. Am. Chem. Soc.*, **62**, 617 (1940).

take place. Comparison of equations (40) and (41) shows that

$$K_a K_b = K_S, \tag{42}$$

and so the dissociation constant of a base is inversely proportional to that of its conjugate acid, and *vice versa;* the proportionality constant is the ionic product of the solvent. This is the quantitative expression of the conclusion reached earlier that the anion of a strong acid, which is its conjugate base, will be weak, while the anion of a weak acid will be a moderately strong base, and similarly for the conjugate acids of strong and weak bases.

For certain purposes it is useful to define the dissociation constant of the solvent itself as an acid or base; by analogy with the conventional method of writing the dissociation constant of any acid or base, the activity of the solvent molecule taking part in the equilibrium is assumed to be unity. In the equilibrium

$$SH + SH \rightleftharpoons SH_2^+ + S^-$$

one molecule of SH may be regarded as functioning as the acid or base, while the other is the solvent molecule; the conventional dissociation constant of either acid or base is then

$$K_a = K_b = \frac{a_{SH_2^+} a_{S^-}}{a_{SH}}$$

$$= \frac{K_S}{a_{SH}}. \tag{43}$$

For most purposes a_{SH} may be replaced by the molecular concentration of solvent molecules in the pure solvent; with water, for example, the concentration of water molecules in moles per liter is 1000/18, i.e., 55.5, so that the dissociation constant of H_2O as an acid or base is equal to the ionic product of water divided by 55.5.

The Ionic Product of Water.—An ionic product of particular interest is that of water: the autoprotolytic equilibrium is

$$H_2O + H_2O \rightleftharpoons H_3O^+ + OH^-,$$

and hence the ionic product K_w may be defined by either of the following equivalent expressions, viz.,

$$K_w = a_{H_3O^+} a_{OH^-} \tag{44}$$

$$= m_{H_3O^+} m_{OH^-} \cdot \gamma_{H_3O^+} \gamma_{OH^-} \tag{44a}$$

$$= c_{H_3O^+} c_{OH^-} \cdot f_{H_3O^+} f_{OH^-}. \tag{44b}$$

By writing the ionic product in this manner it is tacitly assumed that the activity of the water is always unity; in solutions containing dissolved substances, however, the activity is diminished and K_w as defined above will not be constant but will increase. The activity of water in any solution may be taken as equal to p/p_0, where p is the vapor pressure of the solution and p_0 that of the pure water at the same temperature; in a solution containing 1 g.-ion per liter of solute, which is to be regarded as relatively concentrated, the activity of the water is about 0.98. The effect on K_w of the change in the activity of the water is thus not large in most cases.

The equilibrium between H_3O^+ and OH^- ions will exist in pure water and in all aqueous solutions: if the ionic strength of the medium is low, the ionic activity coefficients may be taken as unity, and hence the ionic product of water, now represented by k_w, is given by

$$k_w = c_{H_3O^+}c_{OH^-} \text{ (or } c_{H^+}c_{OH^-}). \tag{45}$$

As will be seen later, the value of k_w is approximately 10^{-14} at ordinary temperatures, and this figure will be adopted for the present.

In an exactly neutral solution, or in perfectly pure water, the concentrations of hydrogen (H_3O^+) and hydroxyl ions must be equal; hence under these conditions,

$$c_{H^+} = c_{OH^-} = 10^{-7} \text{ g.-ion per liter,}$$

the product being 10^{-14} as required. The question of the exact significance of the experimental value of pH will be considered in Chap. X, but for the present the pH of a solution may be defined, *approximately*, by

$$\text{pH} \approx -\log c_{H^+}.$$

It follows, therefore, that in pure water or in a neutral solution at ordinary temperatures, the pH is 7. If the quantity pOH is defined in an analogous approximate manner, as $-\log c_{OH^-}$, the value must also be 7 in water.

By taking logarithms of equation (45), it can be shown that for any dilute aqueous solution

$$\text{pH} + \text{pOH} = \text{p}k_w = 14 \tag{46}$$

at ordinary temperatures, where $\text{p}k_w$ is written for $-\log k_w$. If the hydrogen ion concentration of a solution exceeds 10^{-7} g.-ion per liter, the pH is less than 7 and the solution is said to be acid; the pOH is correspondingly greater than 7. Similarly, in an alkaline solution, the hydrogen ion concentration is less than 10^{-7} g.-ion per liter, but the hydroxyl ion concentration is greater than this value; the pH is greater than 7, but the pOH is smaller than this figure. The relationships between pH, pOH, c_{H^+} and c_{OH^-}, at about 25°, may be summarized in the manner represented below.

c_{H^+}	1	10^{-1}	10^{-2}	10^{-3}	10^{-4}	10^{-5}	10^{-6}	10^{-7}	10^{-8}	10^{-9}	10^{-10}	10^{-11}	10^{-12}	10^{-13}	10^{-14}
c_{OH^-}	10^{-14}	10^{-13}	10^{-12}	10^{-11}	10^{-10}	10^{-9}	10^{-8}	10^{-7}	10^{-6}	10^{-5}	10^{-4}	10^{-3}	10^{-2}	10^{-1}	1
pH	0	1	2	3	4	5	6	7	8	9	10	11	12	13	14
pOH	14	13	12	11	10	9	8	7	6	5	4	3	2	1	0
	←- - - - - - - - - - - - Acid						- - - - - - - - - - - -→	Neutral	←- - - - - - - - - - Alkaline						- - - - - - - - -→

It is seen that the range of pH from zero to 14 covers the range of hydrogen and hydroxyl ion concentrations from a N solution of strong acid on the one hand to a N solution of a strong base on the other hand. A solution of hydrogen ion concentration, or activity, exceeding 1 g.-ion per liter would have a negative pH, but values less than about − 1 in water are uncommon.

Determination of Ionic Product: Conductance Method.—Since it contains a certain proportion of hydrogen and hydroxyl ions, even perfectly pure water may be expected to have a definite conductance; the purest water hitherto reported was obtained by Kohlrausch and Heydweiller[26] after forty-eight distillations under reduced pressure. The specific conductance of this water was found to be 0.043×10^{-6} ohm^{-1} cm.$^{-1}$ at 18°, but it was believed that this still contained some impurity and the conductance of a 1 cm. cube of perfectly pure water was estimated to be 0.0384×10^{-6} ohm^{-1} cm.$^{-1}$ at 18°. The equivalent conductances of hydrogen and hydroxyl ions at the very small concentrations existing in pure water may be taken as equal to the accepted values at infinite dilution; these are 315.2 and 173.8 ohms^{-1} cm.^{2}, respectively, at 18°, and hence the total conductance of 1 equiv. of hydrogen and 1 equiv. of hydroxyl ions, at infinite dilution, should be 489.0 ohms^{-1} cm.^{2} It follows, therefore, that 1 cc. of water contains

$$\frac{0.0384 \times 10^{-6}}{489.0} = 0.78 \times 10^{-10} \text{ equiv. per cc.}$$

of hydrogen and hydroxyl ions; the concentrations in g.-ion per liter are thus 0.78×10^{-7}, and hence

$$\begin{aligned} k_w = c_{H^+}c_{OH^-} &= (0.78 \times 10^{-7})^2 \\ &= 0.61 \times 10^{-14}. \end{aligned}$$

Since the activity coefficients of the ions in pure water cannot differ appreciably from unity, this result is probably very close to K_w, the activity ionic product, at 18°. The results in Table LX give the observed specific conductances and the values of K_w at several temperatures from 0° to 50°.

TABLE LX. SPECIFIC CONDUCTANCE AND IONIC PRODUCT OF WATER

Temp.	0°	18°	25°	34°	50°	
κ	0.015	0.043	0.062	0.095	0.187	$\times 10^{-6}$ ohm^{-1} cm.$^{-1}$
K_w	0.12	0.61	1.04	2.05	5.66	$\times 10^{-14}$

Conductance measurements have been used to determine the ionic products of the amphiprotic solvents ethyl alcohol, formic acid and acetic acid.

[26] Kohlrausch and Heydweiller, *Z. physik. Chem.*, **14**, 317 (1894); Heydweiller, *Ann. Physik*, **28**, 503 (1909).

Electromotive Force Methods.—The earliest E.M.F. methods for evaluating the ionic product of water employed cells with liquid junction;[27] the E.M.F. of the cell

$$H_2(1 \text{ atm.}) \mid KOH(0.01\ N) \parallel HCl(0.01\ N) \mid H_2(1 \text{ atm.}),$$

from which it is supposed that the liquid junction potential has been completely eliminated, is given by

$$E = \frac{RT}{F} \ln \frac{a'_{H^+}}{a''_{H^+}}, \tag{47}$$

where a'_{H^+} and a''_{H^+} represent the hydrogen ion activities in the right-hand and left-hand solutions, i.e., in the 0.01 N hydrochloric acid and 0.01 N potassium hydroxide, respectively. If a''_{OH^-} is the hydroxyl ion activity in the latter solution, then

$$a''_{H^+}a''_{OH^-} = K_w,$$

and substitution of K_w/a''_{OH^-} for a''_{H^+} in equation (47) gives

$$E = \frac{RT}{F} \ln \frac{a'_{H^+}a''_{OH^-}}{K_w}. \tag{48}$$

By measuring each of the electrodes separately against a calomel reference electrode containing 0.1 N potassium chloride, and estimating the magnitude of the liquid junction potential in each case, the E.M.F. of the complete cell under consideration was found to be + 0.5874 volt at 25°. The ionic activity coefficients were assumed to be 0.903 in the 0.01 N solutions, so that a'_{H^+} and a''_{OH^-}, representing the activities of hydrogen and hydroxyl ions in 0.01 N hydrochloric acid and 0.01 N potassium hydroxide, respectively, were both taken to be equal to 0.0093; insertion of these figures in equation (48) gives K_w as 1.01×10^{-14} at 25°. This result is almost identical with some of the best later data, but the close agreement is probably partly fortuitous.

The most satisfactory method for determining the ionic product of water makes use of cells without liquid junction, similar to those employed for the evaluation of dissociation constants (cf. p. 314).[28] The E.M.F. of the cell

$$H_2(1 \text{ atm.}) \mid MOH(m_1)\ \ MCl(m_2)\ \ AgCl(s) \mid Ag,$$

where M is an alkali metal, e.g., lithium, sodium or potassium, is

$$E = E^0 - \frac{RT}{F} \ln a_{H^+}a_{Cl^-}. \tag{49}$$

[27] Lewis, Brighton and Sebastian, *J. Am. Chem. Soc.*, 39, 2245 (1917); Wynne-Jones, *Trans. Faraday Soc.*, 32, 1397 (1936).

[28] Roberts, *J. Am. Chem. Soc.*, 52, 3877 (1930); Harned and Hamer, *ibid.*, 55, 2194 (1933); for reviews, with full references, see Harned, *J. Franklin Inst.*, 225, 623 (1938); Harned and Owen, *Chem. Revs.*, 25, 31 (1939).

Since $a_{H^+}a_{OH^-}$ is equal to K_w, the activity of the water being assumed constant, it follows that

$$E = E^0 - \frac{RT}{F}\ln K_w - \frac{RT}{F}\ln\frac{a_{Cl^-}}{a_{OH^-}}$$
$$= E^0 - \frac{RT}{F}\ln K_w - \frac{RT}{F}\ln\frac{m_{Cl^-}}{m_{OH^-}} - \frac{RT}{F}\ln\frac{\gamma_{Cl^-}}{\gamma_{OH^-}},$$

and rearrangement gives

$$E - E^0 + \frac{RT}{F}\ln\frac{m_{Cl^-}}{m_{OH^-}} = -\frac{RT}{F}\ln K_w - \frac{RT}{F}\ln\frac{\gamma_{Cl^-}}{\gamma_{OH^-}},$$

$$\therefore\quad \frac{F(E - E^0)}{2.303RT} + \log\frac{m_{Cl^-}}{m_{OH^-}} = -\log K_w - \log\frac{\gamma_{Cl^-}}{\gamma_{OH^-}}. \qquad (50)$$

The activity coefficient fraction $\gamma_{Cl^-}/\gamma_{OH^-}$ is unity at infinite dilution, and so the value of the right-hand side of equation (50) becomes equal to $-\log K_w$ under these conditions. It follows, therefore, that if the left-hand side of this equation, for various concentrations of alkali hydroxide and chloride, is plotted against the ionic strength, the intercept for infinite dilution gives $-\log K_w$. The value of E^0 is known to be $+0.2224$ volt at 25°, and by making the assumption that MOH and MCl are completely dissociated, as will be the case in relatively dilute solutions, m_{OH^-} and m_{Cl^-} may be identified with m_1 and m_2, respectively. The results shown in Fig. 92 are for a series of cells containing cesium (I), potassium (II), sodium (III), barium (IV), and lithium (V) chlorides together with the corresponding hydroxides; the agreement between the values extrapolated to infinite dilution is very striking. The value of $-\log K_w$ is found to be 13.9965 at 25°, so that K_w is 1.008×10^{-14}.

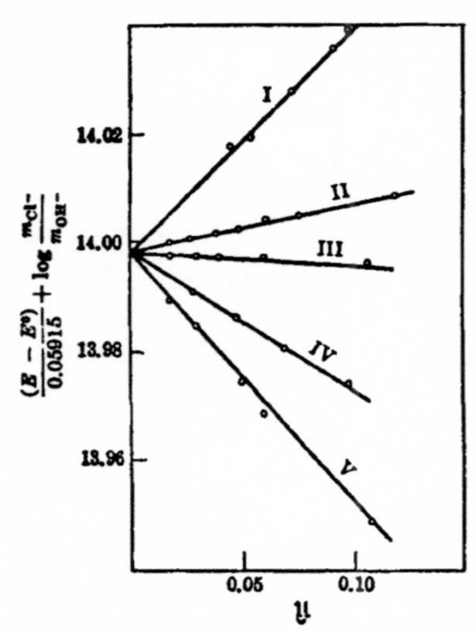

FIG. 92. Determination of the ionic product of water (Harned, *et al.*)

Another method of obtaining the ionic product of water is to combine the E.M.F. of the cell

$$H_2(1\text{ atm.}) \mid HCl(m_1') \; MCl(m_2') \; AgCl(s) \mid Ag$$

with that just considered; the E.M.F. of this cell is given by the same general equation,

$$E' = E^0 - \frac{RT}{F} \ln a'_{H^+}a'_{Cl} . \tag{51}$$

Combination of equations (49) and (51) gives

$$\begin{aligned} E - E' &= \frac{RT}{F} \ln \frac{a'_{H^+}a'_{Cl^-}}{a_{H^+}a_{Cl^-}} \\ &= \frac{RT}{F} \ln \frac{m'_{H^+}m'_{Cl^-}}{m_{H^+}m_{Cl^-}} + \frac{RT}{F} \ln \frac{\gamma'_{H^+}\gamma'_{Cl^-}}{\gamma_{H^+}\gamma_{Cl^-}}, \end{aligned} \tag{52}$$

where the primed quantities refer to the cell containing hydrochloric acid whereas those without primes refer to the alkali hydroxide cell.

If the ionic strengths in the two cells are kept equal, then provided the solutions are relatively dilute the activity coefficient .actor will be virtually unity, and the second term on the right-hand side of equation (52) is zero; hence under these conditions

$$E - E' = \frac{RT}{F} \ln \frac{m'_{H^+}m'_{Cl^-}}{m_{H^+}m_{Cl^-}},$$

and making use of the fact that K_w is equal to $m_{H^+}m_{OH^-} \cdot \gamma_{H^+}\gamma_{OH^-}$, this becomes

$$E - E' = \frac{RT}{F} \ln \frac{m'_{H^+}m'_{Cl^-}m_{OH^-}}{m_{Cl^-}} + \frac{RT}{F} \ln \gamma_{H^+}\gamma_{OH^-} - \frac{RT}{F} \ln K_w. \tag{53}$$

According to the extended Debye-Hückel equation, the value of $\log \gamma_{H^+}\gamma_{OH^-}$ may be represented by $-A\sqrt{\mu} + C\mu$, where A is a known constant for water at the experimental temperature; hence, equation (53), after rearrangement, becomes

$$\begin{aligned} E - E' - \frac{RT}{F} \ln \frac{m'_{H^+}m'_{Cl^-}m_{OH^-}}{m_{Cl^-}} + 2.303 \frac{RT}{F} A\sqrt{\mu} \\ = -\frac{RT}{F} \ln K_w + 2.303C\mu, \end{aligned}$$

$$\therefore \quad \frac{F(E - E')}{2.303RT} - \log \frac{m'_{H^+}m'_{Cl^-}m_{OH^-}}{m_{Cl^-}} + A\sqrt{\mu} = -\log K_w + \frac{F}{RT} C\mu. \tag{54}$$

The plot of the left-hand side of equation (54) against the ionic strength μ should be, at least approximately, a straight line whose intercept for μ equal to zero gives $-\log K_w$. As before, the values of m'_{H^+}, m'_{Cl^-}, m_{OH^-} and m_{Cl^-} are estimated on the assumption that the electrolytes HCl, MCl and MOH are completely dissociated.

A large number of measurements of cells of the types described, containing different halides, have been made by Harned and his collaborators over a series of temperatures from 0° to 50°; the excellent agreement between the results obtained in different cases may be taken as evidence of their accuracy. A selection of the values of the ionic product of water, derived from measurements of cells without liquid junction, is quoted in Table LXI; the data in the last column may be taken as the most reliable values of the ionic product of water.

TABLE LXI. IONIC PRODUCT FROM CELLS CONTAINING VARIOUS HALIDES

t	NaCl	KCl	LiBr	$BaCl_2$	Mean
0°	0.113	0.115	0.113	0.112	0.113×10^{-14}
10°	0.292	0.293	0.292	0.280	0.292
20°	0.681	0.681	0.681	0.681	0.681
25°	1.007	1.008	1.007	1.009	1.008
30°	1.470	1.471	1.467	1.466	1.468
40°	2.914	2.916	—	2.920	2.917
50°	5.482	5.476	—	5.465	5.474

Effect of Temperature on the Ionic Product of Water.—The values of the ionic product in Table LXI are seen to increase with increasing temperature; at 100°, the ionic product of water is about 50×10^{-14}. According to Harned and Hamer[29] the values between 0° and 35° may be expressed accurately by means of the equation

$$\log K_w = -\frac{4787.3}{T} - 7.1321 \log T - 0.010365T + 22.801.$$

From this expression it is possible, by making use of the reaction isochore, i.e.,

$$\frac{d \ln K}{dT} = \frac{\Delta H}{RT^2},$$

to derive the heat change accompanying the ionization of water; the results at 0°, 20° and 25° are as follows:

0°	20°	25°
14.51	13.69	13.48 kcal.

These values are strictly applicable at infinite dilution, i.e., in pure water.

It was seen on page 12, and it is obvious from the considerations discussed in the present chapter, that the neutralization of a strong acid by a strong base in aqueous solution is to be represented as

$$H_3O^+ + OH^- = H_2O + H_2O,$$

which is the same reaction as is involved in the ionization of water, except that it is in the opposite direction. The heats of neutralization obtained experimentally are 14.71, 13.69 and 13.41 kcal. at 0°, 20° and 25°, respectively; the agreement with the values derived from K_w is excellent.

Although the relationship given above for the dependence of K_w on temperature is only intended to hold over a limited temperature range,

[29] Harned and Hamer, *J. Am. Chem. Soc.*, 55, 4496 (1933); see also, Harned and Geary, *ibid.*, 59, 2032 (1937).

it shows nevertheless that the ionic product of water, like the dissociation constants of acids, to which reference has already been made, should pass through a maximum at a relatively high temperature and then decrease. Although the temperature at which the maximum value of K_w is to be expected lies beyond the range of the recent accurate work on the ionic product of water, definite evidence for the existence of this maximum had been obtained several years ago by Noyes (1910). The temperature at which the maximum ionic product was observed is about 220°, the value of K_w being then about 460×10^{-14}.

The Ionization of Water in Halide Solutions.—The cells employed for the determination of the ionic product of water have also been used to study the extent of dissociation of water in halide solutions.[30] Since K_w is equal to $a_{H^+}a_{OH^-}$ and $a_{H^+}a_{OH^-}/\gamma_{H^+}\gamma_{OH^-}$ is equal to $m_{H^+}m_{OH^-}$, equation (53) becomes, after rearrangement,

$$-\frac{RT}{F}\ln m_{H^+}m_{OH^-} = E - E' - \frac{RT}{F}\ln\frac{m'_{H^+}m'_{Cl^-}m_{OH^-}}{m_{Cl^-}},$$

and so the **molal ionization product** $m_{H^+}m_{OH^-}$ in the halide solution present in the cells may be evaluated directly from the E.M.F.'s E and E', and the molalities of the electrolytes. The amounts of hydrogen and hydroxyl ions are equal in the pure halide solution; consequently, the square-root of $m_{H^+}m_{OH^-}$ gives the concentration of these ions, in g.-ions

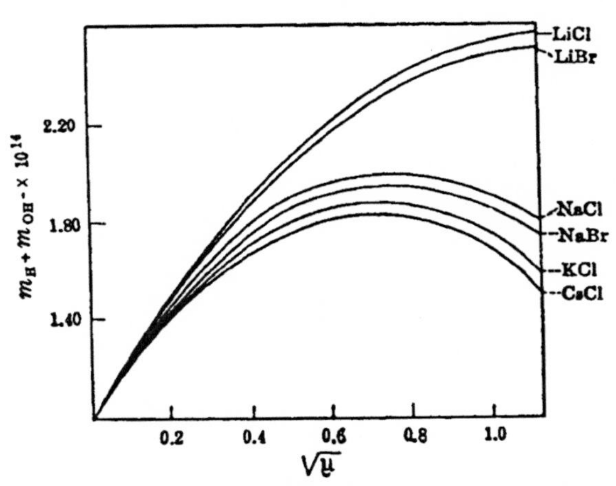

FIG. 93. Variation of molal ionization product of water (Harned, *et al.*)

per 1000 g. of water, produced by the ionization of the water in the halide solution. The results for a number of alkali halides at 25° are shown in Fig. 93; it will be seen that, in general, the extent of the ionization of water increases at first, then reaches a maximum and decreases with

[30] For reviews with full references, see Harned, *J. Franklin Inst.*, 225, 623 (1938); Harned and Owen, *Chem. Revs.*, 25, 31 (1939).

increasing ionic strength of the medium. With lithium salts the maximum is attained at a higher concentration than is shown in the diagram. The explanation of this variation is not difficult to find: the quantity $a_{H^+}a_{OH^-}/a_{H_2O}$, i.e., $m_{H^+}m_{OH^-} \times \gamma_{H^+}\gamma_{OH^-}/a_{H_2O}$, which includes the activity of the water, must remain constant in all aqueous solutions, and since the activity coefficients always decrease and then increase as the ionic strength of the medium is increased (cf. Fig. 46), while a_{H_2O}, i.e., p/p_0,* decreases steadily, it follows that the variation of $m_{H^+}m_{OH^-}$ must be of the form shown in Fig. 93. In spite of the dependence of $m_{H^+}m_{OH^-}$ on the ionic strength of the solution, it is still satisfactory, for purposes of approximate computation, to take the ionic concentration product of water (k_w) to be about 10^{-14} at ordinary temperatures, provided the concentration of electrolyte in the solution is not too great.

PROBLEMS

1. Show that according to equation (10) the plot of Λc against $1/\Lambda$ should be a straight line; test the accuracy of this (approximate) result by means of the data for acetic acid on page 165 and for α-crotonic acid in Problem 7 of Chap. III.

2. Utilize the data referred to in Problem 1 to calculate the dissociation functions of acetic and α-crotonic acids at several concentrations by means of equation (10); compare the results with the thermodynamic dissociation constants obtained in Chap. V.

3. In their measurements of the cell

$$H_2(1 \text{ atm.}) \mid HP(m_1) \; NaP(m_2) \; NaCl(m_3) \; AgCl(s) \mid Ag,$$

where HP represents propionic acid, Harned and Ehlers [*J. Am. Chem. Soc.*, 55, 2379 (1933)] made m_1, m_2 and m_3 equal and obtained the following E.M.F.'s at 25°:

m	E	m	E
4.899×10^{-3}	0.64758	18.669×10^{-3}	0.61311
8.716	0.63275	25.546	0.60522
12.812	0.62286	31.793	0.59958

Evaluate the dissociation constant of propionic acid.

4. Walpole [*J. Chem. Soc.*, 105, 2501 (1914)] measured the pH's of a series of mixtures of x cc. of 0.2 N acetic acid with $10 - x$ cc. of 0.2 N sodium acetate, and obtained the following results:

x	8.0	7.0	6.0	5.0	4.0	3.0	2.0 cc.
pH	4.05	4.27	4.45	4.63	4.80	4.99	5.23

Calculate the dissociation constant of acetic acid by the use of equation (27), the activity coefficients of the acetate ions being obtained by means of the simple Debye-Hückel equation. Derive the dissociation constant by means of the graphical method described on page 324.

* Since pure water, vapor pressure p_0, is taken as the standard state, the activity of water in any solution of aqueous vapor pressure p will be p/p_0.

5. Bennett, Brooks and Glasstone [*J. Chem. Soc.*, 1821 (1935)] obtained the following results in the titration of *o*-fluorophenol in 30 per cent alcohol at 25°; when x cc. of 0.01 N sodium hydroxide was added to 50 cc. of a 0.01 N solution of the phenol the pH's were:

x	10	15	20	25	30	40 cc.
pH	8.73	9.01	9.20	9.37	9.56	10.00

Calculate the dissociation constant of *o*-fluorophenol, using the expression $\log f = -0.683\sqrt{\mu} + 2.0\mu$ to obtain the activity coefficient of the anion. The activity coefficient of the undissociated acid may be taken as unity.

6. The following pH values were obtained by German and Vogel [*J. Am. Chem. Soc.*, **58**, 1546 (1936)] in the titration of 100 cc. of 0.005 molar succinic acid with x cc. of 0.01 N sodium hydroxide at 25°:

x	pH	x	pH
20.0	4.00	60.0	5.11
30.0	4.28	70.0	5.39
40.0	4.56	80.0	5.68
50.0	4.84	90.0	6.03

Determine the two dissociation constants of succinic acid by the graphical method described on page 326.

7. The E.M.F. of the cell

$$H_2(1\ \text{atm.})\ |\ NaOH(m)\ \ NaCl(m)\ \ AgCl(s)\ |\ Ag,$$

with the sodium hydroxide and chloride at equal molalities, was found by Roberts [*J. Am. Chem. Soc.*, **52**, 3877 (1930)] to have a constant value of 1.0508 volt at 25° when the solutions were dilute. Calculate the ionic product of water from this result.

8. The following E.M.F.'s were obtained at 25° by Harned and Copson [*J. Am. Chem. Soc.*, **55**, 2206 (1933)] for the cells

$$(A)\quad H_2(1\ \text{atm.})\ |\ LiOH\ (0.01)\ \ LiCl(m)\ \ AgCl(s)\ |\ Ag$$

$$(B)\quad H_2(1\ \text{atm.})\ |\ HCl\ \ (0.01)\ \ LiCl(m)\ \ AgCl(s)\ |\ Ag.$$

m	E_A	E_B
0.01	1.04979	0.4[illegible]779
0.02	1.03175	0.43855
0.05	1.00755	0.42282
0.10	0.98883	0.40917
0.20	0.96[illegible]57	0.39453
0.50	0.94277	0.37235
1.00	0.91992	0.35191
2.00	0.89203	0.32352
3.00	0.87154	0.29959
4.00	0.85407	0.27754

Utilize the method given on page 343 to derive the ionic product of water from these data. Plot the variation of the molal ionization product with the ionic strength of the solution.

CHAPTER X

THE DETERMINATION OF HYDROGEN IONS

Standardization of pH Values.—The **hydrogen ion exponent,** pH, was originally defined by Sørensen (1909) as the "negative logarithm of the hydrogen ion concentration," i.e., as $-\log c_{H^+}$; most determinations of pH are, however, based ultimately on E.M.F. measurements with hydrogen electrodes, and the values obtained are, theoretically, an indication of the hydrogen ion activity rather than of the concentration. For this reason, it has become the practice in recent years to regard the pH as defined by

$$\text{pH} \equiv -\log a_{H^+}, \tag{1}$$

where H^+ stands for the hydrogen ion, i.e., lyonium ion, in the particular solvent. This definition, however, involves the activity of a single ionic species and so can have no strict thermodynamic significance; it follows, therefore, that there is no method available for the precise determination of pH defined in this manner. It is desirable, nevertheless, to establish, if possible, an arbitrary pH scale that shall be reasonably consistent with certain thermodynamic quantities, such as dissociation constants, which are known exactly, within the limits of experimental error. The values obtained with the aid of this scale will not, of course, be actual pH's, since such quantities cannot be determined, but they will at least be data which if inserted in equations involving pH, i.e., $-\log a_{H^+}$, will give results consistent with those determined by strict thermodynamic methods not involving individual ion activities.

The E.M.F. of a cell free from liquid junction potential, consisting of a hydrogen electrode and a reference electrode, should be given by

$$E = E_{\text{ref.}} - \frac{RT}{F} \ln a_{H^+},$$

or, introducing the definition of pH according to equation (1),

$$E = E_{\text{ref.}} + 2.303 \frac{RT}{F} \text{pH}$$

where $E_{\text{ref.}}$ is the potential of the reference electrode on the hydrogen scale. It follows, therefore, that

$$\text{pH} = \frac{F(E - E_{\text{ref.}})}{2.303RT}. \tag{2}$$

If the usual value for $E_{ref.}$ of the reference electrode is employed in this equation to derive pH's, the results are found to be inconsistent with other determinations that are thermodynamically exact. A possible way out of this difficulty is to find a value for $E_{ref.}$ such that its use in equation (2) gives pH values which are consistent with known thermodynamic dissociation constants. For this purpose use is made of equation (29) of Chap. IX, viz.,

$$\text{pH} = \text{p}K_a + \log \frac{B}{a - B} - A\sqrt{\mu} + C\mu, \tag{3}$$

which combined with equation (2) gives

$$\frac{F(E - E_{ref.})}{2.303RT} = \text{p}K_a + \log \frac{B}{a - B} - A\sqrt{\mu} + C\mu,$$

$$\therefore\ E - \frac{2.303RT}{F}\left(\text{p}K_a + \log \frac{B}{a - B} - A\sqrt{\mu}\right) = E_{ref.} + \frac{2.303RT}{F} C\mu. \tag{4}$$

A series of mixtures, at different total concentrations, of an acid, whose dissociation constant is known exactly, e.g., from observations on cells without liquid junction, and its salt are made up, thus giving a series of values for B and $a - B$. The E.M.F.'s of the cells consisting of a hydrogen electrode in this solution combined with a reference electrode are measured; a saturated solution of potassium chloride is used as a salt bridge between the experimental solution and the one contained in the reference electrode. The E values obtained in this manner, together with B and $a - B$, calculated from the known composition of the acid-salt mixture (cf. p. 324), and the pK_a of the acid, permit the left-hand side of equation (4) to be evaluated for a number of solutions of different ionic strengths. The results plotted against the ionic strength should fall on a straight line, the intercept for zero ionic strength giving the required quantity $E_{ref.}$. In order for this result to have any significance it should be approximately constant for a number of solutions covering a range of pH values and involving different acids; this has in fact been found to be the case in the pH range of 4 to 9, and hence a pH scale consistent with the known pK_a values for a number of acids is possible.[1]

The conclusions reached from this work may be stated in terms of the potentials of the reference electrodes; for example, the value of $E_{ref.}$ of the 0.1 N KCl calomel electrode for the purpose of determining pH's by means of equation (2) is 0.3358 volt * at 25°. In view of possible variations in the salt bridge from one set of experiments to another, it is preferable to utilize these potentials to determine the pH values of a number of reproducible buffer solutions (cf. p. 410) which can form a

[1] Hitchcock and Taylor, *J. Am. Chem. Soc.*, **59**, 1812 (1937); **60**, 2710 (1938); MacInnes, Belcher and Shedlovsky, *ibid.*, **60**, 1094 (1938); see also, Cohn, Heyroth and Menkin, *ibid.*, **50**, 696 (1928).

* This may be compared with 0.3338 volt, given on page 232, employed in earlier pH work.

scale of reference. The results obtained in this manner are recorded in Table LXII for temperatures of 25° and 38°; they are probably correct

TABLE LXII. STANDARDIZATION OF pH VALUES OF REFERENCE SOLUTIONS

Solution	25°	38°
0.1 N HCl	1.085	1.082
0.1 M Potassium tetroxalate	1.480	1.495
0.01 N HCl and 0.09 N KCl	2.075	2.075
0.05 M Potassium acid phthalate	4.005	4.020
0.1 N Acetic acid and 0.1 N Sodium acetate	4.640	4.650
0.025 M KH_2PO_4 and 0.025 M Na_2HPO_4	6.855	6.835
0.05 M $Na_2B_4O_7 \cdot 10H_2O$	9.180	9.070

to ± 0.01 pH unit. With this series of reference solutions it is possible to standardize a convenient combination of hydrogen and reference electrodes; the required pH of any solution may thus be determined. The pH's obtained in this way are such that if inserted in equation (3), they will give a pK_a value which should not differ greatly from one obtained by a completely thermodynamic procedure. These pH values can then be used in connection with equations (29) and (34) of Chap. IX to give reasonably accurate dissociation constants.

Reversible Hydrogen Electrodes.—In previous references to the hydrogen electrode it has been stated briefly that it consists of a platinum electrode in contact with hydrogen gas; the details of the construction of this electrode will be considered here. In addition to the hydrogen gas electrode, a number of other electrodes are known which behave reversibly with respect to hydrogen ions. Any one of these can be used for the determination of pH values, although the electrode involving hydrogen gas at 1 atm. pressure is the standard to which others are referred.

I. The Hydrogen Gas Electrode.—The hydrogen gas electrode consists of a small platinum sheet or wire coated with finely divided platinum black by electrolysis of a solution of chloroplatinic acid containing a small proportion of lead acetate (cf. p. 35). The platinum foil or wire, attached to a suitable connecting wire, is inserted in the experimental solution through which a stream of hydrogen is passed at atmospheric pressure. The position of the electrode in the solution is arranged so that it is partly in the solution and partly in the atmosphere of hydrogen gas. A number of forms of electrode vessel, suitable for a variety of uses, have been employed for the purpose of setting up hydrogen gas electrodes; some of these are depicted in Fig. 94. A simple and convenient type of hydrogen electrode is that, usually associated with the name of Hildebrand,[2] shown in Fig. 95; a rectangular sheet of platinum,

[2] Hildebrand, *J. Am. Chem. Soc.*, **35**, 847 (1913); for further details concerning hydrogen electrodes, see Clark, "The Determination of Hydrogen Ions," 1928; Britton, "Hydrogen Ions," 1932; Glasstone, "The Electrochemistry of Solutions," 1937, p. 375. See also, Hamer and Acree, *J. Res. Nat. Bur. Standards*, **23**, 647 (1939).

of about 1 to 3 sq. cm. exposed area, which is subsequently platinized, is welded to a short length of platinum wire sealed into a glass tube containing mercury. This tube is sealed into another, closed at the top, but widening out into a bell shape in the region surrounding the platinum

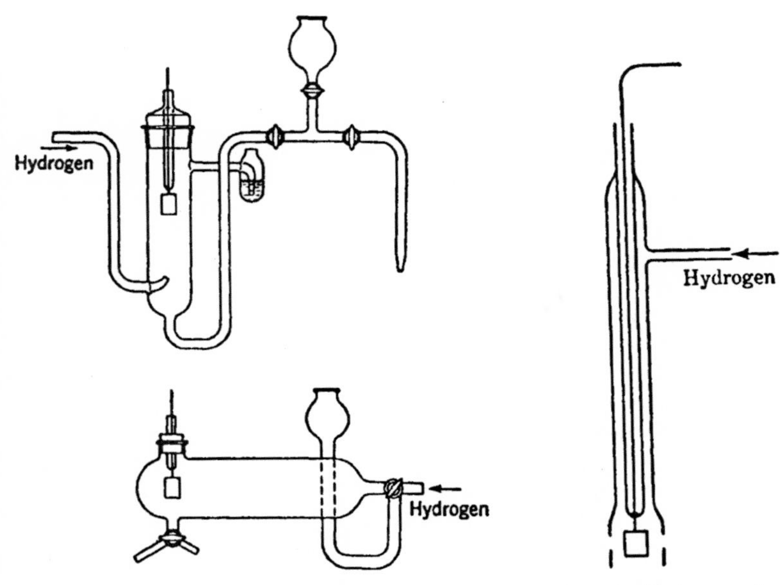

FIG. 94. Forms of hydrogen electrode

FIG. 95. Hydrogen electrode: Hildebrand type

sheet; a side connection is provided for the inlet of hydrogen gas. A number of holes, or slits, are made in the bell-shaped portion of the tube at a level midway up the platinum, so that when the electrode is inserted in a solution and hydrogen passed in through the side-tube the platinum sheet is half immersed in liquid and half surrounded by gas. This arrangement permits the rapid attainment of equilibrium between the electrode material, the hydrogen gas and the solution. The time taken to reach this state of equilibrium depends, among other factors, on the nature of the solution, the thickness of the deposit, and on the previous history of the electrode. As a general rule, an electrode that is functioning in a satisfactory manner will give a steady potential within five or ten minutes of commencing the passage of hydrogen. The use of a platinum sheet in the Hildebrand electrode is not essential, and many workers prefer to use a simple wire of 2 or 3 cm. in length, straight or coiled, for such an electrode attains equilibrium rapidly, although it has a somewhat higher resistance than the form represented in Fig. 95. The hydrogen gas should be purified by bubbling it through alkaline permanganate and alkaline pyrogallol solutions to remove oxygen and other impurities which may influence the functioning of the hydrogen electrode.

Whatever form of electrode vessel is employed, the fundamental principle of the operation is always the same. The hydrogen gas is adsorbed by the finely divided platinum and this permits the rapid establishment of equilibrium between molecular hydrogen on the one hand, and hydrogen ions in solution and electrons, on the other hand, thus

$$\tfrac{1}{2}H_2(g) \rightleftharpoons \tfrac{1}{2}H_2(Pt) + H_2O \rightleftharpoons H_3O^+ + \epsilon.$$

This equilibrium can be attained rapidly from either direction, and so the electrode behaves as one that is reversible with respect to hydrogen ions.

The hydrogen gas electrode behaves erratically in the presence of arsenic, mercury and sulfur compounds, which are known to be catalytic poisons; they probably function by being preferentially adsorbed on the platinum, thus preventing the establishment of equilibrium. An electrode whose operation is affected in this manner is said to be "poisoned"; if it cannot be regenerated by heating with concentrated hydrochloric acid, the platinum black should be removed by means of aqua regia and the electrode should be replatinized. The hydrogen gas electrode cannot be employed in solutions containing oxidizing agents, such as nitrates, chlorates, permanganates and ferric salts, or other substances capable of reduction, e.g., unsaturated and other reducible organic compounds, alkaloids, etc. The electrode does not function in a satisfactory manner in solutions containing noble metals, e.g., gold, silver and mercury, since they tend to be replaced by hydrogen (cf. p. 253), neither can it be used in the presence of lead, cadmium and thallous salts. In spite of these limitations the hydrogen gas electrode has been extensively employed for precise measurements in cells with or without liquid junction, such as those mentioned in Chaps. VI and IX. The electrode has also been found to give fairly satisfactory results in non-aqueous solvents such as alcohols, acetone, benzene and liquid ammonia.

Since the standard state of hydrogen is the gas at 760 mm. pressure, it would be desirable to employ the gas at this pressure; even if the hydrogen were actually passed in at this pressure, which would not be easy to arrange, the partial pressure in the electrode vessel would be somewhat less because of the vapor pressure of the water. A correction for the pressure difference should therefore be made in accordance with equation (50) of Chap. VI; the correction is, however, small as is shown by the values calculated from this equation and recorded in Table LXIII. The results are given for a series of temperatures and for three gas pressures; the corrections are those which must be added, or subtracted if marked by a negative sign, to give the potential of the electrode with hydrogen gas at a partial pressure of 760 mm.

TABLE LXIII. PRESSURE CORRECTIONS FOR HYDROGEN ELECTRODE IN MILLIVOLTS

Temperature	15°	20°	25°	30°
Vapor Pressure	12.8	15.5	23.7	31.7 mm.
Gas Pressure				
740 mm.	0.54	0.61	0.75	0.92
760 mm.	0.20	0.26	0.38	0.56
780 mm.	− 0.13	− 0.08	0.04	0.20

II. The Oxygen Electrode.—The potential of an oxygen electrode, expressed in the form of equation (9b) of Chap. VII, is

$$E = E^0_{O_2, OH^-} + \frac{RT}{F} \ln a_{OH^-}, \tag{5}$$

and since a_{OH^-} may be replaced by K_w/a_{H^+}, where K_w is the ionic product of water, it follows that

$$E = E^0_{O_2, H^+} - \frac{RT}{F} \ln a_{H^+}. \tag{6}$$

The oxygen electrode should thus, in theory, function as if it were reversible with respect to hydrogen ions.

Attempts have been made to set up oxygen electrodes in a manner similar to that adopted for the hydrogen gas electrode, as described above; the results, however, have been found to be unreliable. The potential rises rapidly at first but this is followed by a drift lasting several days. The value reached finally is lower than that expected from the calculated standard potential of oxygen (cf. p. 243) and the known pH of the solution. The use of either iridium or smooth platinum instead of platinized platinum does not bring the potential appreciably nearer the theoretical reversible value, although the use of platinized gold has been recommended. It is evident that the oxygen gas electrode in its usual form does not function reversibly; the difference of potential when the equilibrium

$$\tfrac{1}{2}O_2 + H_2O + 2\epsilon \rightleftharpoons 2OH^-$$

is attained is less than would be expected, and this means that the direct reaction, as represented by this equation, is retarded in some manner not yet clearly understood.

In spite of its irreversibility, the oxygen electrode was at one time used for the approximate comparison of pH values in solutions containing oxidizing substances, in which the hydrogen gas electrode would not function satisfactorily. In order for the results to have any significance the particular oxygen electrode employed was standardized by means of a hydrogen electrode in a solution in which the latter could be employed. The oxygen electrode, with air as the source of oxygen, has also been used for potentiometric titration purposes; in work of this kind the actual potential or pH is immaterial, for all that is required is an indication of the point at which the potential undergoes rapid change.[3] In recent years the difficulty of measuring pH's in solutions containing reducible substances has been largely overcome by the wide adoption of the glass electrode which is described below.

III. The Quinhydrone Electrode.—It was seen in Chap. VIII that a mixture of quinone (Q) and hydroquinone (H_2Q) in the presence of hydrogen ions constitutes a reversible oxidation-reduction system, and the potential of such a system is given by equation (4), page 270, as

$$E = E^0 - \frac{RT}{2F} \ln \frac{a_Q}{a_{H_2Q}} - \frac{RT}{F} \ln a_{H^+}. \tag{7}$$

It is seen, therefore, that the potential of the quinone-hydroquinone system depends on the hydrogen ion activity of the system. For the purpose of pH determination the solution is saturated with **quinhydrone,** which is an equimolecular compound of quinone and hydroquinone; in this manner the ratio of the concentrations c_Q to c_{H_2Q} is maintained at unity, and if the ionic strength of the solution is relatively low the ratio of the activities, i.e., a_Q/a_{H_2Q}, may be regarded as constant. The first two terms on the right-hand side of equation (7) may thus be combined to give

$$E = E_Q^0 - \frac{RT}{F} \ln a_{H^+} \tag{8}$$

$$= E_Q^0 - 2.303 \frac{RT}{F} \log a_{H^+} \tag{8a}$$

$$= E_Q^0 + 2.303 \frac{RT}{F} \text{pH}. \tag{8b}$$

By using the method of standardization described at the beginning of this chapter, the value of E_Q^0 is found at t^0 to be

$$E_Q^0 = -0.6994 + 0.00074(t - 25).$$

This method of expressing the results is of little value for practical purposes; the particular reference electrode and salt bridge employed should be standardized by means of equation (2) using one of the reference solutions in Table LXII. If the reference electrode is a calomel electrode with 0.1 N potassium chloride, and a bridge of a saturated solution of this electrolyte is employed, it has been found possible to express the experimental data by means of the equation

$$E_{Q(\text{cal.})}^0 = -0.3636 + 0.0070(t - 25).$$

This is the potential of the quinhydrone electrode against the Hg, Hg_2Cl_2,

[3] Furman, *J. Am. Chem. Soc.*, **44**, 12 (1922); *Trans. Electrochem. Soc.*, **43**, 79 (1923); Britton, *J. Chem. Soc.*, **127**, 1896, 2148 (1925); Richards, *J. Phys. Chem.*, **32**, 990 (1928).

KCl(0.1 N) electrode when the former contains a solution of hydrogen ions of unit activity, i.e., its pH is zero.[4]

The quinhydrone electrode is easily set up by adding a small quantity of the sparingly soluble quinhydrone, which can be obtained commercially, to the experimental solution so as to saturate it; this solution is shaken gently and then an indicating electrode of platinum or gold is inserted. The surface of the electrode metal should be clean and free from grease; it is first treated with hot chromic acid mixture, washed well with distilled water, and finally dried by heating in an alcohol flame. Gentle agitation of the solution by means of a stream of nitrogen gas is sometimes advantageous. The electrode gives accurate results in solutions of pH less than 8; in more alkaline solutions errors arise, first, because of oxidation of the hydroquinone by oxygen of the air, and second, on account of the ionization of the hydroquinone as an acid (cf. p. 291). Oxidizing or reducing agents capable of reacting rapidly with quinone or hydroquinone are liable to disturb the normal ratio of these substances, and so will affect the potential. The quinhydrone electrode can be used in the presence of the ions of many metals which have a deleterious effect on the hydrogen gas electrode, but ammonium salts exert a harmful influence. The potential of the quinhydrone electrode is affected to some extent by all salts and even by non-electrolytes; this "salt effect" is to be attributed to the varying influence of the salts, etc., on the activities of the quinone and hydroquinone; although the ratio c_Q/c_{H_2Q} remains constant, therefore, this is not necessarily true for a_Q/a_{H_2Q} upon which the electrode potential actually depends. The "salt error" is proportional to the concentration of electrolyte, within reasonable limits; its value, which may be positive or negative, according to the nature of the "salt," is about + 0.02 to − 0.05 pH unit per equiv. per liter of electrolyte. Provided the solution is more dilute than about 0.1 N, the "salt error" is therefore negligible for most purposes. The quinhydrone electrode has an appreciable "protein error," and so cannot be employed to give reliable pH values in solutions containing proteins or certain of their degradation products.[5]

The quinhydrone electrode has been adapted for pH measurements in non-aqueous media, such as alcohols, acetone, formic acid, benzene and liquid ammonia. For the determination of hydrogen ion activities in solutions in pure acetic acid a form of quinhydrone electrode involving tetrachloroquinone (chloranil) and its hydroquinone has been used.[6]

[4] Harned and Wright, *J. Am. Chem. Soc.*, **55**, 4849 (1933); Hovorka and Dearing, *ibid.*, **57**, 446 (1935).

[5] For general references, see Glasstone, "The Electrochemistry of Solutions," 1937, p. 378.

[6] Conant *et al.*, *J. Am. Chem. Soc.*, **47**, 1959 (1925); **49**, 3047 (1927); Heston and Hall, *ibid.*, **56**, 1462 (1934).

IV. The Antimony Electrode.—The so-called "antimony electrode" is really an electrode consisting of antimony and its trioxide, the reaction being

$$2Sb(s) + 3H_2O = Sb_2O_3(s) + 6H^+ + 6\epsilon,$$

so that the potential is given by

$$E = E^0_{Sb,\,Sb_2O_3,\,H^+} - \frac{RT}{F} \ln a_{H^+}, \qquad (9)$$

the activities of the solid antimony and antimony trioxide, and of the water, being taken as unity. The potential of the Sb, Sb_2O_3 electrode should thus depend on the hydrogen ion activity of the solution in which it is placed. The electrode is generally prepared by casting a stick of antimony in the presence of air; in this way it becomes sufficiently oxidized for the further addition of oxide to be unnecessary. A wire is attached to one end of the rod of antimony obtained in this manner, while the other is inserted in the experimental solution; its potential is then measured against a convenient reference electrode. As the potentials differ from one electrode to another, it is necessary that each antimony electrode should be standardized by means of one of the solutions in Table LXII. The antimony electrode behaves, at least approximately, according to equation (9) over the range of pH from 2 to 7, but in more acid or more alkaline solutions deviations occur; these discrepancies are probably connected with the solubility of the antimony oxide in such solutions. Since no special technique is required for setting up or measuring the potential of the antimony electrode, and it is not easily poisoned, it has advantages over other forms of hydrogen electrode. It is, therefore, very convenient where approximate results are adequate, but it is not recommended for precision work.[7]

V. The Glass Electrode.—One of the most important advances of recent years in connection with the determination of pH's is the development which has taken place in the use of the glass electrode. It has long been known that a potential difference is set up at the interface between glass and a solution in contact with it which is dependent on the pH of the latter;[8] this dependence has been found to correspond to the familiar equation for a reversible hydrogen electrode, viz.,

$$E = E^0_G - \frac{RT}{F} \ln a_{H^+}, \qquad (10)$$

[7] Kolthoff and Hartong, *Rec. trav. chim.*, **44**, 113 (1925); Roberts and Fenwick, *J. Am. Chem. Soc.*, **50**, 2125 (1928); Parks and Beard, *ibid.*, **54**, 856 (1932); Perley, *Ind. Eng. Chem.* (Anal. Ed.), **11**, 316 (1939); Hovorka and Chapman, *J. Am. Chem. Soc.*, **63**, 955 (1941).

[8] For references to experimental methods, see Glasstone, *Ann. Rep. Chem. Soc.*, **30**, 283 (1933); Muller and Dürichen, *Z. Elektrochem.*, **41**, 559 (1935); **42**, 31, 730 (1936); Schwabe, *ibid.*, **41**, 681 (1935). For complete review, see Dole, "The Glass Electrode," 1941.

where E^0_G is the "standard potential" for the particular glass employed, i.e., the potential when in contact with a solution of hydrogen ions at unit activity. It is evident, therefore, that measurements of the potential of the so-called "glass electrode" can be utilized for the determination of pH values.

In its simplest form the glass electrode consists of a tube terminating in a thin-walled bulb, as shown at *A*, in Fig. 96; the glass most suitable for the purpose (Corning 015) contains about 72 per cent SiO_2, 22 per cent Na_2O and 6 per cent CaO; it has a relatively low melting point and a high electrical conductivity. The bulb contains a solution of constant hydrogen ion concentration and an electrode of definite potential; a silver chloride electrode in 0.1 N hydrochloric acid or a platinum wire inserted in a buffer solution, e.g., 0.05 molar potassium acid phthalate, saturated with quinhydrone, is generally used. The bulb is inserted in the experimental solution (*B*) so that the glass electrode consists of the system

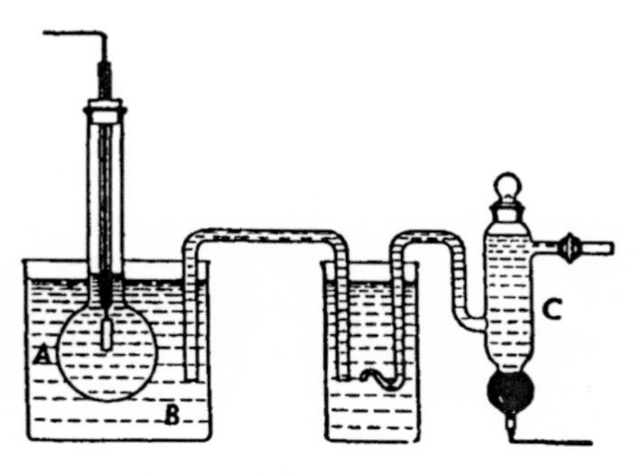

FIG. 96. Glass electrode cell

$$Ag \mid AgCl(s)\ 0.1\ \text{N}\ HCl \mid \text{glass} \mid \text{experimental solution},$$

if silver-silver chloride is the inner electrode of constant potential. The potential of the glass electrode is then measured by combining it with a suitable reference electrode, such as the calomel electrode *C* in Fig. 96, the inner electrode of the glass electrode system serving to make electrical connection.

Owing to the very high resistance of the glass, viz., 10 to 100 million ohms, special methods have to be employed for determining the E.M.F. of the cell; these generally involve the use of an electrometer or of vacuum-tube circuits, as described on page 192. Some workers have successfully prepared thin-walled glass electrodes of relatively large area and hence of comparatively low resistance; in these cases it has been found possible to make E.M.F. measurements without special apparatus, by using a reasonably sensitive galvanometer as the indicating instrument in the potentiometer circuit. Various forms of glass electrode have been employed for different purposes, but the simple bulb type described above can easily be made in a form that is not too fragile and yet has not a very high resistance. Several commercial forms of apparatus are now available which employ robust glass electrodes; by using some form of electrometer triode vacuum tube (p. 193), it is possible to measure the potential to about 0.0005 volt, i.e., 0.01 pH unit, without difficulty. An accuracy of $\pm$ 0.002 pH unit has been claimed for special measuring circuits, but it is doubtful whether the pH scale has been established with this degree of precision.

If both internal and external surfaces of the glass electrode were identical, it is obvious from equation (10) that the potential of the electrode system would be determined simply by the difference of pH of the solutions on the two sides of the glass membrane, apart from the potential of the inner electrode, e.g., Ag, AgCl. This expectation can be tested by measuring the E.M.F. of a cell in which the solution is the same inside and outside the glass bulb and the reference electrode is the same as the inner electrode; thus

$$\text{Ag} \mid \text{AgCl}(s)\ 0.1\ \text{N HCl} \mid \text{glass} \mid 0.1\ \text{N HCl AgCl}(s) \mid \text{Ag}.$$

The E.M.F. of this cell should be zero, but in practice the value is found to be of the order of $\pm$ 2 millivolts, for a good electrode. This small difference is called the **asymmetry potential** of the glass electrode; it is probably due to differences in the strain of the inner and outer surfaces of the glass membrane. It is necessary, therefore, to standardize each glass electrode by means of a series of buffer solutions of known pH; in this way the value of E^0_G in equation (10) for the particular electrode is found.

Before use the glass electrode should be allowed to soak in water for some time, following its preparation, and should not be allowed to become dry subsequently; if treated in this manner equilibrium with the solution in which it is placed is attained rapidly. The potential satisfies equation (10) for a reversible hydrogen electrode very closely in the pH range of 1 to 9, and with fair accuracy up to pH 12,* provided there is no large concentration of salts in the solution. At pH's greater than 9 appreciable salt effects become evident which increase with increasing pH, i.e., increasing alkalinity, of the solution; the magnitude of the salt effects in such solutions depends primarily on the nature of the cations present, but it is of the order of 0.1 to 0.2 unit in the vicinity of pH 11 for 0.1 to 1 N solutions of the salt. In very acid solutions, of pH less than unity, other salt effects, determined mainly by the anions, are observed. Apart from these limitations, the glass electrode has the outstanding advantage that it can be employed in aqueous solutions of almost any kind; the electrode cannot be poisoned, neither is it affected by oxidizing or reducing substances or by organic compounds. It can be used in unbuffered solutions and can be adapted for measurements with very small quantities of liquid. The glass electrode does not function satisfactorily in pure ethyl alcohol or in acetic acid, but it has been employed in mixtures of these substances with water.[9]

* The accuracy may be improved by the use of a special glass now available.

[9] Hughes, *J. Chem. Soc.*, 491 (1928); MacInnes and Dole, *J. Am. Chem. Soc.*, **52**, 29 (1930); MacInnes and Belcher, *ibid*, **53**, 3315 (1931); Dole, *ibid.*, **53**, 4260 (1931); **54**, 3095 (1932); for reviews with references, see Schwabe, *Z. Elektrochem.*, **41**, 681 (1935); Kratz, *ibid.*, **46**, 259 (1940).

There is no completely satisfactory explanation of why a glass electrode functions as a reversible hydrogen electrode, but it is probable that the hydrogen ions in the solution exchange, to some extent, with the sodium ions on the surface of the glass membrane. The result is that a potential, similar to a liquid junction potential, is set up at each surface of the glass; if no ions other than hydrogen ions, and their associated water molecules, are able to enter the glass, the free energy change accompanying the transfer of 1 g.-ion of hydrogen ion from the solution on one side of the membrane, where the activity is a'_{H^+}, to the other side, where the activity is a''_{H^+}, is then

$$\Delta G = RT \ln \frac{a''_{H^+}}{a'_{H^+}} + xRT \ln \frac{a''_{H_2O}}{a'_{H_2O}},$$

where x is the number of molecules of water associated with each hydrogen ion in the transfer; a'_{H_2O} and a''_{H_2O} are the activities of the water in the two solutions. The potential across the glass membrane is consequently given by

$$E_G = \frac{RT}{F} \ln \frac{a'_{H^+}}{a''_{H^+}} + \frac{xRT}{F} \ln \frac{a'_{H_2O}}{a''_{H_2O}}. \tag{11}$$

If the solutions are sufficiently dilute, the activities of the water are the same on both sides of the membrane; the second term on the right-hand side of equation (11) then becomes zero. By retaining the hydrogen ion activity, e.g., a'_{H^+}, constant on one side of the membrane, equation (11) reduces to the same form as (10). If the activity of the water is altered by the addition of alcohol or of appreciable amounts of salts or acids, equation (10) is no longer applicable, and deviations from the ideal reversible behavior of the glass electrode are observed. The salt errors found in relatively alkaline solutions, of pH greater than 9, are probably due to the fact that at these low hydrogen ion concentrations other cations present in the solution are transferred across the glass membrane to some extent. Under these conditions equation (11) is no longer valid, and so the glass electrode cannot behave in accordance with the requirements of equation (10).[10]

Acid-Base Indicators.—An acid-base **indicator** is a substance, which, within certain limits, varies in color according to the hydrogen ion concentration, or activity, of its environment; it is thus possible to determine the pH of a solution by observing the color of a suitable indicator when placed in that solution. Investigation into the chemistry of substances which function as acid-base indicators has shown that they are capable of existing in two or more tautomeric forms having different structures and different colors. In one or other of these forms the molecule is capable of functioning as a weak acid or base, and it is this property,

[10] Dole, *J. Am. Chem. Soc.*, **53**, 4260 (1930); **54**, 2120, 3095 (1932); "Experimental and Theoretical Electrochemistry," 1935, Chap. XXV; "The Glass Electrode," 1941; Haugaard, *J. Phys. Chem.*, **45**, 148 (1941).

together with the difference in color of the tautomeric states, that permits the use of the given compound as an acid-base indicator.[11]

If HIn_1 represents the un-ionized, colorless form of an indicator that is acidic in character, its ionization will be represented by

$$\underset{\text{colorless}}{HIn_1} + H_2O \rightleftharpoons H_3O^+ + \underset{\text{colorless}}{In_1^-},$$

the anion In_1^- having the same structure and color as the molecule HIn_1. Application of the law of mass action to this equilibrium gives the dissociation constant of the acid as

$$K_1 = \frac{a_{H^+}a_{In_1^-}}{a_{HIn_1}}. \tag{12}$$

The colorless ion In_1^- will be in equilibrium with its tautomeric form In_2^-, thus

$$\underset{\text{colorless}}{In_1^-} \rightleftharpoons \underset{\text{colored}}{In_2^-},$$

but the latter, having a different structure from that of In_1^-, will have a different color, and the constant of the tautomeric equilibrium (K_t) will be given by

$$K_t = \frac{a_{In_2^-}}{a_{In_1^-}}. \tag{13}$$

Finally, the colored In_2^- ions will be in equilibrium with hydrogen ions and the colored un-ionized molecules HIn_2, thus

$$\underset{\text{colored}}{HIn_2} + H_2O \rightleftharpoons H_3O^+ + \underset{\text{colored}}{In_2^-};$$

the dissociation constant of the acid HIn_2 is then

$$K_2 = \frac{a_{H^+}a_{In_2^-}}{a_{HIn_2}}. \tag{14}$$

Combination of equations (12), (13) and (14) gives

$$\frac{a_{H^+}(a_{In_1^-} + a_{In_2^-})}{(a_{HIn_1} + a_{HIn_2})} = \frac{K_1K_2(1 + K_t)}{K_2 + K_1K_t} = K_{In}, \tag{15}$$

where K_{In} is a composite constant involving K_1, K_2 and K_t; it follows, therefore, from equation (15) that

$$a_{H^+} = K_{In}\frac{(a_{HIn_1} + a_{HIn_2})}{(a_{In_1^-} + a_{In_2^-})}. \tag{16}$$

If the ionic strength of the medium is relatively low, the activities of HIn_1, HIn_2, In_1^- and In_2^- may be replaced by their respective concen-

[11] For a full discussion of the properties of indicators, see Kolthoff and Rosenblum, "Acid-Base Indicators," 1937.

trations, so that equation (16) becomes

$$a_{H^+} = k_{In} \frac{c_{HIn_1} + c_{HIn_2}}{c_{In_1^-} + c_{In_2^-}}, \tag{17}$$

where the approximate "constant" k_{In}, known as the **indicator constant,** replaces K_{In}.

If a particular compound is to be satisfactory as an acid-base or pH indicator, the numerator and denominator in equation (17) must correspond to two distinct colors: a change in the hydrogen ion activity must clearly be accompanied by an alteration in the ratio of numerator to denominator, and unless these represent two markedly different colors the system as a whole will undergo no noticeable change of color. Since HIn_1 and HIn_2 have different colors, on the one hand, and In_1^- and In_2^- are also different, but the same as HIn_1 and HIn_2, respectively, it is evident that in order to satisfy the condition given above it is necessary that the un-ionized molecules must be almost completely in the form HIn_1, or HIn_2, and the ions must be almost exclusively in the other form. It follows from equation (13) that if the tautomeric constant K_t is small the ions In_1^- will predominate over In_2^-; further, if K_1/K_2 is large, so that HIn_1 is a much stronger acid than HIn_2, it follows that the un-ionized molecules HIn_2 will greatly exceed those of HIn_1. These are, in fact, the conditions required to make the substance under consideration a satisfactory indicator. An alternative possibility which is equally satisfactory is that K_t should be large while K_1/K_2 is small; the ionized form will then consist mainly of In_2^- while the un-ionized molecules will be chiefly in the HIn_1 form. For a satisfactory indicator, therefore, equation (17) may be written as

$$a_{H^+} = k_{In} \frac{\text{Un-ionized form}}{\text{Ionized form}} \tag{18}$$

$$= k_{In} \frac{1 - \alpha}{\alpha}, \tag{19}$$

where α is the fraction of the total indicator present in the ionized form. The actual color exhibited by the indicator will, of course, depend on the ratio of the un-ionized to the ionized form, since these have different colors; hence it follows from equation (18) that it will be directly related to the hydrogen ion activity, or concentration, of the medium. In an acid solution, i.e., a_{H^+} is high, the concentration of un-ionized form must increase, according to equation (18), and the indicator will exhibit the color associated with the main HIn form; in an alkaline medium, on the other hand, the ionized form must predominate and the color will be that of the chief In^- species.

A few indicators are bases in the state in which they are normally employed; an example is methyl orange, which is the sodium salt of *p*-dimethylaminoazobenzene sulfonic acid, the indicator action being due to the basic dimethylamino-group, i.e., $-\ N(CH_3)_2$. There is no reason, however, why the conjugate acid, viz., $-\ NH(CH_3)_2^+$, should not be considered as the indicator, although this is not the form in which it is usually supplied. In view of the fact that the properties of aqueous solutions are invariably expressed in terms of pH, and not of pOH, it is convenient to treat all indicators as acids. If the indicator in its familiar form happens to be a base, then the system is treated as if it consisted of its conjugate acid. All indicator systems, of course, consist of conjugate acid and base, e.g., HIn and In^-, and it is in a sense somewhat arbitrary to refer to certain indicators as acids and to others as bases. The particular term employed refers to the nature of the substance in the form in which it is usually encountered; methyl orange is generally employed as the sodium salt of the sulfonic acid of the free base, and hence it is called a basic indicator; but if it were used as the hydrochloride, or other salt, of the base, it would be called an acid indicator. In the subsequent treatment all indicators will for simplicity and uniformity be treated as acids.

Indicator Range.—If, as on page 287, it is assumed that the color of the ionized form In^- is barely visible when 9 per cent of the total indicator is in this form, i.e., when α is 0.09, it follows from equation (19) that the limiting hydrogen ion activity at which the indicator will show its acid color, due to HIn, will be given by

$$a_{H^+} = k_{In}\frac{0.91}{0.09} \approx 10k_{In},$$

$$\therefore \quad pH \approx pk_{In} - 1, \tag{20}$$

where pk_{In} is the indicator exponent, defined in the usual manner as $-\log k_{In}$. On the other hand when 91 per cent of the indicator is in the ionized form, i.e., α is 0.91, the color of the un-ionized form will be virtually undetectable in the mixture, and so the color will be that of the alkaline form; the pH at which the indicator shows its full alkaline color is then obtained from equation (19), thus

$$a_{H^+} = k_{In}\frac{0.09}{0.91} \approx \frac{1}{10}k_{In},$$

$$\therefore \quad pH \approx pk_{In} + 1. \tag{21}$$

It is seen, therefore, that as the pH of a solution is increased by the addition of alkali, the color of an indicator begins to change visibly at a pH approximately equal to $pk_{In} - 1$, and is completely changed, as far as the eye can detect, at a pH of about $pk_{In} + 1$. The effective transition interval of an indicator is thus very roughly two pH units, one on each side of the pH equal to pk_{In} of the indicator. Since various indicators have different values of k_{In}, the range of pH over which the color changes will vary from one indicator to another.

When the indicator is ionized to an extent of 50 per cent, i.e., α is 0.5, it is seen from equation (19) that

$$a_{H^+} = k_{In},$$

$$\therefore \quad pH = pk_{In}. \tag{22}$$

The indicator will thus consist of equal amounts of the ionized and un-ionized forms, and hence will show its exact intermediate color, when the hydrogen ion activity, or concentration, is equal to the indicator constant.

Determination of Indicator Constants.—A simple method of evaluating the constant of an indicator is to make use of equation (22). Two solutions, containing the same amount of indicator, one in the completely acid form and the other in the alkaline form, are superimposed; the net color is equivalent to that of the total amount of indicator with equal portions in the ionized and un-ionized forms. A series of buffer solutions of known pH (see Chap. XI) are then prepared and a quantity of indicator, twice that present in each of the two superimposed solutions, is added; the colors are then compared with that of the latter until a match is obtained. The matching buffer solution consequently contains equal amounts of ionized and un-ionized indicator and so its pH is equal to the required pk_{In}.

The general procedure is to utilize equation (19) and to determine the proportion of un-ionized to ionized form of the indicator in a solution of known pH; the most accurate method is to measure this ratio by a spectrophotometric method similar to that described on page 328. If the substance is a one color indicator, that is to say it is colored in one (ionized) form and colorless in the other (un-ionized) form, e.g., phenolphthalein and *p*-nitrophenol, it generally has one sharp absorption band in the visible spectrum; by measuring the extinction coefficient when the substance is completely in its colored form, e.g., in alkaline solution, it is possible, by utilizing Beer's law (cf. p. 328, footnote), to determine the concentration of colored form in any solution of known pH from the extent of light absorption by the indicator in that solution (cf. Fig. 100). From the total amount of indicator present, the ratio $(1 - \alpha)/\alpha$ can be evaluated and hence k_{In} can be obtained. The principle of this method of determining the indicator constant is identical with that described on page 329 for the dissociation constant of an acid; k_{In} is in fact the *apparent* dissociation constant of the indicator, assuming it to consist of a single un-ionized form HIn and an ionized form In^- with a different color.

A two color indicator will, in general, have two absorption bands, one for each colored form; by studying the extent of absorption in these bands in a solution of definite pH, as compared with the values in a completely acid and a completely alkaline solution, it is possible to calculate directly the ratio of the amounts of un-ionized and ionized forms in the given solution.

Instead of utilizing spectrophotometric devices, the ratio of the amounts of ionized to un-ionized indicator can be estimated, although less accurately, by visual means. With a one color indicator the fraction of ionized, generally colored, form is determined by comparing the color intensity with that of a solution containing various known amounts of indicator which have been completely transformed by the addition of alkali. With a two-color indicator it is necessary to superimpose the acid and alkaline colors in different amounts until a match is obtained. The precision of the measurements can be greatly improved by the use of a commercial form of colorimeter specially designed for the matching of colors.

The values of pk_{In} for a number of useful indicators, together with the pH ranges in which they can be employed and their characteristic colors in acid and alkaline solutions, are recorded in Table LXIV.

TABLE LXIV. USEFUL INDICATORS AND THEIR CHARACTERISTIC PROPERTIES

Indicator	pk_{In}	pH Range	Color Change	
			Acid	Alkaline
Thymol blue	1.51	1.2– 2.8	Red	Yellow
Methyl orange	3.7	3.1– 4.4	Red	Yellow
Bromphenol blue	3.98	3.0– 4.6	Yellow	Blue
Bromcresol green	4.67	3.8– 5.4	Yellow	Blue
Methyl red	5.1	4.2– 6.3	Red	Yellow
Chlorphenol red	5.98	4.8– 6.4	Yellow	Red
Bromphenol red	6.16	5.2– 6.8	Yellow	Red
Bromcresol purple	6.3	5.2– 6.8	Yellow	Purple
Bromthymol blue	7.0	6.0– 7.6	Yellow	Blue
p-Nitrophenol	7.1	5.6– 7.6	Colorless	Yellow
Phenol red	7.9	6.8– 8.4	Yellow	Red
Cresol red	8.3	7.2– 8.8	Yellow	Red
Metacresol purple	8.32	7.4– 9.0	Yellow	Purple
Thymol blue	8.9	8.0– 9.6	Yellow	Blue
Cresolphthalein	9.4	8.2– 9.8	Colorless	Red
Phenolphthalein	9.4	8.3–10.0	Colorless	Red
Thymolphthalein	9.4	9.2–10.6	Colorless	Blue
Alizarine yellow	—	10.0–12.0	Yellow	Lilac
Nitramine	—	11.0–13.0	Colorless	Orange-brown

Determination of pH: With Buffer Solutions.—If a series of buffer solutions of known pH, which must lie in the region of the pH to be determined, is available the estimation of the unknown pH is a relatively simple matter. It is first necessary to choose, by preliminary experiments, an indicator that exhibits a definite intermediate color in the solution under examination. The color produced is then compared with that given by the same amount of the indicator in the various solutions of known pH. In the absence of a "salt error," to which reference will be made later, the pH of the unknown solution will be the same as that of the buffer solution in which the indicator exhibits the same color. Provided a sufficient number of solutions of known pH are available, this method can give results which are correct to about 0.05 pH unit.

When colored solutions are being studied, allowance must be made for the superimposition of the color on to that of the indicator; this may be done by means of the arrangement shown in plan in Fig. 97. The colored experimental solution, to which a definite amount of indicator has been added, is placed in the tube A and pure water is placed in B; the tube C contains the test solution without indicator, and D contains the buffer solution of known pH together with the same amount of indicator as in A. The solution in D is varied until the color of C and D superimposed is the same as that of A and B superimposed. The pH of the solution in A is then the same as that in D.

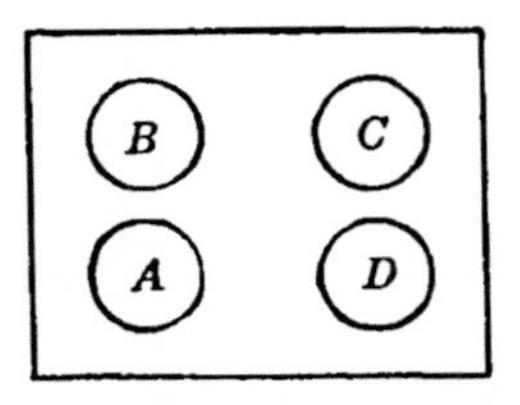

FIG. 97. Indicator measurements with colored solutions

Determination of pH: Without Buffer Solutions.—Provided the constant of an indicator is known, it is possible to determine the pH of an unknown solution without the use of buffer solutions; the methods are the same in principle as those employed for the evaluation of the indicator constant, except that in one case the pH of the solution is supposed to be known while pk_{In} is determined and in the other the reverse is true. For this purpose, equation (18), after taking logarithms, may be written as

$$\mathrm{pH} = pk_{\mathrm{In}} + \log \frac{\text{Ionized form}}{\text{Un-ionized form}}$$

$$= pk_{\mathrm{In}} + \log \frac{\text{Color due to alkaline form}}{\text{Color due to acid form}}. \qquad (23)$$

The problem of determining pH values thus reduces to that of measuring the ratio of the two extreme colors exhibited by a particular indicator in the given solution.

I. Bjerrum's Wedge Method.[12]—A rectangular glass box is divided into two wedge-shaped compartments by the insertion of a sheet of glass diagonally, or two separate wedges are cemented together by Canada balsam to give a vessel of the form shown in plan in Fig. 98. A solution of the indicator which has been made definitely acid is placed in one wedge, and one that is definitely alkaline is placed in the other. By viewing the combination from the front a gradation of colors, from the acid to the alkaline forms of the indicator, can be seen as a result of the superposition of steadily decreasing amounts of acid color on increasing amounts

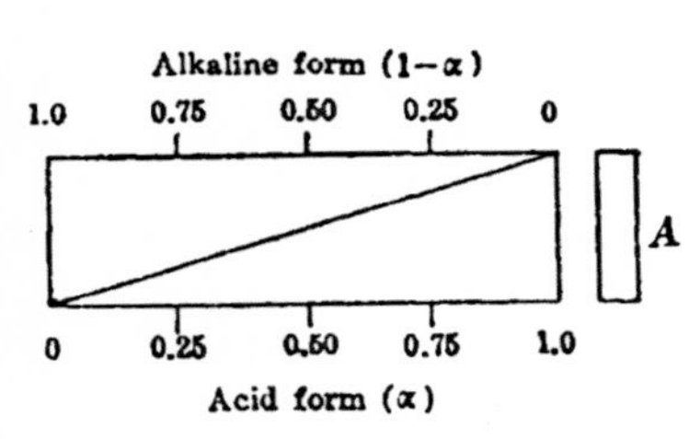

FIG. 98. Representation of Bjerrum wedge

[12] Bjerrum, Ahren's Sammlung, 1914, No. 21; Kolthoff, *Rec. trav. chim.*, **43**, 144 (1924); McCrae, *Analyst*, **51**, 287 (1926).

of the alkaline color. The test solution is placed in a narrow glass box of the same thickness as the combined wedges (Fig. 98, *A*) and the indicator is added so that its concentration is the same as in the wedges. A position is then found at which the color of the test solution matches that of the superimposed acid and alkaline colors; the ratio of the depths of the wedge solutions at this point thus gives the ratio of the colors required for equation (23). If the sides of the box are graduated, as shown, the depths of the two solutions can be obtained and the corresponding pH evaluated. The double-wedge can of course be calibrated so that the logarithmic term, i.e., the second term on the right-hand side, of equation (23) can be read off directly.

II. Colorimeter Method.—One of the simplest forms of colorimeter is shown in Fig. 99; the experimental solution is placed in the vessel *A* and an amount of indicator, giving a known concentration, is added; the fixed flat-bottomed tube *B* contains water to a definite height. The fixed tube *C*, arranged at the same level as *B*, also contains water to the same height as in *B*. Surrounding *C* is a movable tube *D* in which is placed the acid form of the indicator, and this is surrounded by the vessel *E* containing the indicator in its alkaline form; the concentration of the indicator in *D* and *E* is the same as in the test solution in *A*. The inner tube *D* is moved up and down until the color as seen through *C*, *D* and *E* is the same as that seen through *B* and *A*; the ratio of the alkaline color to the acid color in *A* is then given by the ratio of the heights h_1/h_2, so that the pH can be calculated if these heights are measured. If the test solution is colored, the water in *C* is replaced by the test solution to an equal depth; its color is then superimposed on that of the indicator in each case. By the use of special colorimeters it is possible to match the colors with such precision that pH values can be estimated with an accuracy of 0.01 unit.

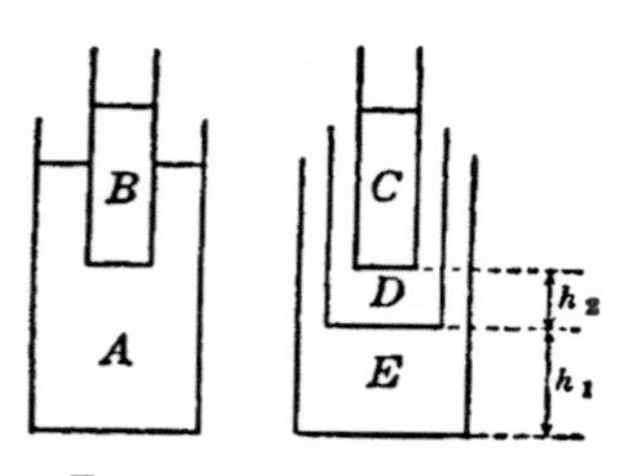

FIG. 99. Colorimeter for pH determinations

III. Spectrophotometric Method.[13]—The use of absorption spectra permits an accurate estimate to be made of the ratio of the amounts of the two colors in a given solution; the method is the same in principle as that already referred to on pages 328 and 363. In order to show the magnitude of the effect on the absorption of light resulting from a change of pH, the transmission curves obtained for bromcresol green in solutions of various pH's are shown in Fig. 100. It is evident that once the extent of the absorption produced by the completely alkaline form of the indi-

[13] Brode, *J. Am. Chem. Soc.*, **46**, 581 (1924); Holmes, *ibid.*, **46**, 2232 (1924); Holmes and Snyder, *ibid.*, **47**, 221, 226 (1925); Vlès, *Compt. rend.*, **180**, 584 (1925); Fortune and Mellon, *J. Am. Chem. Soc.*, **60**, 2607 (1938).

cator is known, the proportion present in a given solution, and hence the pH, can be estimated with fair accuracy.

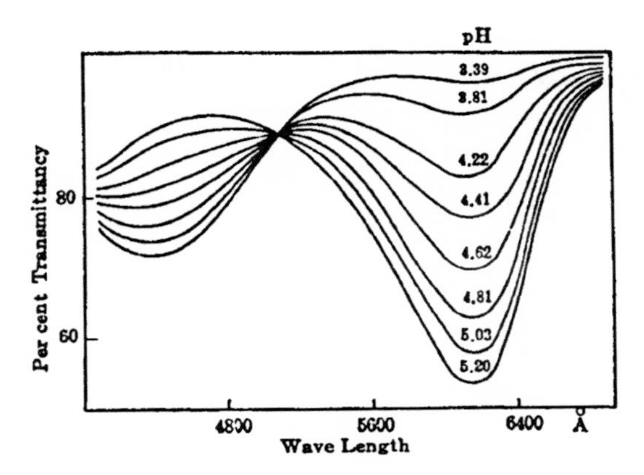

FIG. 100. Light absorption of bromcresol green (Fortune and Mellon)

Errors in Measurements with Indicators.—Three chief sources of error in connection with pH determinations by means of indicators may be mentioned.[14] In the first place, if the test solution is not buffered, e.g., solutions of very weak acids or bases or of neutral salts of strong acids and bases, the addition of the indicator may produce an appreciable change of pH; this source of error may be minimized by employing small amounts of indicator which have been previously adjusted, as a result of preliminary experiments, to have approximately the same pH as the test solution. Such indicator solutions are said to be **isohydric** with the test solution.

The second possible cause of erroneous results is the presence of proteins; as a general rule, indicator methods are not satisfactory for the determination of pH in protein solutions. The error varies with the nature of the indicator; it is usually less for low molecular weight compounds than for complex molecules.

Appreciable quantities of neutral salts produce color changes in an indicator that are not due to an alteration of pH and hence lead to erroneous results. This effect of neutral salts is due to two factors, at least; in the first place, the salt may affect the light absorbing properties of one or both forms of the indicator; and, in the second place, the altered ionic strength changes the activity of the indicator species. In deriving equation (17) the activities of the un-ionized and ionized forms of the indicator were taken to be the same as the respective concentrations; this can only be reasonably true if the ionic strength of the solution is

[14] McCrumb and Kenny, *J. Soc. Chem. Ind.*, 49, 425T (1930); Kolthoff and Rosenblum, "Acid-Base Indicators," 1937, Chap. X.

low, otherwise equation (19) should be written

$$a_{H^+} = K_{In} \frac{1 - \alpha}{\alpha} \cdot \frac{f_{HIn}}{f_{In^-}},$$

where f_{HIn} and f_{In^-} are the activity coefficients of the un-ionized and ionized species, respectively. Taking logarithms, this equation can be put in the form

$$\text{pH} = \text{p}K_{In} + \log \frac{\alpha}{1 - \alpha} + \log \frac{f_{In^-}}{f_{HIn}},$$

and the use of the extended Debye-Hückel equation for $\log (f_{In^-}/f_{HIn})$ then gives

$$\text{pH} = \text{p}K_{In} + \log \frac{\alpha}{1 - \alpha} - A\sqrt{\mu} + C\mu. \quad (24)$$

For a given color tint, i.e., corresponding to a definite value of $\alpha/(1 - \alpha)$, the actual pH will clearly depend on the value of the ionic strength of the solution; at low ionic strengths, e.g., less than about 0.01, the neutral salt error is negligible for most purposes.

The actual neutral salt error is less than estimated by equation (24) because the experimental values of $\text{p}K_{In}$ are generally based on determinations made in buffer solutions of appreciable ionic strength. For equation (24) to be strictly applicable the indicator exponent $\text{p}K_{In}$ should be the true thermodynamic value obtained by extrapolation to infinite dilution.

Universal Indicators.—Since the pH range over which a given indicator can be employed is limited, it is always necessary, as mentioned above, to carry out preliminary measurements with an unknown solution in order to find the approximate pH; with this information available the most suitable indicator can be chosen. For the purpose of making these preliminary observations the so-called **universal indicators** have been found useful:[15] they consist of mixtures of four or five indicators, suitably chosen so that they do not interfere with each other to any extent, which show a series of color changes over a range of pH from about 3 to 11. A convenient and simple form of universal indicator can be prepared by mixing equal volumes of 0.1 per cent solutions of methyl red, α-naphtholphthalein, thymolphthalein, phenolphthalein and bromthymol blue; the colors at different pH values are given below.

pH	4	5	6	7	8	9	10	11
Color	Red	Orange-red	Yellow	Green-yellow	Green	Blue-green	Blue-violet	Red-violet

[15] Carr, *Analyst*, 47, 196 (1922); Clark, "The Determination of Hydrogen Ions," 1928, p. 97; Britton, "Hydrogen Ions," 1932, p. 286; Kolthoff and Rosenblum, "Acid-Base Indicators," 1937, p. 170.

The addition of a small quantity of this universal indicator to an unknown solution permits the pH of the latter to be determined very approximately; it is then possible to choose the most suitable indicator from Table LXIV in order to make a more precise determination of the pH. Universal indicators are frequently employed when approximate pH values only are required, as, for example, in certain processes in qualitative and gravimetric analysis, and for industrial purposes.

PROBLEMS

1. What are the pH values of solutions whose hydrogen ion concentrations (activities) are 2.50, 4.85×10^{-4} and 0.79×10^{-10} g.-ion per liter? Assuming complete dissociation and ideal behavior, evaluate the pH of 0.0095 N sodium hydroxide at 25°.

2. A hydrogen gas electrode in 0.05 molar potassium acid phthalate, when combined with a calomel electrode containing saturated potassium chloride, gave a cell with an E.M.F. of 0.4765 volt at 38°. Calculate the pH of the solution which gave an E.M.F. of 0.7243 volt in a similar cell.

3. The glass electrode cell

$$\text{Pt} \mid \text{Quinhydrone pH 4.00 Buffer} \mid \text{Glass} \mid \text{pH 7.63 Buffer Quinhydrone} \mid \text{Pt}$$

gave an E.M.F. of − 0.2265 volt at 25°; calculate the asymmetry potential of the glass electrode.

4. What are the hydrogen ion activities of solutions of pH 13.46, 5.94 and − 0.5? What are the corresponding hydroxyl ion activities at 25°, assuming the activity of the water to be unity in each case?

5. A quinhydrone electrode in a solution of unknown pH was combined with a KCl(0.1 N), $Hg_2Cl_2(s)$, Hg electrode through a saturated potassium chloride salt bridge; the E.M.F. of the resulting cell was − 0.3394 volt at 30°. Calculate the pH of the solution.

6. If the oxygen electrode were reversible, what change of E.M.F. would be expected when an oxygen gas electrode at 1 atm. pressure replaced (i) a hydrogen electrode at 1 atm., and (ii) a quinhydrone electrode, in a given cell at 25°?

7. An indicator is yellow in the acid form and red in its alkaline form; when placed in a buffer solution of pH 6.35 it was found by spectrophotometric measurements that the extent of absorption in the yellow region of the spectrum was 0.82 of the value in a solution of pH 3.0. Evaluate pk_{In} for the given indicator.

CHAPTER XI

NEUTRALIZATION AND HYDROLYSIS

Types of Neutralization.—The term neutralization is generally applied to the reaction of one equivalent of an acid with one equivalent of base; if the terms "acid" and "base" are employed in the sense of the general definitions given in Chap. IX, the products are not necessarily a salt and water, as in the classical concept of acids and bases, but they are the conjugate base and acid, respectively, of the reacting acid and base. For the reaction between conventional acids, such as hydrochloric, acetic, etc., and strong bases, such as hydroxides in water or alkyloxides in alcohols, there is no difference between the new and the old points of view; it is, however, preferable to discuss all neutralizations from the general standpoint provided by the modern theory of acids and bases. According to this the following reactions are all examples of neutralization:

$$\begin{array}{rlll} HCl + (Na^+)OC_2H_5^- & = (Na^+)Cl^- & + C_2H_5OH \\ CH_3CO_2H + (Na^+)OH^- & = (Na^+)CH_3CO_2^- & + H_2O \\ HCl + RNH_2 & = Cl^- & + RNH_3^+ \\ HCl + (Na^+)CH_3CO_2^- & = (Na^+)Cl^- & + CH_3CO_2H \\ RNH_3^+(Cl^-) + (Na^+)OH^- & = RNH_2 & + H_2O + (Na^+Cl^-). \\ \text{Acid}_1 \qquad \text{Base}_2 & \quad \text{Base}_1 & \quad \text{Acid}_2 \end{array}$$

The last two reactions are of special interest, since they belong to the category usually known as "displacement reactions"; in the first of the two a strong acid, hydrochloric acid, displaces a weak acid, acetic acid, from its salt, while in the second a weak base, e.g., ammonia or an amine, is displaced from its hydrochloride by a strong base. It will be seen later that a much better understanding of these processes can be obtained by treating them as neutralizations, which in fact they are if this term is used in its wider sense.

Incomplete Neutralization: Lyolysis.—The extent to which neutralization occurs, when one equivalent of acid and base are mixed, depends on the nature of the acid, the base and the solvent. If the acid is HA, the base is B and SH is an amphiprotic solvent, i.e., one which can function either as an acid or as a base, the neutralization reaction

$$HA + B \rightleftharpoons BH^+ + A^-$$

takes place, but in addition, since the solvent is amphiprotic, two processes involving it can occur, thus

$$(a) \quad \underset{\text{Acid}}{BH^+} + \underset{\text{Base}}{SH} \rightleftharpoons \underset{\text{Acid}}{SH_2^+} + \underset{\text{Base}}{B}, \qquad (1a)$$

and

$$(b) \qquad \underset{\text{Acid}}{SH} + \underset{\text{Base}}{A^-} \rightleftharpoons \underset{\text{Acid}}{HA} + \underset{\text{Base}}{S^-}. \qquad (1b)$$

In the first of these the free base B is re-formed while in the second the free acid HA is regenerated; it follows, therefore, that the processes (*a*) and (*b*) militate against complete neutralization. This partial reversal of neutralization, or the prevention of complete neutralization, is called by the general name of **lyolysis** or **solvolysis**; in the particular case of water as solvent, the term used is **hydrolysis.**

Conditions for Complete Neutralization.—In order that neutralization may be virtually complete it is necessary that the lyolysis reactions should be reduced as far as possible. For reaction (*a*) to be suppressed it is necessary that B should be a much stronger base than the solvent SH, so that the equilibrium lies to the left. Further, the actual neutralization reaction equilibrium must lie to the right if it is to be practically complete; this means that B must be a stronger base than the anion A^-. For the complete neutralization, therefore, the order of basic strengths must be

$$A^- < B > SH.$$

If B is a weak base, it is necessary that A^- should be still weaker; it has been seen (p. 307) that a strong acid will have a very weak conjugate base, and hence this condition is satisfied if HA is a very strong acid. It is also necessary that the solvent should be a very weak base, and this can be achieved by using a strongly protogenic, i.e., acidic, medium. It has been found, in agreement with these conclusions, that extremely weak bases, e.g., acetoxime, can be neutralized completely by means of perchloric acid, the strongest known acid, in acetic acid as solvent. In water, hydrolysis of the type (*a*) is so considerable that neutralization of acetoxime, even by means of a strong acid, occurs to a negligible extent only.

By similar arguments it can be shown, from a consideration of the lyolytic equilibrium (*b*), that if an acid HA is to be neutralized completely, the condition is that the order of acid strengths must be

$$BH^+ < HA > SH.$$

To neutralize completely a weak acid HA it is necessary, therefore, to use a very strong base, so that its conjugate acid BH^+ is extremely weak, and to work in a protophilic medium, such as ether, acetonitrile or, preferably, liquid ammonia.

It will be evident from the conclusions reached that the lyolysis process (*a*) is due primarily to the weakness of the base B, whereas the process (*b*) results from the weakness of the acid HA. If both acid and base are weak in the particular solvent, then both types of lyolysis can occur, and complete neutralization is only possible in an aprotic solvent, provided the proton donating tendency of the acid HA is considerably greater than that of BH^+, or the proton affinity of the base B is greater than that of A^- (cf. p. 331). If the medium is exclusively protophilic, e.g., acetonitrile, then only the (*a*) type of lyolysis, namely that involving a weak base, is possible; weak acids should be completely neutralized provided a strong base is used. Similarly, in an exclusively protogenic solvent, e.g., hydrogen fluoride, the (*b*) type of lyolysis only can occur; a weak base can thus be completely neutralized in such a medium if a sufficiently strong acid is employed.

Hydrolysis of Salts.—The subject of lyolysis, or hydrolysis, in the event of water being the solvent, can be treated from two angles; in the general treatment already given it has been considered from the point of view of incomplete neutralization, and a return will be made later to this aspect of the subject. Another approach to the phenomena of hydrolysis is to study the equilibria resulting when a salt is dissolved in the given solvent; the situation is, of course, exactly the same as that which arises when an equivalent of the particular acid constituting the salt is neutralized by an equivalent of the base. This particular aspect of the subject of hydrolysis will be treated here; it is convenient to consider the material with special reference to the salt of (*a*) a weak acid, (*b*) a weak base, and (*c*) a weak acid and weak base. The first two of these are often referred to as salts of "one-sided" weakness, and the latter as a salt of "two-sided" weakness. Salts of strong acids and strong bases do not undergo hydrolytic reaction with the solvent, because the conjugate base and acid, respectively, are extremely weak; such salts, therefore, will not be discussed in this section, but reference will be made below to the neutralization of a strong acid by a strong base.

I. Salt of Weak Acid and Strong Base.—When a salt, e.g., NaA, of a weak acid HA is dissolved in water, it may be regarded as undergoing complete dissociation into Na^+ and A^- ions, provided the solution is not too concentrated. Since HA is a weak acid the conjugate base A^- will be moderately strong; hence the latter will react with the solvent molecules (H_2O) giving the type of hydrolytic equilibrium represented by equation (1*b*); in the particular case of water as solvent, this may be written

$$\underset{\text{Unhydrolyzed salt}}{A^-} + H_2O \rightleftharpoons \underset{\text{Free acid}}{HA} + \underset{\text{Free base}}{OH^-}.$$

The hydrolysis of the salt thus results in the partial reformation of the free weak acid HA and of the strong base $(Na^+)OH^-$ from which the salt was constituted. As a consequence of the weakness of the acid HA, therefore, there is a partial reversal of neutralization, and the term hydrolysis is often defined in this sense. It will be observed that the hydrolytic process results in the formation of OH^- ions, and this must obviously be accompanied by a decrease of hydrogen ion concentration (cf. p. 339);

the salt of a weak acid and a strong base thus reacts alkaline on account of hydrolysis. This accounts for the well-known fact that such salts as the cyanides, acetates, borates, phosphates, etc., of the alkali metals are definitely alkaline in solution.

Application of the law of mass action to the hydrolytic equilibrium gives the **hydrolysis constant** (K_h) of the salt as

$$K_h = \frac{a_{HA}a_{OH^-}}{a_{A^-}}, \tag{2}$$

the activity of the water being, as usual, taken as unity. The ionic product of water (K_w) and the dissociation constant (K_a) of the acid and HA are defined by

$$K_w = a_{H^+}a_{OH^-} \quad \text{and} \quad K_a = \frac{a_{H^+}a_{A^-}}{a_{HA}};$$

hence, it follows immediately from these expressions and equation (2) that

$$K_h = \frac{K_w}{K_a}. \tag{3}$$

The hydrolysis constant is thus inversely proportional to the dissociation constant of the weak acid; * the weaker the acid the greater is the hydrolysis constant of the salt.

If the activities are replaced by the product of the concentration and activity coefficient in each case, equation (2) becomes

$$K_h = \frac{c_{HA}c_{OH^-}}{c_{A^-}} \cdot \frac{f_{HA}f_{OH^-}}{f_{A^-}}. \tag{4}$$

In solutions of low ionic strength the activity coefficient f_{HA} of the undissociated molecules is very close to unity, and, further, the ratio of the activity coefficients of the two univalent ions, i.e., f_{OH^-}/f_{A^-}, is then also unity, by the Debye-Hückel limiting law; equation (4), therefore, reduces to the less exact form

$$k_h = \frac{c_{HA}c_{OH^-}}{c_{A^-}}, \tag{5}$$

which is particularly applicable to dilute solutions. As in other cases, the thermodynamic constant K_h has been replaced by the approximate "constant," k_h.

The **degree of hydrolysis** (x) is defined as the fraction of each mole of salt that is hydrolyzed when equilibrium is attained. If c is the stoichiometric, i.e., total, concentration of the salt NaA in the solution, the concentration of unhydrolyzed salt will be $c(1 - x)$; since this may be regarded as completely dissociated into Na^+ and A^- ions, it is possible

* It should be noted that the hydrolysis constant is equal to the dissociation constant of the base A^- which is conjugate to the acid HA.

to write

$$c_{A^-} = c(1 - x).$$

In the hydrolytic reaction, equivalent amounts of OH^- and HA are formed, and if the dissociation of the latter is neglected, since it is likely to be very small especially in the presence of the large concentration of A^- ions, it follows that c_{OH^-} and c_{HA} must be equal; further, both of these must be equal to cx, where x is the fraction of the salt hydrolyzed; hence,

$$c_{OH^-} = c_{HA} = cx.$$

Substitution of these values for c_{A^-} and c_{OH^-} in equation (5) gives

$$k_h = \frac{cx^2}{1 - x}, \tag{6}$$

$$\therefore \quad x = -\frac{k_h}{2c} + \sqrt{\frac{k_h^2}{4c^2} + \frac{k_h}{c}}. \tag{7}$$

From equation (7) it is possible to calculate the degree of hydrolysis at any desired concentration, provided the hydrolysis constant of the salt, or the dissociation constant of the acid [cf. equation (3)], is known. If k_h is small, e.g., for the salt of a moderately strong acid, at not too small a concentration equation (7) reduces to

$$x \approx \sqrt{\frac{k_h}{c}}, \tag{8}$$

so that the degree of hydrolysis is approximately proportional to the square-root of the hydrolysis constant and inversely proportional to the square-root of the concentration of the salt solution. The result of equation (8) may be expressed in a slightly different form by making use of equation (3) which may be written, for the present purpose, as $k_h = k_w/k_a$; thus,

$$x \approx \sqrt{\frac{k_w}{k_a c}}. \tag{9}$$

If two salts of different weak acids are compared at the same concentration, it is seen that

$$\frac{x_1}{x_2} = \sqrt{\frac{(k_a)_2}{(k_a)_1}}, \tag{10}$$

so that the degree of hydrolysis of each is inversely proportional to the square-root of the dissociation constant of the acid; hence the weaker the acid the greater the degree of hydrolysis at a particular concentration. For a given salt, equation (9) shows the degree of hydrolysis to increase with decreasing concentration.

By making use of equation (7) or (8) it is possible to calculate the degree of hydrolysis of the salt of a strong base and a weak acid of known dissociation constant at any desired concentration. The results of such calculations are given in Table LXV; the temperature is assumed to be

TABLE LXV. DEGREE OF HYDROLYSIS OF SALTS OF WEAK ACIDS AND STRONG BASES AT 25°

k_a	k_h	Concentration of Solution			
		0.001 N	0.01 N	0.1 N	1.0 N
10^{-4}	10^{-10}	3.3×10^{-4}	10^{-4}	3.2×10^{-5}	10^{-5}
10^{-6}	10^{-8}	3.2×10^{-3}	10^{-3}	3.2×10^{-4}	10^{-4}
10^{-8}	10^{-6}	3.2×10^{-2}	10^{-2}	3.2×10^{-3}	10^{-3}
10^{-10}	10^{-4}	0.27	0.095	3.2×10^{-2}	10^{-2}

about 25°, so that k_w can be taken as 10^{-14}. It is seen that the degree of hydrolysis increases with decreasing strength of the acid and decreasing concentration of the solution. In a 0.001 N solution, the sodium salt of an acid of dissociation constant equal to 10^{-10}, e.g., a phenol, is hydrolyzed to an extent of 27 per cent. It may be noted that equations (7) and (8) give almost identical values for the degree of hydrolysis in Table LXV, except for the two most dilute solutions of the salt of the acid of k_a equal to 10^{-10}. In these cases the approximate equation (8) would give 0.32 and 0.10, instead of 0.27 and 0.095 given in the table.

It has been seen above that c_{OH^-} is equal to cx, and since the product of c_{H^+} and c_{OH^-} is k_w, it follows that

$$c_{H^+} = \frac{k_w}{cx}, \tag{11}$$

and introducing the value of x from equation (9), the result is

$$c_{H^+} = \sqrt{\frac{k_w k_a}{c}}.$$

Taking logarithms and changing the signs throughout, this becomes

$$-\log c_{H^+} = -\tfrac{1}{2}\log k_w - \tfrac{1}{2}\log k_a + \tfrac{1}{2}\log c. \tag{12}$$

As an approximation, $-\log c_{H^+}$ may be replaced by pH, and using the analogous exponent forms for $-\log k_w$ and $-\log k_a$, it follows that

$$\text{pH} = \tfrac{1}{2}\,\text{p}k_w + \tfrac{1}{2}\,\text{p}k_a + \tfrac{1}{2}\log c. \tag{12a}$$

It is seen, therefore, that the pH, or alkalinity, of a solution of the salt of a weak acid and strong base increases with decreasing acid strength, i.e., increasing pk_a, and increasing concentration. Attention may be called to the fact that although the degree of hydrolysis decreases with increasing concentration of the salt, the pH, or alkalinity, increases. The pH values in Table LXVI have been calculated for dissociation constants and salt concentrations corresponding to those in Table LXV; equation (12) is satisfactory in all cases for which (8) is applicable, but

TABLE LXVI. VALUES OF pH IN SOLUTIONS OF SALTS OF WEAK ACIDS AND STRONG BASES AT 25°

		Concentration of Solution			
k_a	k_h	0.001 N	0.01 N	0.1 N	1.0 N
10^{-4}	10^{-10}	7.5	8.0	8.5	9.0
10^{-6}	10^{-8}	8.5	9.0	9.5	10.0
10^{-8}	10^{-6}	9.5	10.0	10.5	11.0
10^{-10}	10^{-4}	10.4	11.0	11.5	12.0

in the others use has been made of the x values in Table LXV together with equation (11). Since the pH of a neutral solution is about 7.0 at 25°, it follows that the solutions of salts of weak acids can be considerably alkaline in reaction.

It was seen in Chap. IX that the dissociation constant of an acid undergoes relatively little change with temperature between 0° and 100°; on the other hand the ionic product of water increases nearly five hundred-fold. It is evident, therefore, from equation (3) that the hydrolysis constant will increase markedly with increasing temperature; the degree of hydrolysis and the pH at any given concentration of salt will thus increase at the same time.

II. Salt of Weak Base and Strong Acid.—When the base B is weak, the conjugate acid BH^+ will have appreciable strength and hence it will tend to react with the solvent in accordance with the hydrolytic equilibrium (1*a*). It follows, therefore, that if the salt of a weak base and a strong acid is dissolved in water there will be a partial reversal of neutralization, some of the acid H_3O^+ and the weak base B being regenerated; in other words, the salt is hydrolyzed in solution. If the weak base is of the type RNH_2, e.g., ammonia or an amine, the conjugate acid is RNH_3^+, and when the salt, e.g., RNH_3Cl, is dissolved in water it dissociates virtually completely to yield RNH_3^+ and Cl^- ions, the former of which establish the hydrolytic equilibrium

$$RNH_3^+ + H_2O \rightleftharpoons H_3O^+ + RNH_2.$$

When the weak base is a metallic hydroxide, it is probable that the conjugate acid is the hydrated ion of the metal, e.g., $Fe(H_2O)_6^{+++}$ or $Cu(H_2O)_4^{++}$, which may be represented in general by $M(H_2O)_m^+$; the hydrolysis must then be expressed by

$$M(H_2O)_m^+ + H_2O \rightleftharpoons H_3O^+ + M(H_2O)_{m-1}OH,$$

where $M(H_2O)_{m-1}(OH)$ is the weak base. The formation of H_3O^+ ions shows that the solutions react acid in each case.

Writing the hydrolytic equilibrium in the general form

$$\underset{\text{Unhydrolyzed salt}}{BH^+} + H_2O \rightleftharpoons \underset{\text{Free acid}}{H_3O^+} + \underset{\text{Free base}}{B},$$

application of the law of mass action gives, for the hydrolytic constant,

$$K_h = \frac{a_{H^+}a_B}{a_{BH^+}}, \tag{13}$$

and since

$$K_w = a_{H^+}a_{OH^-} \quad \text{and} \quad K_b = \frac{a_{BH^+}a_{OH^-}}{a_B},$$

it follows that

$$K_h = \frac{K_w}{K_b}, \tag{14}$$

where K_b is the dissociation constant of the base B. It is seen that equation (14) is exactly analogous to (3), except that K_b now replaces K_a. By making the same assumptions as before, concerning the neglect of activity coefficients in dilute solution, equation (13) reduces to

$$k_h = \frac{c_{H^+}c_B}{c_{BH^+}}, \tag{15}$$

and from this, since c_{H^+} is now equal to c_B, both of which are equal to cx, while c_{BH^+} is equal to $c(1 - x)$, it follows that

$$k_h = \frac{cx^2}{1 - x}, \tag{16}$$

which is identical in form with equation (6). The degree of hydrolysis in this case is, consequently, also given by equation (7) which reduces to (8) provided the base is not too weak or the solution too dilute. Replacing k_h now by k_w/k_b, by the approximate form of equation (14), it follows that

$$x \approx \sqrt{\frac{k_w}{k_b c}}. \tag{17}$$

The same general conclusions concerning the effect of the dissociation constant of the weak base and the concentration of the salt on the degree of hydrolysis are applicable as for the salt of a weak acid. The results in Table LXV would hold for the present case provided the column headed k_a were replaced by k_b. Further, since the dissociation constants of bases do not vary greatly with temperature, the influence of increasing temperature on the hydrolysis of the salt of a weak base will be very similar to that on the salt of a weak acid.

The hydrogen ion concentration c_{H^+} in the solution of a salt of a weak base is given by cx, as mentioned above, and if the value of x from equation (17) is employed, it follows that

$$c_{H^+} = \sqrt{\frac{k_w c}{k_b}}.$$

This result may be expressed in the logarithmic form

$$pH = \tfrac{1}{2}pk_w - \tfrac{1}{2}pk_b - \tfrac{1}{2}\log c. \tag{18}$$

It is evident that the pH of the solution must be less than $\tfrac{1}{2}pk_w$, i.e., less than 7.0, and so solutions of salts of the type under consideration will exhibit an acid reaction. It was seen on page 339 that in any aqueous solution

$$pH + pOH = pk_w,$$

hence in this particular case

$$pOH = \tfrac{1}{2}pk_w + \tfrac{1}{2}pk_b + \tfrac{1}{2}\log c, \tag{19}$$

which is exactly analogous to equation (12*a*), except that pOH and pk_b replace pH and pk_a, respectively. It follows, therefore, that the results in Table LXVI give the pOH values in solutions of salts of a weak base, provided the column headed pk_a is replaced by pk_b.

III. Salt of Weak Acid and Weak Base.—If both the acid and base from which a given salt is made are weak, the respective conjugate base and acid will have appreciable strength and consequently will tend to interact with the amphiprotic solvent water. When a salt such as ammonium acetate is dissolved in water, it dissociates almost completely into NH_4^+ and Ac^- ions, and these acting as acid and base, respectively, take part in the hydrolytic equilibria

$$NH_4^+ + H_2O \rightleftharpoons H_3O^+ + NH_3,$$

and

$$Ac^- + H_2O \rightleftharpoons HAc + OH^-.$$

Combining the two equations, the complete equilibrium is

$$NH_4^+ + Ac^- + 2H_2O \rightleftharpoons H_3O^+ + OH^- + NH_3 + HAc,$$

or, representing the weak base in general by B and the acid by HA,

$$BH^+ + A^- + 2H_2O \rightleftharpoons H_3O^+ + OH^- + B + HA.$$

Since the normal equilibrium between water molecules and hydrogen and hydroxyl ions, viz.,

$$2H_2O \rightleftharpoons H_3O^+ + OH^-,$$

exists in any event, this may be subtracted from the hydrolytic equilibrium; the result may thus be represented by

$$NH_4^+ + Ac^- \rightleftharpoons NH_3 + HAc$$

for ammonium acetate or, in the general case, by

$$\underset{\text{Unhydrolyzed salt}}{BH^+ + A^-} \rightleftharpoons \underset{\text{Free acid}}{HA} + \underset{\text{Free base}}{B}.$$

The law of mass action then gives for the hydrolysis constant

$$K_h = \frac{a_{HA}a_B}{a_{BH^+}a_{A^-}}, \tag{20}$$

and introduction of the expressions for K_a and K_b leads to the result

$$K_h = \frac{K_w}{K_aK_b}. \tag{21}$$

The hydrolysis constant equation (20) may also be written as

$$K_h = \frac{c_{HA}c_B}{c_{BH^+}c_{A^-}} \cdot \frac{f_{HA}f_B}{f_{BH^+}f_{A^-}}, \tag{22}$$

and since this expression involves the product of the activity coefficients of two univalent ions, instead of their ratio as in the previous cases, it is less justifiable than before to assume that the activity coefficient fraction will become unity in dilute solution. Nevertheless, this approximation can be made without introducing any serious error, and the result is

$$k_h = \frac{c_{HA}c_B}{c_{BH^+}c_{A^-}}. \tag{23}$$

If the original, i.e., stoichiometric, concentration of the salt is c moles per liter, and x is the degree of hydrolysis, then c_{HA} and c_B may both be set equal to cx, whereas c_{BH^+} and c_{A^-} are both equal to the concentration of unhydrolyzed salt $c(1 - x)$, the salt being regarded as completely dissociated. Insertion of these values in equation (23) then gives

$$k_h = \frac{x^2}{(1 - x)^2}, \tag{24}$$

$$\therefore \quad x = \frac{\sqrt{k_h}}{1 + \sqrt{k_h}}. \tag{25}$$

If $\sqrt{k_h}$ is small in comparison with unity, it may be neglected in the denominator so that equation (25) becomes

$$x \approx \sqrt{k_h}, \tag{26}$$

or, introducing the approximate form of equation (21) for k_h,

$$x \approx \sqrt{\frac{k_w}{k_ak_b}}. \tag{27}$$

It appears from equations (25), (26) and (27) that the degree of hydrolysis of a given salt of two-sided weakness is independent of the concentration of the solution; this conclusion is only approximately true, as will be seen shortly.

The hydrogen ion concentration of the solution of hydrolyzed salt may be calculated by using the expression for the dissociation function of the acid, k_a; thus,

$$k_a = \frac{c_{H^+}c_{A^-}}{c_{HA}},$$

$$\therefore \quad c_{H^+} = k_a \frac{c_{HA}}{c_{A^-}}$$

$$= k_a \frac{cx}{c(1-x)} = k_a \frac{x}{1-x}.$$

By equation (24), the fraction $x/(1-x)$ is equal to $\sqrt{k_h}$,

$$\therefore \quad c_{H^+} = k_a\sqrt{k_h}$$

$$= \sqrt{\frac{k_w k_a}{k_b}}, \tag{28}$$

or, expressed logarithmically,

$$\text{pH} = \tfrac{1}{2}\text{p}k_w + \tfrac{1}{2}\text{p}k_a - \tfrac{1}{2}\text{p}k_b. \tag{29}$$

If the dissociation constants of the weak base and acid are approximately equal, i.e., $\text{p}k_a$ is equal to $\text{p}k_b$, it follows that pH is $\frac{1}{2}\text{p}k_w$; the solution will thus be neutral, in spite of hydrolysis. If, on the other hand, k_a is greater than k_b, the salt solution will have an acid reaction; if k_a is less than k_b the solution has an alkaline reaction. As a first approximation the pH of a solution of a salt of a weak acid and weak base is seen to be independent of the concentration.

The conclusion that the degree of hydrolysis and pH of a solution of a salt of double-sided weakness is independent of the concentration is only strictly true if c_{BH^+} is equal to c_{A^-} and if c_B is equal to c_{HA}, as assumed above. This condition is only realized if k_a and k_b are equal, but not otherwise. If the dissociation constants of HA and B are different, so also will be those of the conjugate base and acid, i.e., A^- and BH^+, respectively. The separate hydrolytic reactions

$$A^- + H_2O \rightleftharpoons HA + OH^-$$

and

$$BH^+ + H_2O \rightleftharpoons H_3O^+ + B,$$

will, therefore, take place to different extents, so that the equilibrium concentrations of A^- and BH^+, on the one hand, and of HA and B, on the other hand, will not be equal. The assumptions made above, that c_{BH^+} is equal to c_{A^-} and that c_B is equal to c_{HA}, are consequently not justifiable, and the conclusions drawn are not strictly correct. The problem may be solved in principle by writing

$$c = c_{A^-} + c_{OH^-} = c_{BH^+} + c_B,$$

where the total concentration c is divided into the unhydrolyzed part, i.e., c_{A^-} or c_{BH^+}, and the hydrolyzed part, i.e., c_{OH^-} or c_B, respectively. Further, by the condition of electrical neutrality,

$$c_{H^+} + c_{BH^+} = c_{OH^-} + c_{A^-},$$

and if these equations are combined with the usual expressions for k_a, k_b and k_w, it is possible to eliminate c_{A^-}, c_{OH^-}, c_B and c_{BH^+}, and to derive an equation for c_{H^+} in terms of c and k_a, k_b and k_w. Unfortunately, the resulting expression is of the fourth order, and can be solved only by a process of trial and error. The calculations have been carried out for aniline acetate ($k_a = 1.75 \times 10^{-5}$, $k_b = 4.00 \times 10^{-10}$): at concentrations greater than about 0.01 N the result for the hydrogen ion concentration is practically the same as that obtained by the approximate method given previously. In more dilute solutions, however, the values differ somewhat, the differences increasing with increasing dilution.[1]

Hydrolysis of Acid Salts.—The acid salt of a strong base and a weak dibasic acid, e.g., NaHA, will be hydrolyzed in solution because of the interaction between the ion HA^-, functioning here as a base, and the solvent, thus

$$HA^- + H_2O \rightleftharpoons H_2A + OH^-.$$

The ion HA^- can also act as an acid,

$$HA^- + H_2O \rightleftharpoons H_3O^+ + A^{--},$$

and the H_3O^+ ions formed in this manner may interact with HA^- to form H_2A, thus

$$HA^- + H_3O^+ = H_2A + H_2O.$$

If it were not for this latter reaction $c_{A^{--}}$ would have been equal to c_{H^+}, but since some of the hydrogen ions are removed in the formation of an equivalent amount of H_2A, it follows that

$$c_{A^{--}} = c_{H^+} + c_{H_2A}.$$

Further, if the salt NaHA is hydrolyzed to a small extent only, c_{HA^-} will be almost equal to c, the stoichiometric concentration of the salt. With these expressions for $c_{A^{--}}$ and c_{HA^-}, together with the equations for k_1 and k_2, the dissociation functions of the first and second stages of the acid H_2A, viz.,

$$k_1 = \frac{c_{H^+}c_{HA^-}}{c_{H_2A}} \quad \text{and} \quad k_2 = \frac{c_{H^+}c_{A^{--}}}{c_{HA^-}},$$

it is readily possible to derive the result

$$c_{H^+} = \sqrt{\frac{k_1 k_2 c}{k_1 + c}}. \tag{30}$$

[1] Griffith, *Trans. Faraday Soc.*, 17, 525 (1922).

If k_1 is small in comparison with the concentration, so that it may be neglected in the denominator, equation (30) reduces to the simple form

$$c_{H^+} = \sqrt{k_1 k_2}, \tag{31}$$

$$\therefore \quad pH = \tfrac{1}{2}pk_1 + \tfrac{1}{2}pk_2. \tag{32}$$

In this case, therefore, the pH of the solution is independent of the concentration of the acid salt.

The difference between the results given by equations (30) and (31) increases with increasing dilution, as is to be expected. If k_1 is less than about 0.01 c, however, the discrepancy is negligible.

Displacement of Hydrolytic Equilibrium.—When a salt is hydrolyzed, the equilibrium

$$\text{Unhydrolyzed salt} + \text{Water} \rightleftharpoons \text{Free acid} + \text{Free base}$$

is always established; this equilibrium can be displaced in either direction by altering the concentrations of the products of hydrolysis. The addition of either the free acid or the free base, for example, will increase the concentration of unhydrolyzed salt and so repress the hydrolysis; this fact is utilized in a method for investigating hydrolytic equilibria (p. 383).

If, on the other hand, the free acid or base is removed in some manner, the extent of hydrolysis of the salt must increase in order to maintain the hydrolytic equilibrium. For example, if a solution of potassium cyanide is heated or if a current of air is passed through it, the hydrogen cyanide formed by hydrolysis can be volatilized; as it is removed, however, more is regenerated by the continued hydrolysis of the potassium cyanide. When a solution of ferric chloride is heated, the hydrogen chloride is removed and hence the hydrolytic process continues; the hydrated ferric oxide which is formed remains in colloidal solution and imparts a dark brown color to the system.

Determination of Hydrolysis Constants: I. Hydrogen Ion Methods.—A number of methods of varying degrees of accuracy have been proposed for the estimation of the degree of hydrolysis in salt solutions or of the hydrolysis constant of the salt. One principle which can be used is to evaluate the hydrogen ion concentration of the solution; for a salt of a weak acid c_{H^+} is equal to k_w/cx, where c is the stoichiometric concentration of the salt, and hence it follows from equation (16) that

$$k_h = \frac{k_w^2}{c c_{H^+}^2 - k_w c_{H^+}} \approx \frac{k_w^2}{c c_{H^+}^2}. \tag{33}$$

If c_{H^+} is known, the hydrolysis constant can be calculated. For a salt of a weak base, on the other hand, c_{H^+} is equal to cx; hence

$$k_h = \frac{c_{H^+}^2}{c - c_{H^+}} \approx \frac{c_{H^+}^2}{c}. \tag{34}$$

If the salt is one of two-sided weakness the hydrogen ion concentration alone is insufficient to permit k_h to be evaluated; it is necessary to know, in addition, k_a or k_b.

The hydrogen ion concentration of a hydrolyzed salt solution can be determined by one of the E.M.F. or indicator methods described in Chap. X; it is true that the results obtained in this manner are not actual concentrations, but in view of the approximate nature of equations (33) and (34), the k_h values are approximate in any case.

II. Conductance Method.[2]—In a solution containing c equiv. per liter of a salt of a weak base and a strong acid, for example, there will be present $c(1 - x)$ equiv. of unhydrolyzed salt and cx equiv. of both free acid and base. If the base is very weak, it may be regarded as completely un-ionized, and so it will contribute nothing towards the total conductance of the solution of the salt. The conductance of 1 equiv. of a salt of a very weak base is thus made up of the conductance of $1 - x$ equiv. of unhydrolyzed salt and x equiv. of free acid, i.e.,

$$\Lambda = (1 - x)\Lambda_c + x\Lambda_{HA}. \tag{35}$$

In this equation Λ is the apparent equivalent conductance of the solution, which is equal to $1000\,\kappa/c$, where κ is the observed specific conductance and c is the stoichiometric concentration of the salt in the solution; Λ_c is the hypothetical equivalent conductance of the unhydrolyzed salt, and Λ_{HA} is the equivalent conductance of the free acid in the salt solution. It follows from equation (35) that

$$x = \frac{\Lambda - \Lambda_c}{\Lambda_{HA} - \Lambda_c}, \tag{36}$$

and so the calculation of x involves a knowledge of Λ, Λ_{HA} and Λ_c. As mentioned above, Λ is derived from direct measurement of the specific conductance of the hydrolyzed salt solution; the value of Λ_{HA} is generally taken as the equivalent conductance of the strong acid at infinite dilution, since its concentration is small, but it is probably more correct to use the equivalent conductance at the same total ionic strength as exists in the salt solution. The method is, however, approximate only, and this refinement is hardly necessary.

The evaluation of Λ_c for the unhydrolyzed salt presents a special problem. As already seen, the addition of excess of free base will repress the hydrolysis of the salt, and in the method employed sufficient of the almost non-conducting free base is added to the salt solution until the hydrolysis of the latter is almost zero. For example, with aniline hydrochloride, free aniline is added until the conductance of the solution reaches a constant value; at this point hydrolysis is reduced to a negligible

[2] Bredig, *Z. physik. Chem.*, **13**, 213, 221 (1894); Kanolt, *J. Am. Chem. Soc.*, **29**, 1402 (1907); Noyes, Sosman and Kato, *ibid.*, **32**, 159 (1910); Kameyama, *Trans. Electrochem. Soc.*, **40**, 131 (1921); Gulezian and Müller, *J. Am. Chem. Soc.*, **54**, 3151 (1932).

amount. The conductance of the solution is virtually that of the unhydrolyzed salt, and so Λ_c can be calculated. The data in Table LXVII are taken from the work of Bredig (1894) on a series of solutions

TABLE LXVII. HYDROLYSIS OF ANILINE HYDROCHLORIDE AT 18° FROM CONDUCTANCE MEASUREMENTS *

c	Λ	Λ'_c	Λ''_c	x	$k_h \times 10^5$
0.01563	106.2	96.0	95.9	0.036	2.1
0.00781	113.7	98.2	98.1	0.055	2.5
0.00391	122.0	100.3	100.1	0.077	2.5
0.00195	131.8	101.5	101.4	0.109	2.6
0.000977	144.0	103.3	103.3	0.147	2.5

* Bredig's measurements are not accurate because they were based on an incorrect conductance standard; the values of x and k_h derived from them are, however, not affected.

of aniline hydrochloride of concentration c equiv. per liter and observed equivalent conductance Λ; the columns headed Λ'_c and Λ''_c give the measured equivalent conductances in the presence of N/64 and N/32, respectively, added free aniline. Since the values in the two columns do not differ appreciably, it is evident that N/64 free aniline is sufficient to repress the hydrolysis of the aniline hydrochloride almost to zero; hence either Λ'_c or Λ''_c may be taken as equal to the required value of Λ_c. Taking Λ_{HA} for hydrochloric acid as 380 at 18°, the degree of hydrolysis x has been calculated in each case; from these the results for k_h in the last column has been derived. The values are seen to be approximately constant at about 2.5×10^{-5}.

For the salt of a weak acid, the method would be exactly similar to that described above except that excess of the free acid would be added to repress hydrolysis. The equation for the degree of hydrolysis is then

$$x = \frac{\Lambda - \Lambda_c}{\Lambda_{MOH} - \Lambda_c},$$

where Λ_{MOH} is the equivalent conductance of the strong base. The conductance method has also been used to study the hydrolysis of salts of weak acids and bases, but the calculations involved are somewhat complicated.

The determinations of hydrolysis constants from conductance measurements cannot be regarded as accurate; the assumption has to be made that the added free acid or free base has a negligible conductance. This is reasonably satisfactory if the acid or base is very weak, e.g., a phenol or an aniline derivative, but for somewhat stronger acids or bases, e.g., acetic acid, an appreciable error would be introduced; it is sometimes possible, however, to make an allowance for the conductance of the added acid or base.

III. Distribution Method.[3]—Another approximate method for studying hydrolysis is applicable if one constituent of the salt, generally the weak acid or base, is soluble in a liquid that is not miscible with water, while the salt itself and the other constituent are not soluble in that liquid. Consider, for example, the salt of a weak base, e.g., aniline hydrochloride; the free base is soluble in benzene, in which it has a normal molecular weight, whereas the salt and the free hydrochloric acid are insoluble in benzene. A definite volume (v_1) of an aqueous solution of the salt at a known concentration (c) is shaken with a given volume (v_2) of benzene, and the amount of free aniline in the latter is determined by analysis. If m is the concentration in equiv. per liter of the aniline in benzene found in this manner, then the concentration of free aniline in the aqueous solution (c_B) should be m/D, where D is the "distribution coefficient" of aniline between benzene and water; the value of D must be found by separate experiments on the manner in which pure aniline distributes itself between benzene and water, in the absence of salts, etc. The *amounts* of free aniline in the benzene and aqueous layers are mv_2 and mv_1/D respectively; hence, the amount of free acid in the aqueous solution, assuming none to have dissolved in the benzene, must be the sum of these two quantities, i.e., $mv_2 + mv_1/D$. Since this amount is present in a volume v_1, it follows that the concentration of free acid in the aqueous solution (c_{H^+}) is $mv_2/v_1 + m/D$. The concentration of unhydrolyzed salt (c_{BH^+}) is equal to the stoichiometric concentration (c) less the concentration of free acid, since the latter is equivalent to the salt that has been hydrolyzed; hence, c_{BH^+} is equal to $c - mv_2/v_1 - m/D$. The results derived above may then be summarized thus:

$$c_{H^+} = \frac{mv_2}{v_1} + \frac{m}{D},$$

$$c_B = \frac{m}{D}$$

and

$$c_{BH^+} = c - \frac{mv_2}{v_1} - \frac{m}{D},$$

and so it follows from equation (15) that

$$k_h = \frac{\left(\frac{mv_2}{v_1} + \frac{m}{D}\right)\frac{m}{D}}{c - \frac{mv_2}{v_1} - \frac{m}{D}}. \tag{38}$$

By determining m, therefore, all the quantities required for the evaluation of k_h by means of equation (38) are available, provided D is known

[3] Farmer, *J. Chem. Soc.*, 79, 863 (1901); Farmer and Warth, *ibid.*, 85, 1713 (1904); Williams and Soper, *ibid.*, 2469 (1930).

from separate experiments. The results in Table LXVIII, taken from the work of Farmer and Warth (1904), illustrate the application of the method to the determination of the hydrolysis of aniline hydrochloride; the non-aqueous solvent employed was benzene, for which D is 10.1, and the volumes v_1 and v_2 were 1000 cc. and 59 cc., respectively. The value of k_h is seen to be in satisfactory agreement with that obtained for aniline hydrochloride by the conductance method (Table LXVII).

TABLE LXVIII. HYDROLYSIS OF ANILINE HYDROCHLORIDE FROM DISTRIBUTION MEASUREMENTS

c	m	$c_B = \frac{m}{D}$	c_{H^+}	c_{BH^+}	$k_h \times 10^5$
0.0997	0.0124	0.00123	19.6×10^{-4}	0.0978	2.4
0.0314	0.00628	0.000622	9.9×10^{-4}	0.0304	2.0

The distribution method for studying hydrolysis can be applied to salts of a weak acid, provided a suitable solvent for the acid is available; the hydrolysis constant is given by an equation identical with (38), except that m now represents the concentration of free acid in the non-aqueous liquid. The same principle can be applied to the investigation of salts of two-sided weakness provided a solvent can be found which dissolves either the weak acid or the weak base, but not both.

IV. Vapor Pressure Method.[4]—If the free weak acid or weak base is appreciably volatile, it is possible to determine its concentration or, more correctly, its activity, from vapor pressure measurements. In practice the actual vapor pressure is not measured, but the volatility of the substance in the hydrolyzed salt solution is compared with that in a series of solutions of known concentration. In the case of an alkali cyanide, for example, the free hydrogen cyanide produced by hydrolysis is appreciably volatile. A current of air is passed at a definite rate through the alkali cyanide solution and at exactly the same rate through a hydrogen cyanide solution; the free acid vaporizing with the air in each case is then absorbed in a suitable reagent and the amounts are compared. The concentration of the hydrogen cyanide solution is altered until one is found that vaporizes at the same rate as does the alkali cyanide solution. It may be assumed that the concentrations, or really activities, of the free acid are the same in both solutions. The concentration of free acid c_{HA} in the solution of the hydrolyzed salt of the weak acid may be put equal to cx (cf. p. 374) and hence x and k_h can be calculated.

V. Dissociation Constant Method.—All the methods described above give approximate values only of the so-called hydrolysis "constant" of the salt; the most accurate method for obtaining the true hydrolysis constant is to make use of the thermodynamic dissociation constants of the weak acid or base, or both, and the ionic product of water. For this

[4] Worley *et al.*, *J. Chem. Soc.*, 111, 1057 (1917); *Trans. Faraday Soc.*, 20, 502 (1925); Britton and Dodd, *J. Chem. Soc.*, 2332 (1931).

purpose equations (3), (4) and (21) are employed. The results derived in this manner are, of course, strictly applicable to infinite dilution, but allowance can be made for the influence of the ionic strength of the medium by making use of the Debye-Hückel equations. The methods I to IV are of interest, in so far as they provide definite experimental evidence for hydrolysis, but they would not be used in modern work unless it were not possible, for some reason or other, to determine the dissociation constant of the weak acid or base.

It is of interest to note that some of the earlier measurements of k_h were used, together with the known dissociation constant of the acid or base, to evaluate k_w for water. For example, k_h for aniline hydrochloride has been found by the conductance method (Table LXVII) to be about 2.5×10^{-5}, and k_b for aniline is 4.0×10^{-10}; it follows, therefore, that k_w, which is equal to $k_b k_h$, should be about 1.0×10^{-14}, in agreement with the results recorded in Chap. IX.

Neutralization Curves.—The variation of the pH of a solution of acid or base during the course of neutralization, and especially in the vicinity of the equivalence-point, i.e., when equivalent amounts of acid and base are present, is of great practical importance in connection with analytical and other problems. It is, of course, feasible to measure the pH experimentally at various points of the neutralization process, but a theoretical study of the subject is possible and the results are of considerable interest. For this purpose it is convenient to consider the behavior of different types of acid, viz., strong and weak, with different bases, viz., strong and weak. For the present the discussion will be restricted to neutralization involving a conventional acid and base in aqueous solution, but it will be shown that the results can be extended to all forms of acids and bases in aqueous as well as non-aqueous solvents.

I. Strong Acid and Strong Base.—The changes in hydrogen ion concentration occurring when a strong base is added to a solution of a strong acid can be readily calculated, provided the acid may be assumed to be completely dissociated. The concentration of hydrogen ion (c_{H^+}) at any instant is then equal to the concentration of un-neutralized strong acid at that instant. If a is the initial concentration of the acid in equiv. per liter, and b equiv. per liter is the amount of base added at any instant, the concentration of un-neutralized acid is $a - b$ equiv. per liter, and this is equal to the hydrogen ion concentration. The results obtained in this manner when 100 cc. of 0.1 N hydrochloric acid, i.e., a is 0.1, are titrated with 0.1 N sodium hydroxide are given in Table LXIX. In order to simplify the calculations it is assumed that the volume of the system remains constant at 100 cc.; this simplification involves a slight error, but it will not affect the main conclusions which will be reached here. The values of pH in the last column are derived from the approximate definition of pH as $-\log c_{H^+}$.

When the solution contains equivalent amounts of acid and alkali the method of calculation given above fails, for $a - b$ is then zero; the

TABLE LXIX. NEUTRALIZATION OF 100 CC. 0.1 N HCL BY 0.1 N NAOH

NaOH added	b	c_{H^+}	pH
0.0 cc.	0.00	10^{-1}	1.0
50.0	0.05	5×10^{-2}	1.3
90.0	0.09	10^{-2}	2.0
99.0	0.099	10^{-3}	3.0
99.9	0.0999	10^{-4}	4.0
100.0	0.1000	10^{-7}	7.0
100.1	0.1001	10^{-10}	10.0

system is now, however, identical with one containing the neutral salt sodium chloride, and so the value of c_{H^+} is 10^{-7} g.-ion per liter and the pH is 7.0 at ordinary temperatures. If the addition of base is continued beyond the equivalence-point, the solution will contain free alkali; the pH of the system can then be calculated by assuming that c_{OH^-} is equal to the concentration of the excess alkali and that the ionic product $c_{H^+}c_{OH^-}$ is 10^{-14}. For example, in Table LXIX the addition of 100.1 cc. of 0.1 N sodium hydroxide means an excess of 0.1 cc. of 0.1 N alkali, i.e., 10^{-5} equiv. in 100 cc. of solution; the concentration of free alkali, and hence of hydroxyl ions, is thus 10^{-4} equiv. per liter. If c_{OH^-} is 10^{-4}, it follows that c_{H^+} must be 10^{-10} and hence the solution has a pH of 10.0.

If the titration is carried out in the opposite direction, i.e., the addition of strong acid to a solution of a strong base, the variation of pH may be calculated in a similar manner to that used above. The hydroxyl ion concentration is now taken as equal to the concentration of un-neutralized base, i.e., $b - a$, and the hydrogen ion concentration is then derived from the ionic product of water. The results calculated for the neutralization of 100 cc. of 0.1 N sodium hydroxide by 0.1 N hydrochloric acid, the volume being assumed constant, are recorded in Table LXX.

TABLE LXX. NEUTRALIZATION OF 100 CC. OF 0.1 N NAOH BY 0.1 N HCL

HCl added	a	c_{OH^-}	pH
0.0 cc.	0.00	10^{-1}	13.0
50.0	0.05	5×10^{-2}	12.7
90.0	0.090	10^{-2}	12.0
99.0	0.099	10^{-3}	11.0
99.9	0.0999	10^{-4}	10.0
100.0	0.1000	10^{-7}	7.0
100.1	0.1001	10^{-10}	4.0

The data in Tables LXIX and LXX are plotted in Fig. 101, in which curve I shows the variation of pH with the extent of neutralization of 0.1 N solutions of strong acid and strong base; the two portions of the curve may be regarded as parts of one continuous curve representing the change of pH as a solution of a strong acid is titrated with a strong base until the system contains a large excess of the latter, or *vice versa*. Attention may be called here to the sudden change of pH, from approximately 4 to 10, as the equivalence-point, marked by an arrow, is attained; further reference to this subject will be made later.

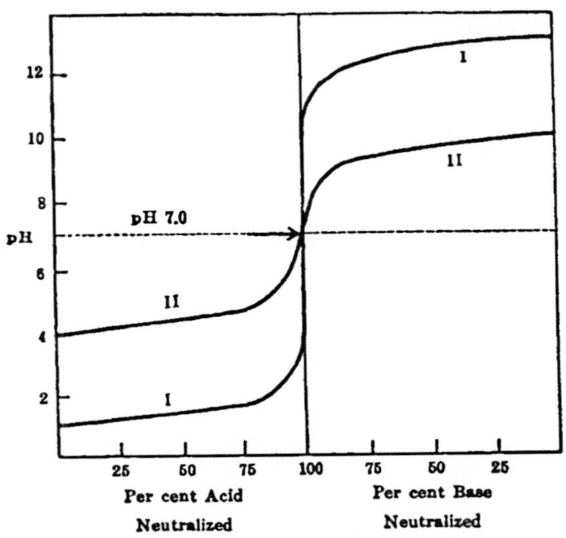

FIG. 101. Neutralization of strong acid and strong base

Similar calculations can be made and analogous pH-neutralization curves can be plotted for solutions of strong acid and base at other concentrations; curve II represents the results obtained for 10^{-4} N solutions. The pH at the equivalence-point is, of course, independent of the concentration, since the pH of the neutral salt is always 7.0. The change of pH at the equivalence-point in curve II is seen to be much less sharp, however, than is the case with the more concentrated solutions.

II. Weak Acid and Strong Base.—The determination of the pH in the course of the neutralization of a weak acid is not so simple as for a strong acid, but the calculations can nevertheless be made with the aid of equations derived in Chap. IX. It was seen on page 323 that if a weak acid, whose initial concentration is a equiv. per liter, is partially neutralized by the addition of b equiv. per liter of base, the activity of the hydrogen ions is given by

$$a_{H^+} = K_a \frac{a - B}{B} \cdot \frac{\gamma_{HA}}{\gamma_{A^-}}, \tag{39}$$

which may be written in the logarithmic form

$$pH = pK_a + \log \frac{B}{a - B} + \log \frac{\gamma_{A^-}}{\gamma_{HA}}, \tag{40}$$

or, utilizing the Debye-Hückel equations,

$$\mathrm{pH} = \mathrm{p}K_a + \log \frac{B}{a - B} - A\sqrt{\mu} + C\mu. \quad (41)$$

The quantity B is defined in this case by

$$B \equiv b + c_{\mathrm{H}^+} - c_{\mathrm{OH}^-}, \quad (42)$$

using volume concentrations instead of molalities: since this involves both c_{H^+} and c_{OH^-}, the latter being equivalent to k_w/c_{H^+}, equation (39) and those derived from it are cubic equations in c_{H^+}, and an exact solution is difficult. The problem is therefore simplified by considering certain special cases.

If the pH of the solution lies between 4 and 10, i.e., c_{H^+} is between 10^{-4} and 10^{-10}, the quantity $c_{\mathrm{H}^+} - c_{\mathrm{OH}^-}$ in equation (42) is negligibly small; under these conditions B is equal to b, and equation (41) becomes

$$\mathrm{pH} = \mathrm{p}K_a + \log \frac{b}{a - b} - A\sqrt{\mu} + C\mu. \quad (43)$$

The partly-neutralized acid system is equivalent to a mixture of unneutralized acid and its salt, the concentration of the former being $a - b$ and that of the latter b; equation (43) can consequently be written as

$$\mathrm{pH} = \mathrm{p}K_a + \log \frac{\text{salt}}{\text{acid}} - A\sqrt{\mu} + C\mu. \quad (44)$$

This relationship, without the activity correction, is equivalent to one derived by L. J. Henderson (1908) and is generally known as the **Henderson equation.** The equation, omitting the activity terms, gives reasonably good results for the pH during the neutralization of a weak base by a strong acid over a range of pH from 4 to 10, but it fails at the beginning and end of the process: under these latter conditions the approximation of setting B equal to b is not justifiable.

For these extreme cases the general equations (39) to (41) are still applicable, and suitable approximations can be made in order to simplify the calculations. At the very beginning of the titration, i.e., when the weak acid is alone present, b is zero and since the solution is relatively acid c_{OH^-} may be neglected; the quantity B is then equal to c_{H^+}, and equation (39) becomes

$$a_{\mathrm{H}^+} = K_a \frac{a - c_{\mathrm{H}^+}}{c_{\mathrm{H}^+}} \cdot \frac{\gamma_{\mathrm{HA}}}{\gamma_{\mathrm{A}^-}}.$$

If the solution has a sufficiently low ionic strength for the activity coefficients to be taken as unity, which is approximately true for the weak acid solution, this equation may be written in the form

$$c_{\mathrm{H}^+} = k_a \frac{a - c_{\mathrm{H}^+}}{c_{\mathrm{H}^+}},$$

$$\therefore \quad c_{\mathrm{H}^+} = -\tfrac{1}{2}k_a + \sqrt{\tfrac{1}{4}k_a^2 + ak_a}. \quad (45)$$

If c_{H^+} or k_a is small, that is for a very weak acid, these equations reduce to

$$c_{H^+} = \sqrt{ak_a}.$$

At the equivalence-point, which represents the other extreme of the titration, a and b are equal, and c_{H^+} may be neglected in comparison with c_{OH^-}, since the solution is alkaline owing to hydrolysis of the salt of the weak acid and strong base. It is seen, from equation (42), therefore, that B is now equivalent to $a - c_{OH^-}$, and, neglecting the activity coefficients, equation (39) becomes

$$c_{H^+} = k_a \frac{c_{OH^-}}{a - c_{OH^-}}.$$

This is a quadratic in c_{H^+}, since c_{OH^-} is equal to k_w/c_{H^+}, and so it can be solved without difficulty, thus

$$c_{H^+} = \frac{k_w}{2a} + \sqrt{\frac{k_w^2}{4a^2} + \frac{k_a k_w}{a}}. \qquad (46)$$

Since $k_w/2a$ is generally very small, it may usually be neglected and so this equation reduces to the form

$$c_{H^+} = \sqrt{\frac{k_w k_a}{a}}, \qquad (47)$$

or

$$\text{pH} = \tfrac{1}{2}\text{p}k_w + \tfrac{1}{2}\text{p}k_a + \tfrac{1}{2}\log a, \qquad (47a)$$

which is identical, as it should be, with the approximate equation (12a) for the hydrogen ion concentration in a solution of a salt of a weak acid and strong base; at the equivalence-point the acid-base system under consideration is, of course, equivalent to such a solution.

It is thus possible to calculate the whole of the pH-neutralization curve of a weak acid by a strong base: equations (45) and (47) are used for the beginning and end, respectively, and equation (43), without the activity corrections, for the intermediate points. The pH values obtained in this manner for the titration of 100 cc. of 0.1 N acetic acid, for which k_a is taken on 1.75×10^{-5}, with 0.1 N sodium hydroxide are quoted in Table LXXI.

When the titration is carried out in the reverse direction, i.e., a strong base is titrated with a weak acid, the pH changes in the early stages of neutralization are almost identical with those obtained when a strong acid is employed. It is true that the salt formed, being one of a weak acid and a strong base, is liable to hydrolyze, but as long as excess of the strong base is present this hydrolysis is quite negligible (cf. p. 382). The hydroxyl ion concentration is then equal to the stoichiometric concentration of un-neutralized base, i.e., c_{OH^-} is equal to $b - a$ where b and a are the concentrations of base and acid which make up the solution, just

TABLE LXXI. NEUTRALIZATION OF 100 CC. OF 0.1 N ACETIC ACID BY 0.1 N NaOH

NaOH added	b	$a - b$	c_{H^+}	pH
0.0 cc.	0.0	0.10	1.32×10^{-3}	2.88
10.0	0.01	0.09	1.60×10^{-4}	3.80
20.0	0.02	0.08	6.93×10^{-5}	4.16
40.0	0.04	0.06	2.63×10^{-5}	4.58
50.0	0.05	0.05	1.75×10^{-5}	4.76
70.0	0.07	0.03	7.42×10^{-6}	5.13
90.0	0.09	0.01	1.95×10^{-6}	5.71
99.0	0.099	0.001	1.75×10^{-7}	6.76
99.9	0.0999	0.0001	1.75×10^{-8}	7.76
100.0	0.10	—	1.32×10^{-9}	8.88

as if the salt were not hydrolyzed. As the equivalence-point is approached closely, however, the concentration of base is greatly reduced and so the hydrolysis of the salt becomes appreciable. The form of the pH curve is then determined by the fact that the hydrogen ion concentration at the equivalence-point is given by equation (47).

The complete curve for the neutralization of 0.1 N acetic acid by 0.1 N sodium hydroxide and *vice versa*, is shown in Fig. 102, I; the right-

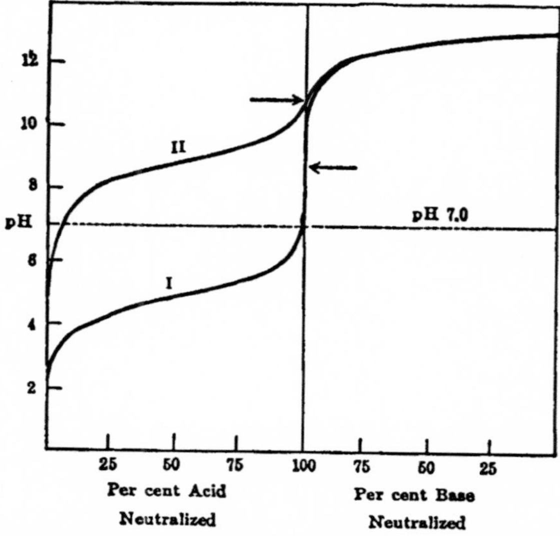

FIG. 102. Neutralization of weak (I) and very weak (II) acid by strong base

hand side is almost identical with that of Fig. 101, I, for a strong base neutralized by a strong acid. It is observed that in this instance there is also a rapid change of pH at the equivalence-point, but it is not so marked as for a strong acid at the same concentration. The equivalence-point itself, indicated by an arrow, now occurs at pH 8.88, the solution of sodium acetate being alkaline because of hydrolysis. If a more dilute acetic acid solution, e.g., 0.01 N, is titrated with a strong base, the main position of the pH-neutralization curve is not affected, as may be seen from an examination of the Henderson equation (44); the pH depends on the *ratio* of salt to un-neutralized acid, and this will be the same at a given stage of neutralization irrespective of the actual concentration. When the neutralization has occurred to the extent of 50 per cent, i.e., at the midpoint of the curve, the ratio of salt to acid is always unity; the pH is then equal to pk_a for the given acid (cf. p. 325), and this does not change markedly with the concentration of the solution. At the beginning and end of the neutralization, when the Henderson equation is not applicable, the pH's, given by equations (45) and (47), are seen to be dependent on the concentration; for 0.01 N acetic acid the values are 3.38 and 8.38, respectively, instead of 2.88 and 8.88 for the 0.1 N solution.

III. Moderately Strong Acid and Strong Base.—If the acid is a moderately strong one, the pH may be less than 4 for an appreciable part of the early stages of the neutralization. The quantity $c_{H^+} - c_{OH^-}$ which appears in the term B cannot then be neglected, but it is more accurate to neglect c_{OH^-} only, so that B becomes $b + c_{H^+}$; under these conditions equation (39), neglecting activity coefficients, becomes

$$c_{H^+} = k_a \frac{a - b - c_{H^+}}{b + c_{H^+}}. \qquad (48)$$

This is a quadratic equation which can be readily solved for c_{H^+}. The pH values for the beginning and end of the titration are derived from equations (45) and (47), as before. The pH-neutralization curve for a moderately strong acid lies between that of a strong acid (Fig. 101) and that of a weak acid (Fig. 102).

IV. Very Weak Acid and Strong Base.—For very weak acids, whose dissociation constants are less than about 10^{-7}, or for very dilute solutions, e.g., more dilute than 0.001 N, of weak acids, the pH of the solution exceeds 10 before the equivalence-point is reached. It is then necessary to include c_{OH^-} in B, although c_{H^+} can be neglected; equation (39) then takes the form

$$c_{H^+} = k_a \frac{a - b + c_{OH^-}}{b - c_{OH^-}}$$

$$= k_a \frac{a - b + k_w/c_{H^+}}{b - k_w/c_{H^+}}. \qquad (49)$$

This equation is also a quadratic in c_{H^+}, and so it can be solved and c_{H^+} evaluated. The results for the neutralization of a 0.1 N solution of an acid of k_a equal to 10^{-9} by a strong base are shown in Fig. 102, II: the equivalence-point, indicated by an arrow, occurs at a pH of 11.0. The inflexion at the equivalence-point is seen to be small, and it is even less marked for more dilute solutions of the acid. It has been calculated that if ak_a is less than about 27 k_w there is no appreciable change in the slope of the pH-neutralization curve as the equivalence-point is attained.

V. Weak Base and Strong Acid.—The equations applicable to the neutralization of weak bases are similar to those for weak acids; the only alterations necessary are that the terms for H^+ and OH^- are exchanged, a and b are interchanged, and k_b replaces k_a. The appropriate form of equation (39), which is fundamental to the whole subject, is

$$a_{OH^-} = K_b \frac{b - B}{B} \cdot \frac{\gamma_B}{\gamma_{BH^+}}, \tag{50}$$

where B is now defined by

$$B \equiv a + c_{OH^-} - c_{H^+}.$$

The Henderson equation, omitting the activity correction, can be written as

$$\text{pOH} = \text{p}k_b + \log \frac{a}{b - a},$$

or

$$\text{pOH} = \text{p}k_b + \log \frac{\text{salt}}{\text{base}},$$

$$\therefore \quad \text{pH} = \text{p}k_w - \text{pOH} = \text{p}k_w - \text{p}k_b - \log \frac{\text{salt}}{\text{base}}. \tag{51}$$

This equation is applicable over the same pH range as before, viz., 4 to 10; outside this range c_{OH^-} may be neglected in more acid solutions, while c_{H^+} can be ignored in more alkaline solutions. At the extremes of the neutralization, i.e., for the pure base and the salt, respectively, the pH values can be obtained by making the appropriate simplifications of equation (50); alternatively, they may be derived from considerations of the dissociation of the base and of the hydrolyzed salt (cf. p. 390).

A little consideration will show that the pH-neutralization curves for weak bases are exactly analogous to those for weak acids, except that they appear at the top right-hand corner of the diagram, with the mid-point, at pH 7, as a center of symmetry. The weaker the base and the less concentrated the solution, the smaller is the change of potential at the equivalence-point, just as in the neutralization of a weak base.

VI. Weak Acid and Weak Base.—The exact treatment of the neutralization of a weak acid by a weak base is somewhat complicated; it is analogous to that for the hydrolysis of a salt of a weak acid and weak base to which brief reference was made on page 381. The result is an equation of the fourth order in c_{H^+} and so cannot be solved easily. The course of the pH-neutralization curve can, however, be obtained, with sufficient accuracy for most purposes, by the use of approximate equations. For the pure weak acid, the pH is given by equation (45) and the values up to about 90 per cent neutralization are obtained by the same equations as were used for the titration of a weak acid by a strong base; as long as there is at least 10 per cent of free excess acid the effect of hydrolysis is negligible. The pH at the equivalence-point is derived from equation (29), based on considerations of the hydrolysis of a salt of a weak acid and weak base. The complete treatment of the region between 90 and 100 per cent neutralization is somewhat complicated, but the general form of the curve can be obtained without difficulty by joining the available points. The variation of the pH in the neutralization of a weak base by a weak acid is derived in an analogous manner; up to about 90 per cent neutralization the behavior is virtually identical

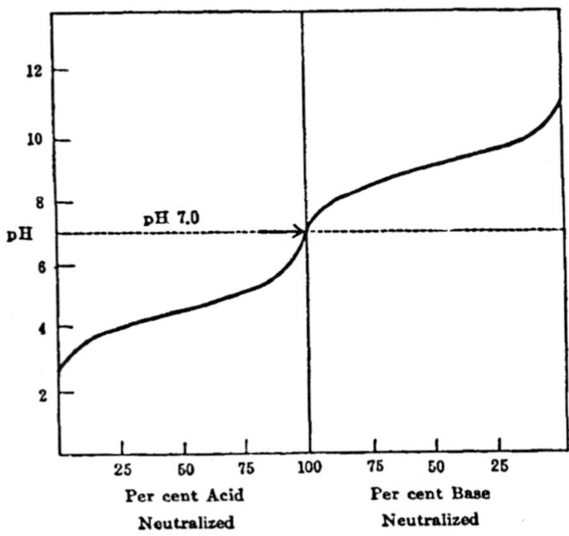

FIG. 103. Neutralization of acetic acid by ammonia

with that obtained for a strong acid. The complete pH-neutralization curve for a 0.1 N solution of acetic acid and 0.1 N ammonia, for which k_a and k_b are both taken to be equal to 1.75×10^{-5}, is shown in Fig. 103; the change of pH is seen to be very gradual throughout the neutralization and is not very marked at the equivalence-point.

Displacement Reactions.—In a displacement reaction a strong acid, or strong base, displaces a weak acid, or weak base, respectively, from one of its salts; an instance which will be considered is the displacement of acetic acid from sodium acetate by hydrochloric acid. Since this process is the opposite of the neutralization of acetic acid by sodium hydroxide, the variation of pH during the displacement reaction will be practically identical with that for the neutralization, except that it is in the reverse direction. In this particular case, therefore, the pH curve is represented by Fig. 102, I, starting from the midpoint, which represents sodium acetate, and finishing at the left-hand end, representing an equivalent amount of free acetic acid. It is evident that there is no sharp change of potential when the equivalence-point is attained. On the other hand, if the salt of a very weak acid, e.g., k_a equal to 10^{-9}, is titrated with hydrochloric acid, the variation of pH is given by Fig. 102, II, also starting from the center and proceeding to the left; a relatively marked inflexion is now observed at the equivalence-point, i.e., at the extreme left of the figure.

The foregoing conclusions are in complete harmony with the concept of acids and bases developed in Chap. IX and of neutralization, in its widest sense, to which reference was made at the beginning of the present chapter. The reaction between sodium acetate and hydrochloric acid, i.e.,

$$\underset{\text{Base}}{(Na^+)Ac^-} + \underset{\text{Acid}}{H_3O^+}(Cl^-) = \underset{\text{Acid}}{HAc} + \underset{\text{Base}}{H_2O} + (Na^+Cl^-),$$

is really the neutralization of the acetate ion base by a strong acid. It was seen on page 338 that the dissociation constant of a conjugate base, such as Ac^-, is equal to k_w/k_a, where k_a is the dissociation constant of the acid HAc; in this case k_a is 1.75×10^{-5} and since k_w is 10^{-14}, it follows that k_b for the acetate ion base is about 5.7×10^{-10}. This represents a relatively weak base and its neutralization would not be expected to be marked by a sharp pH inflexion; this is in agreement with the result derived previously. If the acid is a very weak one, however, the conjugate base is relatively strong; for example, if k_a is 10^{-9} then k_b for the anion base is 10^{-5}. The displacement reaction, which is effectively the neutralization of the anion base by a strong acid, should therefore be accompanied by a change of pH similar to that observed in the neutralization of ammonia by a strong acid.

The arguments given above may be applied equally to the displacement of a weak base, such as ammonia or an amine, from a solution of its salt, e.g., ammonium chloride, by means of a strong base. If the amine RNH_2 has a dissociation constant of about 10^{-5}, its conjugate acid RNH_3^+ will be extremely weak, since k_a will be $10^{-14}/10^{-5}$, i.e., 10^{-9}, and the equivalence-point of the displacement titration will not be marked by an appreciable inflexion. On the other hand, if the base is a very weak one, such as aniline (k_b equal to 10^{-10}), the conjugate anilinium ion acid will be moderately strong, k_a about 10^{-4}, and the equivalence-point will be associated with a definite pH change. It follows, therefore, that only with salts of very weak acids or bases is there any considerable inflexion in the pH curve at the theoretical end-point of the displacement reaction.

Neutralization in Non-Aqueous Media.—As already seen, the magnitude of the inflexion in a pH-neutralization curve depends on the dissociation constant of the acid or base being neutralized; concentration is also important, but for the purposes of the present discussion this will be assumed to be constant. Another important factor, which is less evident at first sight, is the magnitude of k_w; an examination of the equations derived in the previous sections shows that the value of k_w does not affect the pH during the neutralization of an acid, but it has an important influence at the equivalence-point. A decrease of k_w will result in a decrease of hydrogen ion concentration, i.e., the pH is increased, at the equivalence point. When a base is being neutralized, the value of k_w is important, as may be deduced from equation (51); a decrease of k_w, i.e., an increase of pk_w, will be accompanied by a corresponding increase of pH. It may be concluded, therefore, that if the ionic product of water is decreased in some manner, the acid and base parts of the neutralization curve are drawn apart and the inflexion at the equivalence-point is more marked. The two results derived above may be combined in the statement that the smaller k_w/k, where k is the dissociation constant of the acid or base, the greater will be the change of pH as the equivalence-point of a neutralization is approached. The quantity k_w/k is the hydrolysis constant of the salt formed in the reaction; hence, as may be expected, the smaller the extent of hydrolysis the more distinct is the pH inflexion at the end-point of the neutralization. There are thus two possibilities for increasing the sharpness of the approach to the equivalence-point; either k_w may be decreased, while k_a or k_b is approximately unchanged, or k_a or k_b may be increased. The same general conclusions will, of course, be applicable to any other amphiprotic solvent, the quantity k_w being replaced by the corresponding ionic product.

For cation acids, e.g., NH_4^+ or RNH_3^+, or for anion bases, e.g., $CH_3CO_2^-$, the dissociation constants in ethyl alcohol are only slightly less than in water (cf. p. 333), but the ionic product is diminished by a factor of approximately 10^6. It is clear, therefore, from the arguments given above that neutralization of such charged acids and bases will be much more complete in alcoholic solution than in water. The equivalence-points in the neutralization of the anions of acids and of the cations of substituted ammonium salts in alcohol have consequently been found to be accompanied by more marked inflexions than are obtained in aqueous solution.

The dissociation constants of uncharged acids and bases are diminished in the presence of alcohol, and since the ionic product of the solvent is decreased to a somewhat similar extent, the inflexion at the equivalence-point for these substances is similar to that in water.

It was seen on page 371 that lyolysis could be avoided and neutralization made more complete when a weak base was neutralized in a strongly protogenic medium, such as acetic acid. The use of a solvent with a marked proton donating tendency is, effectively, to increase the dissociation constant of the weak base; hence a sharper change of pH is to be expected at the equivalence-point in a strongly protogenic solvent than in water. This argument applies to bases of all types, i.e., charged or uncharged, and the experimental results have been shown to be in accordance with anticipation; the curves in Fig. 104, for example, show the change of pH, as measured by a form of hydrogen electrode, during the course of the neutralization of the extremely weak bases urea and acetoxime by perchloric acid in acetic acid solution.[5] In aqueous solutions these bases would show no detectable change of pH at the equivalence-point. In order to increase the magnitude of the inflexion in the neutralization of a very weak acid it would be necessary to employ a strongly protophilic medium, such as liquid ammonia, or one having no protogenic properties, e.g., acetonitrile.

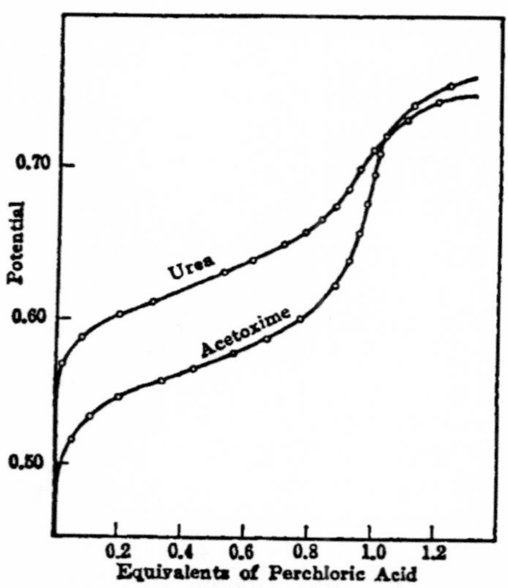

FIG. 104. Neutralization of very weak bases in glacial acetic acid solution

Neutralization of Mixture of Two Monobasic Acids.—An expression for the variation of the pH during the whole course of the neutralization of a mixture of two monobasic acids by a strong base can be derived, but as it is somewhat complicated, simplifications are made which are applicable to certain specific conditions. Let a_{I} and a_{II} be the initial concentrations of the two acids HA_{I} and HA_{II}, whose dissociation constants are k_{I} and k_{II}; suppose that at a certain stage of the neutralization a concentration b of strong base MOH has been added to the mixture of acids. If the salts formed when the acids are neutralized are completely dissociated, then at any instant

$$a_{\text{I}} = c_{\text{HA}_{\text{I}}} + c_{\text{A}_{\text{I}}^-}, \tag{52}$$

and

$$a_{\text{II}} = c_{\text{HA}_{\text{II}}} + c_{\text{A}_{\text{II}}^-}, \tag{53}$$

where c_{HA} represents in each case the concentration of un-neutralized acid while c_{A^-} is that of the neutralized acid, the total adding up to the initial acid concentration. Since the solution must be electrically neutral, the sum of the positive charges must equal that of the negative charges, i.e.,

$$c_{\text{M}^+} + c_{\text{H}^+} = c_{\text{A}_{\text{I}}^-} + c_{\text{A}_{\text{II}}^-} + c_{\text{OH}^-}. \tag{54}$$

The salts MA_{I} and MA_{II} are completely dissociated and so c_{M^+} may be

[5] Hall and Werner, *J. Am. Chem. Soc.*, 50, 2367 (1928); Hall, *Chem. Revs.*, 8, 191 (1931); see also, Nadeau and Branchen, *J. Am. Chem. Soc.*, 57, 1363 (1935).

identified with b, the concentration of added base; further, except towards the end of the neutralization, c_{OH^-} may be neglected, and so equation (54) becomes

$$b + c_{H^+} = c_{A_I^-} + c_{A_{II}^-}. \tag{55}$$

The approximate dissociation constants of the two acids are given by

$$k_I = \frac{c_{H^+}c_{A_I^-}}{c_{HA_I}} \quad \text{and} \quad k_{II} = \frac{c_{H^+}c_{A_{II}^-}}{c_{HA_{II}}},$$

and if these expressions together with equations (52) and (53) are used to eliminate the concentration terms involving A_I^-, A_{II}^-, as well as HA_I and HA_{II}, from (55), the result is

$$c_{H^+} = \frac{a_I k_I}{c_{H^+} + k_I} + \frac{a_{II} k_{II}}{c_{H^+} + k_{II}} - b. \tag{56}$$

This is a cubic equation which can be solved to give the value of the hydrogen ion concentration at any point of the titration of the mixture of acids.

A special case of interest is that arising when the amount of base added is equivalent to the concentration of the stronger of the two acids, e.g., HA_I; under these conditions b may be replaced by a_I, and if all terms of the third order in equation (56) are neglected, since they are likely to be small, it is found that

$$a_I c_{H^+}^2 + k_{II}(a_I - a_{II})c_{H^+} - a_{II}k_I k_{II} = 0.$$

If a_I and a_{II} are not greatly different and k_{II} is small, the second term on the left-hand side in this equation can be omitted, so that

$$c_{H^+} \approx \sqrt{\frac{a_{II}k_I k_{II}}{a_I}}, \tag{57}$$

$$\therefore \quad pH = \tfrac{1}{2}pk_I + \tfrac{1}{2}pk_{II} + \tfrac{1}{2}\log a_I - \tfrac{1}{2}\log a_{II}. \tag{58}$$

This relationship gives the pH at the theoretical first equivalence-point in the neutralization of a mixture of two monobasic acids. If the two acids have the same initial concentration, i.e., a_I is equal to a_{II}, then equation (57) for the first equivalence-point becomes

$$c_{H^+} \approx \sqrt{k_I k_{II}}, \tag{59}$$

$$\therefore \quad pH = \tfrac{1}{2}pk_I + \tfrac{1}{2}pk_{II}. \tag{60}$$

The pH at the equivalence-point for the acid HA_I in the absence of HA_{II} is given by equation (12) as

$$pH = \tfrac{1}{2}pk_w + \tfrac{1}{2}pk_I + \tfrac{1}{2}\log a_I, \tag{61}$$

and comparison of this with the value for the mixture at the first equivalence-point, the latter being designated by $(\text{pH})_m$, shows that in the general case

$$\text{pH} - (\text{pH})_m = \tfrac{1}{2}\text{p}k_w - \tfrac{1}{2}\text{p}k_{\text{II}} + \tfrac{1}{2}\log a_{\text{II}}. \qquad (62)$$

Since $\text{p}k_{\text{II}}$ is generally less than $\text{p}k_w$, the quantity $\text{pH} - (\text{pH})_m$ is positive; the pH at the first equivalence-point of a mixture is thus less than that for the stronger acid alone. This result indicates a flattening of the pH curve in the vicinity of the first equivalence-point, the extent of the flattening being, according to equation (62), more marked the smaller $\text{p}k_{\text{II}}$, i.e., the stronger the acid HA_{II}, and the greater its concentration. If the acid HA_{II} is very weak or its concentration small, or both, the flattening at the first equivalence-point will be negligible, and the neutralization curve of the mixture will differ little from that of the single acid HA_{I}.

Equations for the variation of pH during the course of neutralization beyond the first equivalence-point, similar to those already given, could be derived if necessary, but for most requirements a simpler treatment of the whole neutralization curve is adequate. At the very commencement of the titration the pH is little different from that of the solution of the stronger acid, and during the early stages of neutralization the pH is close to that which would be given by this acid alone. In the vicinity of the first equivalence-point deviations occur, but these can be inferred with sufficient accuracy from the pH at that point, as given by equations (58) or (60). At a short distance beyond the first equivalence-point the pH is close to that for the neutralization of the second acid alone; the pH at the final equivalence-point is the same as that of the salt NaA_{II} and is consequently given by equation (12). A satisfactory idea of the complete neutralization curve can thus be obtained by plotting the curves for the two acids separately side by side, the curve for the stronger acid (HA_{I}) being at the left; the two curves are then joined by a tangent (Fig. 105). The region between the two curves may be fixed more exactly by making use of equation (58) or (60) for the first equivalence-point. The figure shows clearly that if the weaker acid

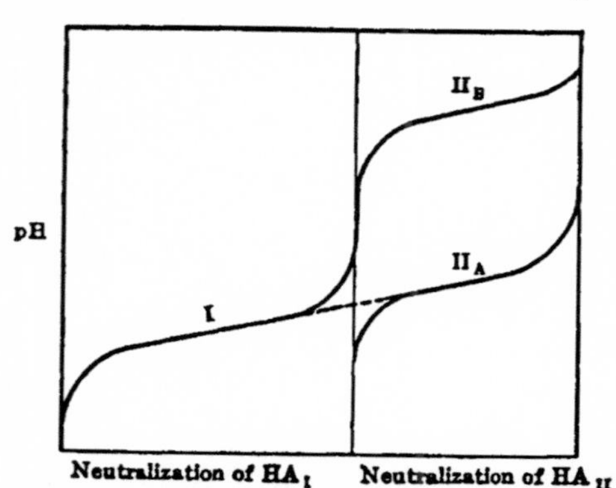

FIG. 105. Neutralization of mixture of acids

HA_{II} is moderately weak, as at II_A, the inflexion at the first equivalence-point will be negligible, but if it is very weak, as at II_B, the inflexion will not differ appreciably from that given by the acid HA_{I} alone. A decrease in the concentration of HA_{II} makes the pH higher at the beginning of the HA_{II} curve and so increases the inflexion to some extent, in agreement with the conclusion already reached. If HA_{I} is a strong acid, e.g., hydrochloric acid, and HA_{II} is a weak acid, the pH follows that for the neutralization of the strong acid alone almost exactly up to the first equivalence-point.

Neutralization of Dibasic Acid by a Strong Base.—If the first stage of the dissociation of the dibasic acid corresponds to that of a strong acid while the second is relatively weak, e.g., chromic acid, the system behaves virtually as two separate acids. The first stage is neutralized as a normal strong acid, then the second stage becomes neutralized independently as a weak acid. When both stages are relatively weak, however, there is some interference between them, and the variation of pH during the course of neutralization may be calculated by means of equations derived in Chap. IX.

For the present purpose, equation (34), page 326, for the hydrogen ion activity of a solution of a dibasic acid, of initial concentration a moles per liter, to which has been added a concentration of b equiv. per liter of strong base, may be written as

$$c_{H^+}^2 \frac{B}{2a - B} = c_{H^+} \frac{a - B}{2a - B} k_1 + k_1 k_2. \tag{63}$$

The activity coefficients have been omitted and the approximate functions k_1 and k_2, for the two stages of dissociation of the dibasic acid, have replaced the corresponding thermodynamic constants. The quantity B is defined as

$$B \equiv b + c_{H^+} - c_{OH^-},$$

and insertion of this value in equation (63) gives a quartic equation for c_{H^+}, which can be solved if necessary. For a considerable range of the neutralization it is possible to neglect c_{OH^-} in the expression for B, and so the equation reduces to a cubic.

At the first equivalence-point, a is equal to b and if c_{OH^-} is neglected, as just suggested, it follows from equation (63) that

$$c_{H^+}^2 = \frac{k_1 k_2 (a - c_{H^+})}{k_1 + a + c_{H^+}}. \tag{64}$$

Since c_{H^+} is generally small in comparison with a, this equation reduces to

$$c_{H^+}^2 = \frac{k_1 k_2 a}{k_1 + a}, \tag{65}$$

which is identical, as it should be, with equation (30), for at the first equivalence-point in the neutralization of the dibasic acid H_2A the system is identical with a solution of NaHA. If k_1 is small, equation (65) becomes, as before,

$$c_{H^+} = \sqrt{k_1 k_2}, \tag{66}$$

$$\therefore \quad \text{pH} = \tfrac{1}{2}\text{p}k_1 + \tfrac{1}{2}\text{p}k_2. \tag{66a}$$

It will be noted that this result is the same as equation (59) for the first equivalence-point in the neutralization of a mixture of equivalent amounts of two weak acids. For a dibasic acid with a very weak first stage dissociation, it may not be justifiable to neglect c_{OH^-}; under these conditions, however, c_{H^+} may be ignored, and the corresponding equations can be derived.

The form of the pH-neutralization curve for a dibasic acid can be represented in an adequate manner by the method used for a mixture of acids; the curves for the two stages are drawn side by side, from the individual dissociation constants k_1 and k_2 treated separately, and then joined by a tangent. The general conclusions drawn concerning the inflexion at the first equivalence-point are similar to those for a mixture of acids; the essential requirement for a dibasic acid to show an appreciable inflexion at the first equivalence-point is that k_1/k_2 should be large. Under these conditions the individual pH-neutralization curves for the two stages of the dibasic acids are relatively far apart and the tangent joining them approaches a vertical direction.

Distribution of Strong Base between the Stages of a Dibasic Acid.— During the course of neutralization of a dibasic acid, the system will consist of undissociated molecules H_2A and of the ions HA^- and A^{--}; the fraction of the total present as HA^- ions, i.e., α_1, is then

$$\alpha_1 = \frac{c_{HA^-}}{c_{H_2A} + c_{HA^-} + c_{A^{--}}}, \tag{67}$$

while that present as A^{--} ions, i.e., α_2, is

$$\alpha_2 = \frac{c_{A^{--}}}{c_{H_2A} + c_{HA^-} + c_{A^{--}}}. \tag{68}$$

Since the HA^- ions arise almost entirely from NaHA, assuming the base to be sodium hydroxide, while the A^{--} ions originate mainly from Na_2A, it follows that α_1 represents, approximately, the fraction of the dibasic acid neutralized in the first stage only, while α_2 is the fraction neutralized in both stages. By using the familiar expressions for the first and second stage dissociation functions (p. 381) to eliminate $c_{A^{--}}$ from equation (67) and c_{HA^-} from equation (68), the results are

$$\alpha_1 = \frac{1}{\frac{c_{H^+}}{k_1} + 1 + \frac{k_2}{c_{H^+}}} \tag{69}$$

and

$$\alpha_2 = \frac{1}{\frac{c_{H^+}}{k_2} + 1 + \frac{c_{H^+}^2}{k_1 k_2}}. \tag{70}$$

It is thus possible, by means of equations (69) and (70), to evaluate the fractions of HA^- and of A^{--} present at any pH for a given dibasic acid, provided k_1 and k_2 are known. As is to be expected, the fraction present as HA^-, i.e., α_1, increases at first as neutralization proceeds; the value then reaches a maximum and falls off to zero when both stages of the acid are completely neutralized. The fraction of A^{--}, on the other hand, increases slowly at first and then more rapidly and finally approaches unity when neutralization is complete and the system consists entirely of Na_2A. Many interesting conclusions can be drawn from the curves for different values of k_1 and k_2 concerning the pH at which the second stage neutralization becomes appreciable, and so on; the main results have, however, already been obtained from a consideration of the pH-neutralization curves.[6]

The point at which the fraction α_1 attains a maximum can be derived by writing $1/\alpha_1$ by means of equation (69) as

$$\frac{1}{\alpha_1} = 1 + \frac{c_{H^+}}{k_1} + \frac{k_2}{c_{H^+}},$$

differentiating with respect to c_{H^+}, thus

$$\frac{d\left(\frac{1}{\alpha_1}\right)}{dc_{H^+}} = \frac{1}{k_1} - \frac{k_2}{c_{H^+}^2},$$

and equating to zero, since $1/\alpha_1$ must be a minimum when α_1 is a maximum. It follows, therefore, that

$$\frac{1}{k_1} = \frac{k_2}{c_{H^+}^2},$$

$$\therefore \quad c_{H^+} = \sqrt{k_1 k_2}, \tag{71}$$

under these conditions. According to equation (66) this is, approximately, the hydrogen ion concentration at the first equivalence-point; hence the fraction of the total acid in the form of HA^- ions is greatest at this point.

Neutralization of Polybasic Acids and Mixtures of Acids.—The treatment of a system consisting of a tribasic or higher acid, or of a mixture of three or more simple acids is complicated, but the general nature of the results can be obtained in the manner already described. The pH-neutralization curve for the whole system is obtained with a fair degree of accuracy by drawing the separate curves for the individual stages of neutralization of the polybasic acid, or for the individual acids in a mixture of acids, in the order of decreasing dissociation constants, and connecting them by means of tangents in the usual way. The pH's at the various equivalence-points can be fixed by using a relationship similar to equation (66); the pH at the nth equivalence-point, i.e., when

[6] Michaelis, "Hydrogen Ion Concentration," translated by Perlzweig, 1926, p. 55.

sufficient strong base has been added to neutralize the first n stages, or n acids, is given by

$$\mathrm{pH} = \tfrac{1}{2}\mathrm{p}k_n + \tfrac{1}{2}\mathrm{p}k_{n+1}, \tag{72}$$

where $\mathrm{p}k_n$ and $\mathrm{p}k_{n+1}$ are the dissociation exponents for the nth and $(n + 1)$th stages, respectively, of a polybasic acid, or of the nth and $(n + 1)$th acids in a mixture arranged in order of decreasing strength.

Another useful method for considering the neutralization of polybasic acids or mixtures of acids, which avoids the necessity of plotting curves, is the following. In general, when an acid is neutralized to the extent of 0.1 per cent, i.e., salt/acid is 1/999, the pH, according to the approximate Henderson equation, is

$$\begin{aligned}\mathrm{pH} &= \mathrm{p}k_a + \log \tfrac{1}{999}\\ &\approx \mathrm{p}k_a - 3.\end{aligned}$$

It follows, therefore, that in a polybasic acid system, or in a mixture of approximately equivalent amounts of different acids, the neutralization of a particular stage or of a particular acid may be regarded as commencing effectively when the pH is equal to $\mathrm{p}k_{n+1} - 3$, where $\mathrm{p}k_{n+1}$ is the dissociation of the $(n + 1)$th stage or acid; at this point the pH-neutralization curve for the mixture will commence to diverge from that of the previous stage of neutralization. Similarly, when an acid is 99.9 per cent neutralized,

$$\begin{aligned}\mathrm{pH} &= \mathrm{p}k_a + \log \tfrac{999}{1}\\ &\approx \mathrm{p}k_a + 3.\end{aligned}$$

The neutralization of any stage may, therefore, be regarded as approximately complete when the pH of the system is equal to $\mathrm{p}k_n + 3$, where $\mathrm{p}k_n$ is the dissociation exponent for the nth stage of a polybasic acid or for the nth acid in a mixture. If this pH is less than $\mathrm{p}k_{n+1} - 3$, the neutralization of the nth stage will be substantially complete before that of the $(n + 1)$th stage commences; if this condition holds, i.e., if $\mathrm{p}k_{n+1} - 3 > \mathrm{p}k_n + 3$, the neutralization of the weaker acid, or stage, will have no appreciable effect on that of the stronger. It is seen, therefore, that if $\mathrm{p}k_{n+1} - \mathrm{p}k_n$ is greater than 6, or k_n/k_{n+1} is greater than 10^6, the pH-neutralization curve for the mixture will show no appreciable divergence, at the nth equivalence-point, from that of the nth acid alone. The inflexion at the nth equivalence-point will then be as definite as for the single nth acid. If $\mathrm{p}k_{n+1} - \mathrm{p}k_n$ is less than 6, the neutralization of the $(n + 1)$th acid, or stage, commences before that of the nth acid is complete, and the result will be a flattening of the pH-curve at the nth equivalence-point; if k_n/k_{n+1} is less than 16, there is no detectable inflexion in the pH-neutralization curve.

Potentiometric Titrations.[7]—The general conclusions drawn from the treatment in the foregoing sections provide the basis for potentiometric,

[7] See general references to potentiometric titrations on page 256.

as well as ordinary volumetric, titrations of acids and bases. The potential E of any form of hydrogen electrode, measured against any convenient reference electrode, is related to the pH of the solution by the general equation

$$E = E_{\text{ref.}} + \frac{RT}{F}\,\text{pH},$$

or, at ordinary temperatures, i.e., about 22°,

$$E = E_{\text{ref.}} + 0.059\,\text{pH},$$

where $E_{\text{ref.}}$ is a constant. It is apparent, therefore, that the curves representing the variation of pH during neutralization are identical in form with those giving the change of hydrogen electrode potential. It should thus be possible to determine the end-point of an acid-base titration by measuring the potential of any convenient form of hydrogen electrode at various points and finding the amount of titrant at which the potential undergoes a sharp inflexion. The underlying principle of the potentiometric titration of a neutralization process is thus fundamentally the same as that involved in precipitation (p. 256) and oxidation-reduction titrations (p. 285). The position of the end-point is found either by graphical determination of the volume of titrant corresponding to the maximum value of $\Delta E/\Delta v$, where ΔE is the change of hydrogen electrode potential resulting from the addition of Δv of titrant, or it can be determined by a suitable adaptation of the principle of differential titration. The apparatus described on page 261 (Fig. 77) can, of course, be employed without modification with glass or quinhydrone electrodes; if hydrogen gas electrodes are used, however, the electrodes are of platinized platinum and the hydrogen must be used for operating the gas-lift, the stream being shut off before each addition of titrant so as to avoid mixing. Any form of hydrogen electrode can be used for carrying out a potentiometric neutralization titration, and even oxygen gas and air electrodes have been employed; since all that is required to be known is the point at which the potential undergoes a rapid change, the irreversibility of these electrodes is not a serious disadvantage. Potentiometric determinations of the end-point of neutralization reactions can be carried out with colored solutions, and often with solutions that are too dilute to be titrated in any other manner.

The accuracy with which the end-point can be estimated obviously depends on the magnitude of the inflexion in the hydrogen potential-neutralization curve at the equivalence-point, and this depends on the dissociation constant of the acid and base, and on the concentration of the solution, as already seen. When a strong acid is titrated with a strong base, the change of potential at the equivalence-point is large, even with relatively dilute solutions (cf. Fig. 101), and the end-point can be obtained accurately. If a weak acid and a strong base, or vice versa, are employed the end-point is generally satisfactory provided the solutions are not too dilute or the acid or base too weak (cf. Fig. 102, I); if c is the concentration of the titrated solution and k_a or k_b the dissociation constant of the weak acid or base being titrated, by a strong base or acid, respectively, then an appreciable break occurs in the neutralization curve at the end-point provided ck_a or ck_b is greater than 10^{-8}. Titrations can be carried out potentiometrically even if ck_a or ck_b is less than 10^{-8}, but the results are less accurate (cf. Fig. 102, II). The potentiometric titration of very weak bases can, of course, be carried out satisfactorily in a strongly protogenic medium (cf. Fig. 104). When a weak acid and weak base are titrated against one another the change of pH at the end-point is never very marked (Fig. 103), but if potential measurements are made carefully, an accuracy of about 1 per cent may be obtained with 0.1 N solutions by determining graphically the position at which $\Delta E/\Delta v$ is a maximum. The principles outlined above apply, of course, to displacement reactions, which are to be regarded as involving neutralization in its widest sense. Such titrations can be performed accurately in aqueous solution if the acid or base that is being displaced is very weak; in other cases satisfactory end-points may be obtained in alcoholic solution.

The separate acids in a mixture of acids, or bases, can often be titrated potentiometrically, provided there is an appreciable difference in their strengths: this condition is realized if one of the acids is strong, e.g., a mineral acid, and the other is weak, e.g., an organic acid. It has been seen that if the ratio of the dissociation constants of two acids exceeds about 10^6, the weaker does not interfere with the neutralization of the stronger acid in the mixture; this conclusion does not take into account the influence of differences of concentration, and it is more correct to say that $c_{I}k_{I}/c_{II}k_{II}$ should be greater than 10^6 where c_I and k_I are the concentration and dissociation constant of one acid and c_{II} and k_{II} that of the other. If this condition is combined with that previously given for obtaining a satisfactory end-point with a single acid, the following conclusions may be drawn: if $c_{I}k_{I}$ and $c_{II}k_{II}$ both exceed 10^{-8} and $c_{I}k_{I}/c_{II}k_{II}$ is greater than 10^6, accurate titration of the separate acids in the mixture is possible. If $c_{I}k_{I}/c_{II}k_{II}$ is less than 10^6 the first equivalence-point cannot be very accurate even if $c_{I}k_{I}$ is greater than 10^{-8}, but an accuracy of about 1 per cent can be achieved by careful titration even if $c_{I}k_{I}/c_{II}k_{II}$ is as low as 10^4. When the first equivalence-point is not detectable, the second equivalence-point, representing neutralization of both acids, may still be obtained provided $c_{II}k_{II}$ exceeds 10^{-8}. The general relationship applicable to mixtures of acids can be extended to polybasic acids, although in the latter case c_I and c_{II} are equal.[8]

In the titration of a strong acid and a strong base the equivalence-point corresponds exactly to the point on the pH-neutralization curve, or the potential-titration curve, at which the slope is a maximum. This

[8] Noyes, *J. Am. Chem. Soc.*, 32, 815 (1910); see also, Tizard and Boeree, *J. Chem. Soc.*, 121, 132 (1922); Kolthoff and Furman, "Indicators," 1926, p. 121.

is not strictly true, however, in the case of the neutralization of a weak acid or a weak base; if $(c_{H^+})_p$ is the hydrogen ion concentration at the potentiometric end-point, i.e., where $\Delta E/\Delta v$ is a maximum, and $(c_{H^+})_s$ is the value at the theoretical, or stoichiometric, equivalence-point, it can be shown that

$$\frac{(c_{H^+})_p}{(c_{H^+})_s} \approx 1 + \frac{3}{2}\sqrt{\frac{k_w}{ak_a}}.$$

Provided ak_a is greater than 10^{-8}, which is the condition for a satisfactory point of inflexion in the titration curve, the ratio of the two hydrogen ion concentrations differs from unity by about one part in 700; this would be equivalent to a potential difference of 0.016 millivolt and hence is well within the limits of experimental error.

Neutralization Titrations with Indicators.—Since, as seen on page 362, an acid-base indicator changes color within a range of approximately one unit of pH on either side of a pH value equal to the indicator exponent (pk_{In}), such indicators are frequently used to determine the end-points of neutralization titrations.[9] The choice of the indicator for a particular titration can best be determined from an examination of the pH-neutralization curve. Before proceeding to consider this aspect of the problem it is useful to define the **titration exponent** (pk_T) of an indicator; this is the pH of a solution at which the indicator shows the color usually associated with the end-point when that indicator is employed in a neutralization titration. It is the general practice in such work to titrate from the lighter to the darker color, e.g., colorless to pink with phenolphthalein and yellow to red with methyl orange; as a general rule a 20 per cent conversion is necessary before the color change can be definitely detected visually, and so if the darker colored form is the one existing in alkaline solution, it follows from the simple Henderson equation (cf. p. 390) that

$$\begin{aligned} pH = pk_T &= pk_{In} + \log\frac{20}{80} \\ &= pk_{In} - 0.6. \end{aligned}$$

This approximate relationship between the titration exponent and pk_{In} is applicable to phenolphthalein and to many of the sulfonephthalein indicators introduced by Clark and Lubs (see Table LXIV, page 364). If the darker color is obtained in acid solution, as is the case with methyl orange and methyl red, then it is approximately true that

$$\begin{aligned} pH = pk_T &= pk_{In} + \log\frac{80}{20} \\ &= pk_{In} + 0.6. \end{aligned}$$

The results quoted in Table LXXII give the titration exponents based

[9] Kolthoff and Furman, "Indicators," 1926, Chap. IV.

TABLE LXXII. TITRATION EXPONENTS OF USEFUL INDICATORS

Indicator	pk_T	End-point Color
Bromphenol blue	4	Purplish-green
Methyl orange	4	Orange
Methyl red	5	Yellowish-red
Bromcresol purple	6	Purplish-green
Bromthymol blue	6.8	Green
Phenol red	7.5	Rose-red
Cresol red	8	Red
Thymol blue	8.8	Blue-violet
Phenolphthalein	9	Pale rose
Thymolphthalein	10	Pale blue

on actual experimental observations, together with practical information, for a number of indicators which may be useful for neutralization titrations; they cover the pH range of from about 4 to 10, since titration indicators are seldom employed outside this range.

In order that a particular indicator may be of use for a given acid-base titration, it is necessary that its exponent should correspond to a pH on the almost vertical portion of the pH-neutralization curve. When the end-point of the titration is approached the pH changes rapidly, and the correct indicator will undergo a sharp color change. The choice of indicator may be readily facilitated by means of Fig. 106 in which the

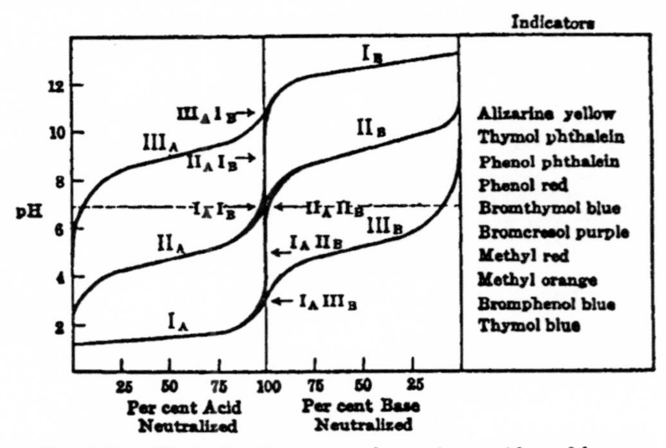

FIG. 106. Neutralization curves for various acids and bases

pH-neutralization curves for a number of acids and bases of different strengths are plotted, while at the right-hand side a series of indicators are arranged at the pH levels corresponding to their titration exponents. The positions of the equivalence-points for the various types of neutralization are marked by arrows. The curves I_A, II_A and III_A show pH changes during the course of neutralization of 0.1 N solutions of a strong acid, a normally weak acid ($k_a = 10^{-5}$) and a very weak acid ($k_a = 10^{-9}$), respectively; curves I_B, II_B and III_B refer to 0.1 N solutions of a strong base, a normally weak base ($k_b = 10^{-5}$) and a very weak base ($k_b = 10^{-9}$), respectively. The complete titration curve for any particular acid and base is obtained by joining the appropriate individual curves.

In the titration of 0.1 N strong acid by 0.1 N strong base (curve I_A–I_B), the pH of the solution undergoes a very sharp change from pH 4 to pH 10 within 0.1 per cent of the equivalence-point (see Table LXIX); any indicator changing color in this range can, therefore, be used to give a reliable indication when the end-point is reached. Consequently, both phenolphthalein, pk_T equal to 9, and methyl orange, pk_T equal to 4, may be employed to give almost identical results in this particular titration. If the solutions are diluted to 0.01 N, however, the change of pH at the equivalence-point is less sharp, viz., from 5 to 9; methyl orange will, therefore, undergo its color change before the end-point is attained, and the titration value would consequently be somewhat too low. When a 0.1 N solution of an acid of k_a equal to 10^{-5} is titrated with a strong base, the equivalence-point is at pH 9, and there is a fairly sharp increase from pH 8 to 10 (curve II_A–I_B); of the common indicators phenolphthalein is the only one that is satisfactory. The less familiar cresolphthalein or thymol blue (second range) could also be used. Any indicator having a titration exponent below 8 is, of course, quite unsatisfactory. In the titration of 0.1 N base of k_b equal to 10^{-5}, the equivalence-point is at pH 5, and the change of potential between pH 4 to 6 is rapid (curve I_A–II_B). Methyl orange is frequently used for such titrations, e.g., ammonia with hydrochloric acid, but it is obvious that the results cannot be too reliable, especially if the solutions are more dilute than 0.1 N; methyl red is a better indicator for a base whose dissociation constant is about 10^{-5}.

It will be evident that if the indicator color is to change sharply at the required end-point, the pH-neutralization curve must rise rapidly at this point. If this curve is not almost vertical, the pH changes slowly and the indicator will show a gradual transition from one color to the other; under these conditions, even if the correct indicator has been chosen, it will be impossible to detect the end-point with any degree of accuracy. In general, the condition requisite for the accurate estimation of a potentiometric end-point, i.e., that ck_a or ck_b should exceed 10^{-8}, is also applicable to titration with an indicator; if ck is less than this value, the results are liable to be in error. They can, however, be improved by using a suitable indicator and titrating to the pH of the theoretical equivalence-point by means of a comparison flask containing a solution of the salt formed at the end-point, together with the same amount of indicator. This procedure may be adopted if it is necessary to titrate a very weak acid or base (curves III_A–I_B and I_A–III_B) or a moderately weak acid by a weak base (curve II_A–II_B); in none of these instances is there a sharp change of pH at the equivalence-point.

Displacement reactions may be treated as neutralizations from the standpoint of the foregoing discussion. If the acid or base displaced is moderately weak, i.e., k_a or k_b is about 10^{-5}, the displacement reaction is equivalent to the neutralization of a very weak base or acid, with k_b or k_a equal to 10^{-9}, respectively; no indicator is likely to give a satisfactory end-point in aqueous solution, although one may possibly be obtained in an alcoholic medium (cf. p. 396). If the acid or base being displaced is very weak, e.g., carbonic acid from a carbonate or boric acid from a borate, there is a marked pH inflexion at the equivalence-point which can be detected with fair accuracy by means of an indicator.

The problem of the detection of the various equivalence-points in a mixture of acids of different concentrations or in a solution of a polybasic acid is essentially the same as that already discussed on page 406 in connection with potentiometric titration, and need not be treated further here. Where the conditions are such that the determination of an accurate end-point appears feasible, the appropriate indicator is the one whose pk_{In} value lies close to the pH at the required equivalence-point.

Buffer Solutions.—It is evident from a consideration of pH-neutralization curves that there are some solutions in which the addition of a small amount of acid or base produces a marked change of pH, whereas in others the corresponding change is very small. A system of the latter type, generally consisting of a mixture of approximately similar amounts of a conjugate weak acid and base, is said to be a **buffer solution;** the resistance to change in the hydrogen ion concentration on the addition of acid or alkali is known as **buffer action.** The magnitude of the buffer action of a given solution is determined by its **buffer capacity;**[10] it is measured by the amount of strong base required to produce unit change of pH in the solution, thus:

$$\text{Buffer capacity } (\beta) \equiv \frac{db}{d(\text{pH})}.$$

An indication of the buffer capacity of any acid-base system can thus be obtained directly from the pH-neutralization curve; if the curve is flat, $d(\text{pH})/db$ is obviously small and the buffer capacity, which is the reciprocal of this slope, is large. An examination of curves IA and IB, Fig. 106, shows that a relatively concentrated solution of strong acid or base is a buffer in regions of low or high pH, respectively. A solution of a weak acid or a weak base alone is not a good buffer, but when an appreciable amount of salt is present, i.e., towards the middle of the individual neutralization curves IIA, IIIA, IIB or IIIB, the buffer capacity of the system is very marked. As the equivalence-point is approached the pH changes rapidly and so the buffer capacity of the salt solution is small. If the acid or base is very weak, or if both are moderately weak, the slope

[10] van Slyke, *J. Biol. Chem.*, **52**, 525 (1922); Kilpi, *Z. physikal. Chem.*, **173**, 223 (1935).

of the pH curve at the equivalence-point is not very great and hence the corresponding salts have moderate buffer capacity.

The buffer action of a solution of a weak acid (HA) and its salt (A^-), i.e., its conjugate base, is explained by the fact that the added hydrogen ions are "neutralized" by the anions of the salt acting as a base, thus

$$H_3O^+ + A^- = H_2O + HA,$$

whereas added hydroxyl ions are removed by the neutralization

$$OH^- + HA = H_2O + A^-.$$

According to the Henderson equation the pH of the solution is determined by the logarithm of the ratio of the concentrations of salt to acid; if this ratio is of the order of unity, it will not be greatly changed by the removal of A^- or HA in one or other of these neutralizations, and so its logarithm will be hardly affected. The pH of the solution will consequently not alter very greatly, and the system will exert buffer action. If the buffer is a mixture of a weak base (B) and its salt, i.e., its conjugate acid (BH^+), the corresponding equations are

$$H_3O^+ + B = H_2O + BH^+$$

and

$$OH^- + BH^+ = H_2O + B.$$

In this case the pH depends on the logarithm of the ratio of B to BH^+, and this will not be changed to any great extent if the buffer contains the weak base and its salt in approximately equivalent amounts.

By the treatment on page 323, the initial concentration of acid, a moles per liter, is equal, at any instant, to the sum of the concentrations of HA and A^-, i.e.,

$$a = c_{HA} + c_{A^-}, \tag{73}$$

and according to the condition for electrical neutrality,

$$b + c_{H^+} = c_{A^-} + c_{OH^-}, \tag{74}$$

where b is the concentration of base added at that instant; since the salt MA is completely dissociated the concentration of M^+ ions, c_{M^+}, has been replaced by b in equation (74). Writing k_a for the dissociation function of the acid, in the usual manner,

$$k_a = \frac{c_{H^+}c_{A^-}}{c_{HA}},$$

and utilizing the value of c_{HA} as $a - c_{A^-}$ given by equation (73), it is found that

$$c_{A^-} = \frac{ak_a}{k_a + c_{H^+}}.$$

Substitution of this expression for c_{A^-}, and k_w/c_{H^+} for c_{OH^-}, in equation

(74), yields the result

$$b = \frac{ak_a}{k_a + c_{H^+}} - c_{H^+} + \frac{k_w}{c_{H^+}}.$$

Remembering that pH is defined, for present purposes, as $-\log c_{H^+}$, differentiation of this equation with respect to pH gives the buffer capacity of the system, thus

$$\beta = \frac{db}{d(\text{pH})} = 2.303\left(\frac{ak_a c_{H^+}}{(k_a + c_{H^+})^2} + c_{H^+} + \frac{k_w}{c_{H^+}}\right). \qquad (75)$$

In the effective buffer region the buffer capacity is determined almost exclusively by the first term in the brackets; hence, neglecting the other terms, it follows that

$$\beta = 2.303a\,\frac{k_a c_{H^+}}{(k_a + c_{H^+})^2}. \qquad (76)$$

The quantity a represents the total concentration of free acid and salt, and so the buffer capacity is proportional to the total concentration of the solution.

To find the pH at which β is a maximum this expression should be differentiated with respect to pH and the result equated to zero; thus

$$\frac{d^2b}{d(\text{pH})^2} = (2.303)^2 ak_a\left(\frac{2c_{H^+}^2}{(k_a + c_{H^+})^3} - \frac{c_{H^+}}{(k_a + c_{H^+})^2}\right) = 0,$$

$$\therefore \quad k_a = c_{H^+}. \qquad (77)$$

It follows, therefore, that the buffer capacity is a maximum when the hydrogen ion concentration of the buffer solution is equal to the dissociation constant of the acid. This condition, i.e., pH is equal to pk_a, arises when the solution contains equivalent amounts of the acid and its salt; such a system, which corresponds to the middle of the neutralization curve of the acid, has the maximum buffer capacity. The actual value of β at this point is found by inserting the condition given by (77) into equation (76); the result is

$$\beta_{\text{max.}} = \frac{2.303}{4}\,a, \qquad (78)$$

and so it is independent of the actual dissociation constant. Exactly analogous results can, of course, be deduced for buffer systems consisting of weak bases and their salts, although it is convenient to consider them as involving the cation acid (BH^+) and its conjugate base (B). The conclusions reached above then hold exactly; the dissociation constant k_a refers to that of the acid BH^+, and is equal to k_w/k_b, where k_b is that of the base B.

Buffer Capacity of Water.—According to equation (74), the condition for electrical neutrality, when a strong base of concentration b has been added to water or to a solution containing a strong acid HA, is

$$b = c_{A^-} - c_{H^+} + c_{OH^-}$$
$$= c_{A^-} - c_{H^+} + k_w/c_{H^+},$$

and differentiation with respect to pH, i.e., $-\log c_{H^+}$, gives the buffer capacity β_{H_2O} of water as

$$\beta_{H_2O} = \frac{db}{d(\text{pH})} = 2.303\left(c_{H^+} + \frac{k_w}{c_{H^+}}\right)$$
$$= 2.303(c_{H^+} + c_{OH^-}). \tag{79}$$

It should be noted that the further addition of base does not affect the concentration of A^- and so its derivative with respect to pH is zero. The buffer capacity of water, as given by equation (79), is negligible between pH values of 2.4 and 11.6, but in more strongly acid, or more strongly alkaline, solutions the buffer capacity of "water" is evidently quite considerable. This conclusion is in harmony with the fact that the pH-neutralization curve of a strong acid or strong base is relatively flat in its early stages.

Preparation of Buffer Solutions.—The buffer capacity of a given acid-base system is a maximum, according to equation (77), when there are present equivalent amounts of acid and salt; the hydrogen ion concentration is then equal to k_a and the pH is equal to pk_a. If the ratio of acid to salt is increased or decreased ten-fold, i.e., to 10 : 1 or 1 : 10, the hydrogen ion concentration is then $10k_a$ or $0.1k_a$, and the pH is p$k_a - 1$ or p$k_a + 1$, respectively. If these values for c_{H^+} are inserted in equation (76), it is found that the buffer capacity is then

$$\beta = \frac{2.303}{12.1}a, \tag{80}$$

which is only about one-third of the value at the maximum. If the pH lies within the range of p$k_a - 1$ to p$k_a + 1$ the buffer capacity is appreciable, but outside this range it falls off to such an extent as to be of relatively little value. It follows, therefore, that a given acid-base buffer system has useful buffer action in a range of one pH unit on either side of the pk_a of the acid. In order to cover the whole range of pH, say from 2.4 to 11.6, i.e., between the range of strong acids and bases, it is necessary to have a series of weak acids whose pk_a values differ by not more than 2 units.

To make a buffer solution of a given pH, it is first necessary to choose an acid with a pk_a value as near as possible to the required pH, so as to obtain the maximum buffer capacity. The actual ratio of acid to salt necessary can then be found from the simple Henderson equation

$$\text{pH} = \text{p}k_a + \log\frac{\text{salt}}{\text{acid}},$$

provided the pH lies within the range of 4 to 10. If the required pH is less than 4 or greater than 10, it is necessary to use the appropriate form of equation (40), where B is defined by (42). Sometimes a buffer solution is made up of two salts representing different stages of neutralization of a polybasic acid, e.g., NaH_2PO_4 and Na_2HPO_4; in this case the former provides the acid $H_2PO_4^-$ while the latter is the corresponding salt, or conjugate base HPO_4^{--}.

In view of the importance of buffer mixtures in various aspects of scientific work a number of such solutions have been made up and their pH values carefully checked by direct experiment with the hydrogen gas electrode. By following the directions given in each case a solution of any desired pH can be prepared with rapidity and precision. A few of the mixtures studied, and their effective ranges, are recorded in Table LXXIII;[11] for further details the original literature or special monographs should be consulted.

TABLE LXXIII. BUFFER SOLUTIONS

Composition	pH Range	Composition	pH Range
Hydrochloric acid and Potassium chloride	1.0–2.2	Potassium dihydrogen phosphate and Sodium hydroxide	5.8– 8.0
Glycine and Hydrochloric acid	1.0–3.7	Boric acid and Borax	6.8– 9 2
Potassium acid phthalate and Hydrochloric acid	2.2–3.8	Diethylbarbituric acid and Sodium salt	7.0– 9.2
Sodium phenylacetate and Phenylacetic acid	3.2–4.9	Borax and Hydrochloric acid	7.6– 9.2
Succinic acid and Borax	3.0–5.8	Boric acid and Sodium hydroxide	7.8–10.0
Acetic acid and Sodium acetate	3.7–5.6	Glycine and Sodium hydroxide	8.2–10.1
Potassium acid phthalate and Sodium hydroxide	4.0–6.2	Borax and Sodium hydroxide	9.2–11.0
Disodium hydrogen citrate and Sodium hydroxide	5.0–6.3	Disodium hydrogen phosphate and Sodium hydroxide	11.0–12.0

Each buffer system is generally applicable over a limited range, viz., about 2 units of pH, but by making suitable mixtures of acids and acid salts, whose pk_a values differ from one another by 2 units or less, it is possible to prepare a **universal buffer mixture**; by adding a pre-determined amount of alkali, a buffer solution of any desired pH from 2 to 12 can be obtained. An example of this type of mixture is a system of citric acid, diethylbarbituric acid (veronal), boric acid and potassium dihydrogen phosphate; this is virtually a system of seven acids whose exponents are given below.

[11] For details concerning the preparation of buffer solutions, see Clark, "The Determination of Hydrogen Ions," 1928, Chap. IX; Britton, "Hydrogen Ions," 1932, Chap. XI; Kolthoff and Rosenblum, "Acid-Base Indicators," 1937, Chap. VIII.

	Citric acid 1st stage	Citric acid 2nd stage	Citric acid 3rd stage	$H_2PO_4^-$	Veronal	Boric acid	HPO_4^{--}
pk_a	3.06	4.74	5.40	7.21	7.43	9.24	12.32

Apart from the last two acids, the successive pk_a values differ by less than 2 units, and so the system, when appropriately neutralized, is capable of exhibiting appreciable buffer capacity over a range of from pH 2 to 12.

Influence of Ionic Strength.—In the discussion so far the activity factor has been omitted from the Henderson equation, and so the results may be regarded as applicable to dilute solutions only. Further, the pH values recorded in the literature for given buffer solutions apply to systems of exactly the concentrations employed in the experiments; if the solution is diluted or if a neutral salt is added, the pH will change because of the alteration of the activity coefficients which are neglected in the simple Henderson equation. In order to make allowance for changes in the ionic strength of the medium, and of the accompanying changes in the activity coefficients, it is convenient to use the complete form of the Henderson equation with the activity coefficients expressed in terms of the ionic strength by means of the Debye-Hückel relationship; as shown on page 326, this may be written as

$$\text{pH} = \text{p}K_n + \log \frac{B}{a - B} - (2n - 1)A\sqrt{\mu} + C\mu, \qquad (81)$$

where $\text{p}K_n$ is the exponent for the nth stage of ionization of the acid, and B has the same significance as before [cf. equation (42)]. If the pH lies between 4 and 10, the fraction $B/(a - B)$ may be replaced by the ratio of "salt" to "acid," as on page 390. For a monobasic acid, e.g., acetic or boric acid, n is unity, and equation (81) reduces to equation (41), but if the acid has a higher basicity, the result is somewhat different. For example, if the buffer consists of KH_2PO_4 and Na_2HPO_4, the concentration of "acid," i.e., $H_2PO_4^-$, may be put equal to that of KH_2PO_4, while that of its "salt" is equal to the concentration of Na_2HPO_4; the dissociation constant of the acid $H_2PO_4^-$ is that for the second stage of phosphoric acid, i.e., K_2, and n is equal to 2; equation (81) thus becomes, in this particular case,

$$\text{pH} = \text{p}K_2 + \log \frac{Na_2HPO_4}{KH_2PO_4} - 3A\sqrt{\mu} + C\mu.$$

The value of A is known to be 0.509 at 25° (cf. p. 146), but that of C must be determined by experiment; to do this two or more measurements of the pH are made in solutions containing a constant ratio of "acid" to "salt" at different ionic strengths. Once C is known, an interpolation formula is available which permits the pH to be calculated at any desired ionic strength.[12]

[12] Cohn *et al.*, *J. Am. Chem. Soc.*, 49, 173 (1927); 50, 696 (1928); Green, *ibid.*, 55, 2331 (1933).

It can be readily seen from equation (81) that the effect of ionic strength is greater the higher the basicity of the "acid" constituent of the buffer solution.

The effect of varying the ionic strength of a buffer solution of constant composition may be expressed quantitatively by differentiating equation (81) with respect to $\sqrt{\mu}$, thus

$$\frac{d(\mathrm{pH})}{d\sqrt{\mu}} = -(2n - 1)A + 2C\sqrt{\mu}.$$

It follows therefore that a change in the ionic strength, resulting from a change in the concentration of the buffer solution or from the addition of neutral salts, results in a greater change in the pH the higher the value of n, i.e., the higher the stage of dissociation of the acid whose salts constitute the buffer system. The change of pH may be positive or negative, depending on the conditions.[13]

PROBLEMS

1. Calculate the degree of hydrolysis and pH of (i) 0.01 N sodium formate, (ii) 0.1 N sodium phenoxide, (iii) N ammonium chloride, and (iv) 0.01 N aniline hydrochloride at 25°. The following dissociation constants may be employed: formic acid, 1.77×10^{-4}; phenol, 1.20×10^{-10}; ammonia, 1.8×10^{-5}; aniline, 4.00×10^{-10}.

2. If equivalent amounts of aniline and phenol are mixed, what proportion, approximately, of salt formation may be expected in aqueous solution? What would be the pH of the resulting mixture?

3. A 0.046 N solution of the potassium salt of a weak monobasic acid was found to have a pH of 9.07 at 25°; calculate the hydrolysis constant and degree of hydrolysis of the salt, in the given solution, and the dissociation constant of the acid.

4. It was found by Williams and Soper [*J. Chem. Soc.*, 2469 (1930)] that when 1 liter of a solution containing 0.03086 mole of *o*-nitraniline and 0.05040 mole of hydrochloric acid was shaken with 60 cc. of heptane until equilibrium was established at 25° that 50 cc. of the heptane layer contained 0.0989 g. of the free base. The distribution coefficient of *o*-nitraniline between heptane and water is 1.790. Determine the hydrolysis constant of the amine hydrochloride.

5. The equivalent conductance of a 0.025 N solution of sodium hydroxide was found by Kameyama [*Trans. Electrochem. Soc.*, 40, 131 (1921)] to be 228.4 ohms^{-1} $\mathrm{cm.}^2$ The addition of various amounts of cyanamide to the solution, so that the molecular ratio of cyanamide to sodium hydroxide was x, gave the following equivalent conductances:

x	1.0	1.5	2.0	4.0
Λ	105.8	94.4	94.1	93.3

Calculate the hydrolysis constant of sodium cyanamide, $NaHCN_2$.

[13] Morton, *J. Chem. Soc.*, 1401 (1928); see also, Kolthoff and Rosenblum, "Acid-Base Indicators," 1937, p. 269.

6. Hattox and De Vries [*J. Am. Chem. Soc.*, **58**, 2126 (1936)] determined the hydrogen ion activities in solutions of indium sulfate, $In_2(SO_4)_3$, at various molalities (m) at 25°; the results were:

$m \times 10^2$	9.99	5.26	2.81	1.58	1.00
pH	2.01	2.20	2.36	2.57	2.69

Evaluate the hydrolytic constants for the two reactions

$$In^{+++} + H_2O = InO^+ + 2H^+$$

and

$$In^{+++} + H_2O = In(OH)^{++} + H^+,$$

and determine from the results which is the more probable. Allowance may be made for the activity of the ions by using the Debye-Hückel equation in the approximate form $\log f_i = -0.5z_i^2\sqrt{\mu}/(1 + \sqrt{\mu})$.

7. The pH of a 0.05 molar solution of acid potassium phthalate is 4.00; the first stage dissociation constant of phthalic acid is 1.3×10^{-3}; what is pk_2 for this acid?

8. Plot the pH-neutralization curves for 0.1 N solutions of (i) formic acid and (ii) phenol, by a strong base. Use the dissociation constants given in Problem 1.

9. Plot the pH-neutralization curves for a mixture of (i) N hydrochloric acid and 0.1 N acetic acid, and (ii) 0.01 N hydrochloric acid and 0.1 N acetic acid. What are the possibilities of estimating the amount of each acid separately by titration?

10. Use the data on page 415 to plot the complete pH-neutralization curve of citric acid in a 0.1 molar solution. Over what range of pH could partially neutralized citric acid be expected to have appreciable buffer capacity?

11. Plot the pH-buffer capacity curve for mixtures of acetic acid and sodium acetate of total concentration 0.2 N. Points should be obtained for mixtures containing 10, 20, 30, 40, 50, 60, 70, 80 and 90 per cent of sodium acetate, the pH's being estimated by the approximate form of the Henderson equation. Plot the buffer capacity curve for water at pH's 1, 2, 3 and 4, and superimpose the result on the curve for acetic acid.

12. Utilize the general form of the acetic acid-acetate buffer capacity curve obtained in Problem 11 to draw an approximate curve for the buffer capacity over the range of pH from 2 to 13 of the universal buffer mixture described on page 415. It may be assumed that the total concentration of each acid and its salt is always 0.2 molar.

13. It is desired to prepare a buffer solution of pH 4.50 having a buffer capacity of 0.18 equiv. per pH; suggest how such a solution would be prepared, using phenylacetic acid ($pK_a = 4.31$) and sodium hydroxide.

CHAPTER XII

AMPHOTERIC ELECTROLYTES

Dipolar Ions.—The term "amphoteric" is applied to all substances which are capable of exhibiting both acidic and basic functions; among these must, therefore, be included water, alcohols and other amphiprotic solvents and a number of metallic hydroxides, e.g., lead and aluminum hydroxides. In these compounds it is generally the same group, viz., $- OH$, which is responsible for the acidic and basic properties; the discussion in the present chapter will, however, be devoted to those **amphoteric electrolytes,** or **ampholytes,** that contain *separate* acidic and basic groups. The most familiar examples of this type of ampholyte are provided by the amino-acids, which may be represented by the general formula NH_2RCO_2H. Until relatively recent times these substances were usually regarded as having this particular structure in the neutral state, and it was assumed that addition of acid resulted in the neutralization of the $- NH_2$ group, viz.,

$$NH_2RCO_2H + H_3O^+ = {}^+NH_3RCO_2H + H_2O,$$

whereas a strong base was believed to react with the $- CO_2H$ group, viz.,

$$NH_2RCO_2H + OH^- = NH_2RCO_2^- + H_2O.$$

It has been long realized, however, that in addition to the uncharged molecules NH_2RCO_2H, a solution of an amino-acid might contain molecules carrying a positive charge at one end and a negative charge at the other, thus constituting an electrically neutral system, viz., ${}^+NH_3RCO_2^-$. These particles have been variously called **zwitterions,** i.e., hermaphrodite (or hybrid) ions, amphions, ampholyte ions, dual ions and dipolar ions. The existence of these dual ions was postulated by Küster (1897) to explain the behavior of methyl orange which, in its neutral form, is an amino-sulfonic acid, but their importance in connection with ampholytic equilibria in amino-carboxylic acids was not clearly realized. The suggestion was made by Bjerrum,[1] however, that nearly the whole of a neutral aliphatic amino-acid is present in solution in the form of the dipolar ion, and that reaction with acids and bases is of a different type from that represented above. A solution of glycine, for example, i.e., $NH_2CH_2CO_2H$, is compared with one of ammonium acetate; if a strong acid is added to the latter, the reaction is with the basic $CH_3CO_2^-$ ion and CH_3CO_2H is formed, but a strong base reacts with the acidic NH_4^+

[1] Bjerrum, *Z. physik. Chem.*, **104**, 417 (1923); see also, Adams, *J. Am. Chem. Soc.*, **38**, 1503 (1916).

ion to yield NH_3. In the same way, the addition of strong acid to glycine, consisting mainly of the dual ions $^+NH_3CH_2CO_2^-$, results in the reaction

$$^+NH_3CH_2CO_2^- + H_3O^+ = {}^+NH_3CH_2CO_2H + H_2O,$$

while reaction with alkali is

$$^+NH_3CH_2CO_2^- + OH^- = NH_2CH_2CO_2^- + H_2O.$$

The products are, of course, the same as in the alternative representation, since there is no doubt that in acid solution the amino-acid forms $^+NH_3CH_2CO_2H$ ions while in alkaline solution the anions $NH_2CH_2CO_2^-$ are formed. It should be noted, however, that the groups exhibiting the acidic and basic functions are the reverse of those accepted in the original treatment of amino-acids; the basic property of the ampholyte is due to the $-CO_2^-$ group whereas the acidic property is that of the $-NH_3^+$ group.

Evidence for the Existence of Dipolar Ions.—The evidence for the presence of large proportions of dipolar ions in solutions of aliphatic amino-acids is very convincing. According to the older treatment the dissociation constants of the $-NH_2$ and $-CO_2H$ groups were extremely small, viz., about 10^{-8} to 10^{-12}; such low values were difficult to understand if they referred to these particular groups, but they are not at all unexpected if they really apply, as just suggested, to the conjugate groups $-NH_3^+$ and $-CO_2^-$, respectively. The ammonium ion acids, e.g., RNH_3^+, and anion bases, e.g., RCO_2^-, are known, from the facts mentioned in previous chapters, to have very low dissociation constants. In changing from water to a medium of lower dielectric constant, such as ethyl alcohol, the dissociation constants of cation acids and of anion bases are not affected appreciably, although the values for carboxylic acids are greatly decreased and those of amines are diminished to a lesser extent (cf. p. 333). It is therefore significant that the acidic and basic dissociation constants of aliphatic amino-acids, as determined from pH measurements in the course of neutralization by alkali and acid, respectively (see Chap. IX), are apparently slightly larger in ethyl alcohol than in water. It is evident that the groups being neutralized cannot be $-CO_2H$ and $-NH_2$, but are probably $-NH_3^+$ and $-CO_2^-$, respectively. Further, if the neutral amino-acid has the structure NH_2RCO_2H, it is to be expected that the basic dissociation constant would be almost the same as that of the corresponding methyl ester NH_2RCO_2Me; actually the two values are of an entirely different order, and hence it appears that the basic groups are not the same in the acid and the ester.

The addition of formaldehyde to an aqueous solution of an amino-acid results in no change in the curve showing the variation of pH in the course of the neutralization by acid, but that for the neutralization by alkali is shifted in the direction of increased acid strength, as shown in Fig. 107. It is known that the formaldehyde reacts with the amino-portion of the amino-acid, and it is evidently this part of the molecule which is neutralized by the alkali. The acidic portion of the electrically neutral ampholyte must consequently be the $-\ NH_3^+$ group.[2]

Important evidence for the dual-ion structure of aliphatic amino-acids has been provided by a study of their Raman spectra; in these spectra each group, or, more exactly, each type of linkage, exhibits a characteristic line. It has been found that neutral amino-acids do not show the line which is characteristic of the carboxylic acid group in aqueous solution, and so the former presumably do not possess this group. When alkali is added to an ordinary carboxylic acid, e.g., acetic acid, the characteristic line of the $-CO_2H$ group disappears, but it appears when a strong acid is added to an amino-acid solution. This is striking evidence for the argument that the basic function of the latter is exercised by the $-\ CO_2^-$ group, for the addition of acid would convert this into $-\ CO_2H$, in harmony with the findings from the Raman spectra. Similarly, free amines have a characteristic Raman line which is absent from the spectrum of an aliphatic amino-acid; the line appears, however, when the latter is neutralized by alkali, implying that reaction takes place with the $-\ NH_3^+$ group.[3]

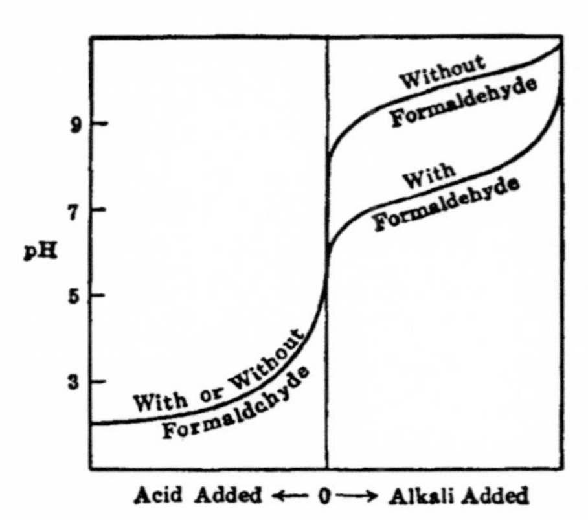

Fig. 107. Titration of amino-acid with and without formaldehyde

There are several other properties of amino-acids which are in agreement with the dipolar-ion type of structure: these are the high melting point, the sparing solubility in alcohol and acetone, and increased solubility in the presence of neutral salts, all of which are properties associated with ionized substances. Examination of crystals of glycine by the method of X-ray diffraction indicates that the substance has the structure $^+NH_3CH_2CO_2^-$ in the solid state. The high dielectric constants of aqueous solutions of aliphatic amino-acids lead to the conclusion that the molecules have very large dipole moments; such large values can only be explained by the presence within the molecule of unit charges of opposite sign separated by several atomic diameters, as would be expected for dipolar ions.[4]

Attention should be called to the fact that the arguments given above apply primarily to aliphatic amino-acids; it is true that aromatic amino-

[2] Harris, *Biochem. J.*, **24**, 1080, 1086 (1930).

[3] Edsall, *J. Chem. Phys.*, **4**, 1 (1936); **5**, 225 (1937).

[4] For summaries of evidence, see Richardson, *Proc. Roy. Soc.*, **115B**, 121 (1934); Neuberger, *ibid.*, **115B**, 180 (1934).

sulfonic acids also exist largely in the dual-ion form, but amino-benzoic acids and amino-phenols consist almost exclusively of neutral, uncharged molecules in aqueous solutions. The properties of these substances are quite different from those of the aliphatic acids.

Dissociation Constants of Amino-Acids.—A very considerable simplification in the treatment of amino-acids can be achieved by regarding them as dibasic acids. Consider, for example, the hydrochloride of glycine, i.e., $Cl^- \cdot {}^+NH_3CH_2CO_2H$; when this is neutralized by an alkali hydroxide, there are two stages of the reaction, corresponding in principle to the two stages of neutralization of a dibasic acid, thus

$$(1) \quad {}^+NH_3CH_2CO_2H + OH^- = {}^+NH_3CH_2CO_2^- + H_2O$$

and

$$(2) \quad {}^+NH_3CH_2CO_2^- + OH^- = NH_2CH_2CO_2^- + H_2O.$$

The two acidic groups are $-\ CO_2H$ and $-\ NH_3^+$, and since the former is undoubtedly the stronger of the two, it will be neutralized first.

It will be noted that the first stage produces the so-called neutral form of the amino-acid which, in this instance, consists almost exclusively of the dual-ion form. If, in the most general case, the dipolar ion, i.e., ${}^+NH_3RCO_2^-$, is represented by $RH^\pm$, the positive ion existing in acid solution, i.e., ${}^+NH_3RCO_2H$, by RH_2^+, and the negative ion present in alkaline solution, i.e., $NH_2RCO_2^-$, by R^-, the two stages of ionization of the dibasic acid ${}^+NH_3RCO_2H$ may be written as

$$(1) \quad RH_2^+ + H_2O \rightleftharpoons H_3O^+ + RH^\pm$$

and

$$(2) \quad RH^\pm + H_2O \rightleftharpoons H_3O^+ + R^-.$$

In the first stage the dissociation of ${}^+NH_3RCO_2H$ is that of the carboxylic acid, while in the second stage the ammonium ion acid dissociates. Applying the law of mass action to these ionization equilibria, the dissociation constants of the two stages are

$$K_1 = \frac{a_{H^+}a_{RH^\pm}}{a_{RH_2^+}} \quad \text{and} \quad K_2 = \frac{a_{H^+}a_{R^-}}{a_{RH^\pm}}, \tag{1}$$

respectively.

The values of these dissociation constants may be determined by means of cells without liquid junction in a manner similar to that described in Chap. IX.[5] For the first stage the acid is the hydrochloride $Cl^- \cdot {}^+NH_3RCO_2H$, i.e., $RH_2^+Cl^-$ and the corresponding "salt" is the electrically neutral form ${}^+NH_3RCO_2^-$, i.e., $RH^\pm$, and so the appropriate cell without liquid junction is

$$H_2(1\ \text{atm.}) \mid RH_2^+Cl^-(m_1)\ RH^\pm(m_2)\ AgCl(s) \mid Ag.$$

[5] Nims and Smith, *J. Biol. Chem.*, 101, 401 (1933); Owen, *J. Am. Chem. Soc.*, 56, 24 (1934); Smith, Taylor and Smith, *J. Biol. Chem.*, 122, 109 (1937).

The E.M.F. of this cell is written in the usual manner, as

$$E = E^0 - \frac{RT}{F} \ln a_{H^+}a_{Cl^-}, \tag{2}$$

and introducing the definition of K_1 given by equation (1), this becomes

$$E = E^0 - \frac{RT}{F} \ln \frac{a_{RH_2^+}a_{Cl^-}}{a_{RH^\pm}} - \frac{RT}{F} \ln K_1,$$

which on rearranging, and replacing the activities by the product of the molalities and the activity coefficients, gives

$$E - E^0 + \frac{RT}{F} \ln \frac{m_{RH_2^+}m_{Cl^-}}{m_{RH^\pm}} + \frac{RT}{F} \ln \frac{\gamma_{RH_2^+}\gamma_{Cl^-}}{\gamma_{RH^\pm}} = -\frac{RT}{F} \ln K_1,$$

$$\therefore \quad \frac{F(E - E^0)}{2.303RT} + \log \frac{m_{RH_2^+}m_{Cl^-}}{m_{RH^\pm}} + \log \frac{\gamma_{RH_2^+}\gamma_{Cl^-}}{\gamma_{RH^\pm}} = -\log K_1. \tag{3}$$

The variation of the activity coefficient of the dipolar ion with ionic strength is given by an expression of the form $\log \gamma_{RH^\pm} = -C'\mu$ (cf. p. 432) and since the values of $\log \gamma_{RH_2^+}$ and $\log \gamma_{Cl^-}$ for the univalent ions RH_2^+ and Cl^- can be written, with the aid of the extended Debye-Hückel equation, as $-A\sqrt{\mu} + C''\mu$, it follows that equation (3) may be put in the form

$$\frac{F(E - E^0)}{2.303RT} + \log \frac{m_{RH_2^+}m_{Cl^-}}{m_{RH^\pm}} - 2A\sqrt{\mu} = -\log K_1 - C\mu. \tag{4}$$

The plot of the left-hand side of equation (4) against the ionic strength should thus be a straight line and the intercept for zero ionic strength should give the value of $-\log K_1$. As in the case treated on page 315, the salt may be taken as completely dissociated so that m_{Cl^-} is equal to m_1; $m_{RH^\pm}$ is equal to $m_2 + m_{H^+}$, and $m_{RH_2^+}$ is $m_1 - m_{H^+}$. The value of m_{H^+}, the hydrogen ion concentration, required for this purpose is best obtained from equation (2) which may be written in the form

$$-\frac{RT}{F} \ln m_{H^+} = E - E^0 + \frac{RT}{F} \ln m_{Cl^-} + \frac{RT}{F} \ln \gamma_{H^+}\gamma_{Cl^-}.$$

The product of the activity coefficients can be estimated from the Debye-Hückel equations, and m_{Cl^-} and E^0 are known; hence m_{H^+} in the given solution can be derived from the measured E.M.F. of the cell.

In order to determine K_1 a series of cells of the type depicted above, in which the ratio of m_1 to m_2 is kept constant but the amounts of $RH_2^+Cl^-$ and of $RH^\pm$ are varied, are set up and the E.M.F.'s (E) measured. The value of E^0 for the hydrogen-silver chloride cell is known, and so the left-hand side of equation (4) can be evaluated; the Debye-Hückel factor A is 0.509 at 25°. In calculating the ionic strength of the solution the dipolar ion $RH^{\pm}$ is treated as a neutral molecule so that it may be regarded as making no contribution to the total. The plot of the left-hand side of equation (4) against the ionic strength is not exactly linear, but it is sufficiently close for an accurate value of K_1 to be obtained by extrapolation.

In the determination of the second dissociation constant (K_2) the "acid" is the neutral form $^+NH_3RCO_2^-$, i.e., $RH^{\pm}$, whereas the corresponding "salt" is the sodium salt $NH_2RCO_2^- \cdot Na^+$, i.e., Na^+R^-; the cell without liquid junction will thus be

$$H_2(1\text{ atm.}) \mid RH^{\pm}(m_1)\ Na^+R^-(m_2)\ NaCl(m_3)\ AgCl(s) \mid Ag.$$

The E.M.F. is given by the general equation (2), and introduction of the value of K_2 from equation (1) results in the expression,

$$E = E^0 - \frac{RT}{F}\ln\frac{a_{RH^{\pm}}a_{Cl^-}}{a_{R^-}} - \frac{RT}{F}\ln K_2,$$

and hence, using the same procedure as before,

$$\frac{F(E - E^0)}{2.303RT} + \log\frac{m_{RH^{\pm}}m_{Cl^-}}{m_{R^-}} = -\log\frac{\gamma_{RH^{\pm}}\gamma_{Cl^-}}{\gamma_{R^-}} - \log K_2. \qquad (5)$$

The activity coefficient term in equation (5) involves a univalent ion in the numerator and denominator, in addition to the dual ion; it follows, therefore, that in dilute solution this term is proportional to the ionic strength. The plot of the left-hand side of equation (5) against μ will thus be linear at low ionic strengths, and the intercept for μ equal to zero gives $-\log K_2$. The experimental procedure is similar to that described for the evaluation of K_1.

Approximate Methods for Dissociation Constants.—Approximate, but more rapid, methods, similar to those used for simple monobasic acids and monoacid bases, have been frequently employed to determine dissociation constants of ampholytes.[6] Upon taking logarithms, the equation for K_1 may be written as

$$\log K_1 = \log a_{H^+} + \log\frac{a_{RH^{\pm}}}{a_{RH_2^+}},$$

$$\therefore\quad pK_1 = pH - \log\frac{c_{RH^{\pm}}}{c_{RH_2}} - \log\frac{f_{RH^{\pm}}}{f_{RH_2}}. \qquad (6)$$

If a solution is made up of c equiv. of neutral amino-acid and a equiv. of a strong acid, $c_{RH^{\pm}}$ is equal to $c - a + c_{H^+}$ and c_{R^+} to $a - c_{H^+}$ (cf. p. 422); inserting these values in equation (6), the result is

$$pK_1 = pH - \log\left(\frac{c}{a - c_{H^+}} - 1\right) - \log\frac{f_{RH^{\pm}}}{f_{RH_2^+}}. \qquad (7)$$

[6] Schmidt, Appleman and Kirk, *J. Biol. Chem.*, 81, 723 (1929); Edsall and Blanchard, *J. Am. Chem. Soc.*, 55, 2337 (1933); Glasstone and Hammel, *ibid.*, 63, 243 (1941).

For the second dissociation constant (K_2) the equation analogous to (6) is

$$pK_2 = pH - \log \frac{c_{R^-}}{c_{RH^\pm}} - \log \frac{f_{R^-}}{f_{RH^\pm}},$$

and if the solution consists of c equiv. of neutral amino-acid and b equiv. of strong base, $c_{RH^\pm}$ is equal to $c - b + c_{OH^-}$ and c_{R^-} to $b - c_{OH^-}$, this becomes

$$pK_2 = pH + \log \left(\frac{c}{b - c_{OH^-}} - 1 \right) - \log \frac{f_{R^-}}{f_{RH^\pm}}. \tag{8}$$

In order to determine pK_1 or pK_2 a solution is made up of known amounts of the neutral amino-acid (c) and either strong acid (a) or strong base (b), and the pH of the solution is determined by means of some form of hydrogen electrode. The values of c_{H^+} or c_{OH^-} are derived from the pH by assuming the activity coefficient of the hydrogen or hydroxyl ions to be equal to the mean values for hydrochloric acid or sodium hydroxide, respectively, at the same ionic strength. Within the pH range of about 4 to 10, however, the terms c_{H^+} and c_{OH^-} may be neglected in equations (7) and (8) respectively, provided the solution is not too dilute. The estimation of the activity coefficient factor presents some difficulty since $-\log f_{RH^\pm}$ is proportional to μ while $\log f_{RH_2^+}$ or $\log f_{R^-}$ is related to $\sqrt{\mu}$; for most purposes, however, the last term in equations (7) and (8) may be taken as zero, provided the ionic strength of the solution is not large. In this event it is necessary to use the symbols pk_1 and pk_2 for the dissociation exponents, or to add a prime, thus pK_1' and pK_2'.

The results of measurements made in this manner with glycine at 20° are given in Table LXXIV; the values of pk_1 and pk_2 are seen to be 2.33

TABLE LXXIV. DETERMINATION OF DISSOCIATION CONSTANTS OF GLYCINE AT 20°

Mixtures of Glycine (c) and Hydrochloric acid (a)

c	a	pH	$c_{H^+} \times 10^3$	$\frac{c}{a - c_{H^+}} - 1$	pk_1
0.0769	0.0231	2.76	2.00	2.650	2.34
0.0714	0.0286	2.58	3.02	1.786	2.33
0.0667	0.0333	2.45	4.17	1.283	2.33
0.0625	0.0375	2.31	5.75	0.972	2.32
0.0588	0.0412	2.21	7.41	0.742	2.34
0.0555	0.0445	2.10	9.55	0.590	2.33

Mixtures of Glycine (c) and Sodium hydroxide (b)

c	b	pH	$c_{OH^-} \times 10^{-5}$	$\frac{c}{b - c_{OH^-}} - 1$	pk_2
0.0833	0.0167	9.22	1.29	3.878	9.82
0.0769	0.0231	9.42	2.09	2.333	9.79
0.0714	0.0286	9.63	3.47	1.500	9.81
0.0667	0.0333	9.78	4.90	1.000	9.78
0.0625	0.0375	9.98	7.95	0.667	9.81
0.0588	0.0412	10.14	10.14	0.250	9.78

and 9.80, which may be compared with 2.37 and 9.75, respectively, derived from cells without liquid junction.

In the methods described above the tacit assumption has been made that the neutralizations of RH_2^+ and of $RH^{\pm}$ do not overlap; this is always true in the early stages of the neutralization of RH_2^+ and in the later stages for $RH^{\pm}$, but it is not necessarily the case in the region of the first equivalence-point, i.e., at $RH^{\pm}$. The problem is, of course, identical with that of an ordinary dibasic acid; provided K_1/K_2 is greater than about 10^6, i.e., $pK_2 - pK_1$ is greater than 6, the two stages may be regarded as independent. If this condition does not hold, the system may be treated as a conventional dibasic acid in the manner described on page 326.

The dissociation constant exponents at 25° of a number of physiologically important amino-acids are recorded in Table LXXV;[7] those

TABLE LXXV. DISSOCIATION CONSTANTS OF AMINO-ACIDS AT 25°

Amino-acid	pK_1	pK_2	Amino-acid	pK_1	pK_2
Alanine	2.340	9.870	Diglycine	3.15	8.10
Arginine	2.02	9.04 12.48	Histidine	1.77	9.18
			Hydroxyproline	1.92	9.73
Aspartic acid	2.09 3.87	9.82	Isoleucine	2.318	9.758
			Leucine	2.328	9.744
Glutamic acid	2.19 4.28	9.66	Norleucine	2.335	9.833
			Valine	2.287	9.719
Glycine	2.350	9.778	Tryptophane	2.38	9.39

given by four significant figures are thermodynamic values, but the others are approximate. The data for K_1 show that the carboxylic acid $^+NH_3RCO_2H$ is a moderately strong acid; the reason is that the positive charge on the nitrogen atom facilitates the departure of the proton from the $-CO_2H$ group, thus increasing the acid strength of the latter. The ammonium ion acid $^+NH_3RCO_2^-$ is relatively weak, however, because the negative charge on the $-CO_2^-$ group has the opposite effect. As the distance of separation increases, the influence of the electrostatic charges becomes less marked. From an examination of the dissociation constants of glycine and diglycine it has been found possible to calculate the distances between the terminal groups.

According to the older ideas concerning amino-acids, neutralization of the electrically neutral form by a strong acid gave the basic dissociation constant k_b of the $-NH_2$ group; this value is now attributed to the $-CO_2^-$ group. Quite apart from any question of theory, however, this dissociation constant must be the value for the base that is conjugate to the acid which gives the experimental dissociation constant K_1; it follows,

[7] For further data, see Hitchcock, Schmidt's "The Chemistry of Amino Acids and Proteins," 1938, Chap. XI.

therefore, that the classical k_b is equal to K_w/K_1, where K_w is the ionic product of water (cf. p. 338). The classical acidic dissociation constant k_a is obtained when the neutral acid is neutralized by a strong base. It was at one time associated with the $-CO_2H$ group but is now attributed to the $-NH_3^+$ group; irrespective of theoretical considerations, it is identical with the quantity to which the symbol K_2 has been given here.

Proportion of Dipolar Ion.[8]—In the foregoing treatment it has been tacitly assumed that the neutral amino-acid is present exclusively in the dual ion form; some indication of the justification of this assumption may be obtained in the following approximate manner. In any solution there will presumably be an equilibrium between the neutral, uncharged, molecules and the dipolar ions, thus

$$NH_2RCO_2H \rightleftharpoons {}^+NH_3RCO_2^-.$$

The equilibrium constant of this system, generally given the symbol K_Z, is then

$$K_Z = \frac{({}^+NH_3RCO_2^-)}{(NH_2RCO_2H)}, \tag{9}$$

where the parentheses are employed to represent activities. Information concerning K_Z may be obtained by considering the two acidic dissociations, whose constants are designated by K_a and K_b, thus

$$(a) \quad {}^+NH_3RCO_2H + H_2O \rightleftharpoons {}^+NH_3RCO_2^- + H_3O^+$$

and

$$(b) \quad {}^+NH_3RCO_2H + H_2O \rightleftharpoons NH_2RCO_2H + H_3O^+.$$

The sum of these two equilibria represents the acidic dissociation of the cationic form ${}^+NH_3RCO_2H$, and it is the sum of the two dissociation constants that is really the quantity K_1 determined above. As a first approximation, which will be justified shortly, K_b may be neglected, so that K_1 may be identified with K_a. The value of K_b cannot be obtained by direct experiment, but it is very probable that the dissociation constant of the ammonium ion acid ${}^+NH_3RCO_2H$ will be almost the same as that of the corresponding acid ${}^+NH_3RCO_2Me$, i.e., of its methyl ester. The dissociation constant of this substance, which is given the symbol K_E, can be determined without difficulty by titration of the hydrochloride of the methyl ester of the amino-acid by means of a strong base. For the aliphatic amino-acids K_E is about 10^{-8}, and since K_1 is of the order of 10^{-3}, the neglect of K_b in comparison with K_a is justifiable.

A comparison of equation (9) with the equilibria (*a*) and (*b*) shows at once that

$$K_Z = \frac{K_a}{K_b},$$

[8] Ebert, *Z. physik. Chem.*, 121, 385 (1926); Edsall and Blanchard, *J. Am. Chem. Soc.*, 55, 2337 (1933).

and in view of the arguments just given

$$K_Z = \frac{K_1}{K_B}, \qquad (10)$$

thus permitting K_Z to be evaluated. For all the aliphatic acids studied so far, K_Z is about 10^5 to 10^6, so that the ratio of $^+NH_3RCO_2^-$ to NH_2RCO_2H in the electrically neutral amino-acid is very large; there is actually less than 0.001 per cent of the acid in the uncharged form, and so the assumption that the amino-acid exists mainly in the dual ion form is justified.

Hydrogen Ion Concentrations in Ampholytes.—In any solution of a *pure* ampholyte there will be present the positive and negative ions $^+NH_3RCO_2H$ and $NH_2RCO_2^-$, represented by RH_2^+ and R^-, to some extent, in addition to hydrogen and hydroxyl ions, and neutral molecules $^+NH_3RCO_2^-$ and NH_2RCO_2H, i.e., $RH^\pm$ and RH. As just seen, however, the proportion of NH_2RCO_2H molecules is negligible for an aliphatic amino-acid. Since the solution is electrically neutral, it follows that

$$c_{H^+} + c_{RH_2^+} = c_{R^-} + c_{OH^-}. \qquad (11)$$

If the solution is sufficiently dilute for activity coefficients to be taken as unity, then the equations for k_1 and k_2, in place of K_1 and K_2, become [cf. equation (1)]

$$k_1 = \frac{c_{H^+}c_{RH^\pm}}{c_{RH_2^+}} \quad \text{and} \quad k_2 = \frac{c_{H^+}c_{R^-}}{c_{RH^\pm}},$$

and if the values of $c_{RH_2^+}$ and c_{R^-} obtained from these relationships are inserted in equation (11), the result is

$$c_{H^+} + \frac{c_{H^+}c_{RH^\pm}}{k_1} = k_2\frac{c_{RH^\pm}}{c_{H^+}} + \frac{k_w}{c_{H^+}},$$

$$\therefore \quad c_{H^+} = \sqrt{\frac{k_1k_2c_{RH^\pm} + k_1k_w}{k_1 + c_{RH^\pm}}}. \qquad (12)$$

If the solution of the ampholyte is relatively acid, as is generally the case, the quantity c_{OH^-} in equation (11) and the corresponding term k_1k_w in equation (12) can be neglected, provided the solution is not too dilute; the value for c_{H^+} then becomes

$$c_{H^+} = \sqrt{\frac{k_1k_2c}{k_1 + c}}, \qquad (13)$$

the quantity $c_{RH^\pm}$ being replaced by c, the equivalent concentration of the ampholyte. It will be noted that equation (13) is identical in form with equation (65), page 401, giving the hydrogen ion concentration of a dibasic acid at the first equivalence-point; this is, of course, as it should be on the basis of the treatment of an amino-acid hydrochloride as a dibasic acid. At the end of the first stage of neutralization of this acid the system consists of the electrically neutral ampholyte. As a general rule k_1 in equation (13) cannot be neglected in comparison with c, and so the simpler form, analogous to (66), page 401, is not applicable.

Isoelectric Points.—An ampholyte is at its **isoelectric point** when the concentration of positive ions is equal to that of the negative ions, i.e., when $c_{RH_2^+}$ is equal to c_{RH^-}. Since these ions have almost the same equivalent conductances, because of their size, equal amounts of positive and negative ions of the ampholyte will migrate in opposite directions. At the isoelectric point, therefore, an amino-acid, or more complex ampholyte, will appear to remain stationary in an electrical field, although the solution may have an appreciable conductance. According to equation (1),

$$c_{RH_2^+} = \frac{a_{H^+}a_{RH^\pm}}{K_1 f_{RH_2^+}} \quad \text{and} \quad c_{R^-} = \frac{K_2 a_{RH^\pm}}{a_{H^+} f_{R^-}},$$

and if these are equated, to give the condition of the isoelectric point, it is found that

$$a_{H^+}^2 = K_1K_2\frac{f_{RH_2^+}}{f_{R^-}},$$

$$\therefore \quad a_{H^+} = \sqrt{K_1K_2\frac{f_{RH_2^+}}{f_{R^-}}}, \tag{14}$$

or

$$\text{pH} = \tfrac{1}{2}\text{p}K_1 + \tfrac{1}{2}\text{p}K_2 - \tfrac{1}{2}\log\frac{f_{RH_2^+}}{f_{R^-}}. \tag{15}$$

By means of equation (15) it is possible to evaluate the pH at the isoelectric point; the activity coefficient term may be neglected since the values for the univalent positive and negative ions will not differ greatly, especially if the solutions are dilute.

It must be pointed out that in general it is not possible to obtain an isoelectric point for a *pure* ampholyte, since equation (13) for the hydrogen ion concentration in the latter will be equal to that given by (14) for the isoelectric point in special cases only. It is usual, therefore, to add a small amount of alkali or acid in order to obtain an isoelectric solution. At the isoelectric point an amino-acid dissociates equally to yield RH_2^+ and R^- ions, but if the hydrogen ion concentration exceeds the theoretical value for this point, there will be a tendency for the hydrogen ions to react with R^- or $RH^\pm$ to yield $RH^\pm$ or RH_2^+, respectively. In a solution of smaller hydrogen ion concentration, however, the RH_2^+ or $RH^\pm$ ions will tend to ionize further. These conclusions have been utilized to determine the isoelectric points of amino-acids: a small amount of ampholyte is added to each of a series of solutions of known pH; these solutions should not be too strongly buffered. If there is an increase of pH, the ampholyte is reacting with hydrogen ions, and if there is a decrease it is dissociating to yield these ions. On the other hand, if the pH of the solution remains unchanged, the RH_2^+ and R^- ions, and hence also the hydrogen and hydroxyl ions, are being formed in equivalent amounts; the pH of the system is then that of the isoelectric point of the ampholyte.

It was seen in connection with the study of dibasic acids (p. 403) that the proportion of acid H_2A present in the HA^- form is a maximum when the hydrogen ion concentration is equal to $\sqrt{k_1k_2}$. Precisely the same arguments apply to an amino-acid system treated as a dibasic acid RH_2^+, and hence the proportion of $RH^\pm$, which is the equivalent of HA^-, must be a maximum at a hydrogen ion activity that is, in fact, identical with that given by equation (14). It follows, therefore, that the proportion of dual ions is a maximum when the ampholyte is at its isoelectric point; the extent of ionization into simple ions must consequently be a minimum at this point. Since the physical properties, e.g., solubility, viscosity, etc., of the dipolar ions probably differ from those of the singly-charged ions, it is to be expected that there may be a maximum or minimum in these properties at the isoelectric point. This has been found especially to be the case with complex ampholytes, such as proteins; the pH of minimum solubility has also been identified with the isoelectric point of sparingly soluble ampholytes.[9]

Attention may be called to the fact that the conclusions reached here concerning the isoelectric point must hold irrespective of whether the neutral ampholyte consists of dual ions or of uncharged molecules. Provided the values of k_1 and k_2 are those derived experimentally they involve no supposition concerning the nature of the groups responsible for them. In the foregoing treatment the assumption has been made that the neutral molecule is $RH^\pm$, as this is true for aliphatic amino-acids, but the results will be quite unchanged if $RH^\pm$ is replaced by RH, as would be the case for an amino-benzoic acid or an amino-phenol.

Neutralization Curves of Ampholytes.—The variation of pH during the course of neutralization of an amino-acid by a strong acid or by a strong base can be calculated by means of equation (7) or (8), respectively, provided pK_1 and pK_2 are known. Alternatively, the pH values determined experimentally may be plotted, as shown by the full curves in Fig. 108 which are for the neutralization of glycine. The addition of acid gives the curve from the center to the left-hand side, while the addition of alkali yields the curve from the middle to the right-hand side. If the system is treated as a dibasic acid, i.e., in the $^+NH_3RCO_2H$ form, the two stages of neutralization by a strong base are represented by the curve starting from the extreme left and proceeding to the extreme right.

[9] Michaelis *et al.*, *Biochem. Z.*, **24**, 79 (1910); **30**, 40 (1910); **47**, 251 (1913); Levene and Simms, *J. Biol. Chem.*, **55**, 801 (1923); Simms, *J. Am. Chem. Soc.*, **48**, 1239 (1926); Hahn and Klockmann, *Z. physik. Chem.*, **157**, 209 (1931); Hitchcock, *J. Biol. Chem.*, **114**, 373 (1936).

The flattening of the curve in the extreme acid and alkaline regions indicates that the electrically neutral form of glycine behaves as a very weak base and as a very weak acid; the corresponding salts, viz., $Cl^- \cdot {}^+NH_3CH_2CO_2H$ and $NH_2CH_2CO_2^- \cdot Na^+$, are, therefore, very con-

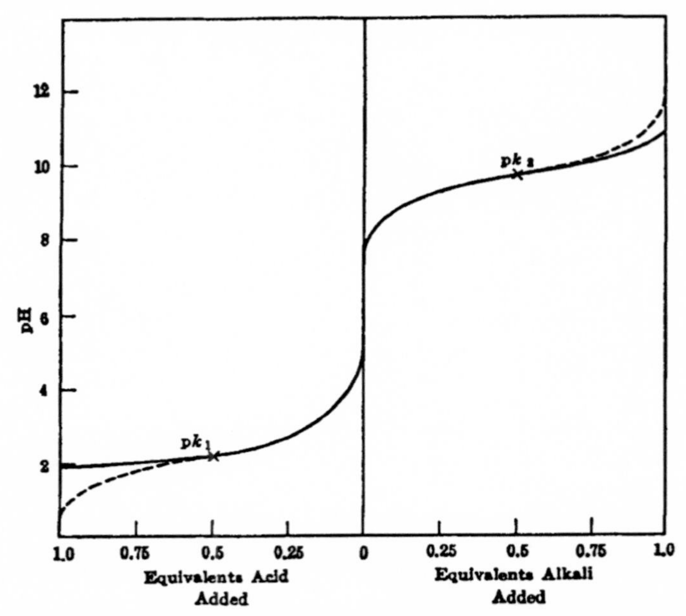

FIG. 108. Uncorrected and corrected neutralization curves for glycine

siderably hydrolyzed, and hence the change of pH as the equivalence-point is approached is very gradual. It is evident from the curves that direct potentiometric or other form of titration of glycine cannot yield satisfactory results. On the other hand, of course, the conjugate acid and base will be relatively strong so that if the titration is commenced either with the hydrochloride or with the sodium salt, a sharp inflexion in the pH curve is obtained in the titration with alkali or strong acid, respectively. The rapid change of pH is seen to occur in the region of pH 4.5 to 7.5 for a 0.1 molar solution of glycine; methyl red and bromcresol purple would consequently be suitable indicators for these titrations.

By making an allowance for the extent of hydrolysis, it is possible to obtain hypothetical "corrected" pH-neutralization curves for glycine and other amino-acids which show appreciable inflexions at the equivalence-points. Two methods of making the necessary corrections have been employed. It will be recalled that hydrolysis represents incomplete neutralization; consequently the free strong acid produced as a result of hydrolysis should be subtracted from the total added in the titration in order to give the amount actually employed in the neutralization. For example, if the original solution contained c equiv. of glycine per liter and if when a equiv. of strong acid, e.g., hydrochloric acid, were added h equiv. of the glycine hydrochloride were hydrolyzed, i.e., h equiv. of free hydrochloric acid were present due to hydrolysis, the actual amount of acid used in neutralization is $a - h$; this represents, therefore, the true extent of neutralization of the glycine. The quantity $100(a - h)/c$, consequently, gives the actual "per cent neutralization" which should be plotted on the pH-neutralization curve in order to eliminate the effect of hydrolysis. Since the acid produced in hydrolysis is a strong one, it may be regarded as completely ionized, so that h is equal to the concentration of hydrogen ions in the solution: the "per cent neutralization" is, therefore, $100(a - c_{H^+})/c$. The value of c_{H^+} can be obtained with sufficient accuracy from the measured pH of the solution by means of the approximate relationship $\text{pH} = -\log c_{H^+}$. An alternative procedure for obtaining h is to add various known amounts of the free strong acid to water, or to a neutral salt solution, until the pH is the same as at any given stage of the neutralization; the quantity of free acid is then assumed to be the same in each case.

An exactly analogous treatment can, of course, be applied to the neutralization by a strong base; in that event h represents the amount of free base at any time and this is practically equal to the hydroxyl ion concentration of the solution. The value of h may also be obtained by the addition of the strong base to water until it gives the desired pH, as described for hydrolysis in the acid solution.

The corrected pH-neutralization curves obtained in this manner are shown by the dotted lines in Fig. 108; the inflexions at the equivalence points are now seen to be sufficiently definite for an accurate estimate of the end-point to be possible. This method has been used for the potentiometric titration of amino-acids.[10]

An alternative, simpler, procedure for improving the inflexion in the neutralization of an amino-acid is to add formaldehyde to the solution; although this does not affect the acid-titration curve, the one for alkaline titration is changed, as seen in Fig. 107. The effect of the formaldehyde is to increase the strength of the ammonium ion acid which is being titrated, and so the pH inflexion at the equivalence-point becomes much more obvious. This is the basis of the "formol" titration of amino-acids discovered by Sørensen (1907): approximately 10 per cent of formaldehyde is added to the solution which is then titrated with standard alkali using phenolphthalein as indicator. In the presence of this concentration of formaldehyde the pH-neutralization curve has a sharp inflexion in the region of pH 9, and so a satisfactory end-point is possible with the aforementioned indicator.

[10] Tague, *J. Am. Chem. Soc.*, **42**, 173 (1920); Harris, *J. Chem. Soc.*, **123**, 3294 (1923); Branch and Miyamoto, *J. Am. Chem. Soc.*, **52**, 863 (1930).

Other methods of titration of amino-acids have been employed that are based on principles already described; as with other very weak bases the neutralization by acid may be carried out in glacial acetic acid solution (cf. p. 398), or use may be made of the fact that charged cation acids and anion bases give sharper pH inflexions when neutralized in media of low dielectric constant, e.g., alcohol or acetone (cf. p. 397).[11]

Activity Coefficients of Ampholytes.—The theoretical treatment of the electrical forces in a solution of an ampholyte containing free single ions, e.g., of added electrolytes, is very difficult; it has, however, been carried out on the assumption that the amino-acid consists essentially of a spherical dipolar ion immersed in a continuous solvent medium.[12] If R is the distance between the charges in the dipolar ion and a is the mean distance of closest approach of the other ions to this ion, the limiting value of the activity coefficient γ of the dipolar ion in dilute solution is given by

$$-\log \gamma = \frac{A}{D^2T^2} \cdot \frac{R^2}{a} \mu, \tag{16}$$

where D is the dielectric constant of the medium of ionic strength μ at the temperature T; the factor A is a constant whose value can be derived theoretically.

It is possible to test equation (16) by means of solubility measurements of amino-acids in the presence of salts; if S_0 is the solubility in a medium of ionic strength zero and dielectric constant D_0, i.e., that in the pure solvent, and S is the solubility in a neutral salt solution of ionic strength μ and dielectric constant D, then since S_0/S is equal to the activity coefficient (cf. p. 177) it follows from equation (16) that

$$\log \frac{S}{S_0} = 0.125 \frac{R^2}{a} \left(\frac{D_0}{D} \right)^2 \mu \tag{17}$$

at 25°, the distances a and R being expressed in Ångström units. This result can also be written in the form

$$\frac{D}{D_0} \log \frac{S}{S_0} = 0.125 \frac{R^2}{a} \cdot \frac{D_0}{D} \mu,$$

and hence the plot of $(D/D_0) \log (S/S_0)$ against $(D_0/D)\mu$ should approach a straight line at high dilution, its slope being equal to $0.125R^2/a$. The value of a can be estimated from the known dimensions of the dipolar ion and of the ions of the neutral electrolyte; consequently R, the distance between the charges in the dipolar ion, can be calculated. From solubility measurements of glycine in aqueous-alcoholic media containing

[11] Jukes and Schmidt, *J. Biol. Chem.*, **105**, 359 (1934); Neuberger, *Proc. Roy. Soc.*, **115B**, 180 (1934); Ogston and Brown, *Trans. Faraday Soc.*, **31**, 574 (1935).
[12] Kirkwood, *J. Chem. Phys.*, **2**, 351 (1934); *Chem. Revs.*, **24**, 233 (1939).

various amounts of sodium chloride the value of R has been estimated as 3.15Å; this may be compared with 3.0Å derived from a model of glycine and in other ways.[13]

It is of interest to note that equation (21) requires the solubility of a dipolar ampholyte molecule to increase in the presence of neutral salts; the solubility of a neutral, uncharged molecule on the other hand decreases with increasing ionic strength of the medium (cf. p. 147). The fact that the solubility of glycine and other aliphatic amino-acids increases under these conditions constitutes, as already mentioned, evidence for their dipolar structure.

PROBLEMS

1. For the determination of the dissociation constants of glycine at 25°, Owen [*J. Am. Chem. Soc.*, **56**, 24 (1934)] measured the E.M.F.'s of the cells consisting of hydrogen and silver chloride electrodes in (*A*) mixtures of glycine (m_1) and glycine hydrochloride (m_2), and (*B*) glycine (m_1), its sodium salt (m_2) and sodium chloride (m_3). The experimental results were as follows:

A		*B*	
$m_1/m_2 = 1.1301$		$m_1/m_3 = 1.0187$; $m_2/m_3 = 0.98487$	
$m_2 \times 10^6$	E	$m_3 \times 10^6$	E
5126	0.52281	4659	0.93750
6833	0.51282	8627	0.92208
10237	0.49967	23135	0.89778
14833	0.48797	34488	0.88796
37740	0.46217	50907	0.87841

Evaluate K_1 and K_2 for glycine by the methods described on pages 422 and 423.

2. Glasstone and Hammel [*J. Am. Chem. Soc.*, **63**, 243 (1941)] determined the pH's of 0.1 N solutions of diglycine during the course of neutralization of 10 cc. by (i) 0.1 N hydrochloric acid, and (ii) 0.1 N sodium hydroxide, at 20°. The following values were obtained:

0.1 N HCl or NaOH added	(i) pH	(ii) pH
2.50 cc.	3.69	7.77
5.00	3.19	8.21
7.50	2.80	8.70

Determine the approximate dissociation constants of diglycine.

3. When 10 cc. of 0.05 N diglycine ester hydrochloride were titrated with 0.05 N sodium hydroxide, the pH values at various stages of neutralization were:

0.05 N NaOH added	3.00	5.00	7.00 cc.
pH value	7.57	7.90	8.27

Utilize the results of Problem 2 to determine the approximate ratio of dipolar ions to uncharged molecules in an aqueous solution of diglycine at 20°.

[13] Cohn *et al.*, *J. Am. Chem. Soc.*, **58**, 2365 (1936); **59**, 2717 (1937); *J. Phys. Chem.*, **43**, 169 (1939); *Chem. Revs.*, **19**, 241 (1936); Keefer, Reiber and Bisson, *J. Am. Chem. Soc.*, **62**, 2951 (1940); see also, Roberts and [illegible], *ibid.*, **63**, 1373 (1941).

4. The value of k_2 for *p*-aminobenzoic acid is 4.93×10^{-5}, whereas the basic dissociation constant (k_b) of ethyl *p*-aminobenzoate is 2.8×10^{-12}. What conclusions may be drawn concerning the structure of *p*-aminobenzoic acid in aqueous solution?

5. Use the dissociation constants obtained in Problem 2 to plot the complete pH-neutralization curve, by a strong acid and by a strong base, of 0.1 N diglycine at 20°; the change in volume during neutralization may be neglected. Assuming the hydrogen or hydroxyl ion concentration, as derived from the pH values, to be equal to the concentration of free strong acid or strong base, respectively, near the end-points, calculate the "corrected" neutralization curve.

6. Utilize equation (69) of Chap. XI to plot the variation with pH of the proportion of dipolar ion ($RH^{\pm}$) in a solution of glycine at 25°. The calculations should be made for integral values of pH from 2 to 11, inclusive. Choose two other sets of values of k_1 and k_2, so as to increase and decrease the ratio k_1/k_2 by a factor of about 10^6 and see how it changes the nature of the curve. What conclusions do you draw from the results?

CHAPTER XIII

POLARIZATION AND OVERVOLTAGE

Electrolytic Polarization.—In the previous chapters the E.M.F.'s of reversible cells and the potentials of reversible electrodes have been considered under such conditions that there is either no net flow of current, or the magnitude of the current passing is so minute that the electrodes are not sensibly disturbed from their equilibrium conditions. In the present and the two succeeding chapters the phenomena accompanying the passage of appreciable currents will be examined. It has been seen (p. 250) that at a reversible electrode in a state of equilibrium the discharge of ions and their re-formation take place at the same rate and there is no net current flow. If the conditions are such that there is an actual passage of current, the electrode is disturbed from its equilibrium condition; this disturbance of equilibrium associated with the flow of current is called **electrolytic polarization,** and the disturbed electrode is said to be **polarized.** It will be seen later that polarization results from the slowness of one or more of the processes occurring at the electrode during the discharge or formation of an ion; the type of polarization depends essentially on the nature of the slow process. Polarized electrodes are often said to behave **irreversibly,** and the phenomena associated with polarization are often referred to as **irreversible electrode phenomena;** the irreversibility arises from the fact that one or other stage in the electrode process requires an appreciable activation energy and is consequently slow.

Dissolution and Deposition Potentials.—If a metal M is placed in a solution of its ions M^+, a reversible electrode represented by M, M^+ is set up; suppose its potential is E. Imagine now that an external source of potential is applied to this electrode so as to make it an anode of an electrolytic cell (p. 8); this will have the effect of increasing the potential, and since the electrode is reversible it will immediately commence to dissolve (cf. p. 184). It follows, therefore, that when a metallic electrode is made an anode, it will begin to dissolve as soon as its potential exceeds the reversible value E by an infinitesimal amount. In other words, the electrolytic **dissolution potential** of a metal when made an anode should be equal to its reversible (oxidation) potential (cf. p. 243) in the given electrolyte. The actual value depends, of course, on the concentration, or activity, in the solution of the ions with respect to which the metal is reversible. On the other hand, if the particular electrode under consideration is made a cathode, so that its potential is reduced below the reversible value, the reverse process, viz., deposition of ions, will commence. Since the cathodic reaction is to be represented by $M^+ + \epsilon \rightarrow M$, which is the opposite to that employed conventionally for electrode potentials (see Table XLIX), the potential of the cathode should be represented by $-E$. The electrolytic **discharge potential** of an ion, or the **deposition potential** of a metal, i.e., the potential at which metal deposition will commence on a cathode, should be equal to the conventional reversible potential in the given solution with the sign reversed, i.e., to the reduction potential. It should be noted that the anodic dissolution (oxidation) and cathodic deposition (reduction) potentials under consideration are those at which the respective processes commence; the reactions occur to a very minute extent and the current strength is very small. If the current is increased, however, the anodic potential becomes more positive and the cathodic potential more negative, as a result of a slow stage in the electrode process; as mentioned above, the electrode is then polarized.

Exactly the same considerations apply with regard to anion discharge at an anode as to cation discharge at a cathode, described above. The condition for a negative ion, e.g., Cl^-, Br^- or OH^-, to be discharged, i.e., oxidized, is just the same as for a positive ion; the anode potential must be at least equal to the reversible potential for the resulting substance, e.g., chlorine, bromine, etc., in the given electrolyte. Appreciable polarization occurs if the current is not extremely small, just as in the case of the discharge or formation of other ions.

Determination of Anode and Cathode Potentials.—Direct measurement of the potentials at which the various processes occur at a cathode or an anode at various current strengths can be made by means of the apparatus depicted in Fig. 109. The electrode A under examination is

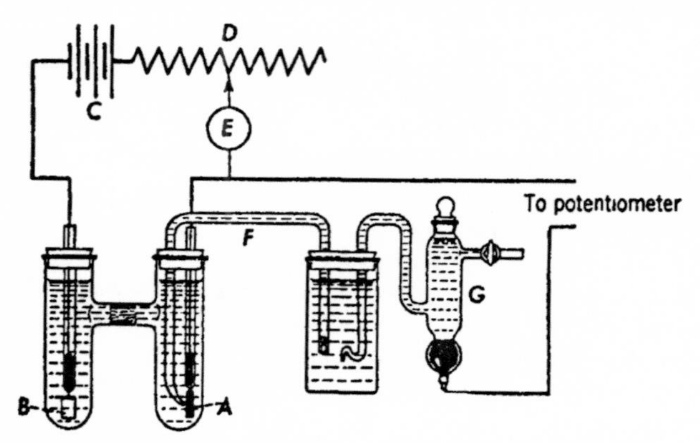

Fig. 109. Measurement of anode or cathode potential

placed in the given solution and combined with another B, so as to make an electrolytic cell; the latter electrode generally consists of an inert material, such as platinum, which does not interfere with the properties of the solution. If the access of oxygen or hydrogen liberated at the electrode *B* should have some influence on the process at *A*, precautions must be taken to prevent this access by using separate electrode compartments. The electrodes *A* and *B* are connected to a source of potential *C*, preferably a high-voltage storage battery, through a variable high resistance *D*; a suitable galvanometer or milliammeter *E* in the circuit indicates the strength of the current passing. The potential of the experimental electrode *A* is determined by connecting it in a convenient manner, e.g., through a salt-bridge *F*, to a reference electrode, e.g., a calomel electrode *G*. The E.M.F. of the cell consisting of the electrodes *A* and *G* is then measured by means of a potentiometer. The reference electrode may consist of the metal to be deposited, or which is being dissolved, at *A* immersed in the same solution as in the electrolytic cell; the difference of potential between *A* and *G* then gives directly the extent of polarization at the particular current employed. By adjustment of the resistance *D*, the current strength is increased gradually and the potential of the electrode measured for each value of the current, after allowing sufficient time for constant conditions to be attained. Observations of this kind with a number of metals, e.g., zinc, cadmium, copper, silver and mercury, have shown that deposition and dissolution do actually commence when the reversible potential of the metal in the given solution is just exceeded. In the deposition of a particular metal, e.g., zinc, on an electrode of another metal, e.g., platinum, the potential is found to change very rapidly as soon as the latter is made a cathode until the value reaches that for zinc in the given solution. The surface of the cathode is then seen to be covered with a thin layer of zinc; subsequently it behaves as a zinc electrode.

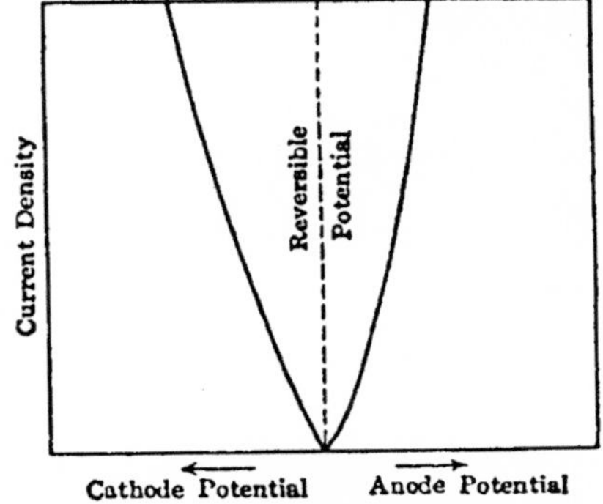

FIG. 110. Variation of electrode potential with current density

As the strength of the current is increased the potential is observed to diverge somewhat from the reversible value in the given solution; typical results are shown in Fig. 110 in which the potential * of a cathode and an anode of the same metal are plotted as a function of the **current density** (C.D.). It is the general practice in electrolytic work to use the current density instead of the current strength: the former is defined as the magnitude of the current per unit area of the electrode, and is expressed in amps. or milliamps. per sq. cm., or per sq. dm. The rates of the processes occurring at an electrode will clearly depend on the exposed area from which the metal can dissolve or to which the ions have access,

* For this purpose the magnitude only of the potential is used; the sign is ignored.

and so it is desirable, especially when comparing the behavior of different electrodes, to consider the current strength per unit area. It is for this reason that the C.D. is employed when plotting electrode potentials.

Concentration Polarization.—For most metals the extent of polarization, as indicated by the departure of the potential from the reversible value in the experimental solution, is relatively small; for reasons which will be considered later it is frequently less for an anode than for a cathode under the same conditions. Most of this polarization is due to concentration changes in the vicinity of the electrode, and is consequently referred to as **concentration polarization.** It arises from the fact that at an anode the dissolution of the metal will result in an increase in the number of ions in contact with it in the solution, and if the rate of diffusion is relatively slow, the concentration of the cations in the immediate vicinity of the anode will be greater than in the bulk of the solution. The result will be a (numerical) increase of potential of the anode which will be greater the higher the C.D., i.e., the higher the rate at which the metal dissolves. At a cathode the opposite condition arises: the discharge of cations results in a diminution of their concentration, and if this is not made up by diffusion, the concentration of cations in contact with the cathode is less than in the bulk of the solution. The potential of the electrode will thus increase in the cathodic direction as the C.D. is increased and the rate of discharge of cations becomes greater. The existence of concentration polarization at an electrode can be readily detected by the fact that it is diminished by any factor which is able to increase the rate of diffusion to or from the electrode; agitation of the electrolyte, increase of its concentration, or raising the temperature all reduce the magnitude of concentration polarization, but they do not eliminate it entirely.

Consecutive Electrode Processes.—If an electrolyte contains a number of different positive and negative ions then, provided there are no disturbing factors, e.g., formation of an alloy, each ionic reaction will take place in turn as the appropriate potential is attained. When the external E.M.F. applied to an electrolytic cell is gradually increased, the potentials of the electrodes change until the ion having the least anodic or cathodic potential is formed at the anode or discharged at the cathode, respectively. Consider, for example, the electrolysis of a solution containing molar zinc and copper sulfates; the deposition (reduction) potentials in this solution are approximately -0.77 and $+0.34$ volt, respectively. As the potential of the cathode, which may be an inert metal such as platinum, is made increasingly negative by means of a gradually increasing external E.M.F., the deposition of copper takes place first, since its deposition potential ($+0.34$) is less cathodic than that of zinc. If the current strength is increased, copper deposition will continue at a greater rate, the potential becoming more negative owing to the effect of concentration polarization. If the electrolysis is prolonged to such an extent that the cupric ions in the solution are almost exhausted, the cathodic potential will then become sufficiently negative, viz., − 0.77 volt, for zinc deposition to commence. Alternatively, should the current be made so large that the cupric ions cannot be brought up to the electrode sufficiently rapidly to satisfy the requirement of the current, the potential will become more negative and simultaneous deposition of copper and zinc will occur. It should be remembered that all aqueous solutions contain hydrogen ions and consequently the liberation of hydrogen is always a possibility to be taken into consideration, especially if the solution is appreciably acid. In evaluating the discharge potential of hydrogen, however, certain complicating factors, to be considered shortly, must be taken into account.

The behavior at an anode is, in general, analogous to that at a cathode: the process requiring the least positive anodic potential, whether it be dissolution of a metal to form cations or the discharge of anions, will take place first. If a copper anode, for example, is placed in an acid solution of copper sulfate, the possible electrode reactions are (*a*) dissolution of the copper to form cupric ions at an anodic potential of − 0.34 volt, (*b*) discharge of hydroxyl ions, which are always present in aqueous solution, at about + 1.2 volts, and (*c*) discharge of SO_4^{--} ions, probably requiring a very high potential. It is evident that when the external E.M.F. is applied to the copper anode the first process to occur will be that the anode will go into solution; under normal conditions, there will be no discharge of hydroxyl ions accompanied by oxygen evolution. The consecutive discharge of anions may be illustrated by reference to the electrolysis of a solution of neutral potassium iodide; such an electrolyte contains iodide and hydroxyl anions whose discharge (oxidation) potentials are about + 0.54 volt and + 0.81 volt, respectively. When this solution is electrolyzed employing platinum, or other inert metal, as anode, it is evident that iodide ions will be discharged preferentially. If the supply of these ions is exhausted, or the current density is so large that the iodide ions are unable to move up to the electrode fast enough, hydroxyl ion discharge and oxygen evolution will occur.

Attention may be called, at this juncture, to an important matter in connection with electrolysis and ionic discharge that is often not clearly understood. It is essential to distinguish clearly between the ions *carrying the current to* the electrodes and those actually *discharged at* the electrodes. The carriage of current is determined by the concentrations and velocities of the various ionic species present in the solution, whereas the discharge potential is determined mainly by the reversible potential in the given solution of the particular ion discharged. The two aspects of the problem are quite independent and should not be confused. As long as the appropriate quantity of electricity is transferred across the solution and at the electrodes, it is immaterial which ions perform the respective functions. In an acid solution of copper sulfate, for example, the current is carried towards the anode almost exclusively by sulfate ions and towards the cathode mainly by hydrogen ions and to some extent by cupric ions; nevertheless, cupric ions only are discharged at the cathode, if the C.D. is not too high, and hydroxyl ions, which play a negligible part in the carriage of current, are discharged at a platinum anode in the same solution.[1]

Decomposition Voltages of Aqueous Solutions.—The foregoing discussion has been based essentially on the assumption that concentration changes are the main source of polarization; this is generally true for the discharge of most metallic cations, and also for certain anions, e.g., the halogen ions. When hydrogen and hydroxyl ions are discharged, however, the flow of an appreciable current is accompanied by polarization of a magnitude far in excess of that to be expected from concentration changes. This fact, and others of interest, can be brought out clearly by measuring the decomposition voltage of an aqueous solution of an acid or a base; such measurements were first made systematically by Le Blanc (1893). The apparatus employed is similar to that represented in Fig. 109; instead of measuring the potential of each electrode separately, however, the voltage across the electrolytic cell is measured, either by placing a voltmeter between the electrodes A and B or by connecting them to a potentiometer. The external E.M.F. is increased gradually and simultaneous measurements are recorded of the current flowing through the cell and the voltage across it. If the solution has an appreciable resistance, allowance should be made for the fall of potential, i.e., the so-called "IR drop," across the solution; this is equal to the product of the current I and the resistance R of the electrolyte between the two electrodes. Proceeding in this manner, the plot of the current density against the cell voltage is of the form shown in Fig. 111; the voltage is seen to increase rapidly at first while the current remains very small, but at a certain point there is a sudden rise in the current and the voltage subsequently changes by only a small amount. The voltage at which the sudden increase of current occurs, viz., D in Fig. 111, is the **decomposition voltage** of the electrolyte in the cell. Actually the nature of the current-voltage curve is such that the point D cannot be determined precisely; in any event, according to modern views it has no exact theoretical significance. The apparent decomposition voltage is nevertheless of interest, for it gives an approximate indica-

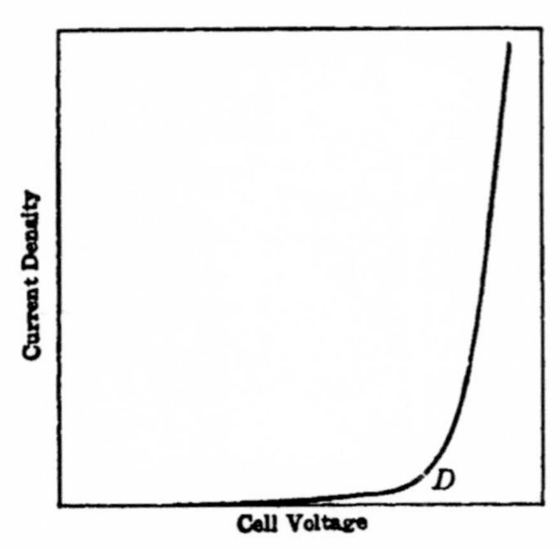

FIG. 111. Variation of current density with cell voltage

[1] For a general treatment, see Glasstone, *School Science Review*, 328 (1935).

tion of the minimum E.M.F. which must be applied to a particular solution in order that currents of appreciable strength shall be able to pass.

When the method just described was applied to the study of aqueous solutions of acids and bases, it was found, provided smooth platinum electrodes were used as anode and cathode, that the decomposition voltage was almost constant, irrespective of the nature of the electrolyte. The results obtained in a number of instances are quoted in Table LXXVI; it is evident that the decomposition voltages of these solutions

TABLE LXXVI. DECOMPOSITION VOLTAGES OF ACIDS AND BASES IN AQUEOUS SOLUTIONS

Acid	Volts	Base	Volts
H_3PO_4	1.70	$N(CH_3)_4OH$	1.74
HNO_3	1.69	NH_4OH	1.74
H_2SO_4	1.67	$NaOH$	1.69
CCl_3CO_2H	1.66	$NH_2(C_2H_5)_2OH$	1.62
$HClO_4$	1.65	KOH	1.67

are all in the vicinity of 1.7 volts. In each case the passage of current is accompanied by the evolution of hydrogen and oxygen gases at atmospheric pressure and the theoretical voltage for these processes to occur is equal to the E.M.F. of the reversible hydrogen-oxygen cell. This E.M.F. for gases at 1 atm. pressure is known to be about 1.23 volts at ordinary temperatures, irrespective of the nature of the electrolyte (cf. p. 242). Since the actual decomposition voltage is much larger than this value, it is evident that the flow of an appreciable current, requiring the discharge of hydrogen and hydroxyl ions *at appreciable rates*, is accompanied by a relatively large polarization.

The fact that the decomposition voltage is the same in solutions of different acids and bases led Le Blanc to conclude that the same electrolytic processes occurred at anode and cathode, respectively, in every case. The only process which could be common to all aqueous solutions is the decomposition of water, resulting in the discharge of hydrogen ions at the cathode, to yield hydrogen gas, and of hydroxyl ions at the anode, leading to the evolution of oxygen gas. In a solution of sulfuric acid, for example, the only cations present are hydrogen ions, and so they must be discharged at the cathode; the most abundant anions in the solution are the sulfate ions, and these undoubtedly carry the current towards the anode, but the hydroxyl ions are actually discharged in spite of their extremely low concentration in the acid solution. Similarly, in a solution of sodium hydroxide it is easy to understand that hydroxyl ions are discharged at the anode, but according to Le Blanc hydrogen ions are discharged at the cathode, although their concentration is very small. A more complete picture of the mechanism of the electrolysis of water will be given later.

If the views proposed above are correct, the decomposition voltage of neutral solutions of sodium sulfate, and sulfates and nitrates of alkali and alkaline earth metals generally, should also be about 1.7 volts, but the actual values are appreciably greater. It must be remembered, however, that since the solutions are not buffered, the discharge of hydrogen ions makes the solution alkaline in the vicinity of the anode, whereas the discharge of hydroxyl ions at the anode causes the solution in its immediate neighborhood to become acid. It is seen, therefore, that hydrogen ions are discharged from a more alkaline solution and hydroxyl ions from a more acid solution than the bulk of the electrolyte; the decomposition voltage is consequently greater than 1.7 volts, for this should be the value when the two ions are discharged from the same solution.

The decomposition voltages of the halogen acids are less than 1.7 volts, as is evident from the values in Table LXXVII for N solutions of

TABLE LXXVII. DECOMPOSITION VOLTAGES OF N-SOLUTIONS OF HALOGEN ACIDS

Hydrochloric acid	1.31 volts
Hydrobromic acid	0.94
Hydriodic acid	0.52

hydrochloric, hydrobromic and hydriodic acids. The products of electrolysis are, however, hydrogen and the appropriate halogen, and not oxygen; the discharge of halide ions evidently occurs more readily than that of hydroxyl ions and the ultimate electrolytic process is the decomposition of the halogen acid, and not of water. On continued electrolysis, especially with dilute solutions, the decomposition voltage rises to about 1.7 volts; the electrolysis of water is now taking place and the gas evolved at the anode contains considerable proportions of oxygen. Attention may be called to the fact that the decomposition voltages of the halogen acids, as given in Table LXXVII, are very close to the theoretical reversible values for the discharge of hydrogen and halide ions. It is evident, therefore, that the polarization accompanying the discharge of these ions at platinum electrodes is small. The relatively high polarization accompanying the electrolysis of other acids must, therefore, be associated with the discharge of hydroxyl ions and the evolution of oxygen. This accounts for the fact that chlorine is liberated in preference to oxygen when N hydrochloric acid is electrolyzed, although theoretically, if there were no polarization, hydroxyl ions should be discharged at about 1.2 volts before chloride ions at 1.3 volts.

Overvoltage.—Although the decomposition voltage of an aqueous solution of an acid or base is about 1.7 volts with smooth platinum electrodes, the value is different if other metals are employed as the electrode material. This dependence of the decomposition voltage and discharge potential on the nature of the electrode had been known for many years, but it was not studied in a systematic manner until 1899 when Caspari observed the potentials at which visible evolution of hydrogen and oxygen gases occurred at cathodes and anodes, respectively, of a number of different metals.[2] The difference between the potential of the electrode

[2] Caspari, *Z. physik. Chem.*, **30**, 89 (1899).

when gas evolution was actually observed and the theoretical reversible value for the same solution was called the **overvoltage.*** It is not certain if the point at which visible evolution of hydrogen or oxygen gas in the form of bubbles has any theoretical significance, although the potential corresponds approximately to that where appreciable current begins to flow (cf. point *D* on Fig. 111); nevertheless the results are of practical interest in connection with the possibility of hydrogen evolution as an alternative process in electrolytic reduction and metal deposition. In order to distinguish between the overvoltage corresponding to the commencement of gas evolution and the value at a definite C.D., the former has been referred to as the "minimum overvoltage." The adjective "minimum" is somewhat misleading, however, since hydrogen evolution is possible at lower overvoltages; the term "bubble overvoltage" will, therefore, be employed as it gives a more exact description of what is measured. Some results obtained with a number of cathodes in N sulfuric acid and anodes † in N potassium hydroxide, for hydrogen and oxygen evolution, respectively, at ordinary temperatures are given in Table LXXVIII. The overvoltage values are not exact because they are difficult to reproduce with any degree of precision; changes often occur with continued electrolysis, and the results are also dependent on the state of the electrode surface. The latter fact is brought out in a striking manner by comparing the overvoltages at smooth and platinized platinum electrodes. The bubble overvoltages for hydrogen are almost independent of pH, although there is sometimes a small decrease in alkaline solution. Anodic overvoltages are even less reproducible than are those accompanying hydrogen evolution, and there is little definite information concerning the influence of pH; it is probable, however, from the fact that

TABLE LXXVIII. CATHODIC AND ANODIC (BUBBLE) OVERVOLTAGES

Electrode	Hydrogen Overvoltage	Oxygen Overvoltage
Platinized Platinum	~0.00 volt	0.25 volt
Palladium	~0.00	0.43
Gold	0.02	0.53
Iron	0.08	0.25
Smooth Platinum	0.09	0.45
Silver	0.15	0.41
Nickel	0.21	0.06
Copper	0.23	—
Cadmium	0.48	0.43
Tin	0.53	—
Lead	0.64	0.31
Zinc	0.70	—
Mercury	0.78	—

* The original word used by Caspari was "Uberspannung"; it is translated as "overvoltage" or "overpotential."

† The anodes are "passive" and do not dissolve to any appreciable extent; see Chap. XIV.

the decomposition voltages of acid and alkaline solutions are the same that the oxygen overvoltage, at least for platinum, is approximately constant. Both cathodic and anodic overvoltages decrease with increasing temperature; the exact extent is not known, but it is generally about 2 millivolts per degree.

It has been seen that, according to the results in Table LXXVII, there is little polarization accompanying the deposition of the halogens at a platinum anode. Experimental work on the overvoltage of halide ion discharge is limited by the fact that many metals are attacked and pass into solution; with the exception of carbon, however, halogen overvoltages appear to be very small at most anodes.

Since the point of bubble evolution represents a more or less indefinite rate of discharge of hydrogen and hydroxyl ions, recent work on overvoltage has been devoted almost exclusively to measurements made at definite C.D.'s; it is then possible to obtain a more precise comparison of the potentials, in excess of the reversible value, which must be applied to different electrodes in order to obtain the same rate of ionic discharge in each case. The details of the methods of measurement and a discussion of the results will be given after the general problem of the mechanism of electrode processes has been considered.

Metal Deposition Overvoltage.—Although concentration changes are probably the most important source of polarization accompanying the deposition of a metal on the cathode, there is evidence of an overvoltage due to other causes. As a general rule, this overvoltage is not large, although it is considerable, e.g., about 0.2 to 0.3 volt, for the deposition of iron, cobalt and nickel at appreciable rates; the behavior of the iron-group metals is exceptional in this respect. It is possible that other elements of Group VIII of the periodic classification require a definite overvoltage for deposition at an appreciable rate, but there are no data in this connection. It may be mentioned that the dissolution of anodes of iron, cobalt and nickel is also accompanied by marked polarization; this is quite large at ordinary temperatures but decreases rapidly as the temperature is raised. Anodes of these metals readily become passive and then they almost completely cease to dissolve; this aspect of anodic behavior is treated more fully in Chap. XIV.

Another form of polarization in metal deposition has been observed in certain instances when the metal is deposited on an electrode of a different material, e.g., of silver on platinum and of mercury on platinum and tantalum. The initial deposition potential exceeds the reversible value by a relatively large amount, but as the C.D. is increased, and the electrode surface becomes covered with the metal being deposited, the potential falls back to normal.

Mechanism of Anodic and Cathodic Phenomena.—Having given a brief review of the simpler phenomena observed at anodes and cathodes, it is opportune, before proceeding to discuss them in further detail, to consider the general problem of the processes occurring at electrodes when there is an appreciable passage of current. When an electrode is at its reversible potential an equilibrium exists between the material of the electrode and its ions in solution; in these circumstances the rate at which the ions are transferred from the electrode to the solution is equal to the rate of passage in the opposite direction (cf. p. 250). Under equilibrium conditions, therefore, there is no *net* flow of current although equivalent currents are actually passing in opposite directions. If the electrode is to be made a cathode or an anode in an electrolytic cell, so that there is a resultant flow of current, the difference of potential between the electrode and electrolyte must be made larger so that the rate of the appropriate reaction is increased; the relation between the current strength and the potential depends essentially on the nature of the slow stage which determines the rate of the electrode process.

The continuous discharge of ions at a cathode involves three main stages: these are (1) transfer of ions from the bulk of the electrolyte to the layer of solution in contact with the electrode by diffusion or other process; (2) discharge of the ions to form atoms on the electrode; and (3) conversion of the atoms to the normal stable form of the deposited substance. In the case of the deposition of a metal, for example, the third stage may represent migration of the atoms from the points at which they are deposited to stable positions in the crystal lattice. When hydrogen is being evolved the corresponding process would be the combination of the atoms to form hydrogen molecules which are evolved in the form of bubbles of gas at atmospheric pressure. Any one of the three stages mentioned above may be the slow, rate-determining one for a given electrode process; different stages may be rate-determining under different conditions even for the same electrode reaction. The consequences of each of the three stages being the slow process in metal deposition will be considered in turn; the problems of hydrogen and oxygen overvoltage are somewhat more involved and so they will be treated separately.

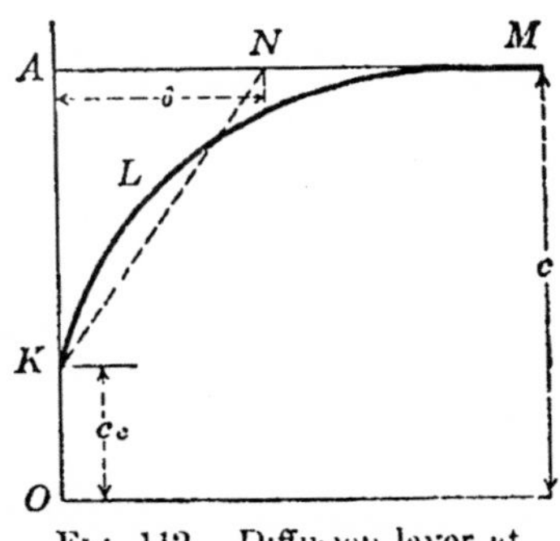

FIG. 112 Diffusion layer at electrode surface

I. Ionic Diffusion as the Slow Process.—When diffusion is the rate-determining stage in the discharge of ions, the conditions are those resulting in the so-called concentration polarization. Owing to the relative slowness of diffusion, the concentration, and activity, of the ions at the electrode surface differs from that in the bulk of the solution; there will consequently be a concentration gradient in passing from the electrode to the electrolyte. This concentration gradient will, theoretically, extend to a considerable distance into the solution, as indicated by the curve *KLM* in Fig. 112, where *K* represents the concentration (c_e) at the electrode and M is the value (c) in the bulk of the solution. It is convenient, however, to consider this gradient as mathematically equivalent to a linear gradient KN; the distance from A to N is then referred to as the thickness of the **diffusion layer.** The actual diffusion layer extends from K to M, but the general practice is to regard it as extending effectively from K to N.[3]

According to Fick's law, the rate of diffusion of a particular ionic species across the layer KN is given by the expression

$$\frac{ds}{dt} = \frac{AD}{\delta}(a - a_e), \tag{1}$$

where A is the exposed area of the electrode surface, D is the diffusion coefficient of the ions, and δ is the thickness of the hypothetical diffusion layer; the activities of the ions under consideration at the two faces of the layer are taken as a on the solution side, and a_e at the electrode surface.* When equilibrium is established at a cathode, the rate of discharge of the ions by the current will be equal to the rate of diffusion to the electrode, as given by equation (1). If I is the current density, i.e., the current strength per unit area, so that A in equation (1) is unity, the rate of discharge of ions is equal to I/nF, where n is the number of electrons involved in the discharge process † and F, the faraday, is the quantity of electricity carried by one equivalent; it follows, therefore, that

$$\frac{I}{nF} = \frac{D}{\delta}(a - a_e), \tag{2}$$

$$\therefore \quad I = \frac{DnF}{\delta}(a - a_e). \tag{3}$$

In this deduction it has been assumed that the ionic species which is being discharged is brought up to the electrode by diffusion only; in general, however, some of the cations will be transported to the electrode by the normal process of transference. If t_+ is the transference number of the discharged cation within the diffusion layer, the rate at which ions are brought up to an electrode of unit area is t_+I/nF; at equilibrium the rate of ionic discharge is equal to the sum of the rates of diffusion and

[3] See Glasstone and Hickling, "Electrolytic Oxidation and Reduction," 1935, Chap. III; Butler, "Electrocapillarity," 1940, Chap. VII.

* Although Fick's law was originally stated in terms of concentrations, it is quite certain that diffusion is determined by the difference of free energy between two points in a solution; it is consequently the difference in activities, rather than that of concentrations, which is employed in equation (1).

† In the case of metal deposition this is, of course, equal to the valence of the ion; the more general form is used, however, because the resulting equations are extended later to other electrode processes.

transference, so that

$$\frac{I}{nF} = \frac{D}{\delta}(a - a_c) + \frac{t_+ I}{nF},$$

$$\therefore \quad I = \frac{DnF}{(1 - t_+)\delta}(a - a_c)$$

$$= \frac{DnF}{t\delta}(a - a_c), \tag{5}$$

where t, written in place of $1 - t_+$, represents the sum of the transference numbers in the diffusion layer of all the ions other than the one being discharged. Replacing the factor $DnF/t\delta$ in equation (5) by k, for brevity, it follows that

$$I = k(a - a_c),$$

$$\therefore \quad a_c = a - \frac{I}{k}. \tag{6}$$

If the electrode processes, other than diffusion, are all rapid, the cathode potential will be almost identical with the reversible value for an ionic activity equal to a_c, i.e., the *actual* value at the electrode surface; hence,

$$E_c = E^0 - \frac{RT}{nF}\ln a_c.$$

When no current is flowing the reversible potential of the electrode is determined by the activity a, i.e., the value in the bulk of the solution, thus

$$E = E^0 - \frac{RT}{nF}\ln a.$$

The numerical magnitude of the concentration polarization ΔE is equal to the difference between E_c and E, i.e.,

$$\Delta E = E_c - E = \frac{RT}{nF}\ln\frac{a}{a_c},$$

and inserting the value of a_c given by equation (6), this becomes

$$\Delta E = \frac{RT}{nF}\ln\frac{ka}{ka - I}. \tag{7}$$

It is seen from this result that the extent of concentration polarization for a given current density is smaller the greater the activity, or concentration, of the ions in the bulk of the solution; further, ΔE can be decreased by increasing the factor k and this can be achieved either by decreasing the thickness of the diffusion layer, e.g., by agitating the solution, or by increasing the diffusion coefficient D of the ions, e.g., by raising the temperature.

The general form of equation (7) is such that for relatively small values of the current I, the concentration polarization ΔE is small and approximately proportional to I; as the current is increased a stage is reached, however, at which ΔE increases rapidly. It is evident from equation (7) that this will occur as the value of I approaches ka, and in fact when I is equal to ka the concentration polarization should, theoretically, increase to infinity. Actually, of course, a finite increase of potential occurs, as shown in Fig. 113, until the discharge of another cation, e.g., hydrogen ion, can take place. The current density at which the rapid increase of potential is observed is called the **limiting current density** (I_d), because it represents the limiting (maximum) rate at which the particular ions can be discharged under the given experimental conditions. If an attempt is made to increase the current above this value, the potential must rise until another cathodic process is possible. The magnitude of the limiting C.D. is given by equating ka to I_d and inserting $DnF/t\delta$ for k, thus

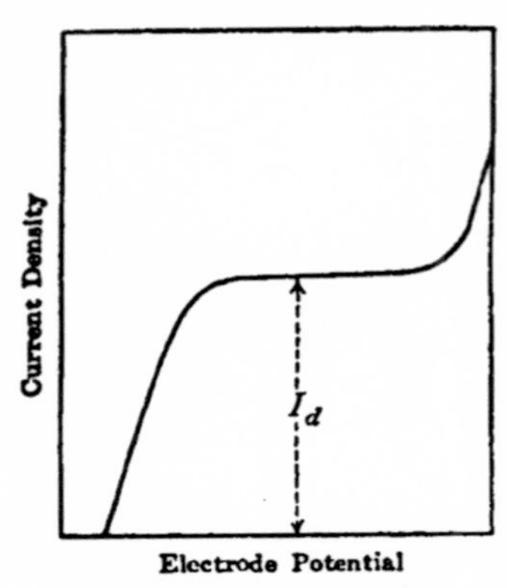

FIG. 113. The limiting current density

$$I_d = \frac{DnF}{t\delta} a; \tag{8}$$

the value is seen to be proportional to the activity, or concentration, of the ions being discharged. Comparison of this equation with (5) shows that when the limiting C.D. is reached the concentration or activity (a_e) of ions at the electrode has fallen to zero; it is evident, therefore, that the rate of diffusion of ions to the electrode must be then a maximum.

If the ka terms in equation (7) are replaced by I_d, the result is

$$\Delta E = \frac{RT}{nF} \ln \frac{I_d}{I_d - I}, \tag{9}$$

which is the form in which the magnitude of the concentration polarization is frequently expressed.[4]

Thickness of the Diffusion Layer.—Since the magnitude of the limiting current can readily be determined experimentally, by gradually increasing the C.D. until the rapid increase of potential (Fig. 113) is observed, it is possible to utilize equation (8) to calculate the thickness δ of the diffusion layer. For this purpose it is desirable to introduce into the solution an excess of an indifferent electrolyte which carries the whole of

[4] Bowden and Agar, *Proc. Roy. Soc.*, **169A**, 206 (1938).

the current; under these conditions t is unity, and so equation (8) becomes

$$I_d = \frac{DnF}{\delta} a. \tag{10}$$

The limiting current as given by equation (10) is often called the **diffusion current,** since it is a measure of the maximum rate at which the discharged ions can be brought up to the electrode by diffusion alone. From this equation it follows that

$$\delta = \frac{DnF}{I_d} a. \tag{11}$$

The diffusion coefficient of an ion, in cm.^{2} sec.$^{-1}$ units, can be expressed in terms of its equivalent conductance λ, in ohms^{-1} cm.^{2}, by means of the relation

$$D = \frac{\lambda RT}{F^2},$$

so that

$$\delta = \frac{\lambda RT}{F} \cdot \frac{nc}{I_d},$$

where the activity a in equation (11) has been replaced by the concentration c of the ions in the bulk of the solution. Inserting the value of R in joules, i.e., 8.313, and F in coulombs, i.e., 96,500, the thickness of the diffusion layer in centimeters is given by

$$\delta = 8.62 \times 10^{-5} \frac{\lambda Tnc}{I_d} \text{ cm.}, \tag{12}$$

the concentration c being in g.-ions *per cc.* of solution.

Although equation (12), and those preceding it, were derived for an electrode process involving the removal of a cation, they apply equally to any cathodic process in which diffusion is the rate determining stage. Such processes of particular interest are electrolytic reduction reactions, especially when the system involved gives a reversible oxidation-reduction potential. For example, if the solution contains ferric ions, they can be reduced at the cathode to ferrous ions and the potential is determined by the ratio of the activities of the two ions in the immediate vicinity of the inert, e.g., platinum, electrode. As reduction proceeds the ferric ions used up are replaced by diffusion, but the concentration at the electrode surface is always less than in the bulk of the solution; if the current density is increased to the limiting value the ferric ion concentration at the cathode is reduced to zero, and the rate of diffusion of the ions to the electrode is then a maximum for the given conditions. It is seen, therefore, that equations (10) and (11) will hold for this particular process, as well as for other analogous electrolytic reductions, provided equilibrium between the oxidation-reduction system and the electrode is established rapidly; in these cases n represents the number of electrons difference between the oxidized and reduced states.

There is nothing in the foregoing discussion that restricts it to reactions at the cathode or to ions; it holds, in fact, for any electrode process, either anodic, i.e., oxidation, or cathodic, i.e., reduction, using the terms oxidation and reduction in their most general sense, in which the concentration of the reactant is decreased by the electrode process, provided the potential-determining equilibrium is attained rapidly. The fundamental equation (10) is applicable, for example, to cases of reversible oxidation of ions, e.g., ferrous to ferric, ferrocyanide to ferricyanide, iodide to iodine, as well as to their reduction, and also to the oxidation and reduction of non-ionized substances, such as hydroquinone and quinone, respectively, that give definite oxidation-reduction potentials.

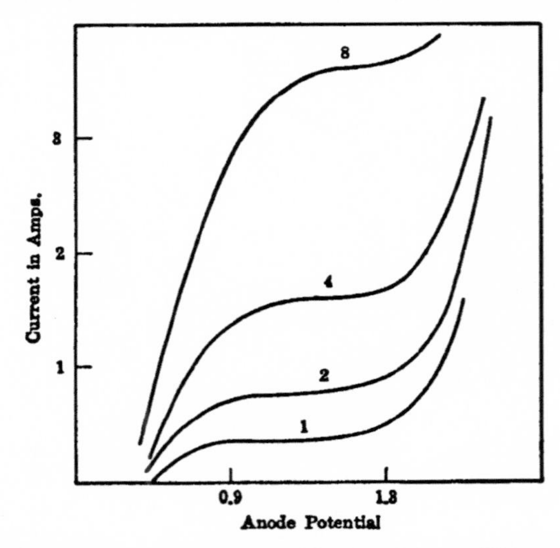

Fig. 114. Limiting current in ferrous chloride solutions (Wilson and Youtz)

Numerous experiments have confirmed the general accuracy of equation (10); the limiting c.d. has been found to be proportional to the concentration of the species that is removed at the electrode. This fact is illustrated by the curves in Fig. 114 which show the variation of the potential of a platinum anode with increasing c.d. in a series of hydrochloric acid (1.5 N) solutions of ferrous chloride with concentrations in the ratio of 1 : 2 : 4 : 8; the limiting current, represented by the horizontal portion of the curve in each case, is seen to increase in almost the same proportion.[5] The measurements were made at 27°, and at this temperature the ionic conductance of the ferrous ion in 1.5 N hydro-

[5] Wilson and Youtz, *Ind. Eng. Chem.*, **15**, 603 (1923); see also, Glasstone and Reynolds, *Trans. Faraday Soc.*, **29**, 399 (1933).

chloric acid is about 40 $ohms^{-1}$ $cm.^2$; it is, therefore, possible to calculate the thickness of the diffusion layer for the various solutions of ferric chloride by means of equation (12). The results obtained are recorded in Table LXXIX; it is seen that in spite of the eight-fold change of con-

TABLE LXXIX. THICKNESS OF THE DIFFUSION LAYER IN FERROUS CHLORIDE SOLUTIONS

$FeCl_3$ concn. Moles per cc.	I_d amp.	δ cm.
0.250×10^{-3}	0.355	0.050
0.125	0.177	0.050
0.0625	0.081	0.054
0.0312	0.038	0.058

centration, the effective thickness of the diffusion layer is approximately 0.05 cm. The electrode used in the work was 66 sq. cm. in area, and similar values for δ have been obtained with very much smaller electrodes; some workers are of the opinion that the thickness of the diffusion layer depends to a certain extent on the size of the electrode, but the effect is certainly not large. Further, a value close to 0.05 cm. for δ has been found for electrolytic processes of various types, viz., deposition of metals, discharge of anions, and electrolytic oxidation and reduction in stationary aqueous solutions at ordinary temperatures.

For most ions the equivalent conductance in solutions of appreciable concentration is about 50 $ohms^{-1}$ $cm.^2$, and if δ is taken as 0.05 cm. at 20°, it follows from equation (12), after rearranging, that

$$I_d = 0.025cn, \tag{13}$$

where c is now the concentration of the electrolytic reactant in moles or g.-ions *per liter*, the limiting current density I_d being in amps. per sq. cm.[6] Strictly speaking, equation (13) is applicable to any ion or molecule whose diffusion coefficient is 1.3×10^{-5} $cm.^2$ $sec.^{-1}$, i.e., 1.1 $cm.^2$ day^{-1}, which is a mean value for a number of substances at ordinary temperatures. If the diffusion coefficient is greater or less than this value, the factor 0.025 in equation (13) is increased or decreased proportionately. For quinone, for example, D is about 0.8 $cm.^2$ day^{-1}, and so the appropriate form of equation (13) would be $I_d = 0.018cn$; in this case n is 2, so that the limiting C.D. is $0.036c$.

Although equation (13) is approximate only, it is of considerable practical value, for it can be employed to calculate the limiting C.D. for metal deposition or for electrolytic oxidation or reduction in a solution of known concentration. Provided the potential at which the particular process occurs is below that of an alternative process, e.g., hydrogen evolution at a cathode or oxygen or halogen evolution at an anode, the current efficiency will be 100 per cent as long as the limiting C.D. is not exceeded. If the current is increased above the limiting value for the given conditions, the efficiency of the main process must inevitably fall

[6] Glasstone, *Trans. Electrochem. Soc.*, 59, 277 (1931).

below 100 per cent, and part of the current will be devoted to another reaction. It should be remembered, of course, that as electrolysis proceeds the concentration in the bulk of the solution of the substance used up at the electrode must inevitably decrease, unless it is replaced in some manner. It follows, therefore, that the limiting C.D. for 100 per cent efficiency will gradually fall off.

Influence of Temperature and Agitation.—If the solution is agitated in any manner or the temperature is raised, the effective thickness of the diffusion layer is diminished. In the former case the decrease depends not only on the rate of agitation, but also on the type of device employed for stirring the solution; no matter what form of agitation is used, however, the diffusion layer cannot be entirely eliminated and it appears to approach a limiting thickness of about 0.001 cm. The question of the influence of temperature on the thickness of the diffusion layer is somewhat uncertain, but in the experience of the present writer the value at 70° is about one-third of that at 20°, in a stationary solution. The diffusion coefficient of most substances in aqueous solution increases by about 2.5 per cent per degree, this being the coefficient of decrease of viscosity; it follows, therefore, that the limiting current is just over four times as great at 70° as at 20°. This increase is small compared with that which results from agitation of the solution. It is the practice, therefore, in metal deposition, and in other aspects of electrolytic work, to maintain efficient circulation of the solution; the current density can then be raised quite considerably and 100 per cent current efficiency maintained.

The Dropping Mercury Cathode.—An interesting application of the principle of the limiting, or diffusion, current is to be found in the **dropping mercury cathode,** which was first adapted to the purpose described below by Heyrovský in 1923.[7] The procedure employed in connection with this form of cathode can be explained with reference to the diagrammatic representation of the apparatus in Fig. 115. The flask *A* contains the experimental solution which can be saturated with hydrogen by passing the gas through the tube *B*; mercury from a reservoir *C* falls through the solution from the end of the capillary tube *D* at a rate of 20 to 30 drops per minute. These drops act as a cathode which is being continuously renewed. A mercury layer *E* at the bottom of the vessel acts as the anode, electrical connection being made by means of a sealed-in wire. The anode (*E*) and cathode (*C*) are joined to the positive and negative poles, respectively, of a battery *F*, and the applied voltage is varied by moving the contact *G* along the potentiometer wire *HI*; the current strength is indicated by the galvanometer *J* across the terminals of which is connected a shunt *K* so as to permit the sensitivity of the instrument to be varied at will. Since the anode has a large area and the

[7] For complete review with references, see Kolthoff and Lingane, *Chem. Revs.*, **24**, 1 (1939); "Polarography," 1941; also, Müller, *Chem. Revs.*, **24**, 95 (1939); von Stackelberg, *Z. Elektrochem.*, **45**, 466 (1939).

current is generally very small, viz., of the order of 10^{-6} amp., the polarization at this electrode is negligible and its potential may be regarded as constant. If this is measured, by comparison with a standard reference electrode, the potential of the cathode can be determined from the total

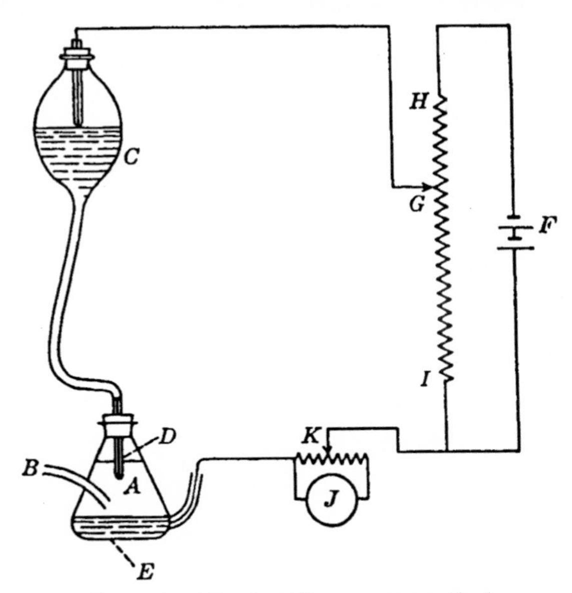

Fig. 115. The dropping mercury cathode

E.M.F. across the cell. The latter includes the IR drop (cf. p. 440), due to the resistance of the electrolyte, which varies with the current strength; as a general rule, however, the variation is small, but allowance for it can be made if necessary. In order to eliminate the possibility of polarization at the anode, some workers employ a separate non-polarizable anode consisting of a large calomel electrode; the potential of this electrode has a known constant value, and so the evaluation of the cathode potential is facilitated.

As the E.M.F. applied to the cell is increased, the potential rises rapidly at first until a cathodic process, e.g., deposition of a metal or electrolytic reduction, that is, reduction in its widest sense, can occur; as with a stationary electrode (cf. Fig. 113), the current can now increase with relatively little increase of potential, until the limiting current for the particular species undergoing reduction is attained. The steady increase of the applied E.M.F. causes the cathode potential to increase until another reduction process can take place, when, once again, the current increases to the limiting value. It is seen, therefore, that as the contact G is moved from I to H, thus increasing the E.M.F. between the electrodes C and E, the current increases in a series of "waves," each wave representing the electro-reduction of a particular species. The complete current-voltage curve is of the form shown in Fig. 116 which was obtained with a solution of 0.1 N calcium chloride * containing copper, lead, cadmium, zinc, manganese and barium ions at concentrations of 10^{-4} g.-ion per liter; the waves for the deposition of the respective ions are indicated. This curve was obtained by means of an automatic register-

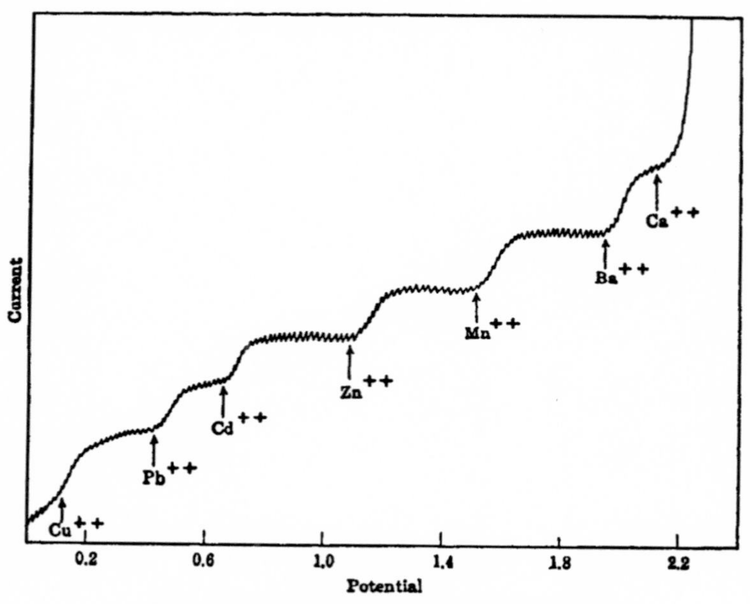

FIG. 116. Polarographic waves (Heyrovský)

ing apparatus known as a **polarograph**; the potentiometer *HI* consists of a number of turns of wire wound around a rotating drum, the contact *G* being fixed. In this way the applied E.M.F. is varied in a regular, known manner. The corresponding current strength is registered photographically by the light reflected from a mirror galvanometer on to a sheet of sensitized paper attached to a rotating drum which is synchronized with the one carrying the potentiometer wire. The resulting curve is known as a **polarogram**; the main waves represent successive cathodic reduction processes while the small oscillations are caused by the changes in the area of the cathode during the growth and detachment of the mercury drops. By using a galvanometer with a relatively long period of swing, the effect of changes in drop size is greatly minimized.

The conditions at a dropping mercury cathode are clearly different from those at a stationary electrode, and the limiting current, or diffusion

* This is the so-called "ground solution" that serves to carry the current.

current (I_d) as it is more frequently called in this work,* is given by the more complicated expression

$$I_d = 0.627nFm^{2/3}t^{1/6}D^{1/2}c, \qquad (14)$$

where n, F, D and c have the same significance as before, m is the mass of mercury dropping per second and t is the average time per drop.[8] The values of m and t vary to some extent with the actual cathode potential because of the corresponding changes in surface tension, but in the vicinity of the potential for any particular electro-reduction process they may be regarded as constant. It then follows from equation (14) that, as in the case of a stationary cathode, the magnitude of the diffusion current, i.e., the height of the polarographic wave, is proportional to the concentration of the species being reduced. It is seen, therefore, that the dropping mercury cathode can be utilized for quantitative analytical purposes. It is not necessary actually to determine m, t and D and to use equation (14) directly; the apparatus is generally calibrated by solutions of the reducible species at known concentrations. The polarographic method of analysis can be used with solutions as dilute as 10^{-5} mole per liter. Since the area of the mercury drop cathode is so small, the actual current is only about 10^{-5} to 10^{-6} amp.; the concentration changes of the bulk of the experimental solution are negligible, and so several polarograms can be recorded in succession without any difference being detectable.

One great advantage of the polarographic method of analysis is that a series of reducible substances, whether they are ions or molecules, can be estimated in one solution. The method has also been applied to the determination of a number of organic compounds, e.g., aldehydes, ketones and nitro-compounds. Dissolved oxygen gives a definite polarographic wave which can be used for analytical purposes. It is because of the reducibility of oxygen that precautions must, in general, be taken to remove traces of this gas from the experimental solution by the passage of a current of hydrogen or nitrogen.

The Half-Wave Potential.—In the preceding description of the analytical applications of the dropping mercury cathode it has been supposed that the nature of the reducible substance has been determined and that the position of the corresponding wave on the current-potential curve is known. If the substance has not been previously identified, however, it is possible to do so by means of the polarographic curve. The reducible material is characterized by its **half-wave potential**;[9] this is the potential

* The diffusion current as defined here is the *actual* current, and not the current density.

[8] Ilkovič, *Coll. Czech. Chem. Comm.*, **6**, 498 (1934); **10**, 249 (1938); *J. chim. phys.*, **35**, 129 (1938); McGillavry and Rideal, *Rec. trav. chim.*, **56**, 1013 (1937); McGillavry, *ibid.*, **57**, 33 (1938); Kolthoff and Lingane, *J. Am. Chem. Soc.*, **61**, 825 (1939).

[9] Heyrovský and Ilkovič, *Coll. Czech. Chem. Comm.*, **7**, 198 (1935); Kolthoff and Lingane, *Chem. Revs.*, **24**, 1 (1939); Lingane, *J. Am. Chem. Soc.*, **61**, 2099 (1939); see also, Müller and Baumberger, *Trans. Electrochem. Soc.*, **71**, 181 (1937); Müller, *Chem. Revs.*, **24**, 95 (1939); *Cold Spring Harbor Symposia Quant. Biol.*, **7**, 59 (1939).

at the point of inflection of its current-potential curve, i.e., half way up its polarographic wave, as shown in Fig. 117. The significance of the half-wave potential can best be understood by considering the behavior of an oxidation-reduction system. Suppose, for example, the solution, which has a definite pH, contains quinone; at the dropping mercury cathode this is reduced to hydroquinone, and the reversible potential of the system as it exists in the immediate vicinity of the drop will be registered on the polarogram. At the bottom of the current-potential wave the solution at the electrode surface consists entirely of quinone while at the top of the wave the concentration of quinone is reduced to zero and the solution must contain hydroquinone only. The polarographic wave in this case is consequently identical in nature with the oxidation-reduction curves shown in Fig. 81 on page 280; at the midpoint of the wave the solution in contact with the mercury drop must contain equivalent amounts of the quinone and hydroquinone and so the half-wave potential is clearly the standard potential of the oxidation-reduction system. This conclusion must hold irrespective of the concentration of the bulk of the solution, for the oxidation-reduction potential depends only on the *ratio* of the oxidized to the reduced state, and this must always be unity at the middle of the current-potential wave. The half-wave potential is thus a characteristic property of the given oxidation-reduction system and can be used for its identification, as well as for other purposes, e.g., determination of pH's, estimation of standard potentials, etc.

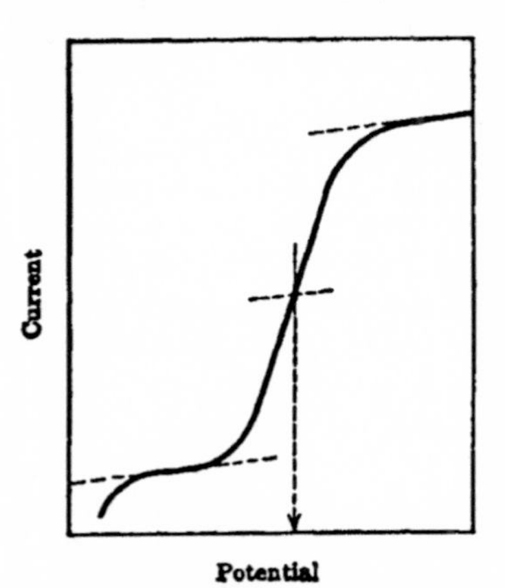

FIG. 117. The half-wave potential

Although it can be readily seen that the half-wave potential is characteristic of an oxidation-reduction system, this is not so obvious when metal deposition is occurring. At the midpoint of the polarographic wave the concentration of the ions being discharged is half the value in the bulk of the solution, and hence it would appear, at first sight, that the potential should depend on the actual ionic concentration of the electrolyte. This would be true if the cathode consisted of the pure metal that is being deposited, but with the dropping mercury cathode an alloy is formed whose concentration varies from zero at the beginning of the wave to a value at the end of the wave that depends on the magnitude of the diffusion current and hence on the concentration of the electrolyte. The potential of a metal in contact with a solution of its ions should be expressed (cf. p. 228) in the form

$$E = \text{const.} - \frac{RT}{zF} \ln \frac{a_{M^+}}{a_M},$$

where a_M represents the activity of the metal and a_{M^+} that of the ions in solution; for approximate purposes this may be written

$$E = \text{const.} - \frac{RT}{zF} \ln \frac{c_{M^+}}{c_M}, \tag{15}$$

where c_M and c_{M^+} are the corresponding concentrations. Suppose a solution of M^+ ions of concentration c_{M^+} is reduced at the dropping mercury cathode, and the maximum concentration of the metal M in the amalgam formed in the drops is c_M; at the half-wave point, the respective concentrations at the drop surface will be $\frac{1}{2}c_{M^+}$ and $\frac{1}{2}c_M$, respectively, so that the potential is given by equation (15). Imagine, now, that the solution is diluted to a fraction x of its original value, i.e., to xc_{M^+}; the concentration in the vicinity of the mercury drop at the midpoint of the polarographic wave is $\frac{1}{2}xc_{M^+}$. Since the concentration of the bulk of the solution is reduced, the height of the wave will be decreased to the same extent, and consequently so also will be the maximum concentration of metal in the amalgam; the latter is thus xc_M, and the concentration at the half-wave potential is now $\frac{1}{2}xc_M$. Since the concentrations of both the ions in solution and the metal in the cathode have been reduced in the same proportion, the potential is still given by equation (15). The half-wave potential should thus be a definite property of the deposited ions and not of their concentration in the solution: this expectation is borne out by the results in Table LXXX for solutions of different con-

TABLE LXXX. POLAROGRAPHIC HALF-WAVE POTENTIALS IN 0.1 N POTASSIUM CHLORIDE

Concentration Moles per liter	Tl^+	Pb^{++}	Cd^{++}	Zn^{++}
0.1×10^{-3}	−0.462	−0.396	—	—
0.2	—	—	−0.594	−0.990
0.5	−0.457	−0.396	−0.593	−0.989
1.0	−0.460	−0.392	−0.594	—
2.0	−0.456	−0.397	−0.601	−0.999
5.0	−0.459	−0.394	−0.598	−0.992
10.0	—	−0.398	−0.605	−1.003

centrations. The potentials recorded are with reference to the calomel electrode containing saturated potassium chloride; the "ground" electrolyte employed in each case was 0.1 N potassium chloride and the observations were made at 25°.

Diffusion at the Anode.—Concentration polarization at an anode produced by the dissolution of the metal under the influence of current is similar in many respects to that due to the reverse process, i.e., ionic discharge, at a cathode; there is, however, one important difference, which makes the effect of a given current smaller at an anode than at a cathode. Provided the thickness of the diffusion layer remains unchanged, a current density which in a particular electrolyte is large enough to reduce the concentration of M^+ ions at a cathode almost to zero, will merely double the concentration if applied to an anode of the metal M in the same solution, assuming the anode of M to pass into solution as M^+ ions. Since the magnitude of the concentration polarization depends on the ratio of the concentration of the ions at the electrode surface to that in the bulk of the electrolyte, as seen above, it follows that it will be very much greater at a cathode under these conditions than at an anode.

The same conclusion may be reached more precisely by following the same treatment as was adopted in the discussion of concentration changes at a cathode. Since the ionic concentration at the electrode c_e is now greater than in the bulk of the solution, it is possible to derive an equation which is exactly analogous to (6), except that a and a_e are reversed, viz.,

$$a = a_e - \frac{I}{k},$$

where k has the same significance as before. If the diffusion layer has the same thickness as for a cathode under the given conditions, k may be replaced by I_d/a where I_d represents the *cathodic* limiting current for the electrolyte being employed. It follows, therefore, that

$$a = a_e - \frac{Ia}{I_d},$$

and since the *numerical* value of the concentration polarization is now given by

$$\Delta E = E - E_e = \frac{RT}{nF} \ln \frac{a_e}{a},$$

the result is

$$\Delta E = \frac{RT}{nF} \ln \frac{I_d + I}{I_d}, \tag{16}$$

which may be compared with the corresponding equation (9) for concentration polarization at a cathode. It is at once apparent that when I is equal to I_d, the anodic polarization according to equation (16) is relatively small, viz., $(RT/nF) \ln 2$, whereas at a cathode the polarization becomes very large.[10]

Ideally, there should be no limit to the magnitude of the C.D. for the dissolution of a metallic anode; in actual practice, however, such a limit is, in fact, observed. The reason is that the solution in the vicinity of the anode becomes saturated with the particular salt that is formed, and this begins to crystallize on the electrode, thus reducing its exposed area and increasing the effective current density. The phenomenon of pas-

[10] Bowden and Agar, *Proc. Roy. Soc.*, 169A, 206 (1938).

sivity will then generally be established and there is a complete change in the behavior of the anode, as described in Chap. XIV.

Limiting currents at the anode will, of course, be observed, just as at a cathode, when ions, e.g., ferrous ions, or other oxidizable substances, e.g., hydroquinone, present in the solution are used up at the electrode during the passage of current. Another instance of the occurrence of anodic limiting currents is when the dissolution of the metal results in a reaction which uses up anions from the electrolyte; a simple example is a silver anode in a cyanide solution. The maximum rate at which the anode can dissolve to form $Ag(CN)_2^-$ ions depends on the maximum rate of diffusion of the cyanide ions to the electrode; the limiting C.D. will be given by an equation exactly the same as (10) or (13). The magnitude of the limiting current for a given cyanide solution will depend on the formula and charge of the complex ion, and consequently will be different for a silver anode than for one of cadmium, since $Cd(CN)_3^{--}$ ions are formed to some extent in the latter case. By measuring the anodic limiting current density and the quantity of the metal dissolved by one faraday of electricity, it should be possible to determine the formula of the complex ion.[11]

If the ions produced when the anode passes into solutions forms an insoluble salt with anions present in the electrolyte, a definite limiting current will be observed provided the precipitate does not adhere to the electrode surface. If the adhesion does occur and the surface is blocked, the anode will, of course, cease to dissolve immediately; this occurs, for example, with a silver anode in a halide solution.

II. Ionic Discharge is the Slow Process.—In considering the possibility that ionic discharge is the rate-determining process at a cathode, it must not be assumed that the slow stage is necessarily the actual neutralization of the ion. It is true that some writers have suggested that the combination of an ion with an electron may occur slowly, but it is doubtful if this is likely to be true for a metal since it consists of a lattice of fixed ions and mobile electrons. It is nevertheless possible that one or other of the stages required to convert a hydrated ion in solution close to the electrode to a neutral atom on the cathode surface may be slow. In this event, the C.D. may be assumed to be such that diffusion of ions to the electrode and the attainment of equilibrium of the atoms on the electrode can be regarded as rapid.

It was seen on page 251 that the rate of the cathodic reaction

$$M^+ + \epsilon \rightarrow M,$$

which results in the discharge of ions, is given by $k_1 a_+ e^{\alpha FE/RT}$, where k_1 is a constant for the given electrode material, a_+ is the activity of the cations in the solution, E is the difference of potential at the electrode, and α lies between zero and unity. Similarly the rate of the anodic

[11] Kolthoff and Miller, *J. Am. Chem. Soc.*, **63**, 1405 (1941).

reaction

$$M \rightarrow M^{+} + \epsilon,$$

leading to the formation of ions in solution, is $k_2e^{-(1-\alpha)FE/RT}$. If E is the reversible potential for the given system, the rates of the two reactions are equal, and there is no net flow of current. If the electrode is made an anode or a cathode, the potential must be changed so that the rate of one of the processes exceeds that of the other. At a cathode, for example, which will be discussed here, the rate of the cathodic, i.e., discharge, reaction must exceed that of the anodic, i.e., ionization, process: the magnitude of the current will be determined by the difference of the two rates.[12] Assuming, for simplicity, that the ions under consideration are univalent, each g.-ion carries a faraday of electricity (F), and so the cathodic current per unit area is given by

$$I_c = Fk_1a_+e^{\alpha FE'/RT}, \tag{17}$$

while the anodic current for the same area is

$$I_a = Fk_2e^{-(1-\alpha)FE'/RT}. \tag{18}$$

The electrode potential is now represented by E' to indicate that it differs from the reversible value for the given electrode and solution; the activity of the cations (a_+) is assumed to be the same as in the bulk of the solution, since diffusion is no longer rate-determining and it may be supposed that the ions are replaced as fast as they are used up. This can be regarded as approximately true provided the C.D. is well below the limiting value. The net value of the current (I) is given by the difference between I_c and I_a; hence the net cathodic current is

$$I = F[k_1a_+e^{\alpha FE'/RT} - k_2e^{-(1-\alpha)FE'/RT}]. \tag{19}$$

Writing E' as $E + \omega$, where E is the reversible potential for the solution containing ions at activity a_+, and ω is the polarization, or overvoltage, at the current density I, equation (19) becomes

$$I = F[k_1a_+e^{\alpha F\omega/RT}e^{\alpha FE/RT} - k_2e^{-(1-\alpha)F\omega/RT}e^{-(1-\alpha)FE/RT}]. \tag{20}$$

At the reversible potential $k_1a_+e^{\alpha FE/RT}$ is equal to $k_2e^{-(1-\alpha)FE/RT}$, as mentioned above, and hence equation (20) may be written

$$I = Fk_1a_+e^{\alpha FE/RT}[e^{\alpha F\omega/RT} - e^{-(1-\alpha)F\omega/RT}]. \tag{21}$$

The factor preceding the brackets is equal to the magnitude of the cathodic, and hence of the anodic, current at the reversible potential; this quantity is given the symbol I_0, and so

$$I = I_0[e^{\alpha F\omega/RT} - e^{-(1-\alpha)F\omega/RT}]. \tag{22}$$

[12] Glasstone, Laidler and Eyring, "The Theory of Rate Processes," 1941, Chap. X; see also, Kimball, *J. Chem. Phys.*, 8, 199 (1940).

The exponentials can be expanded, and if the polarization ω is small all terms beyond the first may be neglected; equation (22) then reduces to

$$I = I_0 \frac{F\omega}{RT},$$

$$\therefore \quad \omega = \frac{RT}{F} \cdot \frac{I}{I_0}. \tag{23}$$

For small polarizations, therefore, the overvoltage ω is proportional to the current I. Since the actual potential E' is equal to the sum of E, which is a constant, and ω, it follows that when some stage in the discharge process is rate-determining, the cathode potential is a linear function of the current at low polarizations. Behavior of this kind has been observed in connection with the discharge of hydrogen ions at platinum cathodes,[13] and also in the deposition of certain metals (see p. 463).

At relatively large currents the polarization becomes appreciable and it is no longer possible to ignore the terms in the expansion of the exponentials in equation (22). Another simplification is, however, possible since the first term in the brackets increases rapidly while the second term decreases as ω is increased. The latter term can thus be neglected in comparison with the former, and equation (22) becomes

$$I = I_0 e^{\alpha F\omega/RT}, \tag{24}$$

$$\therefore \quad \omega = \text{const.} + \frac{RT}{\alpha F} \ln I$$

$$= \text{const.} + \frac{2.3RT}{\alpha F} \log I, \tag{25}$$

so that the overvoltage, and the cathode potential, will be a linear function of the logarithm of the current density. In general, the slope of the line for an ion of valence z should be $2.3RT/\alpha zF$, and at ordinary temperatures this should be $0.0585/z\alpha$ where α lies between zero and unity.

It is probable that a logarithmic relationship, of the type of equation (25), between current and potential results from slow processes other than the discharge of ions; these may sometimes be distinguished by the fact that α is larger than unity. An instance of this kind is possibly to be found in connection with the evolution of hydrogen at certain cathodes, e.g., platinized platinum; the value of α is approximately 2, and the slow process appears to be concerned with the combination of hydrogen atoms to form molecules rather than with the discharge of ions (cf. p. 471).

Most metals can be deposited at an appreciable rate with relatively small overvoltages, but this is not so for iron, cobalt and nickel at ordinary temperatures; the minimum cathode potentials necessary to produce

[13] Butler, *Trans. Faraday Soc.*, 28, 379 (1932); Butler and Armstrong, *J. Chem. Soc.*, 743 (1934).

a visible deposit of these metals in a reasonable time are recorded in Table LXXXI.[14] The electrolyte was a N-solution of the sulfate of the

TABLE LXXXI. EXPERIMENTAL DEPOSITION POTENTIALS IN N-SULFATE SOLUTIONS

Metal	Reversible Potential	Deposition Potential 15°	55°	95°
Iron	− 0.44	− 0.68	− 0.49	− 0.46
Cobalt	− 0.28	− 0.56	− 0.46	− 0.36
Nickel	− 0.24	− 0.57	− 0.43	− 0.29

metal in each case, and the results were found to be almost independent of the pH. The experimental deposition potentials are given at 15°, 55° and 95°, together with the reversible potentials for the process $M^{++} + 2\epsilon \rightarrow M$ for the respective metals. The overvoltages are seen to be quite large at ordinary temperatures, but to decrease with increasing temperature. It is not certain that the results in Table LXXXI have any exact theoretical significance, but they do show that, as in the case of hydrogen ion discharge at certain cathodes, a considerable overvoltage must be applied in order that the ions may be discharged at an appreciable rate. It is reasonably certain that ionic discharge occurs at smaller (cathodic) potentials than are quoted in the table, but the rate must be so small that a long time is required for a visible deposit of the metal to form. With other metals, e.g., copper, mercury, cadmium, etc., appreciable rates of ionic discharge, accompanied by the formation of visible deposits, occur when the reversible potential is exceeded, cathodically, by only a few millivolts. It is probable that the exceptional behavior in the case of the iron-group metals is to be attributed to the slowness of some stage in the discharge process, i.e., between hydrated ions in the vicinity of the cathode, on the one hand, and stable atoms on the electrode surface, on the other hand. This view finds some support in the observation, mentioned on p. 444, that the reverse process, namely the anodic solution of iron, cobalt and nickel, is accompanied by considerable polarization which decreases rapidly as the temperature is raised. It is probable that in both the discharge and the formation of ions a relatively high energy barrier must be surmounted; approximate calculations, based on the data in Table LXXXI, suggest that the value in the forward direction is about 12 to 18 kcal.

Among the possibilities which have been suggested for the rate-determining process in the discharge of the ions of the iron-group are that dehydration of the ions is slow or that the actual neutralization is slow. The present writer, however, considers it improbable that these processes require a high activation energy, and favors the view that the metal is first deposited in a metastable state; the conversion of this form of the metal to the stable state is regarded as the slow rate-determining process.

[14] Glasstone, *J. Chem. Soc.*, 2887 (1926); see also, Foerster and Georgi, *Z. physik. Chem.* (Bodenstein Festband), 453 (1931).

III. Establishment of Equilibrium on the Electrode is the Slow Process.—The fact that an appreciable excess potential must be applied in certain instances before metal deposition can take place on a cathode of another material (cf. p. 444), suggests the possibility that there may be some hindrance to the discharged atoms finding their way to stable lattice positions. The consequences of the rate of formation of nuclei on the cathode surface, at which crystal growth can occur, being the slow process have been considered by Erdey-Grúz and Volmer[15] who reached the following conclusions:

(1) if the slow process is the transfer of atoms from the points at which they are deposited to their final lattice positions, the overvoltage ω is proportional to the current I;

(2) if the slow process is the formation of two-dimensional nuclei, $\log I$ is a linear function of $1/\omega$;

(3) if the slow process is the formation of three-dimensional nuclei, $\log I$ is a linear function of $1/\omega^2$.

At low C.D.'s the deposition of many metals satisfies the requirements of condition (1) above; it must be pointed out, however, that the same result would be obtained at low overvoltages if a stage in the discharge of the ion were the slow process [cf. equation (23)]. According to Erdey-Grúz the variation of overvoltage with the current in the deposition of silver depends on the nature of the electrolyte; for example, condition (1) is satisfied by solutions of silver chloride in ammonium chloride and silver cyanide in potassium cyanide; condition (2) is satisfied by silver iodide in potassium iodide and silver oxide in ammoniacal solution; while in the deposition of silver from silver chloride or bromide in ammonia the overvoltage is a linear function of $\log I$, suggesting that ionic discharge is the slow process. It is difficult to accept these deductions, for it appears hardly probable that the same metal will be deposited in three different ways depending on the conditions. It must be admitted that the results obtained hitherto in connection with the overvoltages of metal deposition do not lead to any definite conclusions. At high current densities, as the limiting value is approached, diffusion will always become the rate-determining stage in the discharge of cations.

Hydrogen Overvoltage.—In view of the foregoing discussion of the dependence of polarization phenomena on the nature of the slow stage of the electrode process, it is evident that in order to give a satisfactory quantitative interpretation of hydrogen overvoltage it is necessary to know the exact nature of its variation with current density.

For the measurement of overvoltage at definite C.D.'s, the method adopted is, in principle, the same as that described on page 436, employing a cathode of known area. The solution should be completely free from dissolved oxygen or other reducible material; for this reason it

[15] Erdey-Grúz and Volmer, *Z. physik. Chem.*, **157A**, 165, 182 (1931); see also, Brandes, *ibid.*, **142A**, 97 (1929).

has become the practice to saturate the solution with hydrogen at atmospheric pressure. The anode and cathode compartments must be kept separate in order to prevent access to the latter electrode of oxygen liberated at the former. The neglect of this precaution by earlier workers on decomposition voltage and overvoltage accounts for some of the phenomena observed which have no connection with the fundamental polarization problem. A current of definite strength is passed through the electrolytic cell, sufficient time being allowed for the potential to become constant; the value of the cathode potential is then measured by combining it with a reference electrode, e.g., a calomel electrode, in the usual manner. If a hydrogen gas electrode, with platinized platinum as the electrode metal, is employed for reference purposes in the experimental solution, the difference in potential between it and the cathode gives the overvoltage directly. When another form of reference electrode is used, its potential against a reversible hydrogen electrode can either be measured directly or the value may be calculated from the known potential of the reference electrode and the pH of the solution.

If the actual current passing through the cell is relatively high, e.g., more than about 0.01 amp., an appreciable ohmic fall of potential, i.e., the *IR* potential, due to the resistance at the surface of the electrode, will be included in the measured potential. The error due to this cause, which is always likely to be present, can be minimized by drawing out the end of the salt-bridge connecting tube (*T* in Fig. 109, p. 436) to a moderate capillary and pressing it close to the surface of the electrode. A method which is sometimes used to eliminate the potential due to the resistance at the electrode surface is to employ special devices which permit the measurement of the cathode potential at a number of very short intervals after the current is cut off; the results are then extrapolated to zero time so as to give the potential, free from the effect of ohmic drop, during the passage of current. This method is no doubt satisfactory with currents of about 0.1 amp., and perhaps more, but for still higher currents the initial rate of fall of potential, after the current is cut off, may be so rapid as to make extrapolation uncertain. An alternative procedure, which has been frequently adopted in recent work, is to use small electrodes, so that relatively high current densities can be obtained although the actual current is low; since the potential fall due to the resistance at the electrode surface is equal to the product of the current strength and the resistance, the consequent error is negligible.

Influence of C.D. on Overvoltage.—Provided the conditions are such that the hydrogen is not removed by reaction with oxygen, or other oxidizing agent, or by diffusion away from the cathode, the overvoltage ω increases with increasing current density I in accordance with the equation

$$\omega = a + b \log I, \tag{26}$$

where a and b are constants; this relationship has been found to hold for a number of cathodes at current densities from 10^{-6} to 10^{-1} amp. per sq. cm.* The plot of overvoltage against log I should be a straight line; the results for a number of cathodes in dilute sulfuric acid, as shown in Fig. 118, are in harmony with expectation.[16] It will be observed that

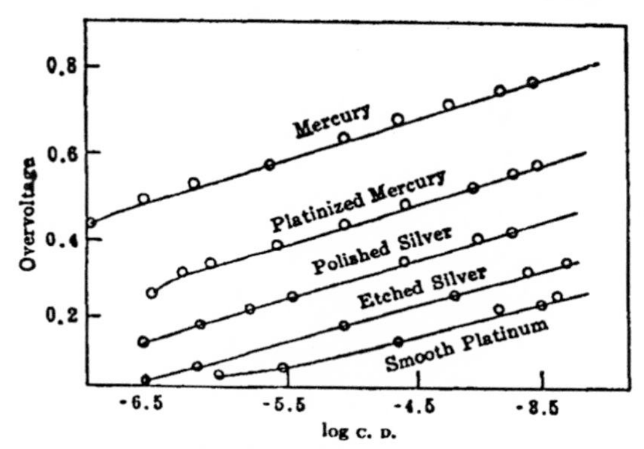

FIG. 118. Variation of overvoltage with current density (Bowden and Rideal)

the lines in the figure are approximately parallel; this means that b in equation (26) is almost the same for a series of metals. The results of the investigations of a number of workers[17] have shown that the constant b is approximately 0.12 for a variety of cathodes, e.g., mercury, gallium, silver, nickel, palladium, aluminum, copper and gold; the value 0.12 for the quantity b can be represented by

$$b = 2 \times \frac{2.30RT}{F},$$

which is equal to 0.118 at ordinary temperatures. That b actually has the significance implied by this expression is shown by the measurements made with a mercury cathode in dilute sulfuric acid between 0° and 72°; the same relationship for b applies over the whole temperature range. The importance of this result may be seen by comparing equations (25) and (26), which are of the same form; if the expression for b is inserted in the latter, it is immediately apparent that α in equation (25) is equal to 0.5 for a number of different cathodes.

Although values of b lying between 0.10 and 0.14 have been obtained in many instances, there are cases of lower and higher values, and changes

* It has been claimed that the equation holds up to current densities of 10^2 amps. per sq. cm., but the data appear to be open to criticism (cf. Hickling and Salt, *Trans. Faraday Soc.*, **36**, 1226 (1940)).

[16] Bowden and Rideal, *Proc. Roy. Soc.*, **120A**, 59 (1928).

[17] Bowden and Rideal, *Proc. Roy. Soc.*, **120A**, 59 (1928); Lewina and Silberfarb, *Acta Physicochim. U.R.S.S.*, **4**, 275 (1936); Kabanow, *ibid.*, **5**, 193 (1936); Pleskow, *ibid.*, **11**, 305 (1939); Hickling and Salt, *Trans. Faraday Soc.*, **36**, 1226 (1940).

are often observed with continued electrolysis. For example, b for smooth platinum increases from about 0.075 to 0.19 with time; for lead and graphite cathodes values as high as 0.3 have been reported, whereas for platinized platinum b is as low as 0.025, although it tends to increase with prolonged use. Changes of overvoltage with time have been observed in numerous instances; such changes are frequently due to alterations in the nature of the surface, which are often visible to the eye, and sometimes to traces of impurities which are deposited on the cathode as electrolysis is prolonged. It is factors such as these which introduce some uncertainty into overvoltage measurements; it seems to be established, nevertheless, that the slope b of the plot of the overvoltage against the logarithm of the C.D. is, in general, close to $2 \times 2.30RT/F$, although exceptions undoubtedly exist.

Influence of pH on Overvoltage.—The dependence of overvoltage on the hydrogen ion concentration is a matter of some uncertainty; there is little doubt that the "bubble overvoltage" is independent of pH for many cathodes, but the results at appreciable current densities are difficult to determine. In order to obtain solutions of low acidity having definite pH's, it is generally necessary to employ buffer systems, and the accumulation of alkali metal ions in the vicinity of the cathode will affect the results obtained. It appears, however, from an examination of a variety of experimental data, that in the absence of strongly adsorbed ions the overvoltages at most cathodes are independent of hydrogen ion concentration over a large range of pH values.[18] In strongly acid or strongly alkaline solutions deviations sometimes occur; this may be due to the large concentration of hydrogen or hydroxyl ions, respectively, or it may be brought about, at least in alkaline solutions, by the alkali metal ions. It is probable that all substances which alter the electrokinetic (zeta-) potential (see Chap. XVI) at the cathode surface, whether they also change the pH or not, have an influence on the overvoltage.

Influence of Temperature on Overvoltage.—Since overvoltage is due to a slow stage in the ionic discharge process, it is evident that increase of temperature will decrease overvoltage. As a general rule, the change is approximately 2 millivolts per degree, but it is more fundamentally satisfactory to express the influence of temperature in terms of the activation energy of the rate-determining stage. The quantity I_0 referred to on page 460 contains the rate constant k_1, apart from temperature-independent factors, and hence, according to modern theory of reaction rates, it may be represented by $Be^{-\Delta H^{\ddagger}/RT}$, where B is a constant and $\Delta H^{\ddagger}$ is the energy of activation of the slow process responsible for overvoltage.

[18] Glasstone, *J. Chem. Soc.*, **125**, 2414, 2646 (1924); Meunier, *J. chim. phys.*, **22**, 629 (1925); Sand, Lloyd and Grant, *J. Chem. Soc.*, 378 (1927); Bowden, *Trans. Faraday Soc.*, **24**, 473 (1928); Lloyd, *ibid.*, **25**, 525 (1929); see, however, Lewina and Sarinsky, *Acta Physicochim. U.R.S.S.*, **7**, 485 (1937); Lukowzew, Lewina and Frumkin, *ibid.*, **11**, 21 (1939); Legran and Lewina, *ibid.*, **12**, 243 (1940).

It follows, therefore, that equation (24) can be written as

$$I = Be^{-\Delta H^{\ddagger}/RT}e^{\alpha F\omega/RT}, \tag{27}$$

and upon taking logarithms of equation (27), the result is

$$\ln I = \ln B + \frac{\alpha F\omega - \Delta H^{\ddagger}}{RT}.$$

Differentiating with respect to temperature, at constant overvoltage, it follows that

$$\left(\frac{\partial \ln I}{\partial T}\right)_{\omega} = \frac{\Delta H^{\ddagger} - \alpha F\omega}{RT^2}, \tag{28}$$

or, at constant current,

$$\left(\frac{\partial \omega}{\partial T}\right)_{I} = \frac{\alpha F\omega - \Delta H^{\ddagger}}{\alpha FT}. \tag{29}$$

Since the overvoltage ω and the value of α, which is derived from the slope b, may be regarded as known, it is clearly possible to determine the activation energy $\Delta H^{\ddagger}$ from a knowledge of the temperature coefficient of $\log I$ at constant overvoltage, or of ω at constant current. Unfortunately, the data from which the values of $\Delta H^{\ddagger}$ may be derived are limited to a few cathodes, but most of the available results are recorded in Table LXXXII.[19] The magnitude of the current I_0, which flows in each direc-

TABLE LXXXII. ACTIVATION ENERGY FOR DISCHARGE OF HYDROGEN IONS

Cathode	Electrolyte	$\Delta H^{\ddagger}$ kcal.	I_0 amp./cm.^{2}
Mercury	0.2 N Sulfuric acid	18.0	6×10^{-12}
Mercury	0.2 N Sodium hydroxide	8.7	7×10^{-9}
Gallium	0.2 N Sulfuric acid	15.2	1.6×10^{-7}
Wood's alloy	0.2 N Sulfuric acid	16.4	1×10^{-8}
Smooth platinum	0.2 N Sulfuric acid	$\sim$10.0	$\sim 3 \times 10^{-6}$
Smooth platinum	0.2 N Sodium hydroxide	$\sim$ 6.5	$\sim 1 \times 10^{-6}$
Palladium	0.2 N Sulfuric acid	9.0	2×10^{-5}
Palladium	0.2 N Sodium hydroxide	10.0	1×10^{-5}

tion at the reversible potential, is also given in each case; this is obtained by plotting the measured overvoltages for a series of current densities against $\log I$ and extrapolating, in a linear manner, to zero overvoltage. In view of the relationship given above between $\Delta H^{\ddagger}$ and I_0, it is clear that when the former is large the latter will be small; a high energy of activation thus means that the current passing at the reversible potential is extremely minute and hence a relatively large overvoltage has to be applied to obtain a reasonable current.

Rate of Growth of Overvoltage.—The rate at which an electrode attains its overvoltage potential, immediately following the switching

[19] Bowden, *Proc. Roy. Soc.*, **126A**, 107 (1929); Bowden and Agar, *Ann. Rep. Chem. Soc.*, **35**, 90 (1938).

on of the current, has been followed either by photographing the movement of a string galvanometer on a rapidly moving film or by means of an oscillograph, both instruments being capable of following rapid changes of potential. For solutions carefully freed from dissolved oxygen and saturated with hydrogen, the potential has been found to increase in a linear manner with the quantity of electricity passed, as shown in Fig. 119 for a mercury cathode; the current is switched on at the point A and the final steady overvoltage is attained at C. The rate of increase of potential is seen to be constant, falling off only at B when the steady state is approached.[20] This constant rate of potential growth implies that the

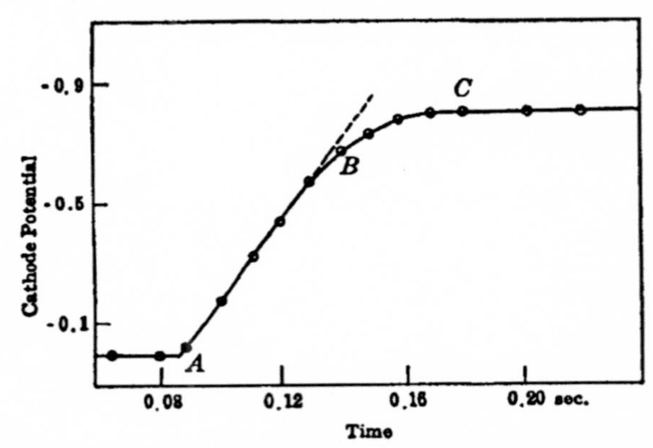

FIG. 119. Growth of overvoltage (Bowden and Rideal)

building up of the cathodic overvoltage is due to the establishment of an electrical double layer, acting virtually as a condenser across which there is no transfer of charge. The capacity of the condenser at the electrode surface is equal to dQ/dE, so that it can be obtained from the slope of the linear plot of the quantity of electricity passed per sq. cm. (Q) against the potential (E), or against the overvoltage. In other words, the capacity of the double layer at the electrode surface is determined by the slope of the portion AB of Fig. 119; the data in this figure lead to a value of $6\mu F$ per sq. cm.,* but this result is probably too low on account of the presence of traces of impurities. By taking careful precautions to exclude all extraneous substances, a value of $20\mu F$ per sq. cm. has been obtained, in agreement with the result derived from other methods of studying the capacity of the cathodic double layer. The magnitude of this capacity appears to be approximately the same for all electrode materials, and this fact has been used to determine the true area of an electrode surface. The ratio of the capacity per *apparent* sq. cm. of

[20] Bowden and Rideal, *Proc. Roy. Soc.*, 120A, 59 (1928); see also, Erdey-Grúz and Volmer, *Z. phys. Chem.*, 150, 203 (1930); Hickling, *Trans. Faraday Soc.*, 36, 364 (1940).

* A microfarad (μF) is equivalent to a capacity of 10^{-6} coulombs per volt.

surface, as determined by the rate of growth of the potential, to the value for an ideally smooth surface, i.e., $20\mu F$, gives the actual or accessible area of the electrode per unit apparent area.

Theories of Overvoltage.—Sufficient general information has now been given to permit a consideration of the theory of overvoltage, and to allow an attempt to be made to elucidate the nature of the slow process involved in the discharge of hydrogen ions. The stages in the reaction at a cathode in aqueous solution are as follows: (1) transfer and diffusion of H_3O^+ ions to the electrode layer; (2) transfer of these ions, or protons, to the electrode; (3) neutralization of the charge of the ions or protons by electrons; (4) combination of the resulting hydrogen atoms to form molecules; and (5) evolution of the hydrogen molecules as bubbles of gas.

I. Bubble Formation as the Slow Process.—It is certain that stage (1) is not rate-determining because, in the first place, the overvoltage depends on the nature of the cathode metal, and, in the second place, the activation energy for transference and diffusion in aqueous solution is known to be much less, viz., about 4 kcal., than that required for overvoltage (cf. Table LXXXII). It has been frequently observed that there is some connection between the bubble overvoltage and the interfacial tensions, or surface forces generally, at a cathode; it was therefore considered at one time that stage (5), i.e., gas-bubble formation, was the essential factor in overvoltage. It was found, for example, that the potential of a platinized platinum cathode fluctuated during the formation and liberation of a bubble of hydrogen gas. This phenomenon is undoubtedly due to supersaturation effects and it probably does not represent any variation of the true overvoltage because the reversible hydrogen potential no doubt alters in a corresponding manner.[21] Alterations in the electrode or electrolyte which influence the conditions required for bubble formation to commence, e.g., the presence of substances which affect the interfacial tensions, undoubtedly alter the bubble overvoltage, but this is partly due to the fact that the effective current density at the instant of bubble formation is also changed. It is also possible, of course, that surface active compounds influence the true overvoltage because of the alteration in the nature of the electrode resulting from adsorption of these substances on the surface. On the whole, however, it must be concluded that ease of bubble formation plays a secondary part and that it cannot generally be the slow process in the discharge of hydrogen ions.

II. Combination of Atoms as the Slow Process.—The view that the formation of molecules from hydrogen atoms, i.e., stage (4), was rate-determining, was proposed by Tafel[22] and received much support at one time. According to this theory the high electrode potential which manifests itself as overvoltage is due to the accumulation of hydrogen atoms

[21] MacInnes and Adler, *J. Am. Chem. Soc.*, **41**, 194 (1919).
[22] Tafel, *Z. phys. Chem.*, **50**, 641 (1905).

at the cathode, and the variation of overvoltage from one metal to another was ascribed to their differing catalytic influence on the rate of combination of these atoms to form molecules. The low overvoltage metals, such as platinum, palladium, nickel and copper, should be good catalysts for the reaction $2H = H_2$, while those of high overvoltage, e.g., lead, mercury, tin and cadmium, would be expected to be poor catalysts. It may be noted in this connection that direct experimental observations on the rate of combination of hydrogen atoms on various metallic surfaces showed their catalytic effect to decrease in the order Pt, Pd, Fe, Ag, Cu, Pb and Hg. Further, low overvoltage metals are the best catalysts for hydrogenation processes in which it is probable that $H_2 = 2H$ is a preliminary step; if the metals catalyze this reaction they should also facilitate the opposite reaction, i.e., the combination of hydrogen atoms. The experimental facts are thus in harmony with the view that the latter is the rate-determining stage in hydrogen evolution at a cathode.

The chief objection to the Tafel theory is that it leads to an incorrect value of the slope b in equation (26). If the electrode surface is sparsely covered with atomic hydrogen, the rate of the reaction $2H = H_2$ will be proportional to n^2, where n is the number of adsorbed atoms per sq. cm. of surface. If the rate of the reverse process, i.e., the dissociation of hydrogen molecules into atoms, is negligible, the current flowing I will be proportional to the rate of the formation of molecular hydrogen, since the latter is assumed to be the slow process at the cathode. It follows, therefore, that I is equal to kn^2, where k is a constant. The potential E of an atomic hydrogen electrode, neglecting activity influences, can be represented by the equation

$$E = \frac{RT}{F} \ln n + \text{constant},$$

where the constant depends on the hydrogen ion concentration of the solution. The potential of an atomic hydrogen electrode corresponding to the reversible (molecular) hydrogen potential is

$$E_0 = \frac{RT}{F} \ln n_0 + \text{constant},$$

where n_0 is the number of hydrogen atoms per sq. cm. of electrode surface in equilibrium with normal hydrogen gas at 1 atm. pressure. The difference between E and E_0 should be equal to the overvoltage, and so

$$\omega = \frac{RT}{F} \ln \frac{n}{n_0} \cdot$$

Since the constant term has disappeared, it follows that according to the Tafel theory the overvoltage should be independent of the hydrogen ion concentration, in general agreement with experiment. Combination of this equation with the result $I = kn^2$, given above, yields

$$\omega = \frac{RT}{2F} \ln I + a, \tag{30}$$

where a is a constant. Although this equation is of the correct form [cf. equation (26)], it requires the slope b of the plot of ω against $\log I$ to be $2.3RT/2F$, i.e., 0.029 at ordinary temperatures. Except for metals of low overvoltage, e.g., platinized platinum, the actual slope is about four times this value, and so it is unlikely that atom-combination is usually the slow stage in hydrogen evolution. At low overvoltages, however, it is probable that this process is actually rate-determining, but it must be replaced by another when the overvoltage is high. At platinized platinum the overvoltage is always low and the slope b is about 0.025, as required by the Tafel theory.

III. Ion Discharge as the Slow Process.—The possibility that the slow step is the neutralization of a hydrogen ion by an electron, i.e., stage (3), has been treated from the quantum-mechanical point of view.[23] It is supposed that an energy barrier exists between the electrons in a metal and a hydrogen ion approaching it; according to classical mechanics the electron cannot cross the barrier to discharge the ion unless it acquires sufficient energy to pass over the top. Quantum mechanics, however, allows of a definite probability of leakage *through* the barrier of an electron from the cathode to an unoccupied level of the same energy in the ion to be discharged. The distribution of electronic energy levels in the metal is given by the Fermi-Dirac statistics whereas that of vacant levels in the hydrogen ions is determined by classical, i.e., Maxwell-Boltzmann, statistics. The current which is able to pass at a given potential is then determined by integrating, over the whole range of unoccupied levels of the ion, the probability of the transition of an electron from any level in the cathode to one of equal energy in the ion. The relationship derived in this manner for the dependence of the current on the overvoltage is of the form of equation (26), and b is found to be $2.3RT/\alpha F$, where α should lie between zero and unity and should probably be about 0.5, in accordance with experiment for a number of metals. Apart from the difficult nature of the quantum-mechanical treatment, the theory suffers from the weakness of requiring the combination of an electron with an ion to be a slow process necessitating an activation energy of the order of 10 kcal. or more. Further, the theory makes no attempt to correlate the overvoltage at a particular cathode with its known physical or chemical properties.

The suggestion that the slow process in the evolution of hydrogen at a cathode is the discharge of the ion was also made independently by

[23] Gurney, *Proc. Roy. Soc.*, **134A**, 137 (1931); see also, Fowler, *Trans. Faraday Soc.*, **28**, 368 (1932).

Smits and by Erdey-Grúz and Volmer.[24] It should be pointed out, however, that the mathematical treatment of the theory by the latter authors and its extension by Frumkin [25] is applicable equally if stage (2), i.e., the transfer of a hydrogen ion (H_3O^+) or of a proton from the solution to the electrode, is the rate-determining process. Although it does not appear to be generally realized, the arguments employed, which are similar to those on page 460, are applicable to any slow stage intervening between the ions in the solution, on the one hand, and hydrogen atoms on the surface on the other hand; it is thus not possible to say whether this process is the transfer of an ion or of a proton to the surface or its actual neutralization by an electron. The treatment of Erdey-Grúz and Volmer is similar to that given previously, but it appears to contain certain errors; the modification made by Frumkin is, however, free from this objection. The latter author utilizes Stern's theory of the structure of the double layer at a solid-liquid interface (see p. 525); it is supposed that this layer consists of a fixed portion, close to the electrode, across which the potential may be represented by ψ, and a diffuse portion, extending into the bulk of the solution, across which the potential fall is ζ, i.e., the electrokinetic (zeta-) potential; the algebraic sum of ψ and ζ is equal to E, the total electrode potential (Fig. 120). The rate of the cathodic reaction is now given by an equation of the same form as (17), but the term a_+ is replaced by the activity, or concentration, of hydrogen ions in the layer AB, instead of that in the bulk of the electrolyte. The latter is in equilibrium with the former across the diffuse portion of the double layer, and so it follows that equation (17) can be written as

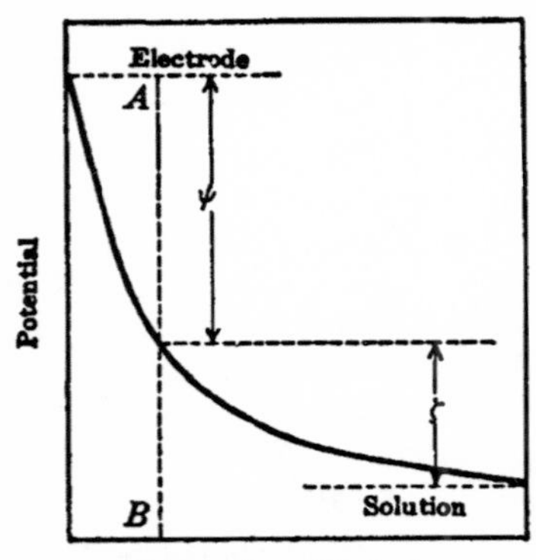

Fig. 120. Double-layer at electrode surface

$$I = Fk_1c_{H^+}e^{F\zeta/RT}e^{\alpha F\psi/RT},$$

the anodic current being neglected since it is very small at appreciable overvoltages. As seen above, $\psi + \zeta$ is equal to the total cathodic potential and this may be represented, as before, by E', and assuming α to be 0.5, in order to give the correct value of the slope b, it follows that

$$I = Fk_1c_{H^+}e^{F\zeta/2RT}e^{FE'/2RT},$$

$$\therefore \quad E' + \zeta = \frac{2RT}{F}\ln I - \frac{2RT}{F}\ln c_{H^+} + \text{const.}$$

[24] Smits, "The Theory of Allotropy," translated by Thomas, 1922, p. 115 *et seq.*; *Z. physik. Chem.*, 172A, 470 (1935); Erdey-Grúz and Volmer, *ibid.*, 150A, 203 (1930); Erdey-Grúz and Wick, *ibid.*, 162A, 53 (1932).

[25] Frumkin *et al.*, *Z. physik. Chem.*, 164A, 121 (1933); *Acta Physicochim. U.R.S.S.*, 7, 475 (1937); 11, 21 (1939); 12, 481 (1940).

The reversible potential E in the given electrolyte is represented by

$$E = -\frac{RT}{F}\ln c_{H^+},$$

and since $E' - E$ is equal to the overvoltage ω, it follows that

$$\omega + \zeta = \frac{2RT}{F}\ln I - \frac{RT}{F}\ln c_{H^+} + \text{const.} \qquad (31)$$

In the absence of excess of extraneous ions, Stern's theory of the double layer leads to the expression

$$\zeta = \text{const.} - \frac{RT}{F}\ln c_{H^+},$$

and hence

$$\omega = \frac{2RT}{F}\ln I + \text{const.} \qquad (32)$$

This equation gives the correct value for the slope b, but this has no significance since α, which must lie between zero and unity, was chosen as 0.5; further, it requires the overvoltage, under the conditions specified, to be independent of the hydrogen ion concentration. According to Frumkin, the diffuse layer potential ζ sometimes remains constant in spite of variations of pH; the overvoltage, according to equation (31), should then increase by 0.059 volt for each unit increase of pH at ordinary temperatures. These conditions are presumed to apply at a nickel cathode in relatively concentrated acid solutions, and at a mercury cathode in the presence of an excess of lanthanum chloride; it appears somewhat improbable, however, that in *both* of these instances the value of ζ would remain constant as the hydrogen ion concentration was changed.

In order to account for the results obtained with a nickel cathode in relatively alkaline solutions Frumkin and his collaborators have postulated that the protons which are discharged originate from water molecules whereas in acid solution they come mainly from H_3O^+ ions.

IV. Proton Transfer as the Slow Process.—A theory of overvoltage having some features in common with that just described, but which definitely regards the slow stage responsible for hydrogen evolution as the transfer of a proton from the solution to the electrode, has been proposed by Eyring, Glasstone and Laidler using the theory of absolute reaction rates.[26] The activation energy for overvoltage is seen from Table LXXXII to be about 10 to 20 kcal., and it is significant that

[26] Eyring, Glasstone and Laidler, *J. Chem. Phys.*, 7, 1053 (1939); *Trans. Electrochem. Soc.*, 76, 145 (1939); see also, Kimball, Glasstone and Glassner, *J. Chem. Phys.*, 9, 91 (1941).

activation energies of the same order have been found for chemical reactions involving the transfer of a proton. Accepting the conclusion, previously mentioned, that under normal conditions, i.e., in the absence of appreciable amounts of neutral salts or when the concentration of hydrogen or hydroxyl ions is not very large, the overvoltage is independent of pH, the simplest assumption to make is that the rate of the cathodic process is given by an equation of the form

$$I = Fk_1ce^{\alpha F\omega/RT}, \tag{33}$$

where c is the concentration in the solution of the species from which the protons originate. If the value of c is the same in all solutions, irrespective of the hydrogen ion concentration, it follows from equation (33) that

$$\omega = \text{const.} + \frac{RT}{\alpha F}\ln I,$$

where ω is independent of the pH of the electrolyte. In addition to the requirement that c is constant, the important point of equation (33) is that the potential across which the proton transfer occurs in the slow process is the overvoltage portion only, and not the whole cathode potential, as implied in the treatment on page 460. The significance of this postulate is that, under such conditions that the overvoltage does not vary with pH, the overvoltage is operative across part only of the double layer at the electrode surface, and that the slow stage in hydrogen evolution involves the transfer of a proton across this portion of the layer. As long as there is no net flow of current, i.e., at the reversible potential, almost the whole potential must be regarded as operating over what may be called the "solution double layer" which is approximately equivalent to Stern's diffuse layer. When there is a resultant cathodic current, however, the electrical equilibrium across this layer remains unchanged, but that across the remainder, called the "electrode double layer," analogous to Stern's fixed layer, is disturbed; in this way, a difference of potential, equal to the overvoltage, is introduced across the latter (cf. Fig. 121). If the slow stage in the discharge of hydrogen ions is the passage of protons across the electrode double layer, then equation (33) follows directly; k_1 is the specific rate of the proton-transfer process when the

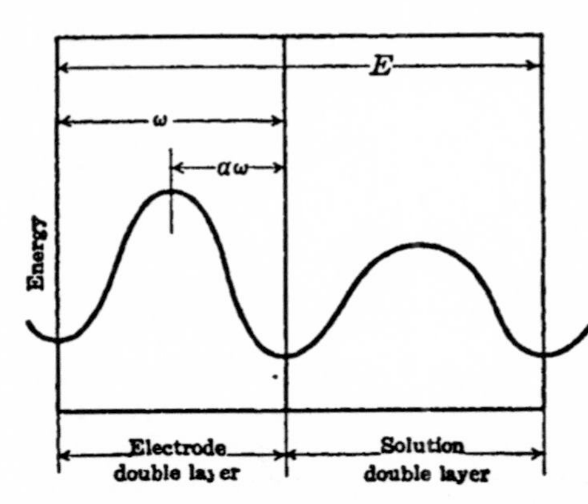

FIG. 121. Potential energy curves at electrode surface (Kimball, Glasstone and Glassner)

overvoltage is zero, i.e., at the reversible potential, and c is the concentration of the species supplying the protons. The fraction α, as implied on page 250, represents the fraction of the potential operative between initial and activated states, and since this is known from experiment to be 0.5, it follows that the energy barrier under consideration must be a symmetrical one; the point is of importance, since none of the theories described hitherto gives an adequate interpretation of this fact.

A comparison of equations (33) and (27) shows that the quantity Fk_1c in the former is equivalent to $Be^{-\Delta H^{\ddagger}/RT}$ and hence to I_0; further, according to the theory of reaction rates k_1 is proportional to the product of $e^{-\Delta H^{\ddagger}/RT}$ and a factor $e^{\Delta S^{\ddagger}/R}$, where $\Delta S^{\ddagger}$ is the standard entropy change of activation for the rate-determining process. It follows, therefore, that the factor B is equivalent to $Fce^{\Delta S^{\ddagger}/R}$ multiplied by a proportionality factor the value of which is known from the theory of absolute reaction rates to be kT/h, where k is the Boltzmann constant, i.e., the gas constant per molecule, and h is Planck's quantum theory constant of action. Since I_0 and $\Delta H^{\ddagger}$ are known for a number of cathodes, according to the data in Table LXXXII, it is possible in these instances to evaluate the factor B: when the necessary calculations were made, the interesting fact emerged that B, and hence $Fce^{\Delta S^{\ddagger}/R}$, was approximately constant for all the electrodes and had almost the same value in acid and alkaline solutions. This result suggests that the same fundamental process is responsible for overvoltage at all the cathodes included in the table on page 467, so that $\Delta S^{\ddagger}$ is the same in each case; further, the species from which the protons originate appear to have the same concentration c in both acid and alkaline electrolytes.

Since water is the species always present in all solutions at an approximately constant concentration, it may be inferred that the rate-determining stage in the evolution of hydrogen is the transfer of a proton across the electrode double layer. For reasons which will appear later, it is suggested that the proton transferred in this way passes to a molecule of water attached to the surface of the electrode; the proton is immediately discharged by an electron to form a hydrogen atom, and a hydroxyl ion is left in solution. The essential discharge mechanism may thus be represented by

$$\mathrm{S} \cdots \mathrm{H_2O} + \mathrm{H_2O} \cdots \overset{\ominus}{\mathrm{M}} \rightarrow \mathrm{S} \cdots \mathrm{OH^-} + \left.\begin{matrix}\mathrm{H_2O}\\ \mathrm{H}\end{matrix}\right\} \cdots \mathrm{M},$$

where the symbol S refers to the solution side and M to the electrode side of the double layer across which the overvoltage operates; the sign $\ominus$ on M indicates that an electron is always available at the cathode. There are two general arguments in favor of this mechanism; first, it accounts for the fact that α in equation (33) is about 0.5, since the energy barrier between the two layers of water molecules will be approximately symmetrical;* and second, it provides an explanation, given by no other theory of overvoltage, of certain unexpected similarities between hydrogen and oxygen overvoltages (see p. 478). In addition, by making some reasonable approximations, it is possible to calculate, from data which are not connected with cathode phenomena, a value of B which is in very good agreement with that derived from the measurements recorded in Table LXXXII.

An important requirement of any theory of overvoltage is that it should account for the dependence of the overvoltage on the nature of the cathode material: this requirement is satisfied by the theory described above. Since B is found to be constant, it follows from equation (27) that the overvoltage at a constant current density increases as the activation energy $\Delta H^{\ddagger}$ becomes larger. If the cathode metal is one which adsorbs hydrogen readily, e.g., platinum, palladium, nickel or copper, it will form relatively strong M—H bonds and consequently the oxygen atom of the water molecule held on such a surface will have an increased tendency to form a bond with hydrogen: the attachment of a proton to such an oxygen atom will, therefore, be facilitated and the energy of activation of the proton transfer process responsible for overvoltage will be low. On the other hand, metals which form weak M—H bonds, e.g., mercury, lead and tin, will be expected to have high activation energies and high overvoltages, in agreement with experiment. If the strength of the M—H bond is very great, the activation energy for proton transfer becomes very small, e.g., at a platinized platinum cathode; another and slower process, e.g., combination of hydrogen atoms, must then be rate-determining.

It has been assumed hitherto that the zeta-potential acting across the solution double layer remains constant; any substance which alters the zeta-potential at constant hydrogen ion concentration will, however, be expected to have some influence on the overvoltage. The effect of strongly adsorbed neutral ions or of hydrogen or hydroxyl ions at high concentrations can be largely accounted for in this manner.

Hydrogen-Deuterium Separation.—When an aqueous solution is electrolyzed, the hydrogen gas evolved always contains relatively more of the lighter isotope than does the electrolyte; the difference is expressed as the **separation factor** (s) defined by

$$s = \frac{(c_{\mathrm{H}}/c_{\mathrm{D}})_{\mathrm{gas}}}{(c_{\mathrm{H}}/c_{\mathrm{D}})_{\mathrm{liq.}}},$$

where c_{H} and c_{D} refer to the concentrations of the lighter, i.e., hydrogen, and of the heavier, i.e., deuterium, isotope, respectively. The value of

* For those cathodes giving high values of the slope b, the fraction α is less than 0.5; this may be due to distortion of the energy barrier, for some reason, or to the operation of an entirely different mechanism. If b lies between 0.118 and 0.059 at ordinary temperatures the value of α is between 0.5 and unity.

the factor s for a number of cathodes under a variety of conditions lies between 3 and 20, and the preferential evolution of the lighter isotope, as indicated by these results, is undoubtedly due to its smaller overvoltage at a given current density.

As a result of experimental studies it appears that the electrolytic hydrogen-deuterium separation factors fall into two distinct classes;[27] in the first s has a value of about 6, viz.,

Ni 6.7; Au 6.4; Ag 6.0; Cu 7.4; Pt (high C.D.) 6.9; Pb (alk. soln.) 6.8,

whereas in the second group it is approximately 3, viz.,

Sn 3.1; Hg 3.1; Pb (acid soln.) 3.0; Pt (platinized) 3.4.

On the basis of these observations, Horiuti and Okamoto concluded that there were two different mechanisms for hydrogen evolution. It was suggested that with cathodes of the first group, for which the so-called "catalytic mechanism" was operative, the rate-determining step is the combination of two hydrogen atoms, adsorbed at adjacent positions on the cathode surface, to form a molecule of gas. For metals of the second group, the "electrochemical mechanism," in which the slow stage is the neutralization of a hydrogen-molecule ion, H_2^+, on the electrode has been proposed.[28] It would appear, at first sight, that the catalytic mechanism is applicable to low overvoltage metals, whereas the electrochemical mechanism holds for high overvoltage cathodes. It will be noted, however, that the lead electrode appears in both groups, according to the pH of the solution, and that platinized platinum has the same separation factor as high overvoltage metals. It is claimed that the postulated mechanisms lead to calculated separation factors for mercury and nickel which are in satisfactory agreement with observation; the theory, however, has little to recommend it over those given previously, and the two values for the separation factor can be explained in an alternative manner.

Oxygen Overvoltage.—Measurements of anodic overvoltages at definite current densities have been made in certain cases, but reliable results are difficult to obtain: one reason for this is that the potential of an anode changes with time even more than does that of a cathode. The most reliable data have been obtained with a smooth platinum anode in dilute sulfuric acid saturated with oxygen;[29] the variation of overvoltage with the current is given by equation (26), just as for hydrogen overvoltage.

[27] Horiuti and Okamoto, *Sci. Papers Inst. Phys. Chem. Res. Tokyo*, **28**, 231 (1936); Walton and Wolfenden, *Trans. Faraday Soc.*, **34**, 436 (1938).

[28] Horiuti and Okamoto, *Sci. Papers Inst. Phys. Chem. Res. Tokyo*, **28**, 231 (1936); Hirota and Horiuti, *Bull. Chem. Soc. Japan*, **13**, 228 (1938); Horiuti, *Sci. Papers Inst. Phys. Chem. Res. Tokyo*, **37**, 274 (1940); see also, Frumkin, *ibid.*, **37**, 473 (1940).

[29] Bowden, *Proc. Roy. Soc.*, **126A**, 107 (1929); Roiter and Jampolskaja, *Acta Physicochim. U.R.S.S.*, **7**, 247 (1937).

Further, the value of b over a range of temperature has been found to be $2 \times 2.3RT/F$ for oxygen as well as for hydrogen evolution; it follows, therefore, that the factor α is approximately 0.5 in each case. The anodic overvoltage at smooth platinum is said to be independent of the concentration of the sulfuric acid used as electrolyte; in sodium hydroxide, however, the value apparently decreases with increasing hydroxyl ion concentration. In very dilute alkaline solutions α is about 0.5, but it apparently increases as the concentration of the electrolyte is increased.

There is ample evidence that many metals become covered with higher oxides, e.g., PtO_3 on platinum, Cu_2O_3 and CuO_2 on copper, and Ni_2O_3 and NiO_2 on nickel, as the result of being employed as anodes; for many years the view was generally accepted that oxygen evolution occurred only through the intermediate formation of such an oxide, and that this oxide was responsible for anodic overvoltage. Although the existence of such oxides is beyond doubt, it is improbable that they are the fundamental cause of overvoltage. At relatively high current densities the measured anode potential is approximately that of an unstable higher oxide, formed by the interaction of the electrode material with atomic oxygen, but the fact that oxygen evolution does not occur until the potential exceeds the reversible value still requires explanation.

The similarity between hydrogen and oxygen overvoltages at platinum electrodes led to the proposal of a mechanism for oxygen liberation which is closely analogous to that suggested on page 475 for the evolution of hydrogen. The slow process in the discharge of hydroxyl ions to liberate oxygen at the anode is believed to be the transfer of a proton from a molecule of water attached to the electrode to one on the solution side of the electrode double layer; in other words, the rate-determining stage is exactly the reverse of that proposed to account for hydrogen overvoltage.[30] The mechanism of the proton transfer process at the anode may be represented in the following manner:

$$S \cdots H_2O + H_2O \cdots \overset{\oplus}{M} \rightarrow S \cdots H_3O^+ + OH \cdots M,$$

the sign $\oplus$ on M implying the deficit of an electron. The hydroxyl radicals formed on the surface react in pairs with the ultimate formation of molecular oxygen, the intermediate stages being relatively rapid. Since the potential barrier is virtually the same as for hydrogen evolution, it is approximately symmetrical, and hence α will be about 0.5, as found. The problem of hydroxyl ion discharge can be treated by the theory of reaction rates in exactly the same manner as hydrogen ion discharge, and since the processes are similar the value of the factor B should be the same in both cases; it is actually somewhat larger for the anodic reaction, but in view of the uncertainties associated with anodic behavior the agreement may be regarded as being reasonably good.

[30] Eyring, Glasstone and Laidler, *J. Chem. Phys.*, 7, 1053 (1939).

If no oxides were formed on an anode surface it is to be expected that metals forming strong M—H bonds should have high oxygen overvoltages, since the attraction would tend to hinder the removal of a proton from a molecule of water attached to the electrode surface. It is true that smooth platinum, palladium and gold, which have the lowest cathodic overvoltages, have the highest values for oxygen evolution (see Table LXXVIII), but the parallelism does not always hold. Nickel and cobalt, for example, have low overvoltages both as anodes and cathodes; the oxidation of the surface is undoubtedly a complicating factor which would probably tend to diminish the oxygen overvoltage.

The Electrolysis of Water.—According to the suggestions made above concerning the mechanism of the evolution of hydrogen and oxygen, the cathodic process may be represented by

$$H_2O + H_2O + \epsilon = OH^- + H_2O + H,$$

while the anodic process is

$$H_2O + H_2O = H_3O^+ + OH + \epsilon.$$

If the electrolyte is acid, i.e., the pH is less than 7, the hydroxyl ions formed at the cathode will be immediately neutralized, thus

$$OH^- + H_3O^+ = 2H_2O,$$

and the over-all reaction, given by the sum of these three equations, is

$$H_2O = H + OH,$$

the H and OH subsequently forming molecular hydrogen and oxygen, respectively. The complete reaction is thus the decomposition of one molecule of water per faraday of electricity. If the electrolyte is alkaline, the H_3O^+ ions remaining at the anode will be neutralized, and the over-all reaction will be exactly the same as before. In other words, when the electrolysis of an aqueous solution yields hydrogen and oxygen, the ultimate process is always the same, viz., the decomposition of water, irrespective of the pH of the solution. If the electrolyte is not well buffered, it is evident that the solution in the vicinity of the cathode will become alkaline, because of the OH^- ions, while in the neighborhood of the anode it will be acid, due to the excess of H_3O^+ ions.

It is opportune to refer here to the problem of why appreciable overvoltages are associated with the discharge of the ions of water, viz., hydrogen and hydroxyl ions, whereas other ions, such as chloride ions, which may also be accompanied by gas evolution, are discharged close to their reversible potentials. The explanation is probably connected with the unusual structure of the water, with the result that the direct discharge of hydrogen and hydroxyl ions is a difficult process. The easiest alternative, as proposed in preceding sections, is the transfer of a proton from one molecule of water to another. It is of interest to note that such a process is somewhat similar to the mechanism which accounts for the high conductance of the hydrogen and hydroxyl ions (cf. p. 66).

PROBLEMS

1. A 0.01 molar solution of quinone is being reduced at a smooth platinum cathode; estimate the limiting current density for 100 per cent efficiency of the process.

2. In the electrolysis of 0.02 N argentocyanide solution, Glasstone [*J. Chem. Soc.*, 690 (1929)] found that 100 per cent efficiency for silver deposition was obtained at C.D.'s up to 5×10^{-4} amp. per sq. cm. in an unstirred solution at 15°, up to 20×10^{-4} in a solution agitated by means of a stirrer rotating 500 times per minute, and up to 22×10^{-4} in an unstirred solution at 70°. Calculate the approximate thickness of the diffusion layer in each case, the diffusion coefficient of the argentocyanide ion being taken as 1.3×10^{-5} cm.^{2} sec.$^{-1}$ at 15°.

3. A 1.0 molar solution of copper sulfate is electrolyzed with an inert, e.g., platinum, anode and a copper cathode of 55 sq. cm. exposed area; the current is maintained constant at 0.040 amp. If the electrolysis vessel contains 1 liter of solution and there is reasonable circulation of the electrolyte, without resort to stirring, estimate approximately how long electrolysis will proceed before hydrogen evolution commences. How much of the original copper will then have been deposited?

4. A solution containing ions of a univalent metal is being electrolyzed; the anode is made of the same material as is being deposited at the cathode. Utilize equations (9) and (16) to plot the variation of the cathodic and anodic polarizations, respectively, with current density at 25°. The limiting current density may be taken as 10^{-3} amp. per sq. cm. for the given solution [Bowden and Agar, *Proc. Roy. Soc.*, **169A**, 206 (1938)].

5. Estimate the separate anode and cathode potentials when appreciable current commences to flow through a 0.1 N solution of sodium hydroxide, the anode being of nickel and the cathode of mercury. The overvoltages may be assumed to be independent of the nature of the electrolyte.

6. When combined with a calomel reference electrode containing saturated potassium chloride solution, a cathode, from which hydrogen was being evolved in a solution of pH 7, gave a cell of E.M.F. equal to 1.25 volt at ordinary temperatures. What is the overvoltage at the given cathode under the experimental conditions?

7. The results recorded below for the variation of the overvoltage with current density at a mercury cathode in dilute sulfuric acid at 25° have been estimated from the data of Bowden and Rideal [*Proc. Roy. Soc.*, **120A**, 59 (1928)].

Overvoltage	C.D. amp./sq. cm.	Overvoltage	C.D. amp./sq. cm.
0.60 volt	2.9×10^{-7}	0.84 volt	250×10^{-7}
0.65	6.3	0.89	630
0.73	28	0.93	1650
0.79	100	0.96	3300

Plot the variation of the overvoltage with log I, where I is equal to the C.D., and determine the value of the slope b; estimate the values of α and I_0 for the mercury cathode.

8. The following data have been derived from the graphical representation of the results obtained by Bowden [*Proc. Roy. Soc.*, **126A**, 107 (1929)] with a mercury cathode in dilute sulfuric acid. At a constant overvoltage of 0.74 volt, the values of log I were $\bar{6}.18$ at 0°, $\bar{5}.00$ at 36° and $\bar{5}.75$ at 72°. At a constant current density of 10^{-5} amp. per sq. cm., the overvoltages were 0.83 at 0°, 0.74 at 36° and 0.63 at 72°. Calculate the heat of activation ($\Delta H^{\ddagger}$) for the rate determining stage of the cathodic process from each set of data.

9. Utilize the value of I_0 from Problem 7 and of $\Delta H^{\ddagger}$ from Problem 8 to calculate the constant B of equation (27). Assuming c, the concentration of the species providing the protons in the discharge of hydrogen, to be 10^{15} molecules per sq. cm., calculate the standard entropy of activation ($\Delta S^{\ddagger}$) of the rate determining stage.

CHAPTER XIV

THE DEPOSITION AND CORROSION OF METALS

Physical Nature of Electrodeposited Metals.—In view of the industrial importance of the electrodeposition of metals, the influence of various factors on the physical appearance of the deposits has been the subject of much investigation. It is generally agreed that electrodeposited metals are crystalline, and the external appearance depends mainly on the rate at which the crystals grow and on the rate of the formation of fresh nuclei. If the conditions are such as to favor the rapid formation of crystal nuclei, the deposit will be fine-grained; if the tendency is for the nuclei to grow rapidly, however, relatively large crystals will form and the deposit becomes rough in appearance. The chief factors influencing the appearance of the electrodeposited metal are (I) the current density of deposition; (II) concentration of the electrolyte; (III) temperature; (IV) presence of colloidal matter; (V) nature of the electrolyte; and (VI) nature of the basis metal.[1]

I. Current Density.—At low current densities the discharge of ions occurs slowly, and so the rate of the growth of nuclei should exceed the rate at which new ones form; the deposits obtained under these conditions should be coarsely crystalline. As the C.D. is raised the rate of formation of nuclei will be greater and the deposit will become more fine-grained. At very high currents the solution in the vicinity of the cathode will be depleted in the ions required for discharge, and, as a result, the crystals will tend to grow outwards towards regions of higher concentration; the deposit then consists of "trees," nodules or protruding crystals. If the C.D. exceeds the limiting value for the given electrolyte, hydrogen will be evolved at the same time as the metal is deposited; bubble formation often interferes with crystal growth, and porous and spongy deposits may be obtained. The discharge of hydrogen ions frequently causes the solution in the vicinity of the cathode to become alkaline, with the consequent precipitation of hydrous oxides or basic salts; if these are included in the deposit, the latter will be fine-grained and dark in appearance.

II. Concentration of Electrolyte.—The effects of electrolyte concentration and of current density are to a great extent complementary: by increasing the concentration or by agitating the solution, higher C.D.'s

[1] For general discussions, see Blum, *Trans. Electrochem. Soc.*, **36**, 213 (1919); Blum and Rawdon, *ibid.*, **44**, 397 (1923); Graham, *ibid.*, **52**, 157 (1927); Hunt, *J. Phys. Chem.*, **35**, 1006 (1931); Blum and Hogaboom, "Electroplating and Electroforming," 1930, Chap. VII; Glasstone, "The Electrochemistry of Solutions," 1937, Chap. XVIII.

can be used before coarse deposits are formed, or before hydrogen evolution occurs with its accompanying spongy or dark deposits. The influence of concentration on the rate of nucleus formation is uncertain; since increase of concentration tends to give firm, adherent deposits, some workers have expressed the opinion that the presence of the large number of ions in a concentrated solution favors the formation of fresh nuclei. Certain experiments, however, indicate that the rate of formation of nuclei is actually decreased by increasing concentration, but the improvement in the deposit is due to an increase in the rate of growth of crystals over the cathode surface, combined with a decrease in the rate of growth in a perpendicular direction.[2]

III. Temperature.—Increase of temperature has two effects which oppose one another: in the first place, diffusion is favored, so that the formation of rough or spongy deposits at relatively high C.D.'s is inhibited. On the other hand, the rate of crystal growth, favoring a coarse deposit, is increased. Further, increase of temperature decreases hydrogen overvoltage, and so facilitates the evolution of the gas, as well as the precipitation of basic salts. At moderate temperatures the first of the three aforementioned factors predominates, so that the deposits are improved, but at higher temperatures deterioration is observed.

IV. Colloidal Matter.—In many instances, e.g., lead from acetate solution or silver from silver nitrate, the metal normally deposits on the cathode in the form of relatively large crystals, but the presence of very small amounts of colloidal matter, or of certain organic compounds, often results in the production of a smooth, fine-grained and microcrystalline deposit. Minute quantities, of the order of 0.05 g. per liter, of the addition agent are sufficient to cause a profound change in the form of the deposit; an excess of the substance may give loose and powdery or brittle deposits. Among the materials used as addition agents the following may be mentioned: gelatin, peptone, agar, glue, various gums, rubber, casein, alkaloids, dyestuffs, sugars, and camphor; the action of any particular substance is often specific and depends on the nature of the metal and the electrolyte.

The addition agents are generally surface-active substances and they are probably adsorbed on the crystal nuclei, thus preventing their growth; the discharged ions are consequently compelled to start new nuclei and the result is that the deposit is fine-grained whereas it might otherwise have been appreciably crystalline. The deposits obtained in the presence of an addition agent have been found to contain a certain proportion of the latter, in agreement with the view that the added substance is adsorbed.

Nickel generally forms smooth deposits in sulfate or chloride solutions even in the absence of colloidal addition agents; since the deposition of nickel is always accompanied by the simultaneous discharge of hydrogen

[2] Glazunov, *Z. physik. Chem.*, 167, 399 (1934).

ions, the electrolyte tends to become alkaline in the vicinity of the cathode, thus leading to the formation of hydrous oxides or basic salts. By interfering with the growth of nuclei, these substances, in a gelatinous or colloidal form, can cause the deposit to become smooth and fine-grained. It is a significant fact that electrodeposits of nickel have been shown on analysis to contain appreciable amounts of oxide.[3]

V. Electrolyte.—The nature of the anion often has a very important influence on the physical form of the deposited metal; for example, lead from lead nitrate solution is rough, but smooth deposits are obtained from silicofluoride and borofluoride solutions. The valence state of the metal may affect the nature of the deposit; thus, from plumbic solutions lead is deposited in a spongy form whereas relatively large crystals are formed in plumbous solutions. In an analogous manner, smooth deposits of tin are obtained from stannate baths, but from stannite solutions the deposits are of poor quality. The difference in the behavior of different electrolytes is sometimes due to the possibility of the formation of colloidal matter which serves to give a fine-grained deposit; this may be the case in the deposition of lead from silicofluoride and borofluoride solutions where a certain amount of colloidal hydrous silica or boron trioxide may be formed by hydrolysis.

A striking fact in connection with metal deposition is that smooth deposits are generally produced from solutions of complex ions, particularly cyanides; an outstanding illustration of this behavior is provided by silver which is obtained as a very coarse deposit from nitrate solutions, except perhaps at very low C.D.'s, while cyanide solutions give the familiar smooth deposits of "electro-plate." Some workers consider that crystal nucleus formation is favored by the extremely small concentration of the simple, e.g., Ag^+, ions in the solution of the complex, e.g., $Ag(CN)_2^-$, anion. Others, however, are of the opinion that some insoluble salt, such as the simple cyanide, is deposited with the metal, thus acting like an addition agent in preventing crystal growth. One possible way in which such solids might be formed at the cathode is by the discharge of complex cations, e.g., $Ag_2(CN)^+$, which would yield silver and silver cyanide in immediate contact; there is some evidence that ions of this type are present in complex cyanide solutions.[4] Another suggestion that has been made is that deposition at the cathode does not occur by the discharge of cations, such as Ag^+, which are present in very small amounts, but by the anion acquiring an additional charge, e.g., $Ag(CN)_2^-$ becoming $Ag(CN)_2^{--}$; the resulting complex is unstable and decomposes more or less rapidly to yield atomic silver and cyanide ions. The formation of the silver is thus in the nature of a secondary process, and consequently

[3] Macnaughtan and Hothersall, *Trans. Faraday Soc.*, 31, 1168 (1935).

[4] Glasstone, *Trans. Faraday Soc.*, 31, 1218 (1935); see also, Glazunov and Schlötter, *J. Electrodepositors' Tech. Soc.*, 13 (1937); Erdey-Grúz, *Z. physik. Chem.*, 172A, 163, 174, 176 (1935).

may not be subject to the conditions applicable in a solution of the simple ions where direct discharge occurs.

The best electrodeposits of chromium are obtained from solutions of chromic acid containing small amounts of certain anions, particularly the sulfate ion. There is little doubt that the nature of the deposit is influenced by the formation of a complex chromic chromate diaphragm on the cathode. The function of the anion appears to be to affect the nature of this diaphragm in such a manner so that it does not block the cathode completely; it thus does not prevent access of ions, yet it is sufficiently adherent to interfere with the growth of crystal nuclei.

VI. Basis Metal.—Although the external form of a deposited metal is rarely affected by the basis metal used as the cathode, there is ample evidence that the latter has some effect on the crystal growth. It is apparent that in many instances the general orientation of the crystals, at least in the first layers of the deposit, is a continuation of that in the basis metal.

Throwing Power.—The property of a solution by virtue of which a relatively uniform deposit of metal may be obtained on a cathode of irregular shape, is known as the **throwing power.** The problem of the causes of good or bad throwing power is very complex and is not at all clearly understood. It appears, from a theoretical standpoint, that throwing power should be influenced by (*a*) the rate of increase of cathode potential with current density, and (*b*) the conductance of the solution. If preferential deposition should take place at any part of the cathode, the effective C.D. will be higher here than on the remainder of the cathode; if this higher density requires a much increased potential, there will be a tendency for the C.D. to be automatically reduced at the part under consideration. In other words, a cathode potential-current density curve with a marked slope should favor uniform current distribution and so should give good throwing power. This appears to be true in certain cases, but there are other important factors to be considered, for a solution of argentocyanide containing excess free cyanide appears to have excellent throwing power, although the polarization does not increase markedly with increasing current density. In general, the addition of colloidal matter improves the throwing power of a solution, but increase of temperature and agitation have the opposite effect.

If the conductance of the electrolyte is low, the current lines will tend to concentrate on the parts of the cathode nearest the anode and the throwing power will be bad. With a solution of good conductance, however, there will be no particular preference, as far as this factor is concerned, for one portion of the cathode over any other: high conductance will thus improve throwing power. The conductance of an electrolytic bath is generally so good, however, that there is no noticeable influence on the throwing power.

Simultaneous Discharge of Cations.—If a solution contains two cations, there is a possibility that simultaneous discharge may occur: this problem is not only of interest in connection with the electrodeposition of alloys, but it is important in the deposition of single metals, since aqueous solutions always contain hydrogen ions. Were it not for a variety of complicating factors, such as the influence of one metal on the deposition potential of the other, the situation would, in principle, be relatively simple; provided the discharge (reduction) potentials of the two ions were the same, simultaneous deposition would occur. For example, the reversible potential of a metal A in a solution of its ions of activity a_{A^+}, i.e., of the electrode A, A^+, would be given by

$$E = E_A^0 - \frac{RT}{z_A F} \ln a_{A^+},$$

and the theoretical discharge potential $E_{dis.}$ is equal to $-E$, i.e.,

$$E_{dis.} = -E_A^0 + \frac{RT}{z_A F} \ln a_{A^+}.$$

In order to obtain the actual discharge potential it is necessary to include the overvoltage ω_A, and so it follows that

$$E_{dis.} = -E_A^0 - \omega_A + \frac{RT}{z_A F} \ln a_{A^+}.$$

If the solution contains two cations, one of which may be the hydrogen ion, then provided there is no interaction in solution or in the deposit, simultaneous deposition will occur when the two discharge potentials, e.g., of A and B, become equal; that is, when

$$-E_A^0 - \omega_A + \frac{RT}{z_A F} \ln a_{A^+} = -E_B^0 - \omega_B + \frac{RT}{z_B F} \ln a_{B^+}. \quad (1)$$

The subject of simultaneous discharge may be divided into two aspects: first, when it is desired to deposit two metals simultaneously, e.g., in alloy deposition, and second, when simultaneous deposition is to be avoided, e.g., in electro-analytical work. These aspects will be considered in turn.

An examination of equation (1) shows that there are, in general, three ways in which the discharge potentials of two cations may be brought together: (i) if the standard potentials are approximately equal and the overvoltages are small; (ii) if the standard potentials are different, but the overvoltages vary sufficiently to compensate for this difference; and (iii) if the differences in reversible potential and overvoltage are compensated for by differences in the activities of the ions. Examples of the three types of behavior are known.

(i) The standard (oxidation) potentials of lead and tin are $+0.126$ and $+0.140$ volt, respectively, and since there is little overvoltage accompanying the deposition of these metals, a small adjustment of ionic concentration is sufficient to allow simultaneous deposition to take place from chloride or fluoborate solutions. If the solution is acid, hydrogen ion discharge should be theoretically possible, since the standard potential is $\pm$ 0.0 volt; on account of the high hydrogen overvoltage at both lead and tin, however, there is no appreciable evolution of hydrogen. Another example of simultaneous ionic discharge that appears to be free from complications occurs when copper and bismuth are deposited from their simple salt solutions; the respective standard potentials are − 0.34 and − 0.23, respectively, and the deposition overvoltage is negligible. The composition of the deposit depends on the concentrations of the two ions present in the solution and on the C.D.; the rates of ionic discharge adjust themselves so that the composition of the layer in immediate contact with the cathode satisfies equation (1) for the particular cathode potential.

(ii) Since the discharge (reduction) potential of zinc from a molar solution of zinc sulfate is about − 0.77 volt, while the theoretical potential for hydrogen ion discharge, even from a neutral solution, is cathodically smaller, viz., − 0.4 volt, it would be expected that hydrogen evolution would occur in preference to zinc deposition. The large hydrogen overvoltage, viz., about 0.7 to 0.8 volt, at a zinc cathode serves, however, to bring the actual discharge potentials close together, and simultaneous deposition of hydrogen and zinc occurs from slightly acid solutions.

The combination of the reversible potential with the overvoltage in each case results in the deposition of hydrogen together with iron, cobalt or nickel. Since it is desired to suppress the hydrogen evolution as much as possible, in order to increase the efficiency of metal deposition, the pH of the solution is raised. There is, however, a limit to this increase because of the danger of the precipitation of basic salts.

Sometimes two metals which have different deposition potentials at one temperature may have similar potentials at another temperature owing to changes of overvoltage. A case in point is that of the deposition of zinc and nickel from an aqueous ammoniacal solution: at 90° the overvoltages for both nickel and zinc deposition are small and the discharge potentials of the ions differ greatly. At 20°, the overvoltage for nickel deposition makes its discharge potential about 0.3 volt more cathodic and it can then be deposited simultaneously with zinc from the solution in aqueous ammonia.

(iii) Although the normal deposition potentials of cadmium and copper and of zinc and copper are far apart, they may be brought together by adjustment of the ionic activities, or concentrations. If potassium cyanide is added to solutions of salts of these metals, complex cyanides are formed in each case, but owing to their different instability constants (cf. p. 173) the concentrations of the simple ions are reduced to different extents so that the discharge potentials now approach one another. This statement is borne out by the data in Table LXXXIII, which gives the deposition potentials of silver, copper, cadmium and zinc from their sulfate solutions, on the one hand, and from complex cyanide solutions containing excess free cyanide, on the other hand.[5]

TABLE LXXXIII. DEPOSITION (REDUCTION) POTENTIALS FOR SIMPLE AND COMPLEX ION SOLUTIONS

Metal	Sulfate	Complex Cyanide
Silver	+ 0.80 volt	− 0.5 volt
Copper	+ 0.34	− 1.0
Cadmium	− 0.40	− 0.9
Zinc	− 0.76	− 1.2

The electrolytes contain approximately 1 g.-atom per liter of the corresponding metal in each case. It is evident that whereas simultaneous deposition of copper, cadmium and zinc is improbable from the solutions containing the simple ions, it should, and in fact does, occur from the complex cyanide solutions.

A special case of the change of ionic concentration arises when the limiting C.D. for one of the ions is exceeded; the concentration of these ions at the cathode falls to a very small value and the cathode potential then rises to such an extent that discharge of the other ions is possible. For example, at relatively low C.D.'s exclusive deposition of copper occurs from a solution containing copper and zinc sulfates; when the limiting C.D. for the copper ions is exceeded, however, simultaneous discharge of copper and zinc ions occurs and an alloy of the two metals is deposited.

Depolarization of Metal Deposition.—Simultaneous discharge of two ions can also be brought about in another manner. When a metal that is being deposited is able to form a solution in the cathode or, better, when it forms a compound with the cathode material which dissolves in the latter, the partial molal free energy of the deposited substance is diminished and deposition can occur at a potential that is less cathodic than the reversible value for the pure metal. A striking example of this type of behavior, referred to as the **depolarization of metal deposition,** is found when the ions of sodium, or of another alkali metal, are discharged from a neutral or alkaline solution at a mercury cathode. The reversible deposition potential of pure sodium from a solution containing 1 equiv. per liter of a sodium salt is about − 2.7 volts (see Table XLIX), but as a consequence of the formation of compounds which are soluble in mercury, appreciable discharge of sodium ions occurs at − 1.2 volts. In a neutral solution of a sodium salt, i.e., at pH 7, the reversible potential for hydrogen ion discharge is − 0.4 volt, but at a mercury cathode there is an overvoltage of about 0.8 volt (see Table LXXVIII); hence the evolution of hydrogen does not commence until the cathode attains a potential of − 1.2 volts. It is apparent, therefore, that owing to the high hydrogen overvoltage at a mercury cathode, and the marked de-

[5] See Glasstone, *J. Chem. Soc.*, 690, 702 (1929); 1237 (1930).

polarizing effect of the latter on the discharge of sodium ions, deposition of sodium, simultaneously with the evolution of hydrogen, can occur from aqueous solution. It may be noted that even if the electrolyte is originally neutral, e.g., sodium chloride, the discharge of hydrogen ions will make it alkaline in the vicinity of the cathode; the potential for the evolution of hydrogen then becomes more cathodic, viz., about − 1.6 volts, and so the deposition of sodium will be favored. The application of these principles makes it possible to prepare dilute alkali amalgams by the electrolysis of alkali chloride solutions; this fact has been utilized in the electrolytic processes for the manufacture of sodium hydroxide.

Depolarization of metal deposition sometimes occurs when two metals which separate simultaneously form compounds or solid solutions. The reversible potential of a solid solution generally lies in between those of the pure constituents; hence an alloy containing both metals may be deposited at a potential that is less cathodic than that necessary for the less noble constituent in the pure state. This probably accounts for the fact that zinc and nickel are deposited simultaneously at a potential of about − 0.6 volt, whereas that required for pure zinc is nearly 0.2 volt more cathodic.[6] The simultaneous deposition of the iron-group metals is partly due to the similarity of the discharge potentials, but the formation of solid solutions also plays an important part. Although the deposition potentials of cobalt and nickel are lower than that of iron, the cathodic deposit almost invariably contains relatively more of the latter metal.[7]

Separation of Metals by Electrolysis.—The complete separation of one metal from another is important in quantitative electro-analysis;[8] the circumstances in which such separation is possible can be readily understood from the preceding discussion of simultaneous deposition of two metals. The conditions must be adjusted so that the discharge potentials of the various cations in the solution are appreciably different. If the standard potentials differ sufficiently and there are no considerable deposition overvoltages, complete separation within the limits of analytical accuracy is possible; this is, of course, contingent upon the metals not forming compounds or solid solutions under the conditions of deposition. Since the concentration of the ions of a deposited metal decreases during electrolysis, the deposition potential becomes steadily more cathodic, and may eventually approach that for the deposition of another metal. For example, if the ionic concentration is reduced to 0.1 per cent of its original value, the potential becomes 3×0.0295 volt more cathodic for a bivalent metal and 3×0.059 volt for a univalent metal, at ordinary

[6] Schoch and Hirsch, *J. Am. Chem. Soc.*, **29**, 314 (1907); Foerster, *Z. Elektrochem.*, **17**, 883 (1911); **22**, 85 (1916).

[7] Glasstone *et al.*, *Trans. Faraday Soc.*, **23**, 213 (1927); **24**, 370 (1928); **26**, 565 (1929); **27**, 29 (1931); **28**, 733 (1932); **29**, 426 (1933).

[8] For review, see Glasstone, "Thorpe's Dictionary of Applied Chemistry," 4th Ed., 1939, Vol. II, p. 695.

temperatures. If the initial deposition potentials of the two metals differ by at least 0.2 volt, therefore, the first will be removed from the solution almost completely, with an error of less than 0.1 per cent, before the second metal commences to be deposited. If a liter of solution contains one mole of each of the simple salts of silver, copper and cadmium, the deposition of the respective metals will commence at + 0.79, + 0.32 and − 0.40 volt, respectively; it is evident that silver can be deposited quantitatively before copper commences to be deposited, and the latter can, in turn, be separated virtually completely from cadmium. It must be pointed out that these arguments presuppose that the C.D. is less than the limiting value for the metal being deposited; this condition is not difficult to realize at the commencement of the electrolysis, but the current should drop to zero at the end of the deposition. Since the actual C.D. must be appreciable, the efficiency of separation is not always quite as complete as the difference of the deposition potentials would imply.

If two metals normally have similar discharge potentials, the conditions can be altered to make them sufficiently different for separation to be possible. For example, in the case of nickel and zinc in ammoniacal solution, to which reference was made previously, the deposition potentials are similar at 20°, but differ at 90°. The two metals can thus be separated satisfactorily at the higher temperature, but not at the lower. Another illustration is provided by the copper-bismuth system, in which simultaneous deposition takes place from simple salt solutions; if cyanide is added, however, the copper ions form the complex cuprocyanide and the discharge potential becomes more negative (cf. Table LXXXIII). If citric or tartaric acid is present to keep the bismuth in solution, the addition of cyanide hardly affects the deposition potential of this metal; quantitative separation from copper is then possible.

In carrying out the electrolytic separation of metals for analytical purposes, it is desirable to maintain control of the cathode potential; this is best achieved by combining the cathode with a reference electrode, somewhat as in Fig. 109, and measuring the E.M.F. approximately by means of the potentiometer-voltmeter arrangement (Fig. 61). A current of 3 to 10 amps. is passed through the solution and the cathode potential is measured at the commencement of electrolysis; as the ions are removed by deposition the current is steadily decreased so as to keep the potential approximately constant. In this manner the C.D. is prevented from exceeding the limiting value. When the current has fallen to about 0.2 amp., the potential of the cathode is allowed to become about 0.2 volt more negative. The concentration of the deposited ions remaining in solution is now negligible. In order to expedite the process of electrolysis the conditions are arranged so as to increase the rate of diffusion of ions to the cathode; it is then possible to use larger C.D.'s without the risk of depositing the second metal. In the so-called "rapid" electro-analytical methods, rapidly rotating wire-gauze electrodes are employed and, if possible, the temperature is raised.

Electrochemical Passivity.—In the discussion of anode potentials in Chap. XIII it was taken for granted that the anode generally commenced to dissolve when the potential was made slightly more positive than the reversible (oxidation) potential in the given electrolyte. As the C.D. is raised, the potential increases somewhat as a result of concentration polarization, but the extent of the increase is usually not large. It was stated on page 444, however, that metals of the iron group do not commence to dissolve anodically until the theoretical reversible potential is exceeded by a relatively large amount, viz., 0.3 to 0.4 volt at ordinary temperatures; this marked polarization, or irreversibility, must be due to the fact that one of the stages involved in the ionization process is a slow one requiring a high energy of activation. Nevertheless, in spite of the large polarization, an anode of iron, cobalt or nickel still dissolves quantitatively in accordance with the requirements of Faraday's laws of electrolysis. If the C.D. is increased, however, a point is reached at which the anode potential rises very rapidly, and there is a corresponding decrease in the current; at the same time the anode practically ceases to dissolve, although it appears, otherwise, to be quite unchanged. The metal is then said to be in the **passive state**, and the phenomenon is

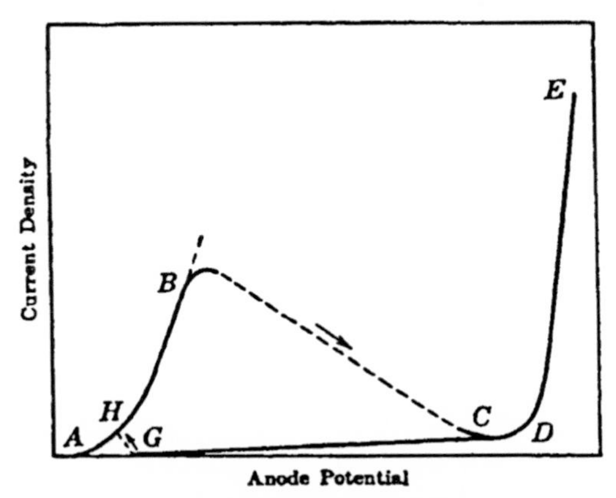

FIG. 122. Anodic passivity

referred to as **passivity**; since the passivity arises in this instance as the result of electrochemical or anodic treatment, one or other of these adjectives is frequently employed to describe this type of passivity.[9]

The general nature of the variation in the potential of an anode of iron, cobalt or nickel as the current is increased is shown in Fig. 122:

[9] For references to recent work, see Hedges, *Ann. Rep. Chem. Soc.*, 31, 127 (1934); see also, Evans, "Metallic Corrosion, Passivity and Protection," 1937.

AB represents the change of anode potential with current density while the electrode is active and is dissolving quantitatively. At *B* the electrode becomes passive and the potential rises rapidly to *C*, the current decreasing at the same time because of the increased voltage now required to pass current through the electrolytic cell. If the current is increased further, the potential becomes more anodic until a new process, usually oxygen evolution, commences at *D*; beyond this point considerable current can flow with relatively little further polarization, as shown by the curve *DE*. In the region *CE* the anode has almost, but not entirely, ceased to dissolve. When the current is decreased from *E*, the active condition of the electrode is not regained at *C*; the current must be reduced to a much lower value, e.g., to *G*, before the potential changes to *H*, on the curve *AB*, and quantitative dissolution recommences. If the current is cut off while the anode is passive, it returns slowly to the active condition on standing. The active state is regained rapidly by making the passive metal cathodic: this can be achieved either by using it as a cathode in an electrolytic cell, or by touching it with a less noble metal or scratching the surface while immersed in an electrolyte. The scratching exposes active metal which together with the passive material forms internal voltaic cells of which the latter constitutes the cathode.

Under given conditions, passivity is attained with increasing readiness in the order: iron, cobalt, nickel; iron is much more difficult to render passive than is nickel in a particular electrolyte. Metals of the iron group become passive more readily in alkaline than in acid solutions, and oxidizing agents, e.g., iodate, bromate, chlorate, chromate and nitrate, favor passivity; chloride ions markedly inhibit the onset of passivity.* Increase of temperature increases the C.D. required for the anode to become passive under a given set of conditions.

A chromium anode can also be rendered passive, although its behavior is somewhat different from that of the iron-group metals. The anode dissolves at low current densities in the form of chromous, i.e., Cr^{++}, ions, but when it becomes passive dissolution does not cease; the potential rises rapidly to about + 1.1 volts, and the chromium now passes into solution in the hexavalent state, forming CrO_4^{--} ions. No oxygen is evolved, however, because the process $Cr \rightarrow CrO_4^{--}$ requires a less positive potential than the evolution of oxygen in the same solution. If the electrolyte contains iodide ions, the discharge of these ions can occur at a still lower potential, viz., + 0.6 volt; the chromium anode then ceases to dissolve when it becomes passive but iodine is liberated. Chromium is passivated more readily in alkaline than in acid solution and temperature has an inhibitive effect, as with the iron-group metals; chloride ions, however, have less influence in preventing the onset of passivity. The chromium anode retains its passivity for some time even in acid solution, but the active state is rapidly regained by cathodic treatment.

* Chlorides are added to nickel and other plating baths in order to prevent the anodes from becoming passive.

Molybdenum and tungsten are rendered passive more readily in acid than in alkaline solution; this is the reverse of the behavior exhibited by chromium and the iron-group metals. Although oxidizing agents generally favor passivity, such is not the case with a tin anode; in this instance, too, chloride ions do not have the inhibiting effect they have in other cases. It is apparent, therefore, that each metal requires its own specific conditions in order that it may be rendered passive.

Platinum and gold are regarded as unattackable metals because they are nearly always in the passive state. A gold anode dissolves in neutral or acid chloride solutions at very low C.D.'s to form Au^{+++} ions at a potential of about + 1.2 volts; as the current is increased, however, the potential rises rapidly to + 1.7 volts, the electrode ceases to dissolve and chlorine is evolved instead. Platinum is actually less noble than gold, but it is rendered passive much more easily, especially in solutions of oxyacids; if the electrolyte contains ammonia or hydrochloric acid, however, platinum suffers appreciable anodic attack, particularly if the metal is in a finely-divided form.

Passivity and Current Density.—It was believed at one time that the onset of passivity required the application of a "critical" current density to the anode; the important work of W. J. Müller,[10] which has thrown much light on the problem of passivity, has shown that this is not the case. Using horizontal electrodes, protected by a glass hood to prevent the products formed at the anode from diffusing away, it was found that passivity could be induced by means of very low C.D.'s, provided sufficient time were allowed. The higher the C.D., however, the smaller the length of time required for the onset of anodic passivity. In N sulfuric acid, for example, iron became passive in 12.4 secs. when the C.D. was 289 milliamps. per sq. cm., but if the C.D. was reduced to 15 milliamps. it was required 848 secs. before passivity was evident. By preventing diffusion from the anode, it has thus been found possible to make metals passive at C.D.'s very much below those which had previously been considered effective. If the solution is stirred or if the temperature is raised, the time required for passivation at a given current is increased; on the other hand, if the electrolyte is previously saturated with the salt formed when the anode dissolves the passivation time is diminished.

It is apparent from these results that the onset of passivity is associated with the saturation of the solution in the immediate vicinity of the anode and the separation of the solid. In several instances deposits of this kind have been removed from the anode and identified. Immediately before the metal becomes passive a visible deposit forms on its surface, but it is thrown off, and dissolves in the bulk of the electrolyte, when the electrode is actually passive and oxygen is evolved; the electrode then appears quite clean.[11]

[10] For reviews, see references on page 491, also Müller, "Bedeckungstheorie der Passivität der Metalle," 1933; *Trans. Faraday Soc.*, 27, 736 (1931).

[11] Müller, *Z. Elektrochem.*, 30, 241 (1924); Hedges, *J. Chem. Soc.*, 2878 (1926).

Chemical Passivity.—If iron is dipped into concentrated nitric acid, of specific gravity greater than 1.25, it may be attacked initially, but the metal does not continue to dissolve; although its appearance is unchanged, the iron is now no longer soluble in dilute nitric acid nor will it displace silver from silver nitrate solution. The iron has thus been rendered passive and since this state has been brought about by chemical action, it is sometimes referred to as **chemical passivity.** Many other oxidizing agents, besides nitric acid, can induce passivity in iron and also in cobalt, nickel and chromium. As in the case of electrochemical passivity, the material recovers its activity if it is subjected to any form of cathodic treatment. The restoration of the active state takes place slowly on standing, but it is facilitated by the presence of chloride ions and increase of temperature. In general, the resemblances between electrochemical and chemical passivity are very marked; it is very probable, therefore, that the fundamental cause of the passivity is the same in each case.

Metals of the iron group and chromium can be made passive by heating in air, and a semi-passive state, in which the metal exhibits an electrode potential between the values for the completely active and completely passive conditions, can be induced in these metals, as well as in molybdenum, tungsten and vanadium, by mere exposure to air. The corrosion resisting properties of stainless steel and of chromium plate are undoubtedly due to passivity resulting from exposure to air.

Theories of Passivity.—A large number of studies have been made of the phenomena of passivity and many theories have been proposed to explain them; the views most widely held at the present time are, in a sense, a modification of those proposed by Faraday in 1836. It was suggested that the surface of a passive metal was oxidized or that the atoms of the "metal were in such a relation to the oxygen of the electrolyte as to be equivalent to an oxidation"; the surface oxide would presumably be so thin as to be invisible. As a result of the metal being completely shielded by the oxide it could no longer dissolve anodically, be attacked by acids or replace other metals from solution. This theory of passivity received general support until Hittorf (1898) showed by his work on chromium that it had to face considerable difficulties, and that the properties of the oxide responsible for passivity would need to be different from those of any known oxides of the metal. Further, the supposed identity in light-reflecting powers of a passive anode and a cathode of the same metal, which is of course in the active condition, suggested that there could be no oxide film on the former.

Most of the aforementioned difficulties have been overcome in recent years. In the first place it is known that if ferric oxide is heated, it becomes almost insoluble in acids, and so it is possible that the oxide on passive iron may be in a similar compact and unattackable form. Second, it has been shown that polarized light is not reflected to the same extent by active and passive forms of a given metal, neither do they exhibit the same photoelectric emission. By gentle anodic action or by the use of iodine, Evans has succeeded in dissolving away the inner portions of iron made passive by anodic treatment, by means of oxidizing agents or by exposure to air; in each case a very thin, transparent and almost colorless film of oxide was left after the basis metal had completely disappeared. The film is quite invisible when attached to the metal, but can be seen when separated from it. The exact nature of the oxide is not known: some investigators consider it to be a form of Fe_2O_3, while others regard it as Fe_3O_4 or FeO_3; it is nevertheless established that it is insoluble in acids.[12]

If oxide films are responsible for passivity, it is to be expected that an anode will become passive most readily in an electrolyte from which the oxide will separate most easily; this expectation is realized in practice. The oxides of iron, cobalt, nickel and chromium are less soluble in alkaline than in acid solution, and passivity sets in more rapidly in the former. The oxides of molybdenum and tungsten, however, are more soluble in alkali than in acid, and so these metals are rendered passive most easily in acid electrolytes.

Although the existence of a thin layer of oxide film on the surface of a passive metal may be admitted, it must be realized that it is still necessary to explain why the material ceases to dissolve and how the oxide film is formed.[13] The following mechanism gives a satisfactory explanation of these and other phenomena of anodic passivity. As a result of the dissolution of the metal the corresponding cations will accumulate in the vicinity of the anode; at the same time the solution will tend to become more alkaline as a consequence of the tendency of hydrogen ions to migrate towards the cathode. If the C.D. is increased or the time of electrolysis is prolonged, the solution will become saturated and, as a result, a normal or basic salt of the metal, depending on the actual acidity, will be formed on the anode surface. By applying mathematical methods, Müller has shown that a porous deposit of salt first grows across the surface in a sideways direction and when the film is complete it commences to thicken so that it becomes visible.[14] When the anode is covered in this manner, the effective C.D. at the exposed portions will be very large and hence the potential will rise rapidly; this gives the region *BC* of the anode potential curve in Fig. 122.

As a result of the increase of potential the discharge of hydroxyl ions becomes possible and the OH radicals, or perhaps atomic oxygen, react with the metal to produce an *adherent and invisible* film of oxide beneath the visible, thick but non-adherent layer. The oxide film spreads rapidly across the anode and when it is covered, oxygen evolution commences. The liberation of gas causes the visible film to be thrown off from the

[12] Evans, *J. Chem. Soc.*, 1020 (1927); Hedges, *ibid.*, 969 (1928); Bancroft and Porter, *J. Phys. Chem.*, 40, 37 (1936).

[13] Hedges, *J. Chem. Soc.*, 969 (1928); Evans, *ibid.*, 109 (1929); 478 (1930).

[14] See references on page 493.

electrode surface, as actually observed, leaving the anode apparently clean. The thick, non-adherent layer has played its part in reducing the effective area of the electrode, and is not necessary for the maintenance of passivity. The thin invisible film that remains adheres firmly to the anode, and is in a form not readily attacked by acids. Since the underlying material is unable to dissolve, it behaves in all respects as a noble, unattackable metal. Cathodic treatment of any kind, however, causes the oxide film to be reduced; the surface of the metal is once more exposed and it becomes active. Increase of temperature increases the solubility of the salt of the metal and thus retards the formation of the visible film; the onset of passivity is thus delayed.

It is important to note that whereas earlier workers were under the impression that a relatively thick film was necessary to ensure passivity, it is now realized that this cannot be so: the fact that the film grows in thickness implies that metal ions are able to pass through it and hence such a film does not prevent the anode from dissolving. A coherent and adherent layer is essential if the metal is to be really passive; such a film does not allow ions to pass through it and so it cannot grow in thickness. A film that is effective for passivity must thus inevitably remain invisible. Chloride ions are apparently able to penetrate certain films, e.g., ferric oxide, thus rendering them porous and unable to prevent the metal dissolving; the presence of chloride ions thus retards the onset of passivity. Even when the anode does eventually become passive in a chloride solution, the passivity is not complete, for the metal continues to dissolve to some extent; it is for this reason that a passive iron anode in an alkaline solution containing chloride ions eventually becomes covered with a visible film of ferric oxide.

When a gold anode is rendered passive in hydrochloric acid solution the conditions appear to be somewhat different from those described above; experiments show that the gold dissolves in the tervalent state to form $AuCl_4^-$ ions in solution. There is consequently a limiting C.D. determined by the maximum rate of diffusion of chloride ions to the anode (cf. p. 459); when this rate is exceeded the potential must rise so as to permit another process to occur. The gold then becomes covered with a layer of oxide, produced either by reaction of Au^{+++} ions with water, or by the direct action of oxygen or hydroxyl radicals, and ceases to dissolve.[15]

It was mentioned above that a chromium anode continues to dissolve in the passive state forming chromate ions; in this case the invisible oxide film may be sufficiently porous to allow ions to penetrate it; alternatively, the oxide film may become oxidized to CrO_3 which dissolves to form chromate, but is immediately regenerated by oxidation of the anode.

[15] Shutt *et al.*, *Trans. Faraday Soc.*, **26**, 635 (1930); **28**, 740 (1932); **29**, 1209 (1933); **30**, 914 (1934); **31**, 636 (1935); Butler *et al.*, *ibid.*, **30**, 1173 (1934); **34**, 806 (1938).

Chemical passivity can be brought into line with the views presented above concerning anodic passivity. If iron is placed in nitric acid there is at first a rapid attack leading to a high local concentration of ferric ions and a depletion of hydrogen ions; the result is that a thin, adherent film of ferric oxide, or of a basic salt, is formed which protects the metal from further attack. The partial or complete passivity of certain metals resulting from exposure to air is ascribed to the formation of a similar continuous film; as already mentioned, this layer can be separated from the surface by suitable treatment. It may be noted that chemical passivity is not restricted to chromium, the iron-group and similar metals; passivity appears to be a property exhibited by most metals under suitable conditions. For example, copper becomes passive in concentrated nitric acid at $-11°$, and even zinc and magnesium show signs of passivity under these conditions; the low temperature is necessary, presumably, in order to prevent the film causing passivity from dissolving.[16]

Mechanical Passivity.—In certain instances the dissolution of an anode is prevented by a visible film, e.g., lead dioxide on a lead anode in dilute sulfuric acid; this phenomenon has been called **mechanical passivity**, but it is probably not fundamentally different from the forms of passivity already discussed. The film is usually not completely impervious, but merely has the effect of decreasing the exposed surface of the electrode to a considerable extent; the effective C.D. is thus increased until another process in which the metal is involved can occur. At a lead anode in sulfuric acid, for example, the lead first dissolves to form plumbous ions which unite with the sulfate ions in the solution to form a porous layer of insoluble lead sulfate. The effective C.D. is increased so much that the potential rises until another process, viz., the formation of plumbic ions, occurs. If the acid is sufficiently concentrated these ions pass into solution, but in more dilute acid media lead dioxide is precipitated and tends partially to close up the pores; the layer of dioxide is somewhat porous and so it increases in thickness until it becomes visible. Such an oxide is not completely protective and attack of the anode continues to some extent; it is, however, a good conductor and so hydroxyl ions are discharged at its outer surface, and oxygen is evolved, in spite of its thickness.

It is of interest to note that if the solution of sulfuric acid or sulfate contains an appropriate amount of chlorate or nitrate, the lead sulfate does not adhere to the anode and the metal continues to dissolve in the bivalent condition. This fact has been utilized in the electrolytic manufacture of lead sulfate and of other insoluble lead salts.

The so-called mechanical passivity, accompanied by the formation of a visible film and oxygen evolution, has been observed with iron, cobalt, nickel, manganese, lead and other anodes in alkali hydroxide solutions, and also with thallium, antimony and bismuth in aqueous sulfuric acid.

[16] Hedges, *J. Chem. Soc.*, 969 (1928); 561 (1930).

The Corrosion of Metals: Hydrogen Evolution Type.[17]—It was seen in Chap. VII that, in general, any metal higher in the electrode potential series should be able to replace one lower in the series from solution; strictly speaking it is more correct to say that any metal with a dissolution potential, i.e., written in the form M, M^+ (cf. p. 435), more positive than that of another, *in the same solution*, should be able to displace the latter. This conclusion applies equally to hydrogen, and so any metal with a more positive dissolution (oxidation) potential in the given solution should liberate hydrogen from that solution. For example, it would be expected that lead ($E_{Pb} \approx + 0.12$) should dissolve in hydrochloric acid with the evolution of hydrogen ($E_H \approx \pm 0.0$), and zinc ($E_{Zn} \approx + 0.75$) should displace hydrogen from a neutral solution, e.g., sodium sulfate, ($E_H \approx + 0.4$).

In actual practice neither of these reactions occurs because the liberation of hydrogen is accompanied by an overvoltage exactly as in the evolution of the gas during electrolysis. The conditions when a metal dissolves to give hydrogen are in this respect identical with those operative at a cathode and the same slow process, whatever it may be, must be an intermediate stage. The hydrogen overvoltages (ω) of lead and zinc are about 0.6 and 0.7 volt, respectively, and so it follows, according to the arguments on page 486, that hydrogen evolution will commence when the respective *discharge* (reduction) potentials are equal to $- E_H - \omega$, i.e., $- 0.6$ volt for lead in hydrochloric acid and $- 1.1$ volts for zinc in a neutral solution. The corresponding hydrogen dissolution potentials would be $+ 0.6$ and $+ 1.1$ volts, and since these are more positive than the potentials of the respective metals, viz., $+ 0.12$ and $+ 0.75$, the evolution of hydrogen will not take place. By increasing the hydrogen ion concentration the hydrogen dissolution potential is reduced, and so the tendency for the metal to dissolve would be increased. Thus, in a 2N solution of a mineral acid the reversible hydrogen potential is approximately ± 0.0; the discharge potential at a zinc cathode will thus be about $- 0.7$, and hence the hydrogen dissolution potential will be $+ 0.7$ volt. This is so close to the dissolution potential of zinc that the metal would be expected to dissolve slowly in a strong acid solution. Pure zinc does, in fact, dissolve in this manner, but impure zinc is attacked more readily, partly because of the lower overvoltage and partly for other reasons to be considered shortly.

If pure zinc is placed in contact with a piece of copper and the combination is immersed in dilute acid, so as to form a short-circuited simple voltaic cell, the rate at which the zinc dissolves is greatly increased; an examination of the system shows, however, that the hydrogen is now being evolved from the copper, instead of from the zinc. It is neverthe-

[17] For general reviews, see Hedges, *Ann. Rep. Chem. Soc.*, **31**, 135 (1934); "Protective Films on Metals," 1932; Evans, "Metallic Corrosion, Passivity and Protection," 1937.

less the zinc, and not the copper, which dissolves. The less noble metal, i.e., the zinc, forms the anode of the voltaic cell; this metal passes into solution as zinc ions, while the hydrogen ions are discharged to form hydrogen gas at the more noble metal, i.e., the copper, which is the cathode of the cell. Since the hydrogen overvoltage at a copper cathode is only about 0.2 volt, compared with 0.7 volt for zinc, hydrogen evolution occurs readily and the zinc dissolves rapidly.

The rate at which the zinc dissolves is, theoretically, governed by Faraday's laws, and hence is greater the larger the current flowing in the short-circuited cell. Increase of current may be obtained by increasing the conductance of the electrolyte and by making the potential difference between the two electrodes, i.e., the E.M.F. of the cell, larger. The E.M.F. of the short-circuited cell is equal to the sum of the dissolution potential of zinc, i.e., about + 0.75 volt, and the discharge potential of hydrogen on the copper, allowing for the overvoltage at the latter; in a neutral solution the reversible discharge potential of hydrogen is − 0.41 while in acid solution it is about ± 0.0 volt, and hence the actual discharge potentials for appreciable rates of gas evolution at copper will be about − 0.61 and − 0.20 volt, respectively. The E.M.F. of the short-circuited zinc-copper cell is thus 0.14 volt in neutral solution and 0.55 volt in acid solution. Assuming the conductances to be the same, the zinc will obviously be attacked more vigorously in an acid than in a neutral solution. If the overvoltage at the cathodic, i.e., more noble, metal were lower than the 0.2 volt at copper, the E.M.F. would be increased and the zinc would dissolve more readily.

It is evident, therefore, that if any metal is in contact with a nobler metal, the former will dissolve more rapidly the lower the hydrogen overvoltage at the latter and the higher the hydrogen ion concentration of the solution. Increasing the conductance of the solution, e.g., by the addition of a neutral salt, will also increase the rate at which the less noble metal dissolves. If the overvoltage of the more noble metal is large, e.g., mercury, the attack on the base metal is hindered; this is the case when zinc is amalgamated, for the overvoltage at the mercury, which is the cathodic part of the cell, is sufficient to reduce the E.M.F. virtually to zero.

The mode of attack on metals described above is referred to as the **hydrogen evolution type of corrosion,** since gas is actually evolved at the more noble portion of the system. This type of corrosion requires a combination of two metals, the nobler one of which has a low overvoltage, but it is not necessary that the metals should be in a massive form. For example, the addition of a small quantity of copper sulfate to an acid solution greatly expedites the rate at which zinc dissolves; copper is deposited, by replacement, on various parts of the zinc and a large number of local short-circuited cells are set up. Another possibility, which frequently arises, is that the base metal should have included in it as a separate phase portions of a more noble conducting material; the latter need not necessarily bé a metal for it may be a sulfide or oxide having a low hydrogen overvoltage.

Corrosion in Presence of a Depolarizer.—In the foregoing discussion the possibility has not been considered that a substance might be present which would react with the hydrogen and would permit a process to occur at the cathode at a lower potential; such a substance is a **hydrogen depolarizer,** e.g., oxygen, from the air, or any oxidizing agent. Unless special precautions are taken to exclude air, a depolarizer is always present, and if this is freely available the conditions of metallic corrosion are changed. Since there is now no actual evolution of hydrogen gas at the cathodic regions, the overvoltage of the noble constituent of the system is unimportant. Whenever two metals are in contact in a solution with free access of air, the baser of the metals is always attacked, irrespective of the hydrogen overvoltage at either. For this reason the iron in imperfect "tin plate" is readily corroded in air, provided moisture is present, but in the absence of oxygen the high overvoltage of tin prevents any attack of the iron. It is a familiar fact that once the underlying iron is exposed, tin plate rusts more rapidly than does iron alone; this corrosion is facilitated by the short-circuited local voltaic cells with atmospheric oxygen as the cathodic depolarizer. In citric and oxalic acid solutions it is the tin, rather than the iron, that is attacked; the reason for this is that the former metal forms complex anions in these particular electrolytes, with the result that its dissolution potential exceeds that of the iron in the same solutions. Under these conditions the tin becomes the anode and the iron the cathode of the cells responsible for corrosion.[18] Even in the presence of a depolarizer, increase of conductance of the liquid will, of course, enhance the attack of the baser metal, and increase of acidity will generally operate in the same direction. The cathodic portions often behave as a species of oxygen electrode and hence the potential, like that of a hydrogen electrode, changes with increasing acidity in such a way as to increase the E.M.F. of the voltaic cells.

Although corrosion is favored by a large difference of potential between the anodic and cathodic portions of a system, even the smallest of such differences is sufficient to stimulate corrosion in the presence of a depolarizer. In an apparently uniform piece of metal, any portion which has been subjected to strain is less noble than an unstrained portion and small crystals are less noble than large ones; further, minute inclusions of noble material are often found in relatively pure metals. These differences permit local voltaic cells to be set up and, in the presence of a depolarizer, corrosion of the baser (anodic) regions will occur.

[18] Hoar, *Trans. Faraday Soc.*, 30, 472 (1934).

The existence of anodic and cathodic parts in an apparently uniform piece of iron can be demonstrated by means of the **ferroxyl indicator:**[19] this consists of a mixture of 100 cc. of 0.1 N sodium chloride, 3 cc. of 1 per cent potassium ferricyanide and 0.5 cc. of a 1 per cent alcoholic solution of phenolphthalein to which sufficient agar is added for the system to form a jelly when cold. The warm solution is poured on a piece of iron, allowed to set and put aside for some hours; it will then be observed that in certain regions the indicator has a blue color while in others it is pink. At the anodic, i.e., baser, portions of the iron the metal passes into solution as ferrous ions which react with the ferricyanide ions to produce a blue color; at the cathodic parts hydrogen ions are removed by the reaction with the ferricyanide, and hence the solution becomes alkaline and gives a pink color with the phenolphthalein. The object of the potassium chloride in the ferroxyl indicator is partly to serve as a conductor and also to prevent the iron becoming passive. A similar indicator, containing the dyestuff alizarine, has been proposed for identifying anodic and cathodic regions of aluminum; at the former a red color is observed while at the latter the indicator turns violet.[20]

The assumption has been made hitherto that the corroded metal is a base one, but this term has been used in a relative sense; any metal can be attacked in the presence of a depolarizer, provided a more noble material is available to form a voltaic cell. It is possible in this manner to bring about the solution of metals which are normally relatively inert: for example, copper is normally insoluble in dilute sulfuric acid, but it will dissolve if air is bubbled through the solution or if hydrogen peroxide is added. In general a second, more noble, metal is not necessary to aid the dissolution of the copper, since irregularities in the metal itself are sufficient to set up voltaic cells; the presence of the cathodic depolarizer serves to keep the potential of these regions below that of the other parts which suffer attack. The vigorous action of nitric acid on copper is to be attributed to the depolarizing action on hydrogen; as is to be expected, reduction products of nitric acid are formed at the same time.

Differential Oxygenation Corrosion.—An interesting type of corrosion results when the depolarizer, generally oxygen, is not uniformly distributed over the surface of the metal; this has been called corrosion due to **differential oxygenation.**[21] If oxygen has access to certain portions of a metal, depolarization of hydrogen will occur more readily there than at other parts; the latter regions will, therefore, tend to become anodic and dissolve, so that hydrogen ions can be discharged at the oxygenated regions which are cathodic. It may appear anomalous, at first sight,

[19] Walker, Cederholm and Bent, *J. Am. Chem. Soc.*, 29, 1256 (1907); Cushman and Gardner, "Corrosion and Preservation of Iron and Steel," 1910; see also, Evans, *Metal Ind.* (*London*), 29, 481, 507 (1926).

[20] Akimow and Aleschko, *Korrosion u. Metallschutz*, 11, 126 (1935).

[21] Aston, *Trans. Electrochem. Soc.*, 29, 449 (1916); Evans, *J. Soc. Chem. Ind.*, 43, 129T, 315T (1924); 45, 38T (1926); *Ind. Eng. Chem.*, 17, 363 (1925); Mears and Evans, *Trans. Faraday Soc.*, 30, 417 (1934).

that the parts of the metal to which oxygen does not have access are those that dissolve: this is because depolarization occurs only at the portions where oxygen is available and hence hydrogen ions will be discharged there to replace the atoms removed by the depolarizer. The development of anodic and cathodic properties as a result of differential oxygenation can be readily demonstrated experimentally; a porous pot and a surrounding vessel are filled with potassium chloride solution, and similar pieces of iron or zinc, cut from the same sheet, are placed in the two solutions. The pieces of metal, i.e., the electrodes, are connected through a galvanometer, and there is no initial flow of current. If one of the solutions is saturated with air, however, the system develops an E.M.F. and current flows; the direction of the current shows that the electrode in the oxygenated solution is the cathode, whereas the other, to which the oxygen does not have access, is anodic and dissolves.

The differential oxygenation mechanism accounts for the fact that the rusting of iron occurs preferentially under portions covered with wet rust: air is less accessible to these parts and hence they become anodic with respect to the others and continue to dissolve as long as the oxygen is available. A similar condition is responsible for the corrosion of any metal covered with a partially protective film, or of a metal that is in contact over a portion of its surface with a non-metal. Corrosion at the bottom of "pits" in a metal surface is also explained by differential oxygenation. The formation of an oxide film, with its accompanying passivity, also helps those parts of the metal exposed to oxygen to become cathodic, thus favoring the attack of the other portions.

In the foregoing discussion different types of corrosion have been considered separately; in practice the situation is complicated by the simultaneous occurrence of two or more forms of corrosion, by the production of adherent films which result in passivity, and by loose deposits, such as rust, which arise from the interaction of the alkali produced at the cathodic portions of the metal with the cations formed at the anodic regions. In addition to these possibilities, due to chemical or electrochemical action, physical factors, such as surface forces, often play a part; a film which would normally be protective may be drawn up into the solution-air interface and thus be prevented from covering the surface of the metal. It is because of these complicating factors that the phenomena of corrosion are sometimes difficult to explain, but it is believed that the principles enunciated in this and the preceding sections represent the fundamentals of electrochemical corrosion.

The so-called "atmospheric corrosion," resulting from exposure of a metal to air, generally containing such gases as hydrogen sulfide and sulfur dioxide as well as carbon dioxide and water vapor, presents a difficult problem; the action has often been regarded as chemical in nature, but it is not impossible that electrochemical phenomena are involved.[22]

[22] Wagner, *Z. physik. Chem.*, 21B, 25 (1933); 32B, 447 (1936).

PROBLEMS

1. A solution containing approximately 0.1 molar cadmium and zinc sulfates is electrolyzed; what proportion of the cadmium can be deposited before zinc separation commences? It may be assumed that the polarization accompanying deposition of the metals is negligible in each case and that no alloy formation occurs.

2. The actual reversible potentials of the metals M_1 and M_2 in a given solution containing simple ions of both metals differ by 0.3 volt. Assuming the metal M_1 to deposit first, determine the proportion remaining in solution when M_2 commences to deposit at ordinary temperatures, if M_1 is (*a*) univalent, (*b*) bivalent.

3. A 0.1 molar solution of copper sulfate in 1.0 N sulfuric acid is electrolyzed; what will be the potential of the cathode when the cupric ion concentration has been reduced to 10^{-7} g.-ion per liter? How much further could the cathode potential be increased and hydrogen evolution still be avoided? The cathodic overvoltage at copper may be taken from Table LXXVIII.

4. An approximately 0.1 molar solution of a silver salt in 1.0 molar ammonia is subjected to electrolysis for analytical purposes; it is desired to reduce the concentration of the silver at least to 10^{-5} molar. Calculate the approximate initial and final potentials of the silver cathode. The instability constant of the $Ag(NH_3)_2^+$ ion may be taken as 7×10^{-8} and the whole of the ammonia may be assumed to exist in the NH_3 state.

5. The (oxidation) potential of copper in a 0.1 molar cuprocyanide solution containing excess of potassium cyanide is 1.0 volt; explain why hydrogen gas is evolved when potassium cyanide solution, pH 11, acts on metallic copper.

6. The metals iron and tin are in contact in a solution of pH 4; utilize the data in Tables XLIX and LXXVIII to calculate the approximate E.M.F. favoring corrosion in the absence of air. The assumption may be made that only simple ions of tin and iron are formed in the given solution. What would be the effect on the magnitude of the E.M.F. and on the nature of the corrosion of the introduction of air, i.e., oxygen at 0.2 atm. pressure?

CHAPTER XV

ELECTROLYTIC OXIDATION AND REDUCTION

Electrolytic Reduction and Oxidation.—Oxidation and reduction, in the special sense of the terms, are equivalent to the removal and addition of electrons, respectively; when a current from an external source of E.M.F. is applied to a pair of electrodes in an electrolytic cell, so as to make them anode and cathode respectively, the former can act as a means for the continuous removal of electrons while the latter serves as a corresponding source of supply. It is apparent, therefore, that there is a possibility of oxidation being brought about electrolytically at an anode, whereas reduction can occur at a cathode. The same conclusion may be reached in a less general manner when it is realized that there is always a tendency for oxygen to be formed at an anode and of hydrogen to be produced at a cathode, thus favoring oxidation and reduction, respectively. Electrolytic oxidation and reduction reactions may be classified as reversible and irreversible; the former refer to systems which normally yield definite reversible potentials, but in the latter are included processes involving systems which for one reason or another do not behave reversibly.

Reversible Oxidation-Reduction Processes.—The fundamental principles concerned in the reduction of a reversible system at a cathode or in the oxidation at an unattackable anode have been already given in Chap. XIII. If the potential of the cathode is made slightly more negative or that of the anode more positive than the reversible potential of the system, reduction or oxidation, respectively, will take place. As the current is increased there will be some polarization due to concentration changes, and eventually the limiting C.D. for the particular process will be attained; any further increase will be accompanied by another reaction, e.g., evolution of hydrogen at a cathode or evolution of oxygen or chlorine at an anode.

In some cases the oxidation-reduction potential is close to the potential at which another process, e.g., hydrogen evolution, can occur; in this event both reactions will take place simultaneously. For example, reduction of the titanic-titanous system in hydrochloric acid commences at about + 0.05 volt, while in the same solution the reversible hydrogen potential is approximately ± 0.0 volt. It follows, therefore, that if a platinized platinum cathode, which has almost zero hydrogen overvoltage, is employed, reduction of the titanic ions and evolution of hydrogen will take place at the same time. The reduction efficiency will then be small. If a high overvoltage cathode is employed, however,

100 per cent efficiency is possible. In the cathodic reduction of stannic ions in hydrochloric acid solution there is simultaneous evolution of hydrogen and deposition of tin; because of the latter occurrence the cathode is inevitably one of tin and so it is not possible to make any change in the overvoltage, except by changing the temperature.

Certain systems which behave reversibly in the equilibrium state exhibit considerable polarization in the course of electrolytic reduction; examples are the conversion of 5-valent vanadium to the 4-valent state, and of the latter to the 3-valent condition, the reduction of 6- to 5- and of 5- to 3-valent molybdenum, and the reduction of 6-valent to 5-valent tungsten. There is reason to believe, however, that in all these cases the abnormal behavior is to be attributed to the presence of oxide films on the cathode: by producing a partial blocking of the surface, these oxide films increase the effective C.D., so that the potential rises.[1] Considerable polarization, accompanied by oxide-film formation, occurs in the reduction of chromate to chromic ions, but it is not certain how far this system is reversible.

The reversible quinone-hydroquinone system also behaves in a somewhat unusual manner: at a platinized platinum electrode there is little polarization other than that due to concentration changes at the electrode, both for oxidation and reduction. With other electrode materials, however, there is marked polarization, especially as the C.D. is increased; at sufficiently high currents diffusion becomes rate determining, but at lower values the nature of the slow process is not at all clearly understood. The polarization appears to be influenced in an unexpected and complex manner by the hydrogen ion concentration of the electrolyte.[2]

Non-Reversible Processes.[3]—Reactions of the non-reversible type, i.e., with systems which do not give reversible equilibrium potentials, occur most frequently with un-ionized organic compounds; the cathodic reduction of nitrobenzene to aniline and the anodic oxidation of alcohol to acetic acid are instances of this type of process. A number of inorganic reactions, such as the electrolytic reduction of nitric acid and nitrates to hydroxylamine and ammonia, and the anodic oxidation of chromic ions to chromate, are also probably irreversible in character. Although the problems of electrolytic oxidation and reduction have been the subject of much experimental investigation, the exact mechanisms of the reactions involved are still in dispute. For example, the electrolytic reduction of the compound RO to R may be represented by

$$RO + 2H^+ + 2\epsilon = R + H_2O,$$

but it is not known with certainty whether (*a*) the protons attach themselves first to RO and then the resulting ion is discharged; (*b*) the elec-

[1] Foerster *et al.*, *Z. physik. Chem.*, **146A**, 81, 177 (1930); **151A**, 321 (1930).

[2] Rosenthal, Lorch and Hammett, *J. Am. Chem. Soc.*, **59**, 1795 (1937).

[3] For further details, see Glasstone and Hickling, "Electrolytic Oxidation and Reduction," 1935.

trons pass from the electrode to RO and the RO^{--} ion combines with the protons; (*c*) the protons are discharged to form atoms and the latter reduce the RO; or (*d*) the hydrogen atoms combine to form molecules which reduce the RO with the cathode material acting as catalyst.[4]

In spite of the considerable confusion which surrounds the subject, it is known that certain factors affect the course and speed of irreversible electrolytic oxidations and reductions; these are as follows: (I) electrode potential, (II) nature and condition of the electrode, (III) concentration of the oxidizable or reducible substance, i.e., the depolarizer, (IV) temperature, and (V) catalysts. In addition, the nature of the electrolyte employed to conduct the current when the depolarizer is a non-conductor often has an important influence. The various factors just enumerated will be considered in turn, first with reference to electrolytic reduction and then to oxidation.

Electrolytic Reduction: I. Electrode Potential.—In the absence of a depolarizer the variation of cathode potential with current density will be represented by a curve similar to I in Fig. 123; the small current flowing between *a* and *b* may be due to the presence of a small amount of reducible material in the solution or to diffusion of hydrogen away from the cathode. Visible hydrogen evolution commences in the vicinity of *b*, which gives the so-called bubble overvoltage potential; beyond this the gas is evolved freely with but little further polarization. If a depolarizer is present, however, another process, i.e., reduction of the depolarizer, will occur in preference to hydrogen evolution; if this reaction is relatively slow, the cathode potential will rise from *a* to *b*′, as shown by curve II. Between *a* and *b*′ the reduction efficiency is 100 per cent, but at the potential *b*′, which is close to that of *b*, appreciable hydrogen evolution will commence and the current efficiency will fall. For a rapid reduction process the cathode potential rises slowly, curve III, until the limiting C.D. is attained; there is then a rapid increase of potential and hydrogen liberation occurs.

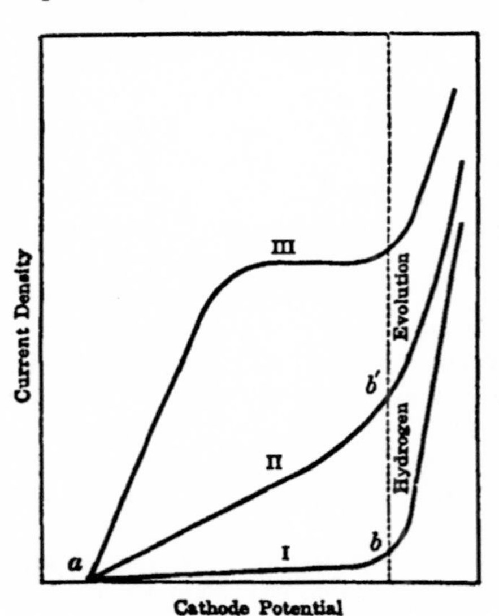

FIG. 123. Cathode potentials in electrolytic reduction

If the reduction process is of the type represented by curve II, it is possible by regulating the current density to maintain a fairly definite

[4] Cf. Leslie and Butler, *Trans. Faraday Soc.*, 32, 989 (1936).

cathode potential. Working in this manner, Haber (1898) found that with a platinum cathode in an alkaline alcoholic solution of nitrobenzene a C.D. of 0.001 amp. per sq. cm. gave a cathodic potential of − 0.6 volt; under these conditions the chief product of electrolytic reduction was azoxybenzene. When the C.D. was increased to 0.0035 amp. per sq. cm., the potential rose (cathodically) to − 1.0 volt and hydrazobenzene was produced. It appears, therefore, since hydrazobenzene represents a more highly reduced state than azoxybenzene, that the higher the cathodic potential the greater the reducing properties of the electrode.

Interesting results were obtained by Löb and Moore (1904) in connection with the electrolytic reduction of nitrobenzene in an alkaline medium. They employed different cathode materials, but adjusted the current density so that the potential was the same, viz., − 1.5 volts, in each case. The main products were azoxybenzene and aniline, and the current efficiencies for these two substances are given in Table LXXXIV:[5]

TABLE LXXXIV. ELECTROLYTIC REDUCTION OF NITROBENZENE AT CONSTANT POTENTIAL

	Current Efficiency	
Cathode	Azoxybenzene	Aniline
Platinum	58.3 per cent	36.9 per cent
Copper	59.7	33.2
Tin	62.6	28.0
Zinc	49.3	40.8
Lead	60.4	30.0
Nickel	62.1	34.9

apart from the results with the zinc cathode, the products at constant potential are almost independent of the nature of the cathode. It must be admitted that this particular instance is somewhat exceptional in the close similarity of the results for the different cathode materials; the latter often exert a catalytic influence, as explained below.

II. Nature of Electrode Material.—When a substance is difficult to reduce electrolytically, e.g., carbonyl compounds and pyridine derivatives, success can often be achieved by the use of a zinc, lead or mercury cathode. These metals all have high hydrogen overvoltages, and it is to this fact that the reducing efficiency is attributed. If the electrodes contain impurities of low overvoltage, or if the electrolyte contains traces of salts of metals with low overvoltages which can be deposited on the cathode, the electrolytic reduction may be entirely inhibited. Similarly, in the reduction of nitro-compounds the use of a zinc, lead, tin or mercury cathode generally results in a good yield of amine; at low overvoltage cathodes, e.g., nickel, platinum or carbon, intermediate products, such as phenylhydroxylamine and benzidine, are obtained.

Since high overvoltage cathodes permit the attainment of a higher (cathodic) potential before hydrogen evolution commences, there is clearly the possibility of a greater degree of reduction, but the actual

[5] Löb and Moore, *Z. physik. Chem.*, 47, 418 (1904).

reason for the increased efficiency is not clear. When the view was widely accepted that overvoltage was due to the slowness of the combination of hydrogen atoms, the reducing properties of high overvoltage cathodes were attributed to the appreciable concentration of atomic hydrogen. Since this theory is now believed to apply only to metals of low overvoltage, the interpretation of electrolytic reduction must be revised. One possibility is that since high overvoltage metals form weak M—H bonds, where M represents the electrode material, the atoms are more readily detached and can consequently react more rapidly with the depolarizer than would be the case if strong M—H bonds were formed. Another possibility is that since the protons have to pass over a high energy barrier before they can reach the electrode in the case of a high overvoltage metal, they will have a higher potential energy, and consequently be more reactive, than if the barrier were a low one, as it is for low overvoltage cathodes. It is also possible that the reduction process may involve the transfer of a proton to the depolarizer adsorbed on the cathode; this represents an alternative to the transfer of the proton to the electrode resulting in hydrogen evolution. If the latter reaction takes place at an appreciable rate at low potentials the former may not be able to occur; on the other hand, if the overvoltage for hydrogen liberation is high, the cathode potential may rise sufficiently for transfer of the proton to the depolarizer to be possible before hydrogen evolution sets in.

Although the overvoltage of the cathode material is important in controlling electrolytic reduction, it is not always the deciding factor. For example, the results in Table LXXXV were obtained in the electrolytic reduction of nitric acid; the current efficiencies for the formation of hydroxylamine and ammonia are recorded.[6] Of the materials employed, spongy copper has the lowest overvoltage, but it gives the greatest reduction efficiency; on the other hand, the least effective cathode, i.e., amalgamated lead, has the highest overvoltage. It is evident that in this instance copper has a catalytic effect in facilitating the reduction of nitric

TABLE LXXXV. CURRENT EFFICIENCIES IN THE ELECTROLYTIC REDUCTION OF NITRIC ACID

Cathode	Condition	Current Efficiency	
		Hydroxylamine	Ammonia
Lead	Amalgamated	70 per cent	17 per cent
Lead	Roughened	24	61
Zinc	Smooth	43	40
Copper	Smooth	11	77
Copper	Spongy	1.5	93

[6] Tafel, *Z. anorg. Chem.*, 31, 289 (1902).

acid to ammonia. It is known that hydroxylamine cannot be reduced electrolytically to ammonia at a copper cathode in acid solution; it follows, therefore, that the two products are obtained from nitric acid by alternative process, the one leading to ammonia being catalyzed by the copper. An analogous example of catalytic deviation of an electrolytic process by the electrode material is found in connection with the reduction of ketones, RCOR. A cadmium cathode favors the formation of the hydrocarbon RCH_2R, although at lead and mercury cathodes, which have larger overvoltages, less highly reduced products, $R_2C(OH)C(OH)R_2$ and $R_2CH(OH)$, are obtained. The hydrocarbon is not a subsequent reduction product of these hydroxylic compounds, and hence the cadmium catalyzes an alternative reaction leading to a completely reduced compound.

Another aspect of the catalytic influence of the cathode is provided by the fact that the reduction of unsaturated to saturated fatty acids does not occur appreciably at smooth platinum, copper or lead cathodes; it does, however, take place readily at electrodes covered with finely divided platinum or nickel. It is probable that the electrolytic reduction of a double bond involves a purely chemical heterogeneous reaction between the reducible substance and molecular hydrogen, produced by electrolysis, with the electrode material as the catalyst.

III. Concentration of Depolarizer.—As a general rule, the concentration of the depolarizer does not affect the nature of the reduction product although it may affect the current efficiency for reduction. By increasing the concentration of the depolarizer or by agitating the electrolyte the rate of the cathodic reaction is increased and it is possible to increase the C.D. without the danger of hydrogen evolution. Any factor which facilitates access of the depolarizer to the cathode increases the slope of the current density-cathode potential curve in Fig. 123, e.g., from II to III in an extreme case.

Another aspect of the advantage of bringing the depolarizer to the cathode is seen by comparing the results in Table LXXXV for smooth and spongy copper electrodes. The much larger effective area of the latter permits more intimate contact of the nitric acid with the cathode. It may be noted in this connection that in his extensive work on the electrolytic reduction of organic compounds, Tafel (1900) frequently used a "prepared" lead electrode, which had been roughened by electrolytic oxidation of the surface to lead dioxide followed by reduction to finely-divided lead by cathodic treatment.

IV. Temperature.—Increase of temperature has three main effects: (*a*) it lowers overvoltage; (*b*) it increases the rate of reaction between the depolarizer and hydrogen; and (*c*) it increases the rate of diffusion of the depolarizer to the cathode. The actual influence on the electrolytic reduction is a balance of the effects of these three factors. If the reduction process does not require too high an overvoltage, an increase of temperature would be expected to improve the reduction efficiency because of the second and third factors; if, on the other hand, the process necessitates a high cathodic potential, an increase of temperature, because it lowers the overvoltage, will be accompanied by a decreased efficiency. If an intermediate stage of reduction, at a lower cathode potential, is possible, it will generally be favored as the temperature is raised.

V. Catalysts.—The addition of certain substances to the electrolyte often produces an increased reduction efficiency; such catalysts are of two main types. The first consists of salts of high overvoltage metals, e.g., zinc, tin or mercury; during the course of the electrolysis the metals are deposited on the cathode, thus raising its overvoltage. The second type are ions capable of existing in two stages of oxidation, e.g., titanium, vanadium, chromium, iron and cerium; these substances are sometimes called **hydrogen carriers,** and they act in the following manner. The higher valence stage of the carrier, e.g., Ti^{++++}, is reduced at the cathode to the lower stage, e.g., Ti^{+++}; the latter, being a powerful reducing agent, reacts with the substance present in the solution and while doing so is re-oxidized to its original state. The resulting ions are once more reduced cathodically, and the process goes on continuously, quite a small amount of the carrier being sufficient for the purpose.

The nature of the electrolyte sometimes has an important influence on the products of electrolytic reduction. The alkalinity or acidity, for example, plays an essential part in determining the nature of the substance obtained in the reduction of nitrobenzene: in this case the effect is mainly due to the influence of the hydrogen ion concentration on various possible side reactions. The formation of azoxybenzene, for example, in an alkaline electrolyte is due to the reaction between phenylhydroxylamine and nitrosobenzene, viz.,

$$C_6H_5NHOH + C_6H_5NO = C_6H_5N : NO \cdot C_6H_5 + H_2O,$$

which is catalyzed by hydroxyl ions. In other cases, however, the cause of the difference in the products obtained in acid and alkaline solutions is not clear; for example, the reduction of aromatic aldehydes in an alkaline electrolyte at a lead cathode yields mainly the corresponding hydrobenzoin, but in acid solution the primary alcohol is formed in addition. It is possible that differences in overvoltage or cathode potential may be of importance in these instances.

Electrolytic Oxidation: I. Electrode Potential.—A series of stable potentials is difficult to obtain at an anode in the presence of a depolarizer; the potential generally rises rapidly from the low value, at which the anode dissolves, to the high value for passivity and oxygen evolution. Since a platinum electrode is nearly always passive, however, it is possible to obtain graded potentials to a limited extent; the data quoted in Table LXXXVI were recorded for the oxidation of an acid solution of ethyl alcohol at a platinum anode at different potentials.[7] It is seen that the proportion of acetic acid in the product increases as the anode potential is raised; by suitable regulation of the potential it is possible to

TABLE LXXXVI. OXIDATION OF ETHYL ALCOHOL AT A PLATINUM ANODE

Current Density amp. per sq. cm.	Anode Potential	Current Efficiency	
		Acetaldehyde	Acetic Acid
0.005–0.015	0.37–0.9 v.	~100 per cent	~0 per cent
0.025	~0.95	85	15
0.06	1.5–1.7	39	61

obtain either acetaldehyde or acetic acid almost exclusively. There are, however, very few instances of this type of behavior in electrolytic oxidation.

II. Nature of Electrode Material.—The current efficiency obtained in the anodic oxidation of (*a*) N sodium formate in 2 N potassium hydroxide at 0.033 amp. per sq. cm.; (*b*) 2 N potassium oxalate in N potassium hydroxide at 0.033 amp. per sq. cm.; and (*c*) N potassium nitrite in N potassium hydroxide at 0.02 amp. per sq. cm., with a number of different electrode materials are quoted in Table LXXXVII;[8] the anodes

TABLE LXXXVII. OXIDATION EFFICIENCIES AT VARIOUS ANODES

Anode	Formate	Oxalate	Nitrite
Platinum	85 per cent	93 per cent	11 per cent
Iridium	86	88	97
Palladium	97	0	96
Iron	18	0	4
Nickel	45	—	4

are given in the order of decreasing overvoltage It is seen that while there is a rough parallelism between the extent of oxidation and the overvoltage of the anode, in no case is the correspondence complete; the difference in behavior of iron and palladium, which have the same overvoltage, in connection with the oxidation of nitrite is very marked, as also is that for the oxidation of nitrite at platinum and iridium anodes.

For reasons which will be evident shortly, it is difficult to lay down definite rules concerning the efficiency of an electrode for anodic oxidation. If the oxidation is brought about by oxygen in an active form or by an oxide of the electrode material, e.g., conversion of chromic to chromate ions and the oxidation of iodate to periodate, the highest effi-

[7] Dony-Hénault, *Z. Elektrochem.*, 6, 533 (1900); Marie and Lejeune, *Compt. rend.*, 187, 343 (1928).
[8] Foerster, "Elektrochemie wässeriger Lösungen," 1922, p. 312.

ciencies are obtained with lead dioxide and platinized platinum anodes. In other cases, such as the oxidation of methyl alcohol and other substances (cf. p. 516), these electrodes give the lowest efficiencies; the best results are then obtained with smooth platinum.

III. Concentration of Depolarizer. IV. Temperature.—The influence of these factors on electrolytic oxidation is similar to their effect in reduction processes.

V. Catalysts.—Electrolytic oxidation reactions are often facilitated by the presence of catalysts capable of existing in two valence stages; examples of such **oxygen carriers** are cerium, chromium, manganese and vanadium ions. Their action is probably similar to that previously described for hydrogen carriers. These catalysts have been used to facilitate the oxidation of toluene to benzaldehyde or benzoic acid, of toluene sulfonamide to saccharin, and of anthracene to anthraquinone.

Small quantities of fluorides often have a beneficial effect in electrolytic oxidation, e.g., of iodate to periodate and of sulfate to persulfate; the part played by the fluoride is not clear, although in most instances its presence is accompanied by an increase of anode potential. Perchlorates have sometimes been added to solutions to improve the oxidation efficiency; they have an effect similar to fluorides.

Polymerization of Anions.—The conversion of thiosulfate to tetrathionate, i.e.,

$$2S_2O_3^{--} = S_4O_6^{--} + 2\epsilon,$$

of sulfite to dithionate, i.e.,

$$2SO_3^{--} = S_2O_6^{--} + 2\epsilon,$$

and of sulfate to persulfate, i.e.,

$$2SO_4^{--} = S_2O_8^{--} + 2\epsilon,$$

are reactions of a similar type which can all be performed anodically: in each instance two ions of reactant combine to form a new anion carrying the same charge as one of the original ions. It was thought at one time that these processes were reversible in the thermodynamic sense, but it is doubtful whether this is actually the case.

Considerable light has been thrown on the mechanism of these ion-doubling processes as a result of a comprehensive study of the oxidation of thiosulfate and of sulfite under a variety of conditions.[9] In the oxidation of thiosulfate, for example, at a smooth platinum anode, two stages of potential were observed, as shown in Fig. 124; the curves in this figure represent the variation of anode potential with time during the passage of current at various strengths* through a 0.025 M solution of thiosulfate buffered at pH 7. It is seen that in the initial stages of the process the

[9] Glasstone and Hickling, *J. Chem. Soc.*, 2345, 2800 (1932); 829 (1933); for review, see *Chem. Revs.*, **25**, 407 (1939).

* The area of the anode was about 50 sq. cm.

potential has a relatively low value but a rise occurs after a certain amount of electricity has been passed through the solution, the quantity being greater the smaller the current. In spite of this change of potential, the over-all current efficiency for the formation of tetrathionate is

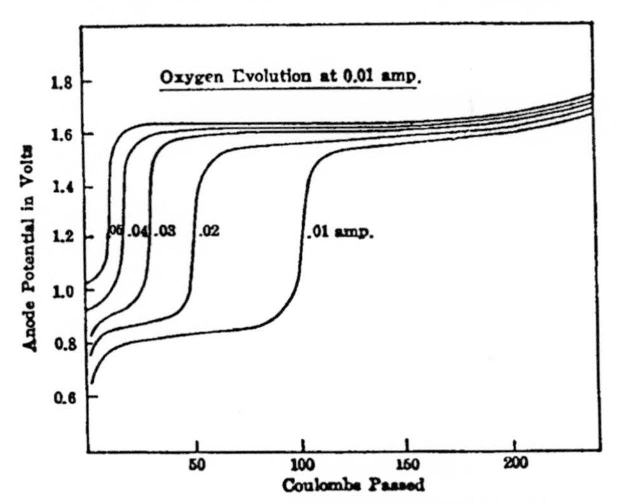

Fig. 124. Variation of potential in anodic oxidation of thiosulfate (Glasstone and Hickling)

not markedly affected by current density, as shown by the results in Table LXXXVIII; the oxidation at a c.d. of 1.0×10^{-3} amp./cm.^{2} was carried out almost entirely at the higher potential while that at 0.2×10^{-3} amp./cm.^{2} was at the lower potential for nearly half the time, as

Table LXXXVIII. Current Efficiency for the Oxidation of 0.025 m Sodium Thiosulfate

Current density $\times 10^3$	1.0	0.8	0.6	0.4	0.2 amp./cm.^{2}
Thiosulfate oxidized	85	85	85	88	90 per cent
Tetrathionate formed	76	76	76	76	79 per cent

seen in Fig. 124. Further, experiments on the variation of the current efficiency during the course of the reaction at the latter c.d. showed no appreciable change, in spite of a rise of 0.7 volt in the anode potential. Among other facts which were elucidated in the course of this work, it was found that the presence of catalysts for the decomposition of hydrogen peroxide, e.g., ferrous, copper, cobalt and manganese salts, in particular, markedly decreased the efficiency of the formation of tetrathionate. The addition of 0.001 m manganous sulfate to 0.025 m thiosulfate reduced the efficiency from 85 per cent to zero; the deposit of manganese dioxide formed on the anode is an active catalyst for the decomposition of hydrogen peroxide.

It is evident from the results just given that the oxidation of $S_2O_3^{--}$ to $S_4O_6^{--}$ cannot be the simple process represented above, involving merely a transfer of two electrons. The reaction must clearly take place in stages, the slow stage, which is responsible for the potential, being different from that in which the oxidation actually occurs; the former changes during the course of time, but the latter remains unchanged. It is known that various oxides are formed on a platinum anode, and hence it is probable that in the lower potential stage the slow process is that involved in the discharge of hydroxyl ions, in some manner, on one oxide of platinum, whereas in the upper stage the ions are discharged on another, higher, oxide. The higher potential is very close to that required for the normal liberation of oxygen at a platinum anode; hence, the electrode surface during the oxidation of thiosulfate is then virtually in the same condition as when oxygen evolution occurs.

In view of the influence of catalysts for hydrogen peroxide on the anodic process, it has been suggested that this substance is formed by the union of two hydroxyl radicals at the anode and that it acts as the effective oxidizing agent. The destruction of the peroxide formed in this manner would then inhibit the formation of tetrathionate from thiosulfate at the anode. It appears to be equally possible that the active oxidizing agent is the hydroxyl radical, for any reaction in which it is involved would undoubtedly take place rapidly. To explain the influence of catalysts it would be necessary to postulate that the catalytic decomposition of hydrogen peroxide involves the formation of free, or almost free, hydroxyl radicals: this is, of course, not improbable. A possible mechanism for the anodic oxidation of thiosulfate might then involve the following stages: first, discharge of a hydroxyl ion from water

$$H_2O = OH + H^+ + \epsilon,$$

which is the slow process (cf. p. 478) followed by

$$2OH = H_2O_2$$

and

$$2S_2O_3^{--} + H_2O_2 = S_4O_6^{--} + 2OH^-,$$

or, possibly, by

$$S_2O_3^{--} + OH = S_2O_3^- + OH^-$$

and

$$2S_2O_3^- = S_4O_6^{--}.$$

In the presence of catalysts for the decomposition of hydrogen peroxide, either the peroxide itself or the hydroxyl radicals are removed in an alternative reaction leading to the formation of molecular oxygen; the efficiency for the formation of tetrathionate would then be decreased. Some investigators have suggested the possibility that the $S_2O_3^{--}$ ions are discharged directly at the anode to form $S_2O_3^-$, and the latter then combine in pairs; this view does not account, however, for the influence of manganous, ferrous, and copper ions, etc., nor for the two potential stages.

The phenomena observed in the anodic oxidation of sulfite to dithionate are essentially similar to those described above for thiosulfate; in the former case, however, there is always a considerable formation of sulfate, and when the conditions are such that the production of dithionate is suppressed, the amount of sulfate increases. Because of the strong reducing properties of sulfite, the oxidation of the platinum anode occurs less readily than in thiosulfate solution and hence larger currents are necessary for the second stage of potential to set in.

Another anodic polymerization process which has been the subject of much study, partly because of its industrial importance, is the oxidation of sulfuric acid to persulfuric acid. The efficiency of this process has been found to run almost parallel with the proportion of HSO_4^- ions present in solutions of different concentration. This fact has been regarded as supporting the view that the first stage in the anodic process is the discharge of HSO_4^- ions, thus

$$HSO_4^- = HSO_4 + \epsilon,$$

followed by the combination in pairs of the resulting radicals,

$$2HSO_4 = H_2S_2O_8.$$

The argument is not convincing, however, because higher efficiencies of persulfate formation can be obtained in ammonium sulfate solution than either in one of sulfuric acid alone or in a mixture of these two substances. The data in Table LXXXIX are of interest in this connection:[10] they

TABLE LXXXIX. EFFICIENCY OF PERSULFATE FORMATION IN MIXTURES OF AMMONIUM SULFATE AND SULFURIC ACID

Ratio of $(NH_4)_2SO_4$ to H_2SO_4	Current Density in Amp. per Sq. Cm.		
	0.1	0.5	1.0
H_2SO_4 alone	1.7 per cent	15.3 per cent	28.5 per cent
1 : 4	5.4	42.4	48.5
1 : 1.5	16.2	55.3	59.6
1.5 : 1	28.4	68.7	69.3
4 · 1	47.3	75.3	79.0
$(NH_4)_2SO_4$ alone	53.1	81.7	84.9

give the efficiency of persulfate formation at a platinum anode in a series of solutions containing different proportions of ammonium sulfate and sulfuric acid, the total sulfate radical concentration being 400 g. per liter in each case. The HSO_4^- ion concentration is probably a maximum when the solution contains approximately equivalent amounts of the two

[10] Elbs and Schönherr, *Z. Elektrochem.*, 2, 245 (1895).

constituents, but the efficiency of persulfate formation shows no such maximum.

Although the conversion of SO_4^{--} to $S_2O_8^{--}$ is undoubtedly a much slower process than the corresponding reactions involving $S_2O_3^{--}$ and SO_3^{--}, it is probable that the fundamental mechanism is the same. Because of experimental difficulties the former of these processes has not been subjected to the same thorough examination as the others, but there is reason for believing that catalysts for hydrogen peroxide decomposition also reduce the efficiency in this case. Since persulfuric acid can be obtained by the action of concentrated hydrogen peroxide on sulfuric acid, and persulfate is produced by the oxidation of sulfate with fluorine, it is not impossible that the anodic oxidizing agent is either hydrogen peroxide or hydroxyl radicals.

Oxidation of Fatty Acids: The Kolbe Reaction.—The phenomena observed in the electrolysis of acetic acid and of acetates have a number of features in common with those found in the reactions just discussed. If a solution of an alkali acetate, alone or mixed with acetic acid, is electrolyzed, the over-all reaction,

$$2CH_3CO_2^- = C_2H_6 + 2CO_2 + 2\epsilon,$$

takes place at a platinum anode; the products, ethane and carbon dioxide, are obtained with a high efficiency. This process, called the **Kolbe reaction,** after the name of its discoverer (Kolbe, 1849), has been very fully investigated, but its interpretation is still a matter for controversy.[11] The reaction can be applied to the salts of most aliphatic acids and by its use paraffin hydrocarbons up to $C_{34}H_{70}$ have been obtained; aromatic acids, however, do not generally undergo the Kolbe reaction. One of the most interesting facts about the process is that although it occurs readily at a smooth platinum anode in aqueous solution, the efficiency for the formation of ethane is very small at platinized platinum, palladium, gold, nickel and iron anodes. If a platinized electrode is poisoned with mercury, however, the Kolbe synthesis takes place to an appreciable extent. The presence of catalysts for hydrogen peroxide decomposition, either those which give deposits on the anode, e.g., manganese or lead salts, or those which do not, e.g., copper and iron salts, brings about a decrease in the formation of ethane when an aqueous solution of an acetate is electrolyzed. The product in these cases consists largely of methyl alcohol, which may be isolated under certain conditions.

If neutral salts are added to the electrolyte, especially if this consists of acetic acid only, with no acetate, the extent of the Kolbe reaction is diminished; under these conditions also, methyl alcohol is formed by the so-called Hofer-Moest reaction (1902). When the conditions are such that the acetate is oxidized to ethane the anode potential is about 2.2

[11] Glasstone and Hickling, *J. Chem. Soc.*, 1878 (1934); for complete review, see "Electrolytic Oxidation and Reduction," 1935, Chap. VIII.

volts, but when this process is suppressed in favor of methyl alcohol formation the potential is lower, viz., about 1.8 volts at pH 5, and corresponds approximately to that for oxygen evolution in the given solution.

Several theories have been proposed to account for the Kolbe electrosynthesis: one, which has been widely held, is that the acetate ions are discharged directly at the anode and the resulting radicals then react in pairs, thus

$$CH_3CO_2^- = CH_3CO_2 + \epsilon,$$

followed by

$$\begin{matrix} CH_3CO_2 \\ + \\ CH_3CO_2 \end{matrix} = \begin{matrix} CH_3 \\ | \\ CH_3 \end{matrix} + 2CO_2.$$

The main argument for this theory is that the formation of ethane does not commence until a definite potential, regarded as the discharge potential of the $CH_3CO_2^-$ ions, is attained. It does not account, however, for the marked influence of electrode material on the Kolbe reaction, nor is it easy to understand why oxygen evolution does not take place in preference, since the latter requires a lower potential. The suppression of the formation of ethane and its replacement by the Hofer-Moest reaction provides the theory with a further difficulty. Another suggestion that has been made is that active anodic oxygen oxidizes the acetate ions or acetic acid to acetyl peroxide, $(CH_3CO_2)_2$, which then decomposes spontaneously to yield ethane and carbon dioxide. The products of the thermal decomposition of acetyl and other peroxides, prepared chemically, have been shown to be similar to those obtained in the Kolbe electrosynthesis. The effect of anode material, of catalysts for hydrogen peroxide decomposition and of added salts cannot, however, be readily explained.

A point of view which brings together some of the essential features of both the foregoing theories,[12] and at the same time accounts for many of the facts which these theories fail to explain, is similar to that previously described in connection with the anodic oxidation of thiosulfate. It is suggested that hydroxyl radicals, or hydrogen peroxide formed by their combination in pairs, react with the acetate ions to form acetate radicals, which then combine, possibly with the intermediate formation of acetyl peroxide, to yield ethane and carbon dioxide; thus,

$$CH_3CO_2^- + OH = CH_3CO_2 + OH^-$$

or

$$2CH_3CO_2^- + H_2O_2 = 2CH_3CO_2 + 2OH^-,$$

followed, possibly, by

$$\begin{matrix} CH_3CO_2 \\ + \\ CH_3CO_2 \end{matrix} = \begin{matrix} CH_3COO \\ | \\ CH_3COO, \end{matrix}$$

[12] Glasstone and Hickling, *Trans. Electrochem. Soc.*, 75, 333 (1939).

and then by

$$\begin{array}{l} CH_3COO \\ \quad\quad\;\, | \\ CH_3COO \end{array} = \begin{array}{l} CH_3 \\ | \\ CH_3 \end{array} + 2CO_2.$$

It is possible that some acetate radicals are formed by the direct discharge of the ions as, it will be seen shortly, is the case in non-aqueous solutions; but an additional mechanism must be introduced, such as the one proposed above, to account for the influence of electrode material, catalysts for hydrogen peroxide decomposition, etc. It is significant that the anodes at which there is no Kolbe reaction consist of substances that are either themselves catalysts, or which become oxidized to compounds that are catalysts, for hydrogen peroxide decomposition. By diverting the hydroxyl radicals or the peroxide into an alternative path, viz., oxygen evolution, the efficiency of ethane formation is diminished. Under these conditions, as well as when access of acetate ions to the anode is prevented by the presence of foreign anions, the reactions mentioned above presumably do not occur, but instead peracetic acid is probably formed, thus,

$$CH_3CO_2^- + 2OH = CH_3CO_2OH + OH^-$$

or

$$CH_3CO_2^- + H_2O_2 = CH_3CO_2OH + OH^-,$$

or even

$$CH_3CO_2H + O = CH_3CO_2OH,$$

followed by its decomposition to give methyl alcohol, thus

$$CH_3CO_2OH = CH_3OH + CO_2.$$

It may be recorded that both ethane and methyl alcohol can be obtained by the chemical oxidation of acetate solutions under suitable conditions.

One of the difficulties of any theory of the Kolbe reaction is to account for the fact that although the potential at which it occurs is so high, oxygen evolution does not take place preferentially. In view of the suggestion made on page 478 that the normal discharge of a hydroxyl ion leading to the anodic evolution of oxygen results from the transfer of a proton from a molecule of water adsorbed on the electrode surface, a possibility is that the water molecules have been largely replaced by acetate ions. The discharge of hydroxyl ions to form oxygen is thus rendered difficult and hence it does not occur until the potential is high enough for the discharge of acetate ions, and their detachment from the anode, to become possible. If the acetate ions are prevented from reaching the electrode, by introducing neutral salts, their adsorption is relatively small and the discharge of hydroxyl ions at the normal potential occurs. The presence of catalysts for hydrogen peroxide decomposition, since they facilitate the formation of oxygen, presumably favors the discharge of hydroxyl ions. The low potentials observed under these two latter conditions can thus be explained.

In non-aqueous solutions the Kolbe electrosynthesis takes place with high efficiency at platinized platinum and gold, as well as at smooth platinum, anodes; increase of temperature and the presence of catalysts for hydrogen peroxide decomposition, both of which have a harmful effect in aqueous solution, have relatively little influence. The mechanism of the reaction is apparently quite different in non-aqueous solutions and aqueous solutions: in the former no hydroxyl ions are present, and so neither hydroxyl radicals nor hydrogen peroxide can be formed. It is probable, therefore, that direct discharge of acetate ions occurs at a potential which is almost independent of the nature of the electrode material in a given solvent. The resulting radicals probably combine in pairs, as in aqueous solution, to form acetyl peroxide, which subsequently decomposes as already described.[13]

The Brown-Walker Electrosynthesis.—The salts of normal dicarboxylic acids do not undergo an oxidation similar to the Kolbe reaction, but alkali metal salts of the semi-esters, viz., $CO_2Et(CH_2)_nCO_2K$, do, however, give a reaction of the same type: this process was discovered by Crum Brown and Walker (1891) and is generally referred to as the **Brown-Walker reaction.** The over-all anodic process may be represented by

$$2\begin{vmatrix} (CH_2)_nCO_2Et \\ CO_2^- \end{vmatrix} = \begin{vmatrix} (CH_2)_nCO_2Et \\ (CH_2)_nCO_2Et \end{vmatrix} + 2CO_2 + 2\epsilon,$$

and the resulting product is the ester of a dibasic acid with double the number of CH_2 groups in the original acid. The Brown-Walker synthesis has been used to prepare dibasic acids containing up to $32CH_2$ groups. The essential phenomena of the Brown-Walker reaction as regards the influence of electrode material, presence of catalysts for hydrogen peroxide decomposition, solvent, temperature, neutral salts, etc., are so similar to those occurring in the Kolbe synthesis that there is little doubt the two processes have the same fundamental mechanism.[14]

Formation of Chromate and Periodate.—In view of the cases just considered, in which hydrogen peroxide catalysts have a harmful influence, it is of interest to record anodic reactions in which the reverse is

TABLE XC. ANODIC OXIDATION EFFICIENCIES

Electrode	Chromic to Chromate	Iodate to Periodate
Smooth platinum	1 per cent	11 per cent
Platinized platinum	53	14
Lead dioxide	100	38

generally true. The results in Table XC,[15] for example, give the current efficiencies at smooth and platinized platinum and lead dioxide anodes

[13] Glasstone and Hickling, *J. Chem. Soc.*, 820 (1936).
[14] Hickling and Westwood, *J. Chem. Soc.*, 1039 (1938); 1109 (1939).
[15] Gross and Hickling, *J. Chem. Soc.*, 325 (1937); Hickling and Richards, *ibid.*, 256 (1940).

for the oxidation of chromic to chromate ions in acid solution (0.1 M chromium potassium sulfate in 0.5 N sulfuric acid; C.D., 0.01 amp. per sq. cm.; 20°) and of iodate to periodate in alkaline solution (0.1 M potassium iodate in N potassium hydroxide; C.D., 0.25 amp. per sq. cm.; 18°). In each case the oxidation is more effective at platinized platinum and lead dioxide electrodes than at smooth platinum; this is the reverse of the type of behavior observed in the oxidation of thiosulfate to tetrathionate, sulfite to dithionate, acetate to ethane, etc. Further, the efficiency for the anodic oxidation of chromic ions and of iodate is greater in alkaline than in acid solution and is, in general, increased by raising the temperature; both these factors decrease the efficiency in the reactions previously considered.

As a result of careful investigations it has been concluded that the effective oxidizing agent for the anodic oxidation of chromic and iodate ions is not hydroxyl radicals or hydrogen peroxide, but an oxide of the metal employed as the anode material. It has been mentioned in Chap. XIII that there is evidence for the formation of higher oxides on metallic anodes during the course of oxygen evolution, and it is probably these oxides which oxidize chromic ions to chromate and iodate ions to periodate. Since chromate and periodate are reduced by hydrogen peroxide to reform chromic and iodate ions, respectively, it is to be expected that conditions which favor the decomposition of hydrogen peroxide or, in general, which facilitate the formation of molecular oxygen from hydroxyl radicals, would increase the anodic efficiency for the formation of chromate and periodate. The fact that increases of temperature and alkalinity increase the efficiency for these processes is, therefore, in agreement with expectation. Platinized platinum is a better catalyst than smooth platinum for the decomposition of hydrogen peroxide, and the results in Table XC show that an anode of the former material is more effective than one of the latter for the reactions under consideration. Lead dioxide is also an excellent catalyst for hydrogen peroxide decomposition, and the higher oxides of lead are very effective in bringing about the chemical oxidation of chromic and iodate ions to chromate and periodate, respectively: it is easy to understand, therefore, the high efficiencies obtained in the electrolytic oxidations with a lead dioxide anode.

CHAPTER XVI

ELECTROKINETIC PHENOMENA

Electrokinetic Effects.—The relative movement, with respect to one another, of a solid and a liquid is accompanied by certain electrical phenomena which are referred to as **electrokinetic effects.**[1] These phenomena are ascribed to the presence of a potential difference at the interface between any two phases at which movement occurs; this potential is known as the **electrokinetic potential** or, frequently, as the **zeta-potential,** from the letter (ζ) of the Greek alphabet by which it is usually represented. If the potential may be supposed to result from the existence of electrically charged layers of opposite sign at the surface of separation between a solid and a liquid, then the application of an electrical field must result in the displacement of one layer with respect to the other. If the solid phase is fixed, e.g., in the form of a diaphragm, while the liquid is free to move, the liquid will tend to flow through the pores of the diaphragm as a consequence of the applied field. The direction in which the liquid flows should depend on the sign of the charge it carries with respect to that of the solid. This movement of a liquid through the pores of a diaphragm under the influence of an E.M.F. constitutes the phenomenon of **electro-endosmosis, or electro-osmosis,** discovered by Reuss in 1809.

If two portions of water are separated by a diaphragm, consisting of a porous non-conducting material, and metal electrodes connected to a source of E.M.F. are placed on each side of the diaphragm, a flow of water will be observed to occur. If the water, or other liquid, is initially at the same level on both sides of the separating membrane, the level will rise on one side and fall on the other, as a result of the application of the electrical field. When the difference of pressure causes the liquid to flow in one direction at the same rate as it passes in the opposite direction as a result of electro-osmosis, a stationary state will be reached; the pressure difference in this state depends on the experimental conditions. If, on the other hand, the liquid is maintained at the same level on both sides of the diaphragm, by means of an overflow tube, there will be a continuous flow of liquid as long as the E.M.F. is applied. A porous diaphragm is actually a mass of fine capillaries, and electro-osmosis is found to occur also in capillary tubes, singly or in groups.

If the solid phase consists of small particles suspended in the liquid, the displacement of one charged layer with respect to the other, conse-

[1] For fuller surveys of the subject, see Abramson, "Elektrokinetic Phenomena and Their Application to Biology and Medicine," 1934; Butler, "Electrocapillarity," 1940.

quent upon the application of the electrical field, now results in the movement of the solid through the liquid; this effect was at one time referred to as **cataphoresis**, but it is now generally called **electrophoresis.** Electrophoretic motion under the influence of an applied E.M.F. has been observed with proteins and other colloidal particles of 5×10^{-6} cm., or less, in diameter, as well as with quartz particles, oil drops and air bubbles. Studies of electrophoresis have proved of great importance, as will be seen later, in the examination of biological fluids, as well as in many aspects of colloid chemistry.

In the two electrokinetic effects just described, the application of an electric field results in the relative movement of the two phases; if, on the other hand, the relative motion is brought about mechanically, the displacement of the charged layers with respect to each other should result in the production of a difference of potential between any two points in the direction of motion. This potential, known as the **streaming potential,** was observed by Quincke (1859) when a liquid was forced through a porous material, e.g., a clay diaphragm, or a capillary tube. The streaming potential may thus be regarded as the reversal of electro-osmosis. In a similar manner, the reversal of electrophoresis results in the so-called **Dorn effect,** or **sedimentation potential;** this potential, first studied by Dorn (1880), arises when small particles are allowed to fall through water under the influence of gravity. A difference of potential is observed between two electrodes placed at different levels in the stream of falling particles.

The Electrical Double Layer and its Structure.—The theoretical treatment of electrokinetic phenomena is based on the concept of the existence of an **electrical double layer** at the solid-liquid boundary. Helmholtz (1879) considered this double layer to consist of two oppositely charged layers at a fixed distance apart, so that it could be regarded as equivalent to an electrical condenser of constant capacity, with parallel plates separated by a distance of the order of a molecular diameter. The original mathematical derivations of Helmholtz are somewhat complicated, and a simpler treatment, based on that of Perrin (1904), is given below. If the electrical double layer at the interface of motion between a solid and a liquid is regarded as a condenser with parallel plates d cm. apart, each carrying a charge σ per sq. cm., then according to electrostatics

$$\zeta = \frac{4\pi\sigma d}{D}, \qquad (1)$$

where ζ is the difference of potential between the plates, which in this case is the electrokinetic (or zeta-) potential, and D is the dielectric constant of the medium. This is the fundamental equation which has been widely employed for the quantitative treatment of electrokinetic phenomena.

The conception of Helmholtz, of the double layer involving a sharp potential gradient, was modified by Gouy[2] and by Chapman[3] who utilized the idea of a **diffuse double layer**; according to this view the solution side of the double layer is not merely one molecule, or so, in thickness but extends for some distance into the liquid phase. In this region thermal agitation permits the free movement of the ions present in the solution, but the distribution of positive and negative charges is not uniform since the electrostatic field arising from the charge on the solid will result in a preferential attraction of particles of opposite sign. The picture of the diffuse electrical double layer at the surface between a solid and a liquid is thus analogous to the Debye-Hückel concept of the oppositely charged ion-atmosphere surrounding a given ion (cf. Chap. III); in fact the mathematical treatment employed by Debye and Hückel, to calculate the thickness of the ion atmosphere and the potential within it, is similar to that used in connection with the diffuse double layer,

If a small planar area, carrying a charge density σ per sq. cm. which is supposed to be fixed to a solid surface, is considered, then a *net* charge of equal magnitude but opposite in sign will be carried by the "ion atmosphere" which extends some distance into the liquid phase. The value of the electrical density ρ per cc. at any point in the solution portion of the diffuse double layer can be derived by the method employed in Chap. III; provided $\epsilon\psi/kT$ is small in comparison with unity, the result is given by equation (8), page 82, where ψ is the average potential at the point under consideration. The electrical density and the potential can also be related by the Poisson equation; in this instance, however, since the charged surface is assumed to be planar, the potential changes only in the direction normal to the surface, and hence it is necessary to consider the variation of potential in one co-ordinate, viz., x, only. The appropriate form of the Poisson equation is then

$$\frac{\partial^2\psi}{\partial x^2} = -\frac{4\pi\rho}{D}, \tag{2}$$

and if the value of ρ, obtained as described above, is introduced, the result is

$$\frac{\partial^2\psi}{\partial x^2} = \kappa^2\psi, \tag{3}$$

where κ has the same significance as given by equation (11), page 83. Integration of equation (3) yields

$$\psi = Ae^{-\kappa x} + A'e^{\kappa x}, \tag{4}$$

where A and A' are integration constants; since ψ must become zero when the distance x from the surface is large, it follows that A' must be

[2] Gouy, *J. de physique*, **9**, 457 (1910).
[3] Chapman, *Phil. Mag.*, **25**, 475 (1913).

zero and hence equation (4) reduces to

$$\psi = Ae^{-\kappa x}. \tag{5}$$

The value of ρ is seen from equations (2) and (3) to be given by

$$\rho = -\frac{D}{4\pi}\kappa^2\psi,$$

and introduction of equation (5) yields

$$\rho = -A\frac{D\kappa^2 e^{-\kappa x}}{4\pi}. \tag{6}$$

Since the charge on the surface must be equal in magnitude, but opposite in sign, to that of the solution, it follows that

$$\sigma = -\int_a^\infty \rho dx, \tag{7}$$

the integration being carried out from a, the distance of closest approach of the ions to the surface, to infinity; combination of equations (6) and (7) then gives

$$\sigma = A\frac{D\kappa^2}{4\pi}\int_a^\infty e^{-\kappa x}dx$$

$$= A\frac{D\kappa e^{-\kappa a}}{4\pi},$$

$$\therefore \quad A = \frac{4\pi\sigma}{D\kappa}e^{\kappa a}.$$

Upon substituting this value of A in equation (5), it follows that

$$\psi = \frac{4\pi\sigma}{D\kappa}e^{\kappa(a-x)}, \tag{8}$$

and if κ is small, i.e., for a dilute solution, this becomes

$$\psi = \frac{4\pi\sigma}{D\kappa}. \tag{9}$$

The quantity κ is known to have the dimensions of a reciprocal length (p. 84), and $1/\kappa$ may be identified with the effective thickness d of the double layer; if, at the same time, the electrical potential ψ in equation (9) is replaced by the electrokinetic potential ζ, this equation becomes

$$\zeta = \frac{4\pi\sigma d}{D}, \tag{10}$$

which is identical with equation (1). With certain approximations, therefore, the diffuse double layer leads to a result of the same form as does the theory of the sharp double layer.

Stern's Theory of the Double Layer.—The variations of capacity of the double layer with the conditions, the influence of electrolytes on the zeta-potential, and other considerations led Stern[4] to propose a model for the double layer which combines the essential characteristics of the Helmholtz and the Gouy theories. According to Stern the double layer consists of two parts: one, which is approximately of a molecular diameter in thickness, is supposed to remain fixed to the surface, while the other is a diffuse layer extending for some distance into the solution. The fall of potential in the fixed layer is sharp while that in the diffuse layer is gradual, the decrease being exponential in nature, as required by equation (5). The potential gradient at the solid-liquid boundary may be represented diagrammatically by Fig. 125 in which the potentials are given by the ordinates and the distances from the surface by the abscissae. The left-hand axis represents the solid phase and the vertical dotted line indicates the extent of the fixed part of the double layer; the relative thickness of this layer is, probably, somewhat exaggerated in the diagram. If the potential of the solid is indicated by A and that of the bulk of the liquid by B, the fall of potential in between may occur in two general ways depending, to a large extent, on the nature of the ions and molecules present in the solution: these are shown by Fig. 125, I and II. In each case, AC is the sharp fall of potential in the fixed portion and CB is the gradual, approximately exponential, change in the diffuse portion of the double layer. The total fall of potential, AB, between the solid and the solution is equal to the reversible potential, in the case of a system that can behave reversibly; this is represented by the symbol E. The electrokinetic or zeta-potential however, which is involved in electro-osmosis, electrophoresis and allied phenomena, is that between the fixed and freely mobile parts of the double layer; this is the potential change from C to B, indicated by ζ in each case.

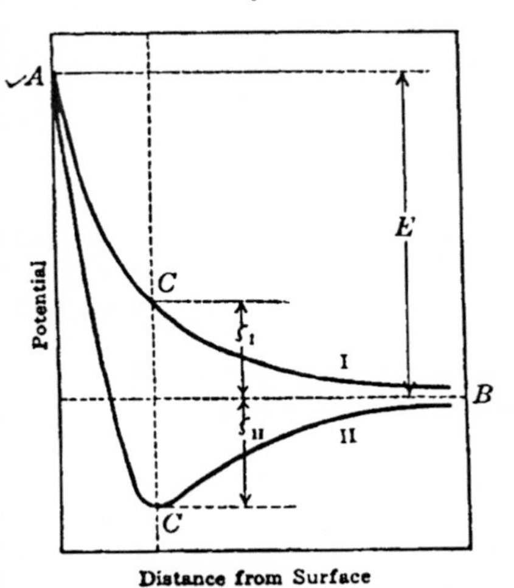

FIG. 125. The Stern double layer

[4] Stern, *Z. Elektrochem.*, **30**, 508 (1924); for a general discussion of the properties of double layers, see Mooney, *J. Phys. Chem.*, **35**, 331 (1931); Verwey, *Chem. Revs.*, **16**, 363 (1935); Mueller, *Ann. New York Acad. Sci.*, **39**, 111 (1939); Crawford, *Trans. Faraday Soc.*, **36**, 85 (1940); Kruyt and Overbeek, *ibid.*, **36**, 110 (1940); Bikerman, *ibid.*, **36**, 154 (1940).

The method given above for calculating the zeta-potential at a diffuse double layer may be applied to the diffuse portion of the Stern double layer. If σ_0 is the charge density on the solid and σ_1 and σ_2 are the corresponding values on the solution sides of the fixed and diffuse layers, respectively, then the condition of electrical neutrality requires that

$$\sigma_0 + \sigma_1 + \sigma_2 = 0,$$

$$\therefore \quad \sigma_0 + \sigma_1 = -\sigma_2.$$

The surface charge density for the diffuse double layer is thus given by $\sigma_0 + \sigma_1$, and this is the quantity which must replace σ in equations (7) to (10). In spite of the slight difference in the significance of σ, equation (10) may, therefore, still be regarded as applicable, provided the solution is dilute. The sign of σ_1 is probably always opposite to that of σ_0, and so $\sigma_0 + \sigma_1$ is numerically less than σ_0, the charge density on the surface of the solid.

It is of interest to note that the combination of fixed and diffuse double layers accounts for the fact that the reversible, or thermodynamic, potential has been found to be independent of the zeta-potential.[5] The addition of capillary active substances, such as dyestuffs, or of ions of high valence, for example, has a marked effect on the zeta-potential at a glass-liquid surface, even to the extent of bringing about a reversal of sign; the thermodynamic potential, as measured by the glass electrode (cf. p. 356), is, however, hardly affected. The added substances do not apparently affect the values of the potentials at A and B, so that the potential E remains unchanged, but the potentials between A and C, and between C and B, the latter giving the electrokinetic potential, may be altered considerably by the adsorption of ions on the solution side of the fixed double layer. It was at one time thought possible that a true null electrode, i.e., with a thermodynamic potential of zero, could be obtained by finding the conditions under which no difference of potential resulted when a metal and a solution of its ions were moved relative to one another. It is now realized, however, that this procedure merely gives a system whose electrokinetic potential is zero, but it is obviously not possible to conclude that the reversible potential is also zero.

Some writers have doubted the necessity of postulating a double layer consisting of two regions;[6] nevertheless, it has been stated that the results of measurements of the electrostatic capacity of the double layer at a mercury-electrolyte interface at various applied potentials can only be accounted for by means of Stern's theory. The variation of capacity with potential was found to consist of two straight lines joined by a non-linear portion, in accordance with the composite double layer of Stern: the Helmholtz theory leads to the expectation of one straight line whereas the Gouy-Chapman theory requires a completely non-linear curve.[7]

[5] Freundlich and Ettisch, *Z. physik. Chem.*, **116**, 401 (1925).
[6] See Urban and White, *J. Phys. Chem.*, **36**, 3157 (1932).
[7] Philpot, *Phil. Mag.*, **13**, 775 (1932).

Electro-osmosis.—When a liquid is forced by electro-osmosis through the fine capillaries constituting a porous diaphragm, or, in general, through any capillary system, the rate of flow will be determined by two opposing factors: these are the force of electro-osmosis, on the one hand, and the frictional force between the moving liquid layer and the wall, on the other hand. When the two forces are equal the flow of liquid will occur at a uniform rate. If d is the effective thickness of the double layer across which the flow takes place, i.e., the one across which the electrokinetic potential is operative, then it may be supposed that the movement of the liquid is confined to this distance. On the wall side of the double layer the velocity of flow is zero whereas on the solution side it may be regarded as having attained the uniform velocity u of the moving liquid; the velocity gradient within the double layer, assuming it to be uniform, is then u/d. The force due to friction, which is the product of the velocity gradient and the coefficient of viscosity (η) of the liquid, is consequently equal to $\eta u/d$. The electrical force causing electro-osmosis is given by $V\sigma$, where V is the applied *potential gradient*, and σ, as before, is the charge density per sq. cm. at the boundary at which the movement occurs; hence, at the stationary state,

$$\frac{\eta u}{d} = V\sigma,$$

and if the value of d obtained from equation (1) or (10) is inserted, the result is

$$\zeta = \frac{4\pi\eta u}{DV}. \qquad (11)$$

If the potential gradient V is unity, i.e., 1 e.s. unit of potential per cm., the uniform velocity, given the symbol u_o, is called the **electro-osmotic mobility**; it is determined by the expression

$$u_o = \frac{\zeta D}{4\pi\eta}. \qquad (12)$$

In the case of electro-osmosis through the pores of a diaphragm, the volume v of liquid transported electro-osmotically per second is equal to qu, where q is the total area of cross section of all the pores in the diaphragm, the assumption being made that the thickness of the double layer is negligible in comparison with the pore diameter. On substituting v/q for u in equation (11) the result is

$$\zeta = \frac{4\pi\eta v}{qDV},$$

or

$$v = \frac{\zeta qDV}{4\pi\eta}. \qquad (13)$$

If the electro-osmosis takes places through a single capillary tube, q is equal to πr^2, where r is the radius of the tube; equation (13) then becomes

$$v = \frac{\zeta r^2 DV}{4\eta}. \tag{14}$$

If liquid is allowed to accumulate on one side of a diaphragm, an excess hydrostatic pressure is set up which eventually counterbalances the electro-osmotic flow. The simplest case to consider is that of electro-osmosis through a single capillary tube, for it is then possible to apply the Poiseuille equation for viscous flow; thus, if v is the volume of liquid of viscosity η flowing per second through a capillary tube of length l and radius r under a difference of pressure P, then

$$v = \frac{\pi r^4 P}{8\eta l}. \tag{15}$$

If this value for v is substituted in equation (14), it follows that

$$P = \frac{2\zeta DVl}{\pi r^2} = \frac{2\zeta DVl}{q}, \tag{16}$$

where P is the maximum pressure difference between the ends of the capillary tube as the result of electro-osmosis, and q is the cross-sectional area of the tube. As already indicated, a diaphragm may be regarded as equivalent to a large number of capillaries, and if q represents their total area of cross section, equation (16) may be regarded as applicable. Since V is the potential gradient and l is the length of the capillary tube, Vl may be replaced by the applied potential E, provided the electrodes are placed near the ends of the tube, or near the sides of the diaphragm; equation (16) then becomes

$$P = \frac{2\zeta DE}{q}. \tag{17}$$

For a given tube or diaphragm, q is constant and so the difference of pressure P maintained as a result of electro-osmosis should be proportional to the applied voltage E and independent of the dimensions of the diaphragm, as has been found by experiment.[8]

In order to utilize equation (17) for the determination of the zeta-potential, it may be written in the form

$$\zeta = \frac{Pq}{2DE} = \frac{P\pi r^2}{2DE}. \tag{18}$$

In this expression the pressure P is in dynes per sq. cm., q is in sq. cm., and ζ and E are in absolute units of potential; if the latter are expressed

[8] Tereschin, *Ann. Physik*, 32, 333 (1887).

in volts, then according to Table I (p. 4),

$$\zeta = \frac{P\pi r^2}{2DE} \times 9 \times 10^4 \text{ volt},$$

and, finally, with the pressure in atmospheres, this becomes

$$\zeta = \frac{P\pi r^2}{2DE} \times 9 \times 1.013 \times 10^{10} \text{ volt}.$$

From experiments on the electro-osmosis of water through a glass capillary tube of radius 0.037 cm., the value of P/E was estimated to be 2×10^{-8} atm. per volt. If the dielectric constant of the medium constituting the double layer is taken as 80, the zeta-potential between water and glass is found to be 0.049 volt.

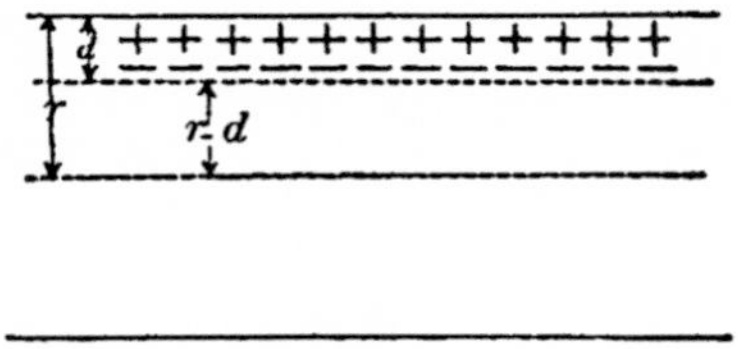

Fig. 126. Calculation of streaming potential

Streaming Potential.—The velocity of a liquid flowing in a capillary tube varies with the distance x from the center of the tube, and according to Poiseuille's treatment it is equal to $P(r^2 - x^2)/4\eta l$, where r is the radius and l is the length of the tube. The moving part of the double layer is at a distance $r - d$ from the center (Fig. 126) and so its velocity u is given by

$$u = \frac{P[r^2 - (r - d)^2]}{4\eta l},$$

and if d^2 is neglected in comparison with $2rd$, i.e., the thickness of the double layer is small relative to the diameter of the tube, the result is

$$u = \frac{Prd}{2\eta l}. \qquad (19)$$

If one side of the double layer is forced past the other, as is done in the measurement of the streaming potential, the strength of the current I produced is

$$I = 2\pi r u \sigma,$$

since $2\pi r u$ represents the area of the double layer moving in unit time and σ is the charge per sq. cm.; on introducing the value of u given by equation (19) it is found that

$$I = \frac{P\pi r^2 \sigma d}{\eta l}.$$

If κ_c is the specific conductance * of the liquid which is streamed through

* In order to avoid confusion with the use of the symbol κ for the reciprocal of the thickness of the double layer or ionic atmosphere, the symbol κ_c is employed here for the specific conductance.

the capillary tube, its resistance (R) is equal to $l/\pi r^2\kappa_c$, since l is the length and πr^2 the cross-sectional area of the tube; if S is the streaming potential produced by the flowing liquid, it follows by Ohm's law that

$$S = IR = \frac{P\pi r^2 \sigma d}{\eta l} \times \frac{l}{\pi r^2 \kappa_c}$$

$$= \frac{P\sigma d}{\eta \kappa_c}, \qquad (20)$$

or, substituting the value of σ obtained from equation (1) or (10),

$$S = \frac{\zeta DP}{4\pi\eta\kappa_c}. \qquad (21)$$

According to equation (20) or (21) the streaming potential should be proportional to the driving pressure P, provided a definite capillary tube or diaphragm is employed with a given electrolyte. This expectation has been verified for the streaming of water and aqueous solutions through glass capillary tubes and also for the passage of water through cellulose diaphragms.[9]

Provided the specific conductance of the liquid is known, the results of streaming potential measurements can be used to evaluate the **zeta-potential** by means of equation (21). In general the values are in satisfactory agreement with those derived from electro-osmotic and other studies, but if a diaphragm of large surface area is employed the results are liable to be in error; the reason for this is that as a consequence of the action of electro-osmotic forces the electrical conductance of a solution in the pores of a diaphragm may differ appreciably from that of the same solution in bulk, especially if it is relatively dilute.[10]

Electrophoresis.—In the derivation of equation (12) for the electro-osmotic mobility under unit potential gradient, it was supposed that the moving liquid was contained in a capillary tube. In other words, the system under consideration was equivalent to a cylinder of moving liquid surrounded by a cylindrical solid wall. The positions of the liquid and wall can be reversed without altering the argument, so that equation (12) will give the velocity with which a solid cylindrical particle moves through a liquid under the influence of an applied electric field of unit potential gradient. This quantity is the **electrophoretic mobility** (u_e) of the par-

[9] Kruyt and Willigen, *Kolloid.-Z.*, **45**, 307 (1928); Briggs, *J. Phys. Chem.*, **32**, 641 (1928); Bull and Gortner, *ibid.*, **35**, 309 (1931); **36**, 111 (1932); Fairbrother and Balkin, *J. Chem. Soc.*, 389 (1931); Bull and Moyer, *J. Phys. Chem.*, **40**, 9 (1936); DuBois and Roberts, *ibid.*, **40**, 543 (1936); Gortner and Lauffer, *ibid.*, **42**, 639 (1938); Gortner, *Trans. Faraday Soc.*, **36**, 63 (1940).

[10] McBain *et al.*, *Trans. Faraday Soc.*, **16**, 150 (1921); *J. Am. Chem. Soc.*, **51**, 3294 (1929); *J. Phys. Chem.*, **34**, 1033 (1930); **39**, 331 (1935); Fairbrother and Mastin, *J. Chem. Soc.*, **125**, 2319 (1924); Briggs, *J. Phys. Chem.*, **32**, 641 (1928); Bull and Gortner, *ibid.*, **35**, 309 (1931); Rutgers, *Trans. Faraday Soc.*, **36**, 69 (1940).

ticle; it follows, therefore, that

$$u_e = \frac{\zeta D}{4\pi\eta} \tag{22}$$

for a cylindrical particle. Although the supposition was made that the particle was solid in nature, equation (22) should be applicable to any particle, solid, liquid or gaseous, suspended in a liquid.

There has been much controversy concerning the applicability of equation (22) to particles of different shapes; according to Smoluchowski's treatment (1903) the equation for the electrophoretic velocity should be independent of the shape of the moving particle. On the other hand, Debye and Hückel [11] find that if the thickness of the double layer, i.e., $1/\kappa$, is large in comparison with the radius of the particle, i.e., for small spherical particles, the velocity of electrophoresis is given by

$$u_e = \frac{\zeta D}{6\pi\eta}, \qquad (23)^*$$

which differs from that derived previously by the replacement of the factor 4π by 6π. The result is analogous to the expression for the so-called "electrophoretic effect" in ordinary ionic conductance: according to equation (29), page 88, this effect is equal to $-\epsilon z\kappa/6\pi\eta$ for a potential gradient of 1 e.s. unit of potential per cm. Since ϵz is the charge carried by the ion, the potential ψ at a distance $1/\kappa$ is $\epsilon z\kappa/D$, so that the electrophoretic contribution to the speed of an ion can be written as $-\psi D/6\pi\eta$. In the case of a relatively large particle, ζ may be employed instead of ψ; the result is then equivalent to equation (23).

The treatment of Debye and Hückel is based on the assumption of an electric field which is constant everywhere and on the supposition that all parts of any spherical shell can move with the same velocity in a given direction. The presence of the particle must, however, distort the electrical field and the hydrodynamic currents, and it is only when the particles are very small in comparison with the thickness of the double layer that the Debye-Hückel result would be expected to hold. Since the value of $1/\kappa$ increases with increasing dilution, it follows that equation (23) should be applicable to small particles in very dilute solutions. For the case of relatively large particles, Henry [12] has derived the modified equation

$$u_e = \frac{\zeta D}{6\pi\eta} \cdot \frac{3\kappa_c}{2\kappa_c + \kappa_c'} \tag{24}$$

[11] Debye and Hückel, *Physik. Z.*, 25, 49 (1924).

* According to Stokes's law, $f = 6\pi\eta r u$, where u is the steady velocity acquired by a particle of radius r under the influence of a force f. If the particle carries a charge Q and moves under a potential gradient of unity, F is equal to Q and u becomes the mobility u_e; hence, $Q = 6\pi\eta r u_e$. The potential at a distance r from the charge Q is equal to Q/Dr, and if this is identified with ζ, equation (23) follows immediately.

[12] Henry, *Proc. Roy. Soc.*, **133A**, 106 (1931); see also, Sumner and Henry, *ibid.*, **133A**, 130 (1931).

where κ_c and κ'_c are the specific conductances of the liquid and of the particle, respectively. If the latter is a non-conductor this equation reduces to equation (22), the same result applying to a cylindrical particle with its axis in the direction of the electrical field; if the particle moves broadside to the field, the factor 8π should replace 4π in equation (22).

As a consequence of the complications associated with the theoretical derivation of the relationship between electrophoretic mobility and the shape of the particle, attempts have been made to solve the problem experimentally. Unfortunately the results in this connection appear to be equally inconclusive. Abramson studied the movement of spherical particles of various oils and of needles of asbestos and of *m*-aminobenzoic acid coated with the same protein, e.g., gelatin or egg albumin, in each case; the results showed that, provided the surface of the particle was completely covered with the protein, the electrophoretic mobility is independent of the shape of the moving particle.[13]

Another method of testing the relative merits of equations (22) and (23) for spherical particles is to compare the electro-osmotic and electrophoretic mobilities, i.e., u_o and u_e, respectively; this can be done by observations on a micro-electrophoresis cell (cf. p. 536) made of the same material as the suspended particles, e.g., glass or quartz. If equation (22) is correct, then it is apparent from equation (12) that u_e/u_o should be unity; on the other hand, if equation (23) applies, the ratio should be 1.5. Experiments with spherical particles and surfaces, coated with adsorbed protein in order to insure identity in the nature of the surface, indicated that u_e/u_o was approximately unity, as required by equation (22). Objection to this conclusion was taken by White, Monaghan and Urban [14] who claimed that the independence of electrophoretic mobility of the shape of the particles and the value of unity for u_e/u_o were due to the use of a liquid medium containing a relatively high concentration of electrolyte. It was stated that if the electrolyte concentration was less than 0.001 M the ratio of electrophoretic to electro-osmotic mobility was not unity. The results of other workers, however, appear to be contrary to this conclusion, and so the situation is somewhat uncertain. It is probable, however, that equation (22) may be regarded as reasonably adequate for particles of any shape.[15]

Determination of Zeta-Potentials.—The measurement of the velocity of electrophoresis of a moving particle, by one of the procedures to be described shortly, provides a convenient method for evaluating the zeta-potential, utilizing equation (22). The value of u_e as given by this equa-

[13] Abramson, *J. Phys. Chem.*, **35**, 289 (1931); *J. Gen. Physiol.*, **16**, 1 (1932).

[14] White, Monaghan and Urban, *J. Phys. Chem.*, **39**, 611 (1935); White and Fourt, *ibid.*, **42**, 29 (1938).

[15] Moyer and Abramson, *J. Gen. Physiol.*, **19**, 727 (1936); *J. Biol. Chem.*, **123**, 391 (1938); Wiley and Hazel, *J. Phys. Chem.*, **41**, 1699 (1937); Moyer, *ibid.*, **42**, 391 (1938); *Trans. Electrochem. Soc.*, **73**, 488 (1938); see also, Briggs, *J. Am. Chem. Soc.*, **50**, 2358 (1928); Bull, *J. Phys. Chem.*, **39**, 577 (1935); Abramson, *Trans. Faraday Soc.*, **36**, 5 (1940).

tion is for a potential gradient of 1 e.s. unit, i.e., 300 volts, per cm.; similarly, ζ is expressed in e.s. units. Consequently if the electrophoretic mobility is for a potential gradient of 1 volt per cm. and the zeta-potential is to be in volts, equation (22) should be written in the form

$$u_e = \frac{\zeta D}{4\pi\eta} \times \frac{1}{9} \times 10^{-4},$$

$$\therefore \quad \zeta = \frac{4\pi\eta u_e}{D} \times 9 \times 10^4 \text{ volt.} \tag{25}$$

The electrophoretic mobilities of small colloidal particles, as well as of larger quartz particles, oil drops and air bubbles, are always about 2 to 4 $\times$ 10^{-4} cm. per sec. in water; hence, in accordance with the requirements of equation (25), η being 0.01 c.g.s. unit (poise) and D approximately 80, the value of the zeta-potential is between 0.03 and 0.06 volt in each case.

Attention may be called to the fact that the mobility of a simple ion under a potential gradient of 1 volt per cm. is about 6 $\times$ 10^{-4} cm. per sec. at ordinary temperatures (cf. p. 60), and hence suspended particles move with a velocity comparable with that of electrolytic ions. The electrical conductance of the suspension, e.g., of a colloidal solution, is, however, quite low because the number of particles carrying the current is very small in comparison with that in a solution of an electrolyte at an appreciable concentration.

Influence of Ions on Electrokinetic Phenomena.—The influence of electrolytes on electrokinetic effects can be described most simply in terms of the apparent changes in the zeta-potential. The results appear to be the same, on the whole, whether they are based on electro-osmotic, electrophoretic or streaming potential measurements. In general, the zeta-potential is made more positive by acids, i.e., by hydrogen ions, and more negative by solutions of alkalis, i.e., by hydroxyl ions. If the pH of the solution is kept constant, the electrokinetic potential is influenced by the addition of salts; the ions having the most appreciable effects are those of sign opposite to that of the solid phase. Further, the higher the valence of these ions the greater is their influence on electrokinetic phenomena. It may be noted that ions of the same sign as the solid do have some effect, but this is relatively small.

These conclusions may be illustrated by reference to the results depicted in Fig. 127, which shows the apparent variation of the electrokinetic potential, calculated by means of equation (21), from streaming potential measurements with different electrolytes in a glass capillary tube.[16] In pure water the zeta-potential at the glass-water interface is negative, the glass being negatively charged with respect to the water; on the addition of potassium or barium chloride there is at first an increase

[16] Freundlich and Ettisch, *Z. physik. Chem.*, 116, 401 (1925); see also, Gortner, *J. Phys. Chem.*, 34, 1509 (1930); 35, 700 (1931).

in the negative charge of the glass, presumably owing to the action of the chloride ions. As the concentration of these electrolytes increases the zeta-potential reaches a maximum and then decreases; the influence of the ions of sign opposite to that of the glass now becomes evident. When positive ions of high valence, e.g., La^{+++} or Th^{++++}, are present in the solution their effect is considerable and hence any influence of the anions is so small in comparison that it cannot be observed. It will be seen that at high concentrations of thorium nitrate the zeta-potential after becoming positive tends to reverse its sign again, presumably because of the effect of the negative ion.

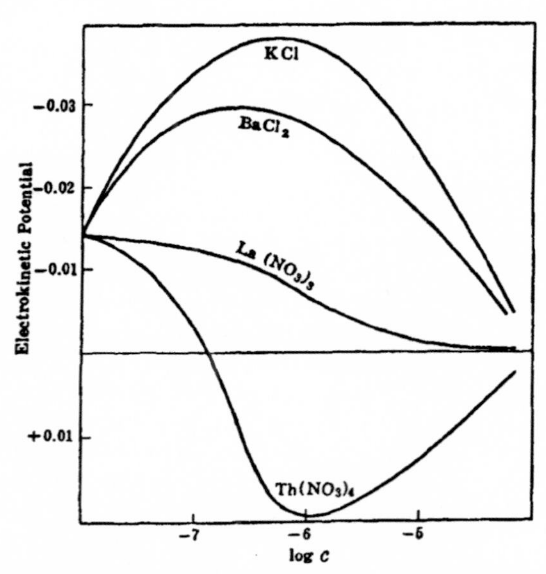

Fig. 127. Zeta-potentials in presence of salts (Freundlich and Ettisch)

The influence of ions on electrokinetic effects can be readily explained with the aid of Stern's concept of the double layer. Substances like silicon carbide, cellulose, sulfur and carbon, which do not ionize, are negatively charged in contact with water and the addition of small amounts of uni-univalent electrolytes tends to increase this charge. It is probable that in these cases the negative zeta-potential is due in the first place to the firm attachment to the surface of hydroxyl ions from the water and possibly also of anions from the electrolyte. An equivalent number of positive ions, some closely held in the fixed part of the double layer and the remainder in the diffuse portion, will be left in the solution. The potential gradient between the solid surface and the bulk of the liquid, which is pure water or a dilute solution, is shown diagrammatically in Fig. 128, I. If the electrolyte concentration is increased, there will be a tendency for the cations to accumulate on the solution side of the fixed double layer, i.e., in the vicinity of the line XY in the diagram: by increasing the positive charge density at XY the zeta-potential becomes less negative, as shown by Fig. 128, II. If the positive ion concentration is made large, especially if these ions have a high valence, the sign of the electrokinetic potential may eventually be reversed (Fig. 128, III).

If the solid material is a substance capable of ionization, then the sign and magnitude of the zeta-potential may be determined by such ionization. For example, silica, tungstic and stannic acids, acid dye stuffs, soaps and glass are negatively charged in contact with water, since in each case there is a tendency for a small cation, e.g., a hydrogen or alkali metal ion, to pass into solution leaving the complex anion to remain part of the positively charged solid. On the other hand, aluminum and other basic hydroxides and basic dyestuffs, which can take up protons from aqueous solution, form positively charged solids. The influence of hydrogen and hydroxyl ions in these cases, and also with amphoteric substances, such as proteins, can be explained in terms of ordinary acid-base equilibria. The presence of other ions, however, may be expected to affect the zeta-potential by changing the charge density on the solution side of the fixed double layer, as explained above.

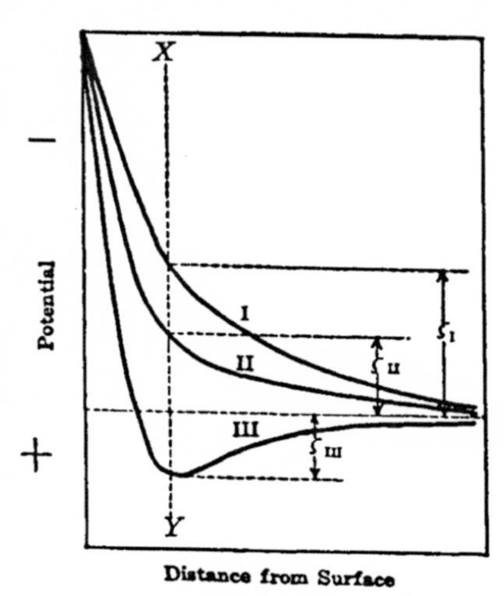

Fig. 128. Interpretation of effect of ions on zeta-potential

In the potential gradients depicted in Fig. 128 the tacit assumption is made that the effective thickness of the double layer remains unchanged in spite of changes in the concentration and nature of the electrolyte; it is probable, however, that the thickness of the diffuse double layer decreases with increasing concentration of ions in the solution, especially if they have high valences. It was seen on page 524 that the effective thickness of the double layer could be regarded as equivalent to the reciprocal of κ, as defined by equation (18), page 84; it is known that this quantity, i.e., $1/\kappa$, is inversely proportional to the square-root of the ionic strength and hence it should be smaller the greater the concentration of the ions and the higher their valence. An examination of equation (1) or (10) shows that a decrease in either σ, the surface charge density, or in d, the thickness of the double layer, will result in a decrease of the zeta-potential. It is thus possible for the apparent zeta-potential to change merely as a result of a change in the effective thickness of the double layer, although it is probable that there is always a simultaneous alteration in the charge density.

The presence of high valence ions of the same sign as the solid surface results in a numerical increase of the charge density which would normally be accompanied by an increase of the zeta-potential; it is possible, however, for the thickness of the double layer to decrease sufficiently at the same time for the zeta-potential to decrease in magnitude. This effect has been observed in certain instances. If the ionic concentration of the liquid phase is increased sufficiently it is possible for the thickness of the double layer to diminish to such an extent that it eventually collapses and reforms with the charges reversed. Even if the collapse does not occur, the double layer will become very thin so that at high ionic concentrations the zeta-potential should be small: this may account for the fact, which is evident from Fig. 127, that the zeta-potential tends to approach zero in the presence of relatively large amounts of electrolyte.

It is opportune to mention here that some writers prefer to avoid the use of the concept of the zeta-potential;[17] it is true that there must be some form of potential across the double layer, but it is so variable in sign and magnitude that its exact significance is regarded as uncertain. The quantity which is called the zeta-potential is, according to equation (1), proportional to the product of the surface charge density and the thickness of the double layer, i.e., to σd; it is, therefore, considered preferable to regard it as a measure of the electric moment per sq. cm. of the double layer.

Electrophoretic Measurements: I. Microscopic Method.—In the microscopic method for the study of electrophoresis the colloidal solution or suspension of the particles under examination is placed in a special micro-electrophoresis cell which is fixed on the stage of a microscope.[18] A shallow flat cell of rectangular cross section has been commonly employed with an electrode sealed in at each side. These electrodes are connected to a source of E.M.F., and the rate of movement of any particle is observed by means of a scale in the eye-piece of the microscope. The potential gradient is calculated from the current strength and the resistance of the solution between the electrodes, the distance between the latter being known. Owing to the fact that particles tend to settle out, a vertical micro-electrophoresis cell has been devised; the effect of gravity is eliminated by making observations first with the applied field in one direction and then repeating the measurements with the field reversed. Cells of cylindrical bore have also been used for the study of electrophoresis, since they are more easily made and cleaned than those having

[17] See, for example, McBain and McBain, *Z. physik. Chem.*, **161**, 279 (1932); Guggenheim, *Trans. Faraday Soc.*, **36**, 139 (1940).

[18] Abramson *et al.*, *J. Phys. Chem.*, **36**, 1454 (1932); *J. Am. Chem. Soc.*, **58**, 2362 (1936); *Ann. New York Acad. Sci.*, **39**, 121 (1939); *Trans. Faraday Soc.*, **36**, 5 (1940); Stern, *Ann. New York Acad. Sci.*, **39**, 147 (1939).

a rectangular cross section; the curvature of the walls, however, makes accurate observation of the moving particle a matter of some difficulty.

The results of measurements by the microscopic method show that the electrophoretic mobility of the particles varies with the distance from the wall of the cell; particles close to the wall move in a direction opposite to that in which those in the center migrate. In any event, the results show an increase in velocity from the walls to the center of the cell. The explanation of this fact lies in the electro-osmotic movement of the liquid; a double layer is set up between the liquid and the walls of the cell and under the influence of the applied field the former exhibits electro-osmotic flow. For the purpose of obtaining the true electrophoretic velocity of the suspended particles it is necessary to observe particles at about one-fifth the distance from one wall to the other. A more accurate procedure is to make a series of measurements at different distances from the side of the cell and to apply a correction for the electro-osmotic flow.[19] The algebraic difference of the corrected electrophoretic velocity and the speed of the particles near the walls gives the electro-osmotic mobility of the liquid in the particular cell. If the solution contains a protein which is adsorbed on the surface of the walls of the vessel and on the particles, it is possible to compare the electrophoretic and electro-osmotic mobilities in one experiment; reference to the significance of such a comparison was made on page 532.

II. Macroscopic Methods.—Although the microscopic procedure has several advantages, in respect of simplicity, time required and the information it gives concerning the size, shape and orientation of the particles, the macroscopic method has attracted much attention in recent years, chiefly because of the development by Tiselius of a particularly useful form of apparatus. This method has been employed for many years in the relatively approximate study of electrophoresis; the apparatus in its simplest form is shown in Fig. 129. The lower part of the U-tube contains the suspension, etc., to be investigated and this is covered in each limb by some pure solvent in which dip two platinum electrodes. When an E.M.F. is applied to these electrodes, the boundaries between the solvent and the suspension move with a speed equal to that of the electrophoresis of the particles. If the suspension is colored, the position of the boundary can be readily observed and its rate of movement measured; from a knowledge of the potential gradient, the mean electrophoretic mobility of the particles can be evaluated. If the boundary is not visible

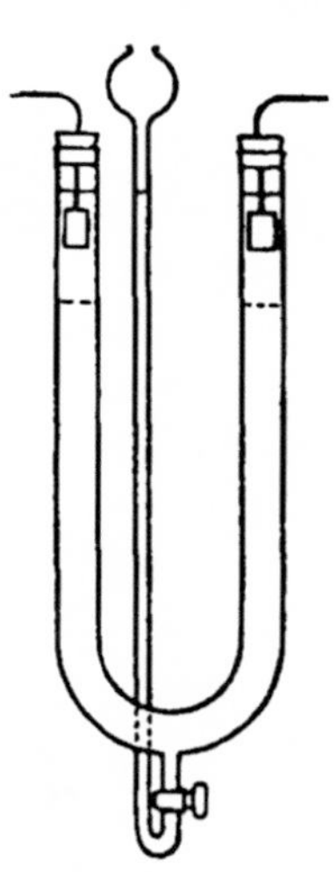

Fig. 129. Study of electrophoresis (Burton)

[19] Ellis, *Z. physik. Chem.*, 78, 321 (1912); Abramson, "Electrokinetic Phenomena," 1934, Chap. III.

to the eye, it can be sometimes rendered visible by causing it to fluoresce as the result of exposure to ultra-violet light; the apparatus must, however, be made of quartz.[20]

The moving boundary method for the study of electrophoresis, as just described, is liable to a number of errors, one of the most important being the disturbance of the boundary. In the determination of transference numbers by observations of a moving boundary, the tendency toward automatic establishment of the conditions necessary for a sharp boundary (cf. p. 118) will compensate to a large extent for the disturbing effects of diffusion. In the study of suspended particles, however, there is virtually no automatic adjustment. The chief difficulties of the macroscopic method have been overcome in a form of apparatus designed by Tiselius; the most important advances are the use of a liquid, generally a buffer solution, of the same composition throughout, so that the environment of the particles is not changed as the boundary moves forward, and the employment of tubes of rectangular cross section maintained at a temperature of about 3°, i.e., the temperature of maximum density of the buffer solution. When an electric current passes through a liquid in a tube, heat is generated, but more is lost by conduction from the liquid near the walls than from the center of the tube. The result is that the latter becomes hotter than the former and the difference of density produces convection currents which will disturb the boundary in a moving boundary apparatus. The use of tubes of rectangular cross section provides a maximum amount of wall area and so facilitates the removal of heat to the surroundings, and the maintenance of the apparatus at somewhat below 4°, where the variation of the density of the liquid with temperature is very small, greatly diminishes convection effects. With these improvements, and a device for obtaining initially sharp boundaries, the macroscopic method has proved a valuable means for the study of electrophoresis, and for its utilization to separate particles moving with different speeds.

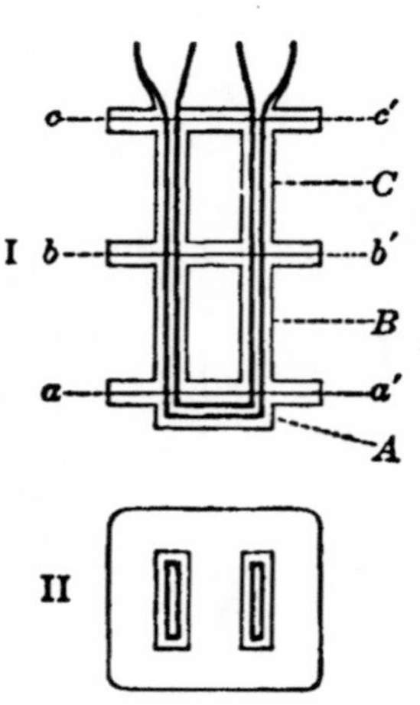

FIG. 130. Portion of Tiselius apparatus in section (I) and plan (II)

The Tiselius Apparatus.—The U-tube portion of the apparatus, shown in section (I) and in plan (II) in Fig. 130, is in several parts which can be slid horizontally across one another along the planes aa', bb' and cc'; the surfaces between the sections are suitably lubricated in order to facilitate the sliding and to prevent leakage of liquid. The channel running through the center of the apparatus, shown by the thicker lines,

[20] Svedberg *et al.*, *J. Am. Chem. Soc.*, 45, 954 (1923); 46, 2700 (1924); 48, 2272 (1926).

is rectangular in cross section with dimensions of 3 mm. by 25 mm. In order to form the boundaries whose motion is to be studied, the channels in the sections *A* and *B*, up to a level just above *bb'*, are filled with the experimental system, e.g., a protein or mixture of proteins in a buffer solution. The section *C* is then slid aside, either by means of small pneumatic pumps, as used by Tiselius,[21] or by means of a screw and ratchet device, as employed in the form of apparatus modified by Longsworth and MacInnes.[22] The excess of liquid in *C* is removed by a pipette, the channel is washed out and filled with the pure buffer solution; the section *C* is then slid back to its original position, thus forming a sharp boundary at *bb'*.

The portion of the apparatus just described is connected, prior to being filled, to two large electrode chambers *D* and *D'*, as shown in Fig. 131; the electrodes *E* and *E'* are silver-silver chloride electrodes, of the

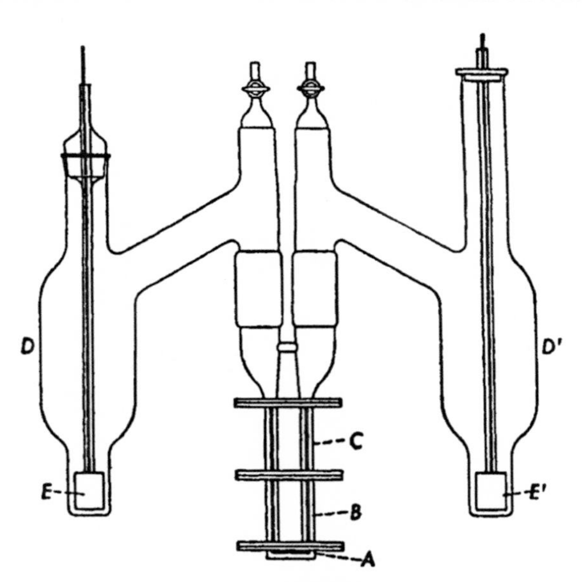

FIG. 131. Complete Tiselius apparatus

type described in Chap. VI but of much larger current carrying capacity. The narrow portions of the electrode vessels, immediately surrounding the electrodes, contain a concentrated solution of an alkali chloride, while the remainder of the apparatus contains the same buffer solution as forms the medium in which the protein particles are suspended. After the

[21] Tiselius, *Trans. Faraday Soc.*, 33, 524 (1937); *Kolloid.-Z.*, 85, 129 (1938); Stern, *Ann. New York Acad. Sci.*, 39, 147 (1939); see also lecture by Tiselius, reported by Krejci, *J. Franklin Inst.*, 228, 797 (1939).

[22] Longsworth and MacInnes, *Chem. Revs.*, 24, 271 (1939); Longsworth, Cannan and MacInnes, *J. Am. Chem. Soc.*, 62, 2580 (1940).

boundaries are set up, as described above, a source of E.M.F. is applied to the electrodes and the electrophoretic movement commences.

Two methods have been used to observe the motion of the boundaries; both depend on changes of refractive index. The one most generally employed is an adaptation of the Toepler "schlieren" (shadow) technique; by adjustment of a diaphragm the boundaries are made to appear as bands of shadow, which can be seen as such on a ground glass screen or they can be recorded photographically. By taking observations after various intervals of time, the rate of movement of the boundary, and hence the electrophoretic mobility, can be calculated. The second procedure is the so-called "scale" method in which a photograph of a graduated scale is taken through the boundary; the displacement of the lines, due to changes of refractive index, gives an indication of the position of the moving boundary at any instant. The apparatus required for the scale method is much less costly than that employed in the "schlieren" procedure, but its use is very tedious.[23]

Electrophoretic mobilities determined by the Tiselius method have been found generally to be in good agreement with those obtained by the microscopic observation of single particles.[24] Measurements have been made by both methods, with horse serum protein and the results, for solutions of various pH's, are plotted in Fig. 132. In this case, although the agreement is not always so good, almost identical electrophoretic mobilities have been obtained by the microscopic method using various particles coated with horse serum albumin. It will be observed from Fig. 132 that the mobility of the protein in relatively acid solution is positive, so that it carries a positive charge; in view of the amphoteric character of proteins this result is to be expected. If the pH exceeds about 4.9 the particles carry a negative charge and move in the direction opposite to that of the applied field. At pH 4.9 the horse serum albumin particles exhibit no electrophoretic mobility; hence pH 4.9 represents the isoelectric point (cf. p. 428) of this

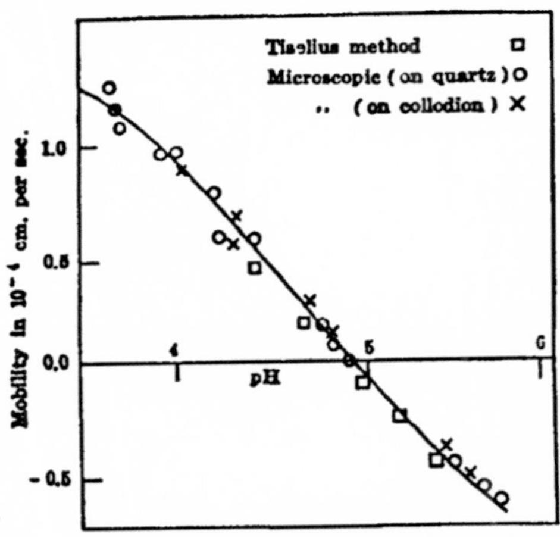

Fig. 132. Mobility of horse serum albumin (Abramson, Gorin and Moyer)

[23] Longsworth, *Ann. New York Acad. Sci.*, 39, 187 (1939); Horsfall, *ibid.*, 39, 203 (1939); Kekwick, *Trans. Faraday Soc.*, 36, 147 (1940).
[24] Moyer, *J. Phys. Chem.*, 42, 71 (1938); *J. Biol. Chem.*, 122, 641 (1938); Abramson, Gorin and Moyer, *Chem. Revs.*, 24, 345 (1939).

substance. Measurements of electrophoresis at a series of pH values is, of course, the most direct method for the determination of the isoelectric points of proteins and allied substances.

Separations by the Tiselius Apparatus.—One of the most valuable features of the Tiselius electrophoresis apparatus is that it permits of the separation of constituents of a mixture provided they have appreciably different electrophoretic mobilities. This aspect of recent work on electrophoresis has found particular application in the study of proteins from various natural sources. Several methods have been adopted for the purpose of indicating whether a particular system is homogeneous or whether it consists of two or more proteins. If the particles are all of one kind, there can never be more than one boundary, and hence the "schlieren" photographs taken at various time intervals will show one shadow only. If two or more different substances with somewhat different electrophoretic mobilities are present, however, two or more boundaries will form in the course of time, each one moving steadily ahead of the succeeding ones (cf. p. 127). This result is brought out clearly by the photographs shown in Fig. 133; they were obtained by Tiselius with human blood serum at pH 8.06 and were taken at intervals of 20 minutes. The fastest moving boundary is due to the albumin constituent, and the others are for α- and β-globulins, respectively; the actual electrophoretic mobilities of these three proteins of human blood serum are about 6.0, 4.3 and 2.8 $\times 10^{-5}$ cm. per sec., respectively, for a potential gradient of 1 volt per cm.

FIG. 133. Series of "schlieren" photographs of blood serum (Tiselius)

The "schlieren scanning" technique, which is a modification of the foregoing, not only provides information concerning the number of proteins with different mobilities, and their electrophoretic mobilities, but also gives the amounts of the various substances present.[25] By adjusting the "schlieren" diaphragm a series of positions can be obtained, and the corresponding "schlieren" photographs indicate the variation of refractive index through the system of boundaries. If the adjustment of the diaphragm is synchronized with the movement of a photographic plate the result, obtained after electrophoresis has taken place for a short time, is of the type shown in Fig. 134; this is a "schlieren scanning" photograph obtained with normal blood plasma.[26] The peak marked A is due to the albumin boundary, whereas α, β and γ represent three globulins,

[25] Longsworth, *J. Am. Chem. Soc.*, **61**, 529 (1939); Longsworth and MacInnes, *ibid.*, **62**, 705 (1940); Longsworth, Shedlovsky and MacInnes, *J. Exp. Med.*, **70**, 399 (1939).

[26] Longsworth *et al.*, *J. Exp. Med.*, **70**, 399 (1939).

and ϕ is produced by fibrinogen; the rapid separation of five protein constituents of blood plasma is very striking. Further, the areas under the respective peaks are proportional to the quantities of the various substances present, and hence a complete analysis is possible. It should be borne in mind, however, that two proteins having the same electrophoretic mobility would always give one boundary, and hence one peak; the two substances would behave as a single one. In such cases it is sometimes possible to effect a separation by changing the pH of the system, especially if the two proteins have different isoelectric points. Change of pH, however, may produce alterations in biological systems which may invalidate the conclusions drawn from the electrophoretic studies.

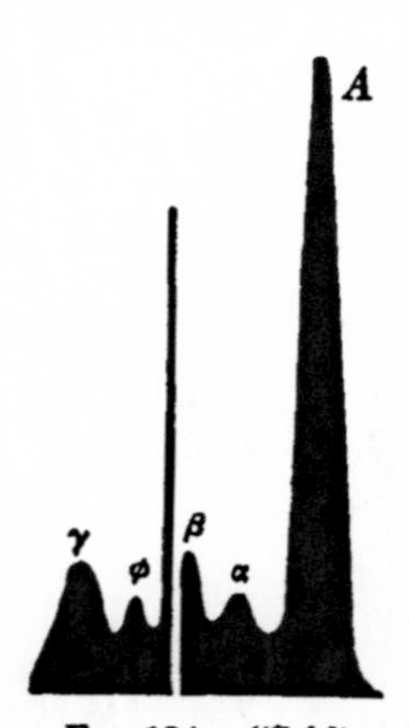

FIG. 134. "Schlieren scanning" photograph of blood plasma (Longsworth, Shedlovsky and MacInnes)

One of the chief difficulties in the interpretation of the results obtained by the Tiselius apparatus is the appearance of "anomalous" boundaries; one example is provided by the very sharp peak near the center of Fig. 134 which is found with descending boundaries. This particular peak is believed to represent an anomalous boundary due to convection; it is supposed to result from a reaction in the neighborhood of the β-globulin boundary following on the electrophoretic separation of the constituents of the system. Other anomalous boundaries have been found which are attributed to protein concentration gradients and to changes in the buffer solution. Such anomalies can often be minimized by the correct choice of the buffer system; it should have a high buffer capacity but, on the other hand, its specific conductance should be relatively low in order to decrease disturbances due to the heating effect of the current. Since high buffer capacity requires a high concentration of the constituents (cf. p. 412), the salts chosen should consist, as far as possible, of ions having relatively small conductances, e.g., sodium salts should be used in preference to potassium salts.

Not only can the Tiselius apparatus be used to indicate the presence of several constituents in a system, but an actual separation of different fractions is sometimes possible. If sufficient time is allowed for the leading boundary to get some distance ahead of the next, the solution between it and the following boundary will consist almost entirely of the faster moving constituent. Since electrophoretic mobilities are, in general, not very different, the boundaries will have moved above the section C or below B in Fig. 131 before there is any appreciable distance between them. If the solution as a whole is given a velocity equal to that of the slower moving boundary but opposite in direction, this boundary will remain stationary at bb' in Fig. 131 while the faster one moves steadily ahead, e.g., from b into section C. After some time the section C can be slid aside and its contents removed; this now contains only the constituent, or constituents, with the highest electrophoretic mobility.

If the system contains three constituents with different mobilities, three boundaries will form; a rate of movement equal in magnitude, but opposite in direction, to that of the middle boundary is now imparted to the whole liquid. The fastest moving constituent moves ahead, whereas the slowest constituent is given an apparent negative velocity; after electrophoresis has proceeded for some time, one limb of the section C contains the former constituent and the other contains the latter in a pure form. Several devices have been employed to impart a movement to the liquid: one method is to withdraw gradually, by means of clockwork, a plunger which fits loosely into one of the electrode vessels, while another is to keep one electrode vessel closed, e.g., the left-hand one in Fig. 131, and to force buffer solution into it at the desired rate by means of a syringe operated by a constant speed motor.

Electrophoretic Mobility and Bound Hydrogen Ion.—If the treatment given on page 523 is applied to a spherical particle, the result for the potential at a distance r from the center of the particle is similar to equation (15), page 83, viz.,

$$\psi = \frac{Q}{D} \cdot \frac{e^{-\kappa r}}{r}, \tag{26}$$

where Q, the total charge carried by the particle, replaces $z_i\epsilon$, the charge of the ion. If r is equal to the sum of the radius of the particle and the thickness of the double layer, ψ may be replaced by the electrokinetic potential ζ; further, if κr is relatively small, i.e., for a dilute solution, $e^{-\kappa r}$ may be replaced by $1 - \kappa r$, or by $(1 + \kappa r)^{-1}$, so that equation (26) takes the form

$$\zeta = \frac{Q}{Dr} \cdot \frac{1}{1 + \kappa r}. \tag{27}$$

If this result is combined with equation (23), it is found that

$$Q = 6\pi\eta u_e r(1 + \kappa r), \tag{28}$$

so that for a given particle and a medium of constant ionic strength, i.e., κ and r are constant, the electrophoretic mobility (u_e) should be proportional to the charge (Q) carried by the particle.[27]

The charge carried by a protein in a solution of a given pH may be regarded as approximately proportional to the number of equivalents of acid or alkali required to bring the protein system from the isoelectric point to its actual pH value. This statement presupposes, among other matters, that the electrolytes in the solution do not affect the charge of the particles, as will be the case if univalent ions only are present; that the protein salts are completely ionized or ionized to a constant extent;

[27] Abramson, *J. Gen. Physiol.*, 15, 375 (1932); 16, 593 (1933).

and that the charges on the protein ions are at or near the surface and are approximately uniformly distributed. If these conditions hold, it follows from equation (28), provided the viscosity of the solvent and the radius of the particle remain unchanged, that the electrophoretic mobility of a protein particle in a solution of given pH, at constant ionic strength, should be proportional to the amount of acid or alkali bound by the protein, i.e., to the amount required to bring the system from the isoelectric point to the given pH. The general accuracy of this conclusion is shown by the results for egg albumin in Fig. 135; the circles give the electrophoretic mobilities while the black dots represent the number of moles of acid or alkali bound per gram of the protein.[28] The two scales, which must of course coincide at the isoelectric point, are adjusted to bring the two sets of points as close together as possible. The fact that the agreement extends over a range of pH values provides support for the deduction made above. It follows, therefore, that electrophoretic measurements may be employed to determine the relative number of protons bound or liberated by a protein at any given pH.

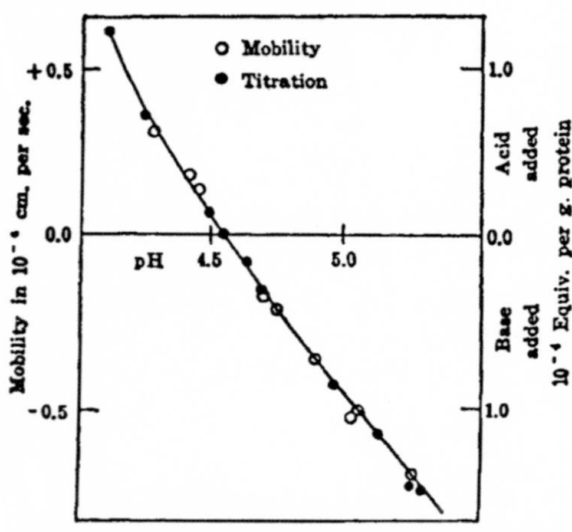

Fig. 135. Bound acid or alkali and mobility (Moyer and Abels)

If the radius of a spherical particle is known, equation (28) may be utilized to estimate the charge Q which it carries; inserting the values of the universal constants in κ (cf. p. 84), it follows that

$$Q = 6\pi\eta u_e r(1 + 0.33 \times 10^8 r\sqrt{\mu}), \qquad (29)$$

where μ is the ionic strength of the medium. Attention may be drawn to the fact that equation (28) is based on the supposition that κr is small, i.e., for a dilute solution; at higher concentrations the influence of ionic strength on the electrophoretic mobility at constant pH, i.e., for a protein particle of constant charge, is much greater than is implied by equation (28) or (29).[29]

[28] Moyer and Abels, *J. Biol. Chem.*, **121**, 331 (1937); Abramson, Gorin and Moyer, *Chem. Revs.*, **24**, 345 (1939).

[29] Gorin, Abramson and Moyer, *J. Am. Chem. Soc.*, **62**, 643 (1940); Davis and Cohn, *ibid.*, **61**, 2092 (1939); *Ann. New York Acad. Sci.*, **39**, 209 (1939).

PROBLEMS

1. A glass particle suspended in water ($\eta = 0.01$ poise) was observed to move with a velocity of 21.0×10^{-4} cm. per sec. under a potential gradient of 6.0 volts per cm. Calculate the zeta-potential at the glass-water interface.

2. Assuming the zeta-potential to be the same as in Problem 1, calculate the rate of electro-osmotic flow of water through a glass capillary tube of 0.05 cm. radius under a potential gradient of 1 volt per cm.

3. If the length of the tube in Problem 2 is 25 cm., what pressure, in terms of mm. of water, could be supported by the electro-osmotic effect of a potential gradient of 1,000 volts per cm.?

4. Using the result obtained by Bull and Gortner [*J. Phys. Chem.*, 36, 111 (1932)], that S/P for the streaming of 2×10^{-4} N sodium chloride through a diaphragm of quartz particles is about 25 millivolts per cm. of mercury pressure, calculate the approximate specific surface conductance of the solution used. Compare the result with the normal value for sodium chloride at the same concentration. The viscosity and dielectric constant of the solution may be assumed to be the same as for water, and the zeta-potential may be taken as 0.05 volt. (Care should be exercised in the matter of units, use being made of the conversion factors in Table I.)

INDEX

Printed in the United Kingdom
by Lightning Source UK Ltd.
125635UK00001B/245/A